Neuromuscular Fundamentals

Neuromuscular Fundamentals

How Our Musculature is Controlled

Nassir H. Sabah

American University of Beirut, Lebanon

CRC Press
Taylor & Francis Group
Boca Raton London New York

CRC Press is an imprint of the
Taylor & Francis Group, an **informa** business

First edition published 2021
by CRC Press
6000 Broken Sound Parkway NW, Suite 300, Boca Raton, FL 33487-2742

and by CRC Press
2 Park Square, Milton Park, Abingdon, Oxon, OX14 4RN

Library of Congress Cataloging-in-Publication Data

Names: Sabah, Nassir H., author.
Title: Neuromuscular fundamentals / Nassir H. Sabah.
Description: First edition. | Boca Raton : CRC Press, 2021. | Includes
bibliographical references and index. | Summary: "The book can serve as
a textbook for a one-semester course on the neuromuscular system or as a
reference in a more general course on neuroscience"-- Provided by
publisher.
Identifiers: LCCN 2020023631 (print) | LCCN 2020023632 (ebook) | ISBN
9780367456924 (hardback) | ISBN 9781003024798 (ebook)
Subjects: MESH: Sensorimotor Cortex | Neuromuscular Junction |
Neuroeffector Junction | Muscles--physiology
Classification: LCC QP383.15 (print) | LCC QP383.15 (ebook) | NLM WL 307
| DDC 612.8/252--dc23
LC record available at https://lccn.loc.gov/2020023631
LC ebook record available at https://lccn.loc.gov/2020023632

ISBN: 978-0-367-45692-4 (hbk)
ISBN: 978-1-003-02479-8 (ebk)

Typeset in Palatino
by Deanta Global Publishing Services, Chennai, India

To those who aspire to advance our understanding of

Nature's greatest marvel: the human brain.

And in fond remembrance of

Sir John Carew Eccles

and his wife

Helena Táboríková.

~

The whole universe is a play of unity in variety, and of variety in unity.

Swami Vivekananda

Brief Contents

Brief Contents

Contents

Preface

This book is rather unique in its approach and coverage. The approach is essentially that of an engineering textbook, emphasizing the quantitative aspects and highlighting the fundamentals and basic concepts involved. The coverage progresses in a logical and systematic manner from the subcellular, starting with the electrophysiology of the cell membrane, then proceeding to synapses, neurons, and muscle, before considering neuronal motor ensembles and the neuromuscular system as a whole. Simple, clear, and comprehensive explanations are given throughout.

The objectives of the book are:

1. To provide a thorough, quantitative understanding of the electrophysiology of the membranes of excitable cells and the characteristic features of the basic functional components of the nervous system, namely, synapses and neurons. These topics are fundamental to neuroscience and are prerequisites to advanced study of any specific area of neuroscience or part of the nervous system.

2. To present a particular, specialized area of neuroscience, namely, the somatomotor system and the control of movement and posture. This is an area that is important in its own right and is distinguished from other areas of neuroscience, such as the visual system, auditory system, and cognition, in that the end-objective, which is skeletal muscle control, is effected outside the central nervous system and, at least in principle, is amenable to precise measurement of the dynamic and kinematic variables involved. In contrast, the end-objective of sensory systems, for instance, is sensory perception, which is intangible and much more elusive. The same is true of cognitive functions.

3. To expose the reader to much of the basic, specialized terminology associated with the topics covered, to explain their meaning and significance, and to demystify some of the concepts, particularly those of a biochemical nature, associated with neuronal function. This is particularly helpful to readers who do not have a rigorous background in biochemistry and is indispensable for keeping abreast with the literature on the topics covered.

The book is intended for undergraduate or graduate students in the natural sciences, mathematics, or engineering who seek a deeper understanding of the fundamentals of neuroscience and the somatomotor system, in accordance with the aforementioned objectives. The book can serve as a textbook for a one-semester course on the neuromuscular system or as a reference in a more general course on neuroscience. The book material has been used by the author for many years, with continuing updates, in a course on the neuromuscular system.

As can be seen from the contents, the topics covered are treated quite exhaustively in their different aspects. In addition, the following highlights illustrate the nature of the book's philosophy and approach:

- The expressions for chemical and electrochemical potentials in one dimension are derived in Chapter 1. The significance of the electrochemical potential as the driving force for the passive diffusion of ions is emphasized and this concept is applied in many instances, as in discussions of the membrane equivalent circuit and the movement of ions through ion channels.

- The membrane equivalent circuit is thoroughly explained, particularly the interpretation of the conductances and battery sources that appear in the branches for each of the ions involved and the implications for the admissible range of membrane voltages.

- The basic features of membrane rectification are simply explained, based on the expression for membrane current given by the Goldman–Hodgkin–Katz current equation derived from first principles. It is also explained quite simply how the nonlinear, time-varying membrane conductance gives rise to a membrane reactance. The analogy between membrane and semiconductor systems is noted, as both systems satisfy the same electrodiffusion equations once certain simplifying assumptions are made.

- The cable equation for an unmyelinated axon is derived step-by-step and solved for two special cases that yield rather non-intuitive results, physically interpreted, and which are useful for understanding some aspects of dendritic responses.

- Active and passive propagation of electric signals are compared in terms of their basic features.

- The basics of second-messenger systems are explained clearly and simply.

- Mechanical models are used to help explain the basic functional properties of skeletal muscle and their receptors.

- The Equilibrium Point Hypothesis is presented in considerable detail. In the author's view, this is an attractive hypothesis on the control of movement, according to which, movement is essentially controlled through modifications of innate reflex responses, and the initial motor command from higher centers is based on an internal physiological variable rather than kinematic or dynamic variables of the intended movement. The end-objective of the movement is then reached using sensory feedback.

The book deals essentially with the human nervous system. Data and figures from the nervous systems of animals are included only for comparison purposes, when appropriate, or where corresponding data on humans is not available, or when necessitated by the nature of the discussion.

Every chapter includes learning objectives at the beginning, a summary of the main concepts at the end, and, starting from Chapter 3, one or more "Spotlight on Techniques" boxes that describe some methodology that is relevant to the topics discussed in the chapter. Problems, with answers, are included for enhancing and extending understanding of the material. An extensive bibliography for each chapter is included at the end of the book that can be used for further reading on the topics covered. A solutions manual is available through the publisher for adopting professors.

The sophisticated engineering of the neuromuscular system is both awesome and fascinating, as is true of biological systems in general. It is hoped that the book gives a good appreciation of the amazing features of the neuromuscular system and will contribute to stimulating some profound thinking about this system.

Acknowledgments

I am, as always, indebted to my students and colleagues for their valuable interactions and discussions relating to the subject matter of this book. I would also like to express my sincere appreciation of the efforts of the staff of CRC Press and their associates in producing and promoting this book, particularly Marc Gutierrez, Editor, Electrical & Biomedical Engineering, for his support and encouragement, Nick Mould, Editorial Assistant, for his professional and effective management of the publishing of the book, and Michelle van Kampen, Account Manager, Deanta Global Publishing Services, for her patience, understanding, and superb handling of the production process. I am also very grateful to the graphic designers, Bassel Fatayri, Luna Akil, and Dalida Raad, for their contributions in preparing the figures in this book.

Acknowledgments for Reproduction of Figures

I am grateful to the following for their kind permission to reproduce figures in this book:

- Oxford Publishing Limited, and Professor U.J. McMahan, for permission to reproduce Figure 5.3.
- John Wiley and Sons – Books – for their permission to reproduce Figures 5.11 and 7.8.

Convention for Symbols

The following convention for symbols is adhered to as much as possible:

- *Voltages and currents*: Small-case letters denote instantaneous values; capital letters denote dc or steady-state values.
- *Conductance, resistance, and capacitance*: capital letters denote per unit area values; small letters denote per unit length values.
- *Subscripts*:
  - Small-case m used with voltages, currents, resistance, conductance, or capacitance, refers to the membrane.
  - Small-case i refers to the intracellular medium.
  - Small-case o refers to the extracellular medium.
  - Small-case r denotes voltages with respect to the resting state as zero reference.

Author Biography

Nassir H. Sabah is professor of biomedical engineering in the Electrical and Computer Engineering department at the American University of Beirut, Lebanon. He received his B.Sc. (Hons. Class I) and his M.Sc. in electrical engineering from the University of Birmingham, U.K., and his Ph.D. in biophysical sciences from the State University of New York (SUNY/Buffalo). He has served as Chairman of the Electrical Engineering Department, Director of the Institute of Computer Studies, and Dean of the Faculty of Engineering and Architecture, at the American University of Beirut. In these capacities, he was responsible for the development of programs, curricula, and courses in electrical, biomedical, communications, and computer engineering. Professor Sabah has extensive professional experience in the fields of electrical engineering, electronics, and computer systems, with more than 35 years' teaching experience in neuroengineering, biomedical engineering, electronics, and electric circuits. He has over 100 technical publications, mainly in neurophysiology, biophysics, and biomedical instrumentation. He has served on numerous committees and panels in Lebanon and the region. He is a Fellow of the Institution of Engineering and Technology (IET, U.K.), a member of the American Association for the Advancement of Science (AAAS), and a member of the American Society for Engineering Education (ASEE).

1

Introduction: Background Material

Objective and Overview

The chapter provides some general background material for subsequent chapters and for following the relevant literature. The first part (Sections 1.1, 1.2, and 1.3) deals with some biological background, starting with living cells and cell organelles that are important for later discussions. Specialized cells, namely neurons and glia, which are the constituents of the nervous system, are considered next, with emphasis on some of their basic features. In order to appreciate the references to various parts of the nervous system in the following chapters, Section 1.3 presents the general organization of this system, including the main subdivisions of the brain.

Section 1.4 is concerned with some basic concepts on diffusion, fluxes, and potentials. The concept of chemical potential is explained and the expression for chemical potential derived. The concept of chemical potential and its expression are then generalized to those of electrochemical potential, and some important conclusions involving electrochemical potential are drawn. The definition of permeability and its relation to diffusion are considered.

The concepts introduced in Section 1.4 are used to discuss ionic equilibriums in Section 1.5, including the well-known case of the Gibbs–Donnan equilibrium. The chapter ends with an introduction to chemical kinetics and their extension to the kinetics of ion-channel gating.

1.1 Living Cells

All living cells have a surrounding envelope referred to as the **cell membrane**, or **plasma membrane**. Animal cells are **eukaryotic**, that is, they have a well-developed nucleus and other membrane-bounded **organelles**, which are cell elements that perform some specialized functions. Figure 1.1 illustrates a typical eukaryotic cell and some of its organelles. The part of the cell that is outside the nucleus and bounded by the cell membrane is the

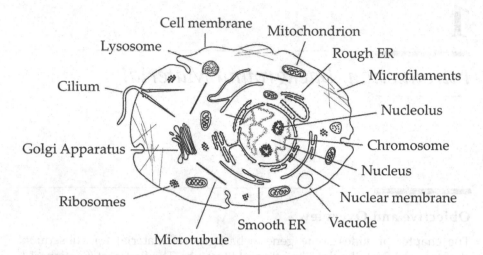

FIGURE 1.1
Eukaryotic cell and cell organelles.

cytoplasm. The **cytosol**, or **intracellular fluid**, is the liquid part of the cyto-plasm, exclusive of organelles. It consists of a complex mixture of substances that are dissolved or suspended in water. The cell membrane is discussed in considerable detail in Sections 2.1 and 2.2. The following cell organelles are particularly relevant for our purposes.

1.1.1 Endoplasmic Reticulum

The **endoplasmic reticulum (ER)** is an elaborate network of vesicles, tubules, and cisternae. It can take the form of a **rough endoplasmic reticulum (RER)** because of the presence of **ribosomes**, which are granular structures that are the sites of protein synthesis. Ribosomes can also be free in the cytosol in addition to being membrane-bound in the RER. The RER is thus involved in the synthesis of integral membrane proteins and proteins that are to be secreted outside the cell, including hormones. The **smooth endoplasmic reticulum (SER)** is almost devoid of ribosomes and is involved in the syn-thesis of lipids and steroids and in Ca^{2+} **homeostasis**, that is, regulation of Ca^{2+} concentration in the cytosol, by acting as the principal store of Ca^{2+} in the cell. A type of SER, the **sarcoplasmic reticulum (SR)**, is found in muscle cells (Section 9.1.2) and is specialized for the uptake and release of Ca^{2+}.

1.1.2 Mitochondria

The mitochondrion, illustrated in Figure 1.2, usually has an elongated shape, ranging between 0.1 and 0.5 μm in diameter and one to several micrometers in length, although mitochondria can be up to about 20 μm long. The number of mitochondria in a cell varies widely between cells of different tissues and

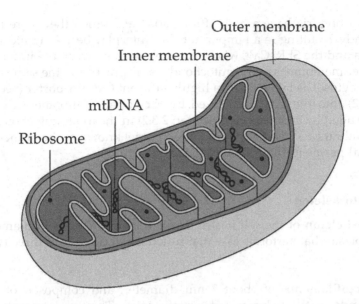

Outer membrane

Inner membrane

mtDNA

Ribosome

FIGURE 1.2
Mitochondrion.

different organisms, from a single mitochondrion to several thousand mitochondria per cell. Like the nucleus, mitochondria have a double membrane, the inner membrane being highly convoluted and forming folds called **cristae**, which greatly increases the surface area of the inner membrane. Mitochondria can fuse and can divide independently of the cell, in response to particular needs of the cell. They have their own DNA (mtDNA) that is used to synthesize mitochondrial RNAs and proteins. mtDNA is passed only by the mother to both her male and female offspring and changes very slowly with time. Although mitochondria perform many essential functions, including involvement in cell death, or **apoptosis**, two functions are of special interest for our purposes:

1. Mitochondria are the "power plants" of the cell because they generate most of the cell's supply of **adenosine triphosphate (ATP)**, which is used to fuel cell activities that consume energy, including ionic pumps (Section 2.3.1) and muscular contraction (Section 9.2). Many cellular processes requiring energy are coupled to the reaction of ATP with water (hydrolysis), to produce **adenosine diphosphate (ADP)** of lower energy. In many cases, proteins are activated, or "primed for action", by **phosphorylation**, which involves the addition of a phosphate group to an amino acid of the protein, usually through ATP donating a phosphate group to the protein and becoming converted to ADP in the process. We will encounter in future chapters many examples of proteins being activated by phosphorylation and deactivated by **dephosphorylation**.

2. Mitochondria have a high affinity for Ca^{2+}, which they store transiently, resulting in an important Ca^{2+} interplay between mitochondria and the SER. Ca^{2+}, whose concentration between the inner and outer membranes of the mitochondrion is practically the same as in the cytosol, is taken up by a highly efficient Ca^{2+} **uniporter** (Section 2.3.2) into the matrix enclosed by the inner membrane. Ca^{2+} are extruded by an **antiporter** (Section 2.3.2) in the inner mitochondrial membrane as well as by a complex channel known as the **mitochondrial permeability transition pore (mPTP)**.

1.1.3 Cytoskeleton

The **cytoskeleton** of the cell is a dynamic network of protein filaments in the cytoplasm that performs essential functions. It consists of three types of fibers:

1. **Microfilaments**, of about 7 nm diameter, and composed of two strands of the polymerized protein actin. They are concentrated mostly adjacent to the cell membrane and are attached to it at many points. They maintain cellular shape in two dimensions and allow the membrane to resist tension. Microfilaments cause dynamic changes in the shape of dendritic spines (Section 6.5.3).

2. **Intermediate filaments**, of about 10 nm diameter, and composed of a variety of proteins that differ between different types of cells. They are more strongly bound to the cell membrane than microfilaments and play a key role in the three-dimensional structure of the cell and in holding in place the Z disks and myofibrils in muscle cells (Section 9.1.2). They are especially abundant in axons of neurons (Section 1.2) and in cells of the epidermis, that is, the outer layer of the skin, where they constitute the major structural components of skin and hair. **Neurofilaments** are intermediate filaments found in nerve cells and are responsible for radial growth of the axon (Section 1.2) and hence determine the axon diameter.

3. **Microtubules** having a hollow tubular structure of about 15 nm inner diameter and about 24 nm external diameter, their length ranging dynamically between a fraction of a μm and hundreds of μms. They are composed of polymers of the protein tubulin. Microtubules are important components of: (i) **cilia** (Figure 1.1) – short hair-like projections from cells, which are capable of a beating movement that, for example, propels mucus along air passageways of the lungs, and (ii) **flagella**, which are long tapering processes from cells, which are responsible for movement of microorganisms as well as sperm cells. Microtubules play a key role in cell division and in intracellular transport, as described later.

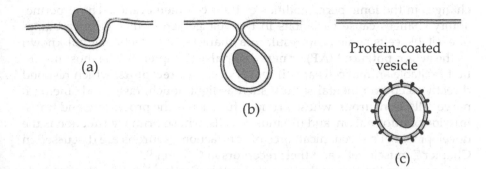

Protein-coated
vesicle

(a)

(b)

(c)

FIGURE 1.3
Endocytosis.

1.1.4 Endocytosis and Exocytosis

Endocytosis is a process by which extracellular material, such as large polar molecules, or molecules inserted in the cell membrane, such as receptors, are **internalized**, that is, brought inside the cell. First, an invagination of the cell membrane is formed at the location of the material to be internalized, as illustrated in Figure 1.3a for an extracellular molecule. The invagination then enlarges into the cell, with a narrow restriction, or neck, adjacent to the cell membrane (Figure 1.3b). Finally, pinch-off occurs at the neck, leaving a membrane-bounded vesicle (Figure 1.3c) of about 100 nm diameter that is coated with the protein clathrin. The type of this endocytic vesicle and the pathway that is subsequently followed depend on the particular endocytic function. Larger, membrane-bounded and protein-coated sacs, referred to as **endosomes** of several types, are usually involved in these pathways. The **early endosome** sorts the vesicles according to their destination. Material to be degraded is sent to another type of endosome, the **late endosome,** on its way to a **lysosome**, which is a spherical organelle that contains enzymes that break up cellular debris and other cellular material to be destroyed (Figure 1.1). Receptors to be recycled back to the cell membrane are sent to a third type of endosome, the **recycling endosome**.

The reverse of endocytosis is **exocytosis**, in which a membrane-bounded vesicle containing some molecules fuses with the cell membrane, thereby releasing the molecules into the extracellular space. Secretion of hormones and release of neurotransmitters (Section 6.1.2.2) is by exocytosis.

1.2 Neurons and Glia

1.2.1 Neurons

An important subset of living cells is **excitable cells**, which, when stimulated by an adequate stimulus of appropriate strength, undergo specific changes in the ionic permeabilities of their cell membranes. These permeability changes cause variations in the voltage across the cell membranes of excitable cells, which can result in a characteristic electric signal known as the **action potential (AP)** or **nerve impulse** (Chapter 3). The most important excitable animal cells are: (i) sensory cells, or **receptors**, which respond directly to environmental stimuli such as light, touch, taste, and smell, (ii) nerve cells, or **neurons**, whose primary function is the processing and transmission of information, and (iii) muscle cells, whose primary function is the development of a mechanical force of contraction. Neurons are discussed in Chapter 7, muscle cells and their receptors in Chapter 9.

Neurons are the core of the nervous system (Section 1.3). They have a wide variety of sizes and shapes that serve their particular functions (Section 7.1). The number of neurons in the brain is estimated at 86 billion, with less than 1 billion in the brainstem (Section 1.3) and the spinal cord. Typically, a neuron has four distinct regions that are specialized in terms of function (Figure 1.4):

1. The cell body, also referred to as the **perikaryon** or **soma** (plural **somas** or **somata**), and which contains the nucleus.
2. The dendritic tree, or **dendrites**, which are a branched, tree-like structure that extends for up to about a few mms from the cell body.

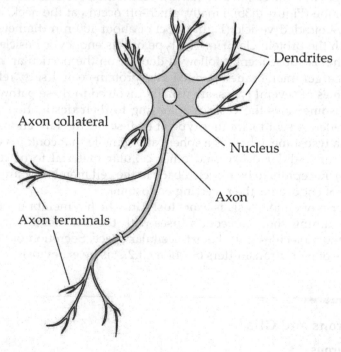

FIGURE 1.4
Typical neuron.

The dendrites are only diagrammatically illustrated in Figure 1.4. They are generally much more elaborate (Figure 7.1).

3. The **axon**, which is a relatively long process that extends from the cell body and which can vary in length from a fraction of a millimeter to about one meter in an adult human. Axons branch mostly near their distal ends (Figure 1.4), the branching being quite extensive in some cases. An axonal branch close to the cell body is an **axon collateral**.

4. Axon terminals at the ends of axonal branches. Axon terminals that lie entirely within the brain and spinal cord form specialized structures, known as **synapses** (Chapter 6), located mainly on the dendrites and cell bodies of other neurons. Axon terminals that lie outside the brain and spinal cord end up in **neuromuscular junctions** on skeletal muscles (Chapter 5), in synapses of the autonomic nervous system (Section 1.3), in glands, or in various tissues and organs whose activity is regulated by the nervous system, such as blood vessels and the heart.

Neurons communicate with one another over the short term, that is, over a time interval less than a fraction of a second, by means of electric signals such as the AP. The cell bodies and dendrites of neurons are specialized for the *reception and processing of electric signals* from other neurons, whereas axons are specialized for the *transmission* of the AP from a given neuron to the axon terminals. In the vast majority of cases, axon terminals are specialized for the release of a characteristic chemical, the **neurotransmitter**.

The neuron illustrated in Figure 1.4 is a particular neuron, the motor neuron, or **motoneuron**, whose cell body is in the anterior, that is, the frontal, part of the spinal cord. The axon of the motoneuron leaves the spinal cord and terminates on skeletal muscle cells in a type of synapse known as the neuromuscular junction (Chapter 5). Motoneuron axons that terminate on the distal parts of the leg, for example, are more than a meter long in an adult human. A motoneuron is shown in Figure 1.5 having an axon collateral terminating on another neuron, the **Renshaw cell**, whose axon terminates, in turn, on the same motoneuron as well as on other motoneurons nearby. Neurons, such as the Renshaw cell, that connect with other neurons over a relatively short distance are **interneurons**, discussed in more detail in Section 7.1.

The term **nerve fiber** covers all slender processes, that is, thin projections from the cell body of a neuron, that conduct electric signals, usually in the form of APs, as in the case of axons. Nerve fibers that conduct APs towards the brain and spinal cord are termed **afferent**, whereas nerve fibers that conduct APs away from the brain and spinal cord, like the motoneuron axon, are termed **efferent**. The terms afferent and efferent are also commonly applied to a particular region of the brain.

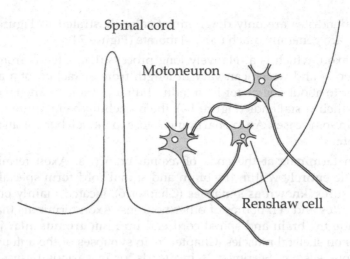

FIGURE 1.5
Motoneurons and Renshaw cells in the spinal cord.

A **peripheral nerve** or simply a **nerve** is a cable-like bundle of nerve fibers in the peripheral nervous system. A **fiber tract**, or a **tract**, is a bundle of nerve fibers in the central nervous system. A **nerve fascicle**, or **fascicle**, is a small fiber tract whose nerve fibers have similar origin, termination, and function. A bundle of one or more nerve fascicles is a **funiculus.**

Functionally related neurons often have their cell bodies grouped together into dense, well-defined groupings. Such a grouping of cell bodies is referred to as a **nucleus**, if it occurs inside the central nervous system, and as a **ganglion**, if it occurs in the peripheral nervous system, although there are exceptions to this terminology, as in the case of the basal ganglia (Section 1.3).

Neural tissue that consists predominantly of cell bodies and their unmyelinated fibers is described as **gray matter** because of its color, whereas neural tissue that consists predominantly of myelinated axons (Section 4.5) is described as **white matter** because of the white appearance of the myelin sheath.

1.2.2 Axonal Transport

In most cells, substances move within the cell by diffusion. In the case of neurons having long axons, however, some other mechanism must be used to exchange substances between the cell body and the distal parts of axons. Specialized motor proteins, fueled by ATP, move substances in the **retrograde** (toward the cell body) and **orthograde** (away from the cell body), directions using microtubules as "tracks". Some substances, such as cytoskeletal proteins, are transported at a slow rate of 1–10 mm/day, whereas other substances, such as enzymes, lipids, and other small-molecular weight material, are transported at a fast rate of up to 400 mm/day. Mitochondria are transported at an intermediate rate. Axonal transport is discussed further in Section 5.2 in connection with the neuromuscular junction.

1.2.3 Glial Cells

The nervous system consists of two types of cells: neurons and **glial cells**, also referred to as **glia** or **neuroglia,** which are intimately associated with neurons. Like typical eukaryotic cells, glia have a nucleus and the usual organelles. They do not have axons, since they do not generate or propagate APs, but some glial cells have processes that resemble dendrites, which are used to transport substances from one end of the cell to the other. The relative numbers of neurons and glia vary between various parts of the brain and between species. In humans, the brain, as a whole, has roughly the same number of neurons and glia. Glia outnumber neurons in the cerebral cortex (Section 1.3) by a factor of roughly 3.7:1, whereas neurons outnumber glia in the cerebellum (Section 1.3) by a factor of roughly 4.3:1. The first function ascribed to glia was of that of "connective tissue of nerve tissue" that holds neurons together, as exemplified by the term neuroglia being derived from "nerve glue". Since then, several types of glial cells have been found to perform a variety of highly important functions.

Microglia are small glial cells of the brain and spinal cord that act as **microphages** by engulfing and digesting damaged neurons, cellular debris, and infectious organisms. They constitute the first and main immune defense of the central nervous system (CNS). **Oligodendrocytes** in the CNS and **Schwann cells** in the peripheral nervous system (PNS) provide the myelin sheath of axons, which very effectively increases the conduction speed of the AP (Section 4.2). **Astrocytes** in the CNS regulate the extracellular environment of neurons by removing excess ions, notably K^+ following APs, and by recycling neurotransmitters involved in synaptic transmission. By having processes that terminate on both blood capillaries and neurons, they regulate blood flow and supply neurons with glucose and oxygen. They seem to play a crucial role in many aspects of neuronal operation, as will be elaborated later in Chapters 6 and 7. **Ependymal cells** line the cavities of the CNS as discussed later (Section 1.3). Glial cells play a key role in the development of the nervous system and in guiding neurons and axons to their destinations. Glia have also been implicated in some pathological conditions, such as chronic pain, motor neuron disease, and Alzheimer's disease. Unlike neurons, they can divide, and may cause a type of brain tumor known as **glioma**. It was once believed that neurons that die cannot be replaced. But it has since been shown that neurons can be regenerated from special types of stem cells, and that astrocytes can sometimes change into neurons.

1.3 Organization of the Nervous System

The nervous system is divided into the **peripheral nervous system (PNS)** and the **central nervous system (CNS)**. The central nervous system consists of the brain, enclosed by the skull, or **cranium**, and the spinal cord, enclosed by the vertebral column. The peripheral nervous system, being the rest of the nervous system outside the brain and spinal cord, mainly comprises:

1. Neuronal aggregations referred to as **ganglia**.
2. Sensory cells and receptors that respond to external stimuli or to changes in the internal state of the body.
3. Nerve fibers found outside the brain and the spinal cord.

The nervous system is essentially, though not exactly, symmetrical with respect to a **midsagittal** plane, that is, a plane that divides the body into right and left halves.

The peripheral nervous system has two main subdivisions (Figure 1.6):

1. The **somatic nervous system**, concerned with sensory input to the central nervous system and with motor output to skeletal muscle.
2. The **autonomic nervous system**, concerned with the control of visceral functions such as heart rate, digestion, respiration, and perspiration. The autonomic nervous system has two main subdivisions: (i) the **sympathetic nervous system**, involved in the "fight-or-flight" response that mobilizes the body to respond to stressful or threatening conditions, and (ii) the **parasympathetic nervous system**, concerned with activities of the body at rest, such as digestion and waste elimination. Most organs and systems of the body receive both sympathetic and parasympathetic stimulation acting in opposition, thereby providing a more effective, finer control.

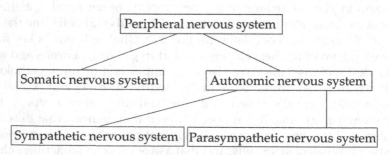

FIGURE 1.6
Subdivisions of the peripheral nervous system.

There are several classifications and names for the various brain structures, which is somewhat confusing. Conventionally, the brain may be divided into four major subdivisions, as indicated in Figure 1.7 and are illustrated in Figure 1.8 in a midsagittal section that divides the brain into right and left halves. These two halves of the brain are interconnected by a massive fiber tract, the **corpus callosum**, having a cross-sectional area of about 700 mm² and consisting of about 200 million fibers. A smaller tract, the **anterior commissure**, also connects the two hemispheres. The four major subdivisions

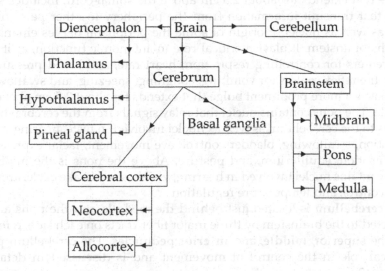

FIGURE 1.7
The four main subdivisions of the brain.

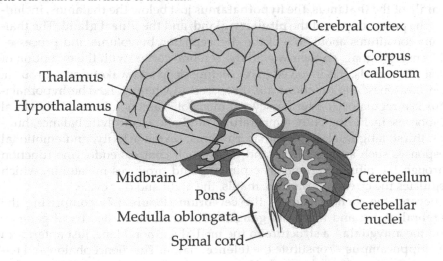

FIGURE 1.8
Midsagittal section through the brain.

will be described very briefly in what follows. More details about the structure of these subdivisions, their substructures, and their functions will be presented in future chapters as needed for our discussion of the neuromuscular system.

As one proceeds upwards from the spinal cord in humans, the first major subdivision encountered is the **brainstem**, consisting of three structures: first, the **medulla oblongata** (or simply the **medulla**), then the **pons**, and finally the **midbrain** or **mesencephalon** (Figure 1.8). The medulla, which is a bulge that extends for about 2.5 cm above the spinal cord, includes some nuclei that transmit information from the periphery to other parts of the brain, as well as nuclei of origin of some the peripheral nerves emanating from the brainstem. It plays a critical role in autonomic function, as it contains centers for controlling respiration, heart rate, and blood pressure as well as the reflex centers for vomiting, coughing, sneezing, and swallowing. The pons is a more prominent bulge that extends for about 2.5 cm above the medulla. The pons contains nuclei that relay signals from the cerebral hemispheres to the cerebellum, as well as nuclei involved in taste, hearing, sleep, respiration, swallowing, bladder control, eye movement, facial expressions and sensation, equilibrium, and posture. Above the pons is the midbrain which contains nuclei involved in hearing, vision, sleep/wake cycle, arousal, motor control, and temperature regulation.

The **cerebellum** is located just behind the medulla and the pons and is connected to the brainstem by three major fiber tracts on each side, referred to as the **superior**, **middle**, and **inferior peduncles**. The cerebellum plays a critical role in the control of movement and is discussed in detail in Chapter 12.

The next major subdivision encountered is the **diencephalon**, consisting mainly of the **thalamus**, the **hypothalamus** just below the thalamus, including the posterior part of the **pituitary gland**, and the **pineal gland**. The thalamus constitutes about 80% of the diencephalon by volume and processes all sensory signals on their way to the cerebral cortex, with the exception of olfactory signals. It is involved in regulating sleep and wakefulness, arousal, and awareness. Its functions are discussed in Chapter 12. The hypothalamus is a regulatory center for several metabolic, autonomic, and behavioral responses, including body temperature, fluid and electrolytic balance, hunger, thirst, fatigue, sleep, circadian rhythms, sexual activity, and emotional responses such as anger, fear, and pleasure. It controls endocrine function through the pituitary gland. The pineal gland produces melatonin, which regulates the **circadian rhythm**, that is, the night and day cycle.

Beyond the diencephalon is the **cerebrum** (Figure 1.7), comprising the **cerebral cortex** and the **basal ganglia**. The cerebral cortex, basal ganglia, and the **amygdala** – a structure in the medial temporal lobe, just anterior to the **hippocampus** – constitute the **telencephalon**. The diencephalon and telencephalon are referred to as the **forebrain**. The basal ganglia and thalamus are more clearly illustrated in the frontal section of Figure 12.6. The cerebral cortex, where the word cortex refers to a rind or outer covering, is divided into the **neocortex**, which in humans is almost the whole of the cerebral cortex, and the phylogenetically older **allocortex**. The cerebral cortex constitutes in humans about 77% of the brain by volume and up to 40% by mass. It is the highly convoluted, outer layer covering the two cerebral hemispheres. Its thickness varies between different regions from less than 2 mm to about 4.5 mm. It contains about 16 billion neurons in a well-defined structure of up to six layers in the neocortex, labeled I–VI, layer I being the most superficial. The large folds, or convolutions, of the cerebral cortex are called gyri (singular, **gyrus**) and are separated by fissures known as sulci (singular, **sulcus**). Some of the deep fissures are used as landmarks that divide the neocortex into four lobes that can be identified on the outer surface of each hemisphere: the **frontal, temporal, parietal,** and **occipital lobes** (Figure 1.9). The **central sulcus** separates the frontal lobe from the parietal lobe, and the **lateral sulcus** separates the temporal lobe from the parietal and frontal lobes. Two other lobes are tucked inside the cerebral hemispheres: (i) the **limbic lobe**, which is an arc-shaped region on the medial surface of each cerebral hemisphere, and is contiguous with parts of the frontal, parietal, and temporal lobes, and (ii) the **insular lobe**, located deep within the lateral sulcus. The cerebral cortex is involved in higher functions, including sensory perception, voluntary movement, conscious thought, and language.

The allocortex areas are in the peripheral parts of the cerebral cortex and have only three or four layers. The main constituents of the allocortex are the **hippocampus** and parts of the olfactory system concerned with the sense of smell. The hippocampus, located in the medial temporal lobe, acts as a memory gateway to the brain by contributing to short-term memory and by helping to consolidate short-term memory to long-term memory. It is also involved in spatial navigation and spatial memory and in some behavioral respects, such as the ability to inhibit previously learnt responses.

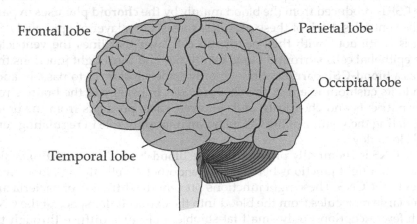

FIGURE 1.9
The four lobes of the brain.

The basal ganglia are a collection of nuclei situated on both sides of the thalamus. They are involved in a variety of functions, including motor function and the generation of voluntary movement in particular. They will be considered in more detail in Chapter 12. In actual fact, some of the nuclei of the basal ganglia are in the cerebrum, whereas other nuclei are in the midbrain.

It should be emphasized that the aforementioned subdivisions of the brain are primarily anatomical, not functional. Functional systems of the brain generally span across these anatomical subdivisions. For example, the **limbic system** supports a variety of functions such as emotion, motivation, learning, and formation of long-term memories. It includes the hippocampus, the amygdala (in the cerebrum), the **cingulate gyrus** (just above the corpus callosum), the hypothalamus, and the anterior thalamic nuclei (in the diencephalon).

The CNS is physically protected by:

1. A bony structure in the form of the skull and vertebral column.

2. A system of enveloping membranes, referred to as **meninges**. In mammals, the meninges are: the **dura mater** (outermost membrane), the **arachnoid mater**, and the **pia mater** (innermost membrane).

3. The cushioning effect of the **cerebrospinal fluid** (**CSF**) that surrounds the brain and spinal cord. The CSF fills the **subarachnoid space** between the arachnoid mater and the pia mater.

The CSF also circulates in four cavities, or **ventricles**, in the brain: two large lateral ventricles on either side in the cerebrum, a small third ventricle along the midline in the diencephalon, and a fourth ventricle of intermediate size along the midline at the back of the pons and the upper half of the medulla. The central canal of the spinal cord is continuous with the fourth ventricle. The CSF is produced from the blood mainly by the **choroid plexuses** in parts of the ventricles. Each of these plexuses consists of a layer of epithelial cells that is continuous with the ependymal cell layer that lines the ventricles. The epithelial cells surround blood capillaries and form tight junctions that act as a **blood-CSF barrier**, allowing only some substances to pass. In addition to its cushioning effect, the CSF provides buoyancy to the brain, circulates nutrients and chemicals, and removes waste products from the brain. Regulating the volume of the CSF plays an important role in regulating cerebral blood flow.

The CNS is chemically protected by the **blood-brain barrier** (**BBB**), which arises from tight junctions between the endothelial cells lining blood capillaries in the CNS. These tight junctions prevent the diffusion of bacteria and other large molecules from the blood into the extracellular space of the CNS. With few exceptions, only small fat-soluble molecules diffuse through the BBB. Substances that can diffuse through the BBB include alcohol, caffeine, nicotine, hormones, O_2, and CO_2, whereas other substances, including glucose, are actively transported through the BBB. The BBB prevents large drug molecules injected into the blood stream from reaching the cells of the CNS. On the other hand, the peripheral nervous system is vulnerable to neurotoxins because it is not chemically protected by a blood barrier.

It is helpful to note the meaning of some terms that are commonly used to designate relative anatomical locations in the CNS:

Rostral: situated toward a rostrum or beaklike projection, such as the oral and nasal region. In humans, it denotes a higher location for areas in the spinal cord, and a more frontal location for areas in the brain.

Caudal: situated toward a tail-like structure. It denotes a lower location in humans and a more posterior location in quadrupeds.

Anterior: located toward the front or facing forward.

Posterior: relating to the dorsal side in humans and to the rear end of the body in quadrupeds.

Ventral: relating to the abdomen, or the frontal direction of the body; often used synonymously with "anterior".

Dorsal: relating to the back in humans or to the upper direction in quadrupeds.

1.4 Diffusion, Fluxes, and Potentials

1.4.1 Chemical Potential

Atomic particles such as ions, atoms, or molecules that are free to move in a given medium will preferentially move from a region of higher concentration of the given particle to a region of lower concentration simply as a result of random thermal motion. If a drop of a water-soluble dye is deposited in a beaker of water, for example, the dye will gradually expand because of random thermal motion of dye molecules, until the dye is uniformly distributed throughout the beaker.

Since freely moving particles move, in general, from regions of higher potential energy to regions of lower potential energy, a type of potential energy, referred to as **chemical potential energy**, is associated with concentration. The motion of a freely moving particle down a concentration gradient can then be described as motion from a region of higher chemical potential energy of the given particle to a region of lower chemical potential energy, just as a free, electrically charged particle will move from a region of higher electric potential energy to a region of lower electric potential energy, or a free body will fall from a region of higher gravitational potential energy to a region of lower gravitational potential energy.

The expression for chemical potential energy per particle can be simply derived by considering a one-dimensional flow of particles along the x-axis. According to Fick's law, the flux U in number of particles per unit area per unit time passing through a plane normal to the x-axis at a point x is:

$$U = -D \frac{d[Y]}{dx} \tag{1.1}$$

where $[Y]$ is the concentration of a given particle Y at point x and D is a **diffusion constant** that depends on the resistance to the motion of the atomic particles in the given medium. D is thus a measure of how easily a given atomic particle can move under the influence of a concentration gradient in the medium under consideration; the larger D the less is the resistance to motion of the given particle and the larger is the flux of the particle per unit concentration gradient. D depends on the size and shape of the diffusing particle, on the viscosity of the medium, and on any interaction between the molecules of the medium and the particles in question.

If the concentration $[Y]$ decreases with x $(d[Y]/dx < 0)$, the flux is in the positive x-direction $(U > 0)$. The negative sign in Equation 1.1 ensures that D is a positive constant. When U is in particles/(cm²s), $[Y]$ is in particles/cm³, and x is in cm, then D is in cm²/s. For ions, typical values of D are 1.33×10^{-5} cm²/s for Na⁺, 1.96×10^{-5} cm²/s for K⁺, 2.03×10^{-5} cm²/s for Cl⁻, 9.31×10^{-5} cm²/s for H⁺, and 0.79×10^{-5} cm²/s for Ca⁺. For macromolecules, typical values are in the range of 10^{-8}–10^{-6} cm²/s.

For charged particles in an applied electric field, the **mobility** μ is defined as the *magnitude* of the average velocity v acquired per unit electric field ξ:

$$\mu = \left| \frac{v}{\xi} \right| = \frac{|v|}{|\xi|} \tag{1.2}$$

μ is thus a measure of how easily a given charged atomic particle can move under the influence of an applied electric field in the medium under consideration; the larger the μ the less is resistance to motion of the given particle and the larger is the velocity acquired by the particle per unit applied electric field.

v and ξ are in the same direction for positively charged particles and in opposite directions for negatively charged particles. Defining mobility in terms of the magnitude of the ratio v/ξ makes μ a positive quantity for both positively and negatively charged particles. When v is in cm/s and ξ is in volts/cm, μ is in cm²/(Vs).

Since both D and μ depend on resistance to the motion of the given particle in the medium under consideration, they are related, as given by the Nernst–Einstein relation:

$$D = \frac{\mu kT}{|z|q} \tag{1.3}$$

where T is the absolute temperature, k is Boltzmann's constant in J/K, q is the *magnitude* of the electronic charge, and z is the valence of the particle, including the sign. Thus, $z = +1$ for Na$^+$ and $z = -1$ for Cl$^-$. In both cases, $|z| = +1$.

The velocity acquired by a particle in a medium per unit driving force can be considered to be independent of the nature of the force, depending only on the particle and the medium. Considering diffusion to be due to some "diffusion force" F_d, the velocity acquired per unit of this force is v/F_d, and is a positive quantity because the velocity acquired is in the direction of force for a particle of positive mass. Under the influence of an electric field, the magnitude of the electric driving electric force is $|z|q|\xi|$. The magnitude of the velocity acquired per unit *driving electric force* is then $|v|/(|z|q|\xi|) = \mu/(|z|q)$, using Equation 1.2, and is a positive quantity as it should be. Equating this expression to v/F_d, on the basis that the velocity acquired per unit driving force is independent of the nature of the driving force, gives:

$$F_d = \frac{|z|qv}{\mu} \tag{1.4}$$

By definition, **flux** is the product of concentration and velocity:

$$U = [Y]v \tag{1.5}$$

Substituting in Equation 1.4, for v from Equation 1.5, for U from Equation 1.1, and for D/μ from Equation 1.3:

$$F_d = -\frac{kT}{[Y]}\frac{d[Y]}{dx} = -\frac{d}{dx}\left(kT \ln\frac{[Y]}{[Y]_{ref}} + \kappa'_{ref}\right) \tag{1.6}$$

where $[Y]_{ref}$ and κ'_{ref} are arbitrary constants, and the argument of the log function is dimensionless, as it should be. Since force is quite generally the negative of a potential energy gradient, the bracketed term in Equation 1.6 is identified as the chemical potential energy per particle κ'. Thus:

$$\kappa' = kT \ln\frac{[Y]}{[Y]_{ref}} + \kappa'_{ref} \ \text{J/particle} \tag{1.7}$$

where κ'_{ref} is the chemical potential energy per particle of a reference state having a concentration $[Y]_{ref}$. This is in accordance with the basic principle that potential energy is generally defined with respect to an arbitrary zero reference. Thus, the zero for electric potential energy is conventionally taken as that of the earth, and the zero for gravitational potential energy is conventionally taken at sea level.

Problem 1.1

Using the relation $\xi = -\dfrac{dv}{dx}$, where ξ is the electric field, and v is voltage, verify that Equation 1.6 applies in the form of electric force being equal to the negative of the gradient of electric potential energy.

In biological systems, it is usually convenient to define **chemical potential** $\tilde{\kappa}'$ for a given substance as chemical potential energy per gram-molecular weight, or **mole**, of the given substance rather than per particle, where a mole, whose symbol is mol, is formally defined as the amount of substance that contains as many atomic particles, such as molecules, atoms or ions, as 12 g of carbon 12. This number is in fact Avogadro's constant N_A (nearly 6.022×10^{23}). One mole of a substance is also the amount in grams that is equal to the molecular weight. The molar concentration, or molarity, of a solute is the number of moles/liter, and is denoted by M. Thus, a concentration of 100 mM of Na^+ is equivalent to a 0.1 of a mole of Na^+ per liter of solution. It is also 0.1 of a millimole per cm^3 of solution.

Since a mole contains N_A particles, $\tilde{\kappa}' = N_A \kappa$. Multiplying both sides of Equation (1.7) by N_A gives:

$$\tilde{\kappa}' = RT \ln \frac{[Y]}{[Y]_{\text{ref}}} + \tilde{\kappa}'_{\text{ref}} \text{ J/mole} \tag{1.8}$$

where $R = N_A k$ is the gas constant.

1.4.2 Electrochemical Potential

When the atomic particle under consideration is electrically charged and in a region where the voltage is v, the particle possesses, in general, electric potential energy because of its electric charge and chemical potential energy because of its concentration. The algebraic sum of these two potential energies is the **electrochemical potential energy**. Since we will be considering ions in solution, the electric charge of an ion Y is conveniently expressed as $z_Y q$, where q is the magnitude of the electronic charge and z_Y is the valence of the ion, including the sign. The electric potential energy per particle is then $z_Y q v$, and is positive for a positive ion and negative for a negative ion, assuming v is positive. The expression for the electrochemical potential energy per particle K is obtained by adding $z_Y q v$ to κ' in Equation 1.7 to obtain:

$$\kappa = kT \ln \frac{[Y]}{[Y]_{\text{ref}}} + z_Y q (v - v_{\text{ref}}) + \kappa_{\text{ref}} \tag{1.9}$$

where κ_{ref} is the electrochemical potential energy per particle of a reference state having a concentration $[Y]_{ref}$ and a voltage v_{ref}.

If we define **electrochemical potential** $\tilde{\kappa}$ as electrochemical potential energy per mole, then multiplying both sides of Equation 1.9 by N_A gives:

$$\tilde{\kappa} = RT \ln \frac{[Y]}{[Y]_{ref}} + z_Y F(v - v_{ref}) + \tilde{\kappa}_{ref} \tag{1.10}$$

where $F = N_A q$ is Faraday's constant and is approximately 96,500 C/mol.

Electrochemical potential is of fundamental importance in considering ionic systems, as will become clear from later discussions. We will next show that when ions move under the influence of both a concentration gradient and an electric potential gradient, the effective driving force is the negative of the electrochemical potential gradient. To do this, we first substitute for v from Equation 1.4 in Equation 1.5 and using Equation 1.3:

$$U_d = \frac{D}{kT}[Y]F_d \tag{1.11}$$

where U_d is identified as the flux due to diffusion.

Under the influence of an electric field, the force on a charged particle is $F_e = z_Y q \xi = -z_Y q dv/dx$. By substituting for μ from Equation 1.3, Equation 1.2 can be written, with due regard to the sign, as:

$$v = \frac{D}{kT}(z_Y q \xi) = \frac{D}{kT}F_e = -\frac{D}{kT}\left(z_Y q \frac{dv}{dx}\right) \tag{1.12}$$

According to Equation 1.12, v is in the same direction, that is having the same sign, as F_e and ξ for a positively charged particle ($z_Y > 0$), and is in the opposite direction for a negatively charged particle ($z_Y < 0$), which is evidently correct.

If the flux due to the electric field is U_e, then substituting for v in Equation 1.5:

$$U_e = -\frac{z_Y q D}{kT}[Y]\frac{dv}{dx} \tag{1.13}$$

This is the same as replacing U_d in Equation 1.11 by U_e and F_d by $F_e = -zqdv/dx$. The total flux, using Fick's law for U_d, is:

$$U_d + U_e = -D\frac{d[Y]}{dx} - \frac{z_Y q D}{kT}[Y]\frac{dv}{dx} \tag{1.14}$$

$$= -\frac{D}{kT}[Y]\left(\frac{kT}{[Y]}\frac{d[Y]}{dx} + z_Y q \frac{dv}{dx}\right)$$

$$= -\frac{D}{kT}[Y]\frac{d}{dx}\left(kT\ln\frac{[Y]}{[Y]_{\text{ref}}} + z_Y q(v - v_{\text{ref}}) + \kappa_{\text{ref}}\right)$$

$$= -\frac{D}{kT}[Y]\frac{d\kappa}{dx} \tag{1.15}$$

Comparing Equations 1.15 and 1.11, it is seen that the effective driving force is $-d\kappa/dx$.

The current density J is related to flux as $J = z_Y q(U_d + U_e)$, so that:

$$J = -\frac{z_Y q D}{kT}[Y]\frac{d\kappa}{dx} \tag{1.16}$$

Equation 1.16 is referred to as **generalized Ohm's law,** since it reduces to Ohm's law in the absence of a concentration gradient (Problem 1.2). Note that if $d\kappa/dx < 0$, the particles move in the positive x-direction (Equation 1.15), and if the particles are positively charged ($z_Y > 0$), the current is in the positive x-direction ($J > 0$). Conversely, if the particles are negatively charged ($z_Y < 0$), they still move in the positive x-direction if $d\kappa/dx < 0$, but the current is in the negative x-direction.

Problem 1.2

Show that in the absence of a concentration gradient, Equation 1.16 reduces to Ohm's law at a point, $J = \sigma\xi$, where the conductivity $\sigma = |z_Y|q\mu[Y]$.

Equation 1.14 in terms of the total flux U can be expressed as:

$$U = -D\left[\frac{d[Y]}{dx} + \frac{z_Y q}{kT}[Y]\frac{dv}{dx}\right] \tag{1.17}$$

Equation 1.17 is known as the **Nernst–Planck equation for electrodiffusion** under steady-state conditions, where the time variation is zero and there is no bulk movement of the medium, that is, only the particle in question is moving.

Another useful relation involving diffusion was derived by Einstein, who showed that the distance x a particle diffuses in time t due to random thermal motion and, in accordance with Fick's law, is given by a Gaussian distribution having a mean square value of:

$$\bar{x}^2 = 2nDt \tag{1.18}$$

where n is the number of dimensions of motion of the particle in the medium. Thus, $n = 3$ for motion in three dimensions. In practice, the time for a particle to reach a target by diffusion depends on the diameter y of the target, the time being independent of y/x for one dimension, proportional to $\log(y/x)$ for two dimensions, and linear in y/x for three dimensions.

It may be noted that the Boltzmann relation for electrodiffusion under equilibrium conditions follows readily from the expression for electrochemical potential by considering that the reference state has an electrochemical potential of zero, that is, $\kappa_{ref} = 0$ in Equation 1.9. Under equilibrium conditions, $\kappa = 0$ as well (Section 1.5.2), it follows from Equation 1.9 that:

$$[Y] = [Y_{ref}]e^{\frac{-z_Y q(v - v_{ref})}{kT}} \tag{1.19}$$

According to Equation 1.19, concentration follows a Boltzmann-type distribution with respect to electric potential energy.

It should be noted that in considering chemical and electrochemical potentials, "potential" denotes potential energy per particle or per mole (N_A particles). Its unit is joules/per unit amount of substance, that is, joules/particle or joules/mole. In Equation 1.9, $z_Y q v$ is electric potential energy per particle assuming $v_{ref} = 0$. Strictly speaking, and in conformity with the interpretation of chemical and electrochemical potentials, "electric potential" should denote electric potential energy per particle and should not be used to denote voltage, as is often done, because voltage is electric potential energy per unit charge, that is, $v = z_Y q v/(z_Y q)$, independently of the sign of the charge. Its unit is joules/coulomb, or volt. In this book, a distinction is made between electric potential as electric potential energy per particle and voltage as electric potential energy per unit charge.

1.4.3 Permeability

Closely associated with the concept of diffusion is that of permeability. Consider a membrane extending from $x=0$ to $x=\delta$ (Figure 1.10). Under steady-state conditions, that is, conditions that do not change with time, the flux of a substance Y through the membrane is constant. Assuming that D_Y is constant within the membrane, Equation 1.1 can be integrated to give:

$$U_Y = \frac{D_Y}{\delta}\left([Y]_0 - [Y]_\delta\right) = \frac{D_Y \beta_Y}{\delta}\left([Y]_i - [Y]_o\right) \tag{1.20}$$

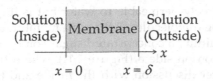

FIGURE 1.10
Diagrammatic illustration of a membrane separating two solutions.

where $[Y]_0$ and $[Y]_\delta$ are the concentrations of the given substance *inside* the membrane at $x=0$ and $x=\delta$, respectively, and $[Y]_i$ and $[Y]_o$ are the corresponding concentrations of the given substance *in the solution* on either side of the membrane. In general, depending on the system under consideration, $[Y]_i \neq [Y]_0$ and $[Y]_o \neq [Y]_\delta$. In such cases it is usually assumed that the concentration ratios on either side of the membrane-solution interface are related by a **partition coefficient** for the given substance β_Y defined as:

$$\beta_Y = \frac{[Y]_0}{[Y]_i} = \frac{[Y]_\delta}{[Y]_o} \tag{1.21}$$

The **permeability** P_Y of the membrane to the given substance is defined as the flux per unit difference of concentrations of the given substance in the solution on the two sides of the membrane. Using Equations 1.20 and 1.3:

$$P_Y = \frac{U_Y}{[Y]_i - [Y]_o} = \frac{D_Y \beta_Y}{\delta} = \frac{\mu k T \beta_Y}{|z_Y| q \delta} \tag{1.22}$$

When U is in particles/(cm²s) and $[Y]$ is in particles/cm³, P is in cm/s, the same units as velocity. Permeability can therefore be considered as a measure of how fast a given substance can move through the membrane. If flux is expressed in mol/(cm²s) and concentration in mol/cm³ (which is $10^{-3} \times$M), P is also in cm/s.

1.5 Ionic Equilibriums

It is convenient to invoke semipermeable membranes in discussing ionic equilibriums, where a **semipermeable membrane** is a membrane that selectively allows certain substances to pass freely through the membrane in either direction while blocking other substances.

1.5.1 Osmotic Equilibrium

Consider, to begin with, a semipermeable membrane that, by virtue of the size of its pores, is freely permeable to water but is impermeable to larger molecules such as sugar molecules. Let such a membrane separate two compartments, one containing a concentrated sugar solution on side i, the other a dilute sugar solution on side o (Figure 1.11), where these two sides may be identified in a later discussion with the *inside* and the *outside* of a cell, respectively. How is equilibrium established?

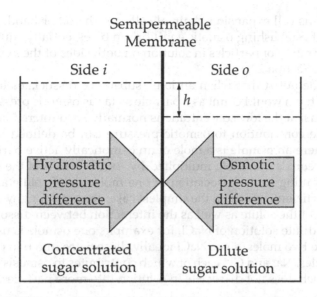

FIGURE 1.11
Equilibrium between concentrated and dilute sugar solutions.

Water is effectively more concentrated in the dilute solution on side *o*, which means that there will be a concentration gradient of water from side *o* to side *i*. Initially, there will be more water molecules striking the membrane on side *o* per unit area per unit time due to random thermal motion, or **Brownian motion**, than on side *i*, which results in a net flow of water down its concentration gradient, from side *o* to side *i*. This net flow raises the level of water on side *i* with respect to side *o*, creating a hydrostatic pressure that opposes the flow of water from side *o* to side *i*. Under equilibrium conditions, the tendency of water to flow down its gradient, from side *o* to side *i*, referred to as **osmotic pressure** difference, is balanced by an equal and opposite tendency of water to flow from side *i* to side *o* under the influence of the hydrostatic pressure difference due to a head *h*. Equilibrium is dynamic, meaning that flow in either direction does not stop but becomes equalized.

A living cell must maintain its volume in order to preserve its integrity. From a functional viewpoint, the concentration of particles in solution inside the cell is ordinarily larger than outside the cell, which means that water would flow down its concentration gradient into the cell, expanding cell volume. If the cell membrane is elastic enough, then as the cell expands in volume, an elastic force is developed that opposes the osmotic force. The elastic force, like that of an elastic spring, increases with the cell volume, so that an equilibrium state is reached when the opposing forces are equal. If, however, the cell membrane is not sufficiently elastic, and its elastic limit is exceeded, the cell would burst and die. Plant cells are surrounded by a rigid cell wall that constrains cell expansion. Animal cells, on the other hand, must have the means of establishing osmotic equilibrium by essentially equalizing the total concentration of particles in solution on both sides of the membrane, as discussed in Section 2.3.

It should be noted that when an ionic substance dissociates into ions in solution, each ion would count as a particle as far as osmotic pressure is concerned, at least at the low concentrations normally encountered in biological systems. The contribution to osmotic pressure can be defined in terms of osmoles, where an **osmole** is a mole of an osmotically active particle. Thus, one osmole equals one mole multiplied by $n \times o$, where n is the number of particles resulting from dissociation of one molecule of solute and o is the **osmotic coefficient**, which, in the simplest case, accounts for any incomplete dissociation of the solute as well as the interaction between dissociated particles. For a dilute solution of NaCl, for example, one osmole is usually considered to be two moles, since NaCl ideally dissociates into two osmotically active particles, Na^+ and Cl^-, each of which contributes to osmosis like a mole of sugar which does not dissociate in solution into more particles.

Problem 1.3

How would equilibrium be established in the system of Figure 1.11 if the membrane were also permeable to sugar molecules? What will be the osmotic pressure difference?

ANS.: Solutions are of equal concentrations on both sides; no osmotic pressure difference.

1.5.2 Basic Ionic Equilibrium

Consider a semipermeable membrane that separates two compartments, one containing a concentrated K^+Cl^- solution on side i, the other a dilute K^+Cl^- solution on side o, the membrane being permeable to K^+ but not to Cl^- (Figure 1.12a). How will such a system reach equilibrium? K^+, being free to diffuse through the membrane, will diffuse from side i to side o under the influence of the concentration gradient. However, as they do so, side o becomes positively charged with respect to side i, creating an electric potential difference that drives K^+ from side o to side i. Clearly, under equilibrium conditions, there will be no net movement of K^+ in either direction, which means that the tendency of K^+ to move from side i to side o under the influence of the concentration gradient is balanced by an equal and opposite tendency for K^+ to move from side o to side i under the influence of the electric potential gradient.

In order to derive a quantitative expression for the electric potential difference between the two sides at equilibrium, we note first that because of differences in total concentration of particles on the two sides, there will be a difference in osmotic pressure between the two sides and a consequent

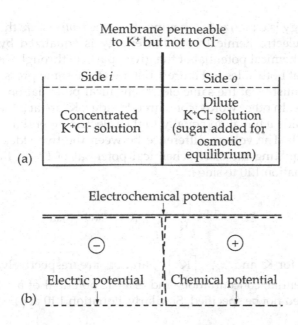

Membrane permeable
to K^+ but not to Cl^-

Side i | Side o

Concentrated
K^+Cl^- solution

Dilute
K^+Cl^- solution
(sugar added for
osmotic
equilibrium)

(a)

Electrochemical potential

$\ominus$ $\oplus$

Electric potential | Chemical potential

(b)

FIGURE 1.12
Ionic equilibrium. (a) two K^+Cl^- solutions separated by a semipermeable membrane; (b) variation of electric potential, chemical potential, and electrochemical potential between the two
sides at equilibrium.

difference in hydrostatic pressure as explained in connection with Figure 1.11.
But in view of the aforementioned argument that osmotic equilibrium is
maintained in animal cells, we will assume that the system of Figure 1.12a is
also in osmotic equilibrium, as by adding to side o an appropriate concentration of an electrically neutral substance, such as sugar, that does not diffuse
through the semipermeable membrane.

A fundamental attribute of the state of equilibrium is that no work is done
in taking a small quantity of a freely moving particle from any part of the
system to any other part, where a freely moving particle is a particle that
is free to move throughout the system, like K^+ in Figure 1.12a but not Cl^-.
Stipulating a small quantity ensures that the movement of such a quantity
throughout the system does not significantly disturb the system. When no
work is done in taking a small quantity of a freely moving particle from any
part of the system to any other part, this means that the freely moving particle
has the same potential energy throughout the system. If this is not the case,
then work is done on the particle in moving it from a region of low potential energy to a region of higher potential energy. Conversely, energy can be
extracted from the system if the particle is allowed to move from a region of
high potential energy to a region of lower potential energy. In either case,
the system, by definition, will not be at equilibrium. The potential energy in
a system can be mechanical, thermal, or electrochemical. Mechanical potential energy is equalized by having the same *pressure* throughout the system.

Thermal energy is equalized by having the same *temperature* throughout the system, and electrochemical potential energy is equalized by having the same electrochemical potential of the given particle throughout the system. Assuming that both sides in Figure 1.12a are at the same pressure and temperature, K^+ must be at the same electrochemical potential on both sides of the membrane. In other words, compared to side i, K^+ are at a lower chemical potential on side o but at an equally higher electric potential, as illustrated in Figure 1.12b. The voltage difference between the two sides can then be derived by equating the electrochemical potential of K^+ on the two sides. Applying Equation 1.10 to side i:

$$\kappa_{Ki} = kT \ln \frac{\left[K^+\right]_i}{\left[K^+\right]_{\text{ref}}} + q(v_i - v_{\text{ref}}) + \kappa_{K\text{ref}} \tag{1.23}$$

where $z_K = +1$ for K^+ and, $\kappa_{K\text{ref}}$, $\left[K^+\right]_{\text{ref}}$, and v_{ref} are, respectively, the electrochemical potential, concentration, and electric potential of a reference state for K^+ that need not be specified. Similarly, Equation 1.10 applied to side o is:

$$\kappa_{Ko} = kT \ln \frac{\left[K^+\right]_o}{\left[K^+\right]_{\text{ref}}} + q(v_i - v_{\text{ref}}) + \kappa_{K\text{ref}} \tag{1.24}$$

where the same reference state is used. Equating κ_{Ki} and κ_{Ko}, the quantities for the reference state cancel out, giving:

$$(v_i - v_{\text{ref}}) = E_K = -\frac{kT}{q} \ln \frac{\left[K^+\right]_i}{\left[K^+\right]_o} = -\frac{RT}{F} \ln \frac{\left[K^+\right]_i}{\left[K^+\right]_o} \tag{1.25}$$

With $\left[K^+\right]_i > \left[K^+\right]_o$, it follows that $v_o > v_i$, as argued previously.

Equation 1.25 is the **Nernst equation** for K^+. It relates the voltage difference between the two sides at equilibrium to the ratio of concentrations of a diffusible ion on both sides of the membrane. As it is derived from equality of electrochemical potential for a given ion, it applies to any diffusible ionic species, under conditions of equilibrium and equality of pressure and temperature throughout the system. E_K is known as the **Nernst potential**, or **equilibrium potential**, for K^+, but it will be referred to in this book as the **equilibrium voltage** because of the distinction between potential and voltage referred to previously.

Summary: *Any ionic species that is diffusible through the membrane will have, under conditions of equal pressure and temperature, the same electrochemical potential at equilibrium on both sides of the membrane. The concentration ratios of the given ions on the two sides are related to the voltage difference between the two sides by the Nernst equation.*

Problem 1.4

Determine the concentration ratio of K^+ that will result in a 20 mV voltage difference in the system of Figure 1.12.

ANS.: $\dfrac{[K]_i}{[K]_o} = 2.618$

In the case of living cells, an important deduction can be made concerning the magnitude of the charge needed to establish the voltage difference involved. Consider a small neuron whose cell body can be approximated as a sphere of diameter d μm. Let the voltage across the cell membrane be 70 mV and the membrane capacitance be 1 μF/cm^2 (Section 2.2). The charge in coulombs (C) across the cell membrane associated with the 70 mV transmembrane voltage is $(70 \times 10^{-3}\ V) \times (1\ \mu F/cm^2) \times \pi \times (10^{-4}d\ cm)^2 = 7 \times 10^{-16} \pi d^2\ C \cong 2.2 \times 10^{-15} d^2$ C. If we consider the concentration of K^+ alone inside the cell to be about 150 mM, or 0.15 moles/liter, the charge due to K^+ inside the cell is $0.15 \times 96{,}485 \times 10^{-3} \times (\pi/6) \times (10^{-4}d\ cm)^3 \cong 7.6 \times 10^{-12} d^3$ C. The ratio of the charge associated with the 70 mV transmembrane voltage to the charge of K^+ alone inside the cell is $2.9 \times 10^{-4}/d$. Even if $d = 1$ μm, this ratio is only about 0.03%. It can be concluded, therefore, that *the quantity of charge needed to establish the voltage difference across biological membranes is very small compared to the charge of either polarity that is present on the two sides of the membrane.*

An implication of the smallness of the charge transferred is that in the system of Figure 1.12, where *the movement of ions of one polarity is involved, the final concentrations of ions on either side of the membrane at equilibrium are practically the same as the initial concentrations.*

Problem 1.5

Determine: (a) the number of ions in 1 cm^3 of a 100 mM solution; (b) the number of monovalent ions that would give a charge of 10^{-7} C, considering the electronic charge to be 1.602×10^{-19} C.

ANS.: (a) 6.022×10^{19} ions per cm^3; (b) 6.24×10^{11} ions.

1.5.3 Equilibrium Voltage

The Nernst equation for K^+ (Equation 1.25) can be generalized for an ion Y as:

$$E_Y = v_i - v_o = -\frac{kT}{z_Y q} \ln \frac{[Y]_i}{[Y]_o} = -\frac{RT}{z_Y F} \ln \frac{[Y]_i}{[Y]_o} \tag{1.26}$$

where $E_Y = v_i - v_o$ is in the assigned positive direction for the voltage across the cell membrane, that is, the voltage of the inside with respect to the outside. When z_Y is positive and $[Y]_i > [Y]_o$, Y will flow from the inside to the outside under the influence of its concentration gradient. The membrane voltage at equilibrium will then have to be outside positive with respect to the inside, so as to oppose the concentration gradient. E_Y will therefore have a negative value, in accordance with Equation 1.26.

If the temperature is 37°C ($T = 310$ K), numerical values are used for R and F ($R = 8.3147$ J/(mol.K), $F = 96,485$ C/mol), and the natural logarithm is converted to logarithm to base 10 (by multiplying it by $\ln(10) = 2.3026$), then $(RT/F) \times 2.3026 = 61.71$ mV. Equation 1.25 for K^+ ($z = 1$) becomes:

$$E_K \cong -62 \log_{10} \frac{\left[K^+\right]_i}{\left[K^+\right]_o} \text{ mV} \qquad (1.27)$$

Problem 1.6

Two ions A and B of valencies z_A and z_B, respectively, are at equilibrium on the two sides 1 and 2 of a membrane. What is the relation between the concentration ratios of the two ions on either side of the membrane? If A is Ca^{++} ions having a concentration ratio of 100, determine the concentration ratio of B if B is: (a) K^+ ions, (b) Cl^-.

ANS.: $\left(\dfrac{[A]_i}{[A]_o}\right)^{1/Z_A} = \left(\dfrac{[B]_i}{[B]_o}\right)^{1/Z_B}$; (a) $\dfrac{\left[K^+\right]_i}{\left[K^+\right]_o} = 10$; (b) $\dfrac{\left[Cl^-\right]_o}{\left[Cl^-\right]_i} = 10$.

1.5.4 Gibbs–Donnan Equilibrium

The Gibbs–Donnan equilibrium is a type of ionic equilibrium that was believed at one time to model the generation of a resting voltage across the cell membrane. Consider a semipermeable membrane that separates two compartments, one containing a solution of K^+Cl^- solution on side o, the other a solution of $K^+Cl^-A^-$ on side i, where A^- is a large anion that does not diffuse through the membrane, unlike K^+ and Cl^- (Figure 1.13). How does the system reach equilibrium?

We can deduce from a simple qualitative argument that at equilibrium, side i will be at a negative voltage with respect to side o. Suppose that the solution on side i did not contain any A^-. Then since the membrane is permeable to K^+ and Cl^-, equilibrium is established with equal concentrations of these ions on the two sides of the membrane. There will be no net charge on either side and, hence, no voltage difference. Suppose now that a number of Cl^- on side i are instantaneously replaced by an equal number of A^-. The charge balance will not be disturbed, but Cl^- will be more concentrated on side o than on side i. There will be a movement of Cl^- from side o to side i under the influence of this concentration gradient, making side i negative with respect to side o. K^+ will also be attracted to the negatively charged

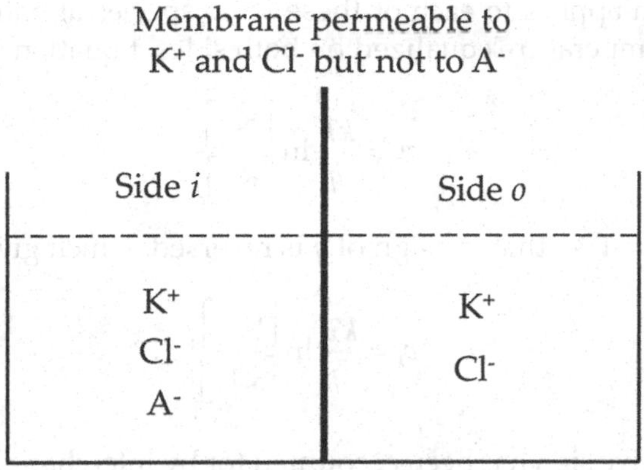

Membrane permeable to
K⁺ and Cl⁻ but not to A⁻

Side *i*	Side *o*
K⁺	K⁺
Cl⁻	Cl⁻
A⁻	

FIGURE 1.13
Gibbs–Donnan Equilibrium.

side *i*. It follows that at equilibrium $[Cl^-]_o > [Cl^-]_i$ and $[K^+]_i > [K^+]_o$. It cannot be argued that the movement of K⁺ completely neutralizes the negative charge on side *i*, thereby resulting in zero voltage across the membrane. For in order to have both Cl⁻ and K⁺ at equilibrium, with $[Cl^-]_o > [Cl^-]_i$ and $[K^+]_i > [K^+]_o$, side *i* must be negative with respect to side *o*, so that the electric potential difference will oppose diffusion of K⁺ from side *i* to side *o*, and diffusion of Cl⁻ from side *o* to side *i*.

Before analyzing the system, it is important to note an argument that is often invoked in the analysis of ionic systems. At distances sufficiently far from the membrane, the effect of the membrane is not significantly felt. Any small volume of solution on either side of the membrane can therefore be considered to be electrically neutral because each solution that was introduced on either side was electrically neutral to begin with, and the system as a whole is electrically neutral, or electroneutral. Because the net charge transferred to establish equilibrium is small and is confined to a distance of the order of atomic dimensions on either side of the membrane, the concentration of a given ionic species can be assumed to be the same throughout a given side of the membrane except in the immediate vicinity of the membrane. Electroneutrality applied to side *i* gives:

$$\left[K^+\right]_i = \left[Cl^-\right]_i + \left[A^-\right]_i \tag{1.28}$$

and for side *o*:

$$\left[K^+\right]_o = \left[Cl^-\right]_o \tag{1.29}$$

To simplify the analysis, it will be assumed that some measure has been taken to establish osmotic equilibrium, as in Figure 1.12, and that the temperature and pressure are the same on both sides. Since K⁺ and Cl⁻ are diffusible, the

Nernst equation applies to each of these ionic species at equilibrium, with pressure and temperature equalized on both sides. Equation 1.25 for K^+ is:

$$v_0 - v_i = \frac{kT}{q} \ln \frac{\left[K^+ \right]_i}{\left[K^+ \right]_o} \qquad (1.30)$$

For Cl^- ions, $z_{Cl} = -1$, so that the sign of q is reversed, which gives:

$$v_0 - v_i = \frac{kT}{q} \ln \frac{\left[Cl^- \right]_o}{\left[Cl^- \right]_i} \qquad (1.31)$$

where the negative sign has been accounted for by interchanging the numerator and denominator in the argument of the log function. It follows from Equations 1.30 and 1.31 that:

$$\left[K^+ \right]_i \left[Cl^- \right]_i = \left[K^+ \right]_o \left[Cl^- \right]_o \qquad (1.32)$$

Note that all the concentrations in the preceding equations are those *at equilibrium*. Since both K^+ and Cl^- are diffusible through the membrane, we cannot, strictly speaking, assume that the final concentrations are nearly the same as the initial concentrations, as in Figure 1.12, where movement of K^+ only is involved.

Dividing both sides of Equation 1.28 by $\left[K^+ \right]_i$:

$$1 = \frac{\left[Cl^- \right]_i}{\left[K^+ \right]_i} + \frac{\left[A^- \right]_i}{\left[K^+ \right]_i} \qquad (1.33)$$

Let $\eta = \dfrac{\left[K^+ \right]_i}{\left[K^+ \right]_o}$, where η is the **Gibbs–Donnan ratio**. From Equation 1.32,

$\eta = \dfrac{\left[Cl^- \right]_o}{\left[Cl^- \right]_i}$. The ratio $\left[Cl^- \right]_i / \left[K^+ \right]_i$ in Equation 1.33 can be expressed as:

$$\frac{\left[Cl^- \right]_i}{\left[K^+ \right]_i} = \frac{\left[Cl^- \right]_i}{\left[Cl^- \right]_o} \times \frac{\left[Cl^- \right]_o}{\left[K^+ \right]_i} = \frac{\left[Cl^- \right]_i}{\left[Cl^- \right]_o} \times \frac{\left[K^+ \right]_o}{\left[K^+ \right]_i} = \frac{1}{\eta^2} \qquad (1.34)$$

where $\left[Cl^- \right]_o / \left[K^+ \right]_i$ has been replaced by $\left[K^+ \right]_o / \left[K^+ \right]_i$ in accordance with Equation 1.29. Substituting for $\left[Cl^- \right]_i / \left[K^+ \right]_i$ in Equation 1.33 and solving for η:

$$\eta = 1 \Big/ \sqrt{1 - \frac{\left[A^- \right]_i}{\left[K^+ \right]_i}} \qquad (1.35)$$

But from equation 1.28, $\left[A^-\right]_i < \left[K^+\right]_i$, so that $\eta > 1$. It follows that $\left[K^+\right]_i > \left[K^+\right]_o$, $\left[Cl^-\right]_o > \left[Cl^-\right]_i$, and $v_o > v_i$, as concluded from the qualitative discussion.

Finally, let us compare the total ionic concentrations on both sides. The total concentration of ions on side i is, from Equation 1.28:

$$[\text{total}]_i = \left[K^+\right]_i + \left[Cl^-\right]_i + \left[A^-\right]_i = 2\left[K^+\right]_i \qquad (1.36)$$

whereas the total concentration of ions on side o is, from Equation 1.29:

$$[\text{total}]_o = \left[K^+\right]_o + \left[Cl^-\right]_o = 2\left[K^+\right]_o \qquad (1.37)$$

But $\left[K^+\right]_i > \left[K^+\right]_o$, which means that $[\text{total}]_i > [\text{total}]_o$. The system will not be in osmotic equilibrium, unless a concentration $\left([\text{total}]_i - [\text{total}]_o\right)$ of some neutral solute is added to side o, as in Figure 1.12.

Problem 1.7
Show that if the indiffusible anions are positively charged, the side containing these ions will be positively charged with respect to the other side.

Problem 1.8
Show that in the case of a mixture of univalent diffusible ions in the Gibbs–Donnan equilibrium, cations X_1, X_2, etc. are distributed as K^+, whereas anions Y_1, Y_2, etc. are distributed as Cl^-,

$$\frac{\left[K^+\right]_i}{\left[K^+\right]_o} = \frac{[X_1]_i}{[X_1]_o} = \frac{[X_2]_i}{[X_2]_o} = \dots = \frac{\left[Cl^-\right]_o}{\left[Cl^-\right]_i} = \frac{[Y_1]_o}{[Y_1]_i} = \frac{[Y_2]_o}{[Y_2]_i} = \dots = \eta.$$

Problem 1.9
Consider that in the system of Figure 1.13, the initial concentrations are:

Inside: $\left[A^-\right]_{i0} = 50\,\text{mM}$, $\left[Cl^-\right]_{i0} = 100\,\text{mM}$, $\left[K^+\right]_{i0} = 150\,\text{mM}$

Outside $\left[Cl^-\right]_{o0} = 150\,\text{mM}$, $\left[K^+\right]_{o0} = 150\,\text{mM}$

Determine the final concentrations of K^+ and Cl^-, assuming electroneutrality.

ANS.: Inside $\left[A^-\right]_{i0} = 50\,\text{mM}$, $\left[Cl^-\right]_{iF} = 113.6\,\text{mM}$, $\left[K^+\right]_{iF} = 163.6\,\text{mM}$

Outside: $\left[Cl^-\right]_{oF} = 136.4\,\text{mM}$, $\left[K^+\right]_{oF} = 136.4\,\text{mM}$

1.6 Chemical Kinetics

1.6.1 Reaction Rates

In a chemical reaction involving a given substance, or species A, where A may be a reactant being consumed by the reaction or a product being produced by the reaction, the instantaneous rate of change of the concentration of A is its time derivative:

$$\text{Rate of change of concentration of } A = \frac{d[A]}{dt} \qquad (1.38)$$

where [A] is the concentration of A at time t and $d[A]/dt$ is the slope of the graph of [A] with respect to time at time t. This derivative has a positive value for a product and a negative value for a reactant.

In general, a reaction cannot proceed unless reacting molecules possess a minimum energy known as the **activation energy,** E_a, of the reaction (Figure 1.14). The horizontal coordinate, representing the progress of the reaction along a reaction pathway, is referred to as the **reaction coordinate**. It is usually a geometric parameter, such as bond length or bond angle, that changes during the conversion of one or more reactants into one or more products. A simple classical example is the breaking of a covalent bond in the dissociation of a hydrogen molecule into two hydrogen atoms.

The rate of reaction under given conditions basically depends on two main factors: (i) the concentrations of the reactants, and (ii) the height of this energy barrier compared to the energies of the reactant molecules, as will be explained later.

Consider the reaction:

$$O_2 + 2H_2 \rightarrow 2H_2O \qquad (1.39)$$

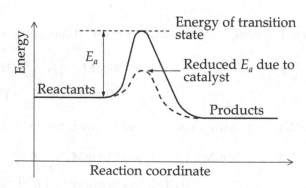

FIGURE 1.14
Energy profile of chemical reaction.

This is a **stoichiometric equation**, where **stoichiometry** is concerned with the relative quantities of reactants and products in a chemical reaction, and a stoichiometric equation shows the quantitative relationship between reactants and products. A proper stoichiometric equation must be balanced, that is, the number of atoms of any element must be the same on both sides of the equation in accordance with conservation of mass. Thus, in Equation 1.39, there are two atoms of oxygen and four atoms of hydrogen on either side. The number multiplying each species in the stoichiometric equation is the **stoichiometric coefficient**.

The **rate of reaction**, r, for any species A is defined as:

$$r = \pm \frac{1}{\varsigma} \frac{d[A]}{dt} \qquad (1.40)$$

where ς is the stoichiometric coefficient of species A, the positive sign is used for products ($d[A]/dt > 0$) and the negative sign is used for reactants ($d[A]/dt < 0$). This makes the rate of reaction positive for both reactants and products. Including the stoichiometric coefficient in Equation 1.40 equalizes the rates of reaction for the various reactants and products. Thus, for the reaction of Equation 1.39,

$$\text{rate of reaction} = \frac{1}{2} \frac{d[H_2O]}{dt} = -\frac{1}{2} \frac{d[H_2]}{dt} = -\frac{d[O_2]}{dt} \qquad (1.41)$$

This means that the rate of increase of $[H_2O]$ is the same, in magnitude, as the rate of decrease of $[H_2]$ and is twice the rate of decrease of $[O_2]$.

Consider the reaction:

$$aA + bB \rightarrow cC \qquad (1.42)$$

The rate of increase of $[C]$ is:

$$\frac{d[C]}{dt} = k[A]^a[B]^b \qquad (1.43)$$

where k is the **rate constant** of the reaction, a is the **order of the reaction** with respect to A, and b is the order of the reaction with respect to B. The overall order of the reaction is $a + b$. It should be noted that the orders of the reaction with respect to the various species are not, in general, the same as the stoichiometric coefficients, as in Equations 1.42 and 1.43 and can only be determined experimentally. However, the orders are the same as the stoichiometric coefficients in the case of an **elementary reaction**, defined as a single-step reaction having a single transition state and no intermediate products, where a transition state is the state of maximum energy during the course of the reaction (Figure 1.14).

That the reaction rates depend on the product of the concentrations of the species involved is an expression of the **law of mass action**. As the concentrations of one or more of the reactants decrease during the course of a reaction, the probability of encounters between reacting molecules is reduced, and the rate of reaction decreases.

In addition, the reaction rate depends on the rate constant, which in turn depends on:

1. Presence of a **catalyst**, which is a substance that increases the reaction rate without it being consumed by the reaction. A catalyst reduces the activation energy, E_a (Figure 1.14) by bringing together reactant molecules in a way that facilitates their reacting with one another. The reaction of Equation 1.39, for example, requires platinum as a catalyst. In cellular reactions, enzymes play a vital role as catalysts. Not only do they increase the rate of reactions which would otherwise be too slow at body temperature, but since enzymes are proteins made by the cell, enzymes are also used to turn reactions on and off as may be needed by the cell. Enzymes can also be activated and deactivated by various molecules in the cytosol.

2. The available areas of catalysts and species involved in the reaction. Clearly, if the available area of the catalyst is limited, this will limit the rate of the reaction. Moreover, if one of the species involved in the reaction is a solid, for example, then having the solid as small-sized particles, increases the surface area and hence the rate of reaction.

3. Temperature of the reaction; generally speaking, the rate doubles for every 10°C rise in temperature. The energies of reacting molecules increase with temperature, so that more molecules will be able to cross the energy barrier E_a, and the rate of the reaction increases. The dependence of the rate constant on temperature is expressed by the **Arrhenius equation**:

$$k = Ae^{-E_a/RT} \tag{1.44}$$

where A is a constant known as the **frequency factor**, R is the gas constant in J/(mol.K) and E_a is in J/mol. Equation 1.44 is of the form of the Maxwell-Boltzmann distribution of statistical mechanics, according to which the number of particles possessing energies greater than say E_x is proportional to e^{-E_x/k_BT}, where k_B is Boltzmann's constant in J/K.

Problem 1.10

Determine E_a in J/mol that will double the reaction rate for a temperature rise from 27°C to 37°C, assuming $R = 8.414$ J/(mol.K).

ANS.: $E_a = 53,594$ J/mol.

1.6.2 Order of Reactions

In a reaction of **zero order,** the rate of the reaction is independent of the concentration of the reactant as may occur, for example, when the surface of a catalyst required for the reaction is saturated by the reactant. An example of a zero-order reaction is the decomposition of ammonia at high pressure on a platinum or tungsten metal surface:

$$2NH_3 \rightarrow 3H_2 + N_2 \tag{1.45}$$

If A is a reactant in a zero-order reaction, the rate of reaction is given by:

$$r = -\frac{d[A]}{dt} = k \tag{1.46}$$

where k is the rate constant. Equation 1.46 can be integrated to give:

$$[A] = -kt + [A]_0 \tag{1.47}$$

where $[A]_0$ is the concentration of A at $t = 0$.

In a reaction of the **first order,** the rate of reaction is proportional to the concentration of only one reactant. Any other reactants that may be present are of zero order. Thus, if A is the reactant in question,

$$r = -\frac{d[A]}{dt} = k[A] \tag{1.48}$$

Integrating Equation 1.48 gives (Problem 1.11):

$$[A] = [A]_0 e^{-kt}, \tag{1.49}$$

A reaction of the **second order** depends on the reaction of two molecules of the same species, that is, a **unimolecular reaction,** or two molecules of two different species of the first order, that is, a **bimolecular reaction.** If A is the reactant in a unimolecular reaction, and A and B are the reactants in a bimolecular reaction, the rate of reaction is given, respectively, by:

$$r = -\frac{d[A]}{dt} = k[A]^2 \quad \text{or} \quad r = -\frac{d[A]}{dt} = k[A][B], \tag{1.50}$$

An example of a second-order unimolecular reaction is:

$$2NO_2 \rightarrow 2NO + O_2 \tag{1.51}$$

The first of Equations 1.50 can be readily integrated by first dividing both sides by $[A]^2$, multiplying by dt, and then integrating with respect to time to give:

$$\frac{1}{[A]} = \frac{1}{[A]_0} + kt \tag{1.52}$$

where $[A]_0$ is $[A]$ at $t=0$. The second of Equations 1.50 can also be integrated to give (Problem 1.13):

$$\frac{[A]}{[B]} = \frac{[A]_0}{[B]_0} e^{([A]_0 - [B]_0)kt} \tag{1.53}$$

assuming $[A]_0 \neq [B]_0$. When $[A]_0 = [B]_0$, it follows that $[A] = [B]$ for all t, and the second of Equations 1.50 reduces to the form of the first equation.

A special case of a second-order reaction is when the concentration of one of the reactants, say B, is so large that it hardly changes during the reaction. The rate equation then becomes:

$$r = -\frac{d[A]}{dt} = k[A][B] = k'[A] \tag{1.54}$$

where $k' = k[B]$. The reaction becomes first order and is referred to as a **pseudo-first-order reaction**.

Unimolecular reactions of the third order are rare because they require three molecules to interact together at the same time and in a particular manner, the probability of such interaction being quite small.

The dimensions of the rate constant k depend on the order of the reaction. These dimensions are $M.s^{-1}$ for a zero-order reaction (Equation 1.46), s^{-1} for a first-order reaction (Equation 1.48), $M^{-1}.s^{-1}$ for a second-order reaction (Equation 1.50), and $M^{1-n}.s^{-1}$ for an nth order reaction.

Problem 1.11

A reaction having only one reactant is a unimolecular reaction and is first order. Its rate is given by Equation 1.48.

(a) Integrate Equation 1.48 by moving $[A]$ to the LHS and dt to the RHS to obtain Equation 1.49. Note that by writing Equation 1.49 as $\ln[A] = -kt + \ln[A]_0$ and plotting $\ln[A]$ vs. t, k can be determined as the slope of the straight-line graph.

(b) The time $t_{1/2}$ that it takes $[A]$ to become one half of the initial concentration $[A]_0$ is known as the **half-life** of the reaction. Show that $t_{1/2} = \dfrac{\ln 2}{k}$ and is independent of $[A]_0$. This means that all half-life periods are equal.

Problem 1.12

(a) Consider the first-order reaction: $2H_2O_2 \longrightarrow 2H_2O + O_2$ having a rate constant of 0.04/min. Determine the concentration of H_2O_2 after 15 min, starting with an initial concentration of 0.50 M.

(b) What is the half-life of the reaction in (a)?

(c) After how long will the concentration of H_2O_2 become 0.15 M?

ANS.: (a) $[H_2O_2] = 0.27$ M; (b) $t_{\frac{1}{2}} = 17.33$ min; (c) 30.1 min.

Problem 1.13

Show that in a first-order reaction, the concentration at the end of n half-life periods is $[A]_0 \left(e^{-kt_{1/2}} \right)^n$.

Problem 1.14

Show that the half-life of the second-order reaction $A + A \longrightarrow$ (Products) is given by the relation $t_{1/2} = \dfrac{1}{k[A]_0}$. Each half-life period is thus twice as long as the preceding one. Note that for the second-order reaction $A + B \longrightarrow$ (Products), with $[A]_0 \neq [B]_0$, the half-life periods of A and B will be different and a general expression for the half-life of the reaction cannot be determined from Equation 1.53.

Problem 1.15

Derive Equation 1.53 (Hint: Assume [Y] to be the concentration that has already reacted at time t, starting at $t = 0$, then express as partial fractions, and integrate).

1.6.3 Reversible Reactions

A reversible reaction proceeds in both the forward direction (reactants $\longrightarrow$ products) and the reverse direction (products $\longrightarrow$ reactants), resulting after a sufficiently long time in a state of dynamic equilibrium consisting of a mixture of reactants and products. For example, consider the simple, first-order reaction:

$$A \underset{k_r}{\overset{k_f}{\rightleftharpoons}} C \tag{1.55}$$

where a species A can exist in another form C, so that at a certain temperature both forms will exist in dynamic equilibrium as $t \longrightarrow \infty$. The differential equations for [A] and [C] at any time t are:

$$\frac{d[A]}{dt} = -k_f [A] + k_r [C] \tag{1.56}$$

and,

$$\frac{d[C]}{dt} = k_f [A] - k_r [C] \tag{1.57}$$

To solve for one of the concentrations, say [A], as a function of t, starting with initial concentrations $[A]_0$ and $[C]_0$, we note that, from conservation of mass, the amount of reactant lost by A, at any time t, which is $[A]_0 - [A]$, must equal the amount of product gained by C, which is $[C] - [C]_0$, that is, $[A]_0 - [A] = [C] - [C]_0$. This is the same as saying that the total initial concentration $[A]_0 + [C]_0$ must equal the total concentration at $[A] + [C]$ at any t. This gives:

$$[C] = [A]_0 + [C]_0 - [A] \tag{1.58}$$

Substituting for [C] in Equation 1.56 gives a differential equation in [A]:

$$\frac{d[A]}{dt} = k_r \left([A]_0 + [C]_0\right) - \left(k_f + k_r\right)[A] \tag{1.59}$$

This equation can be readily integrated by first dividing both sides by the RHS and multiplying by dt:

$$\int_{[A]_0}^{[A]} \frac{d[A]}{k_r \left([A]_0 + [C]_0\right) - \left(k_f + k_r\right)[A]} = \int_0^t dt \tag{1.60}$$

Integrating and making use of Equation 1.58,

$$\ln\left(\frac{k_f[A] - k_r[C]}{k_f[A]_0 - k_r[C]_0}\right) = -\left(k_f + k_r\right)t \tag{1.61}$$

$$\frac{k_f[A] - k_r[C]}{k_f[A]_0 - k_r[C]_0} = e^{-(k_f + k_r)t} \qquad \text{or,} \tag{1.62}$$

As $t \longrightarrow \infty$, $k_f[A] = k_r[C]$. This can also be deduced by writing the rate of the reaction of Equation 1.55 at any t as:

$$r = k_f[A] - k_r[C] \tag{1.63}$$

and noting that at equilibrium $r = 0$. The **equilibrium constant**, defined when $r = 0$ is:

$$K_e = \frac{k_f}{k_r} = \frac{[C]_e}{[A]_e} \tag{1.64}$$

where $[A]_e$ and $[C]_e$ are the equilibrium concentrations. The equilibrium is dynamic in the sense that the forward and reverse reactions do not stop at equilibrium; rather, they occur at equal rates in opposite directions so that concentrations will not change with time.

A first-order, two-stage reversible reaction is described by:

$$A \underset{k_{r1}}{\overset{k_{f1}}{\rightleftharpoons}} B \underset{k_{r2}}{\overset{k_{f2}}{\rightleftharpoons}} C \tag{1.65}$$

The differential equations that characterize the system are:

$$\frac{d[A]}{dt} = -k_{f1}[A] + k_{r1}[B] \tag{1.66}$$

$$\frac{d[B]}{dt} = k_{f1}[A] - k_{r1}[B] - k_{f2}[B] + k_{r2}[C] \tag{1.67}$$

$$\frac{d[C]}{dt} = k_{f2}[B] - k_{r2}[C] \tag{1.68}$$

This set of simultaneous differential equations can be conveniently solved using the Laplace transform including the initial conditions of the problem. The procedure is straightforward but tedious algebraically.

A second-order, reversible reaction is described by:

$$A + B \underset{k_r}{\overset{k_f}{\rightleftharpoons}} C + D \tag{1.69}$$

The rate of reaction is:

$$r = -\frac{d[A]}{dt} = k_f[A][B] - k_r[C][D] \tag{1.70}$$

At equilibrium, $r=0$, $k_f[A]_e[B]_e = k_r[C]_e[D]_e$, where the subscript e refers to the equilibrium state. The equilibrium constant is:

$$K_e = \frac{k_f}{k_r} = \frac{[C]_e[D]_e}{[A]_e[B]_e} \tag{1.71}$$

A large value of K_e indicates that the reaction proceeds almost to completion; that is, the forward reaction is dominant compared to the reverse reaction. Conversely, when K_e is small, the forward reaction does not proceed to any significant extent. K_e is independent of the initial concentrations of reactants, but is temperature dependent because of the temperature dependence of k_f and k_r.

To derive the differential equation for the reaction, let the concentration that has already reacted at time t be $[Y]$, starting at $t=0$, with $[A]=[A]_0$, $[B]=[B]_0$, $[C]=0=[D]$. It follows that: $[A]=[A]_0 - [Y]$, $[B]=[B]_0 - [Y]$, and $[C]=[Y]=[D]$. Substituting in Equation 1.70:

$$\frac{d[Y]}{dt} = k_f([A]_0 - [Y])([B]_0 - [Y]) - k_r[Y]^2 \tag{1.72}$$

Equation 1.70 can be integrated analytically with respect to [Y]. Again, the procedure is straightforward but tedious algebraically.

Problem 1.16

Consider the reversible reaction: $A + B \underset{k_r}{\overset{k_f}{\rightleftharpoons}} 2C$ where a molecule of A reacts with a molecule of B to produce two molecules of C. Derive the expression of the equilibrium constant for this reaction in terms of the equilibrium concentrations of A, B, and C, and determine its value if these concentrations are, respectively, 0.16 M, 0,25 M, and 0.80 M.

ANS.: $K_e = \dfrac{[C]^2}{[A][B]} = 16$ and is dimensionless.

Problem 1.17

Show that in a $A + B \rightleftharpoons C$ and $C + D \rightleftharpoons P + Q$ two-step reaction: the equilibrium constant for the overall reaction is the product of the equilibrium constants of the individual reactions.

ANS.: $K_e = \dfrac{[P][Q]}{[A][B][D]}$ and is the equilibrium constant of the overall reaction:

$$A + B + C \rightleftharpoons P + Q$$

1.6.4 Kinetic Models of Ion Channel Gating

As discussed in Chapter 2, ion channels in membranes can be gated, that is, opened or closed, by the membrane voltage in some cases or by binding of various substances in other cases. Channel gating can be represented by kinetic models involving "reversible reactions". Formally, the transitions between the various states $S_1, S_2, ..., S_n$ of the ion channel can be represented by a sequence of reversible reactions referred to as a **state diagram**:

$$S_1 \underset{k_{21}}{\overset{k_{12}}{\rightleftharpoons}} S_2 \underset{k_{32}}{\overset{k_{23}}{\rightleftharpoons}} \cdots \underset{k_{n(n-1)}}{\overset{k_{(n-1)n}}{\rightleftharpoons}} S_n \qquad (1.73)$$

where $S_1, S_2, ..., S_n$ are the numbers of channels in a given state. If these numbers are divided by the total number of identical channels under consideration, then the $S_1, S_2, ..., S_n$ are replaced by the fraction of channels in each state, denoted by $s_1, s_2, ..., s_n$. When the number of these identical channels is large, the fraction, s_i, $i = 1, 2, ..., n$, becomes the probability of state S_i. The rate constant k_{ij} from state S_i to the next state S_j becomes the probability of the transition from state S_i to state S_j.

Consider the following three-state diagram, where transitions between all states are allowed:

$$k_{13}$$

$$s_1 \underset{k_{21}}{\overset{k_{12}}{\rightleftharpoons}} s_2 \underset{k_{32}}{\overset{k_{23}}{\rightleftharpoons}} s_3 \qquad (1.74)$$

$$\overset{k_{31}}{}$$

The differential equation for s_2 becomes:

$$\frac{ds_2}{dt} = k_{12}s_1 + k_{32}s_3 - k_{21}s_2 - k_{23}s_2 \qquad (1.75)$$

Equation 1.75 is a special case of the general differential equation for s_i, $i = 1$, $2, ..., n$, in an n-state system in which all transitions are allowed. This general equation can be written as follows:

$$\frac{ds_i}{dt} = \sum_{j=1}^{j=n} k_{ji}s_j - \sum_{j=1}^{j=n} k_{ij}s_i \qquad (1.76)$$

It can be readily verified that Equation 1.76 reduces to Equation 1.75 for the three-state diagram of Equation 1.74 (Problem 1.18).

A system defined by a set of equations exemplified by Equation 1.76 is described as a **Markov model**. Transition probabilities in such models are, in general, time dependent. The solution to these types of differential equations is discussed in books on stochastic processes. Software packages are available to facilitate the solutions to these equations.

Problem 1.18

Show that Equation 1.76 reduces to Equation 1.75 for $i = 2$, and $j = 1, 2$, or 3.

Problem 1.19

Consider a four-state system in which all transitions between all states are allowed. Derive the equations for each of the four states and verify that Equation 1.76 applies in each case.

ANS.: $\dfrac{ds_1}{dt} = k_{21}s_2 + k_{31}s_3 + k_{41}s_4 - k_{12}s_1 - k_{13}s_1 - k_{14}s_1$

$\dfrac{ds_2}{dt} = k_{12}s_1 + k_{32}s_3 + k_{42}s_4 - k_{21}s_2 - k_{23}s_2 - k_{24}s_2$

$\dfrac{ds_3}{dt} = k_{13}s_1 + k_{23}s_2 + k_{43}s_4 - k_{31}s_3 - k_{32}s_3 - k_{34}s_3$

$\dfrac{ds_4}{dt} = k_{14}s_1 + k_{24}s_2 + k_{34}s_3 - k_{41}s_4 - k_{42}s_4 - k_{43}s_4.$

Summary of Main Concepts

- The smooth endoplasmic reticulum and mitochondria are involved in the regulation of Ca^{2+} concentration in the cytosol. Most of the cell's ATP is generated in the mitochondria.

- Neurons are the core of the nervous system. Cell bodies and dendrites of neurons are specialized for the reception and processing of electric signals from other neurons, whereas axons are specialized for the transmission of the AP from a given neuron to the axon terminals. Axon terminals are usually specialized for the release of a neurotransmitter.

- The diffusion constant is a measure of how easily a given atomic particle can move under the influence of a concentration gradient in the medium under consideration. Mobility is a measure of how easily a given charged atomic particle can move under the influence of an applied electric field in the medium under consideration. Both the diffusion constant and mobility depend on the resistance to the motion of the given particle in a particular medium, both have positive values, and are related by the Nernst–Einstein relation.

- Chemical potential is chemical potential energy per particle, or per mole, associated with concentration, so that the negative of the gradient of this potential is the "force" that effectively drives the particles from regions of higher concentration to regions of lower concentration.

- Electrochemical potential of a charged atomic particle is the algebraic sum of its chemical and electric potentials. The negative of the electrochemical potential gradient is the net driving force acting on a given atomic particle in the presence of both concentration gradients and electric potential gradients. Ohm's law can be generalized in terms of electrochemical potential to account for the current under the influence of both concentration gradients and electric potential gradients.

- Under equilibrium conditions, temperature, pressure, and electrochemical potential of freely moving particles do not vary with time and are equalized throughout the system, so that no net work is done in taking a small quantity of a freely moving particle from any part of the system to any other part.

- Any ionic species that is diffusible through the membrane will have, under conditions of equal pressure and temperature, the same electrochemical potential at equilibrium on both sides of the membrane. The concentration ratios of the given ions on the two sides are related to the voltage difference between the two sides by the Nernst equation.

- The quantity of charge needed to establish the voltage difference across biological membranes is very small compared to the charge of either polarity that is present on the two sides of the membrane.

- When a semipermeable membrane separates two compartments at equal pressure and temperature, with ions that can freely diffuse through the membrane, then the equilibrium concentrations of any of these ions in the two compartments are related to the voltage across the membrane by the Nernst equation.

- A system consisting solely of ions in a Gibbs–Donnan equilibrium is not in osmotic equilibrium.

- Enzymes play the role of catalysts in cellular reactions. Not only do they increase the rates of these reactions, but they allow turning these reactions on and off by means of availability of enzymes or by their activation and deactivation by various molecules in the cytosol.

- The temperature dependence of the rate constant of a reaction is given by the Arrhenius equation, which is of the form of the Maxwell-Boltzmann distribution of statistical mechanics.

- The dimensions of the rate constant of a reaction of order n are $M^{1-n}.s^{-1}$, where $n = 0, 1, 2$, etc.

- Gating of ion channels can be represented by kinetic models involving a sequence of "reversible reactions" between the various states. The transitions between these reversible reactions are represented by a state diagram. When the states are described in terms of state probabilities, the rate constants become probabilities of transitions between states. Differential equations can be written for the system in terms of state probabilities and transition probabilities. This mathematical formalism is a Markov model.

2

The Cell Membrane in the Steady State

Objective and Overview

The electrical properties of the cell are essentially those of the cell membrane. The chapter begins therefore with a description of the basic structure of the cell membrane, highlighting its electrical properties. As ion channels play a vital role in the electrical behavior of the cell by controlling which ions cross the membrane and under what conditions, ion channels are discussed in general terms. The distribution of some major ions across the cell membrane is then considered, leading to the inference that there must be an ion pump that actively extrudes Na^+. It is shown how the pump plays a vital role in establishing osmotic equilibrium and an electrochemical potential difference across the cell membrane. This electrochemical potential difference is associated with a resting membrane voltage and allows the generation of electric signals by the cell. The origin of the resting membrane voltage is explained, quantified, and interpreted in terms of a useful and commonly invoked equivalent circuit that includes conductances and equilibrium voltages of the various ions. Two interesting electrical properties of the membrane are discussed, namely, rectification and reactance. It is shown that, basically, rectification arises from unequal distribution of ions on the two sides of the membrane, whereas reactance, which could be inductive or capacitive, is manifested by conductances that are nonlinear and time-varying. The chapter ends with a direct analogy between ionic and semiconductor systems.

Learning Objectives

To understand:

- The basic structure of the cell membrane and its impact on the electrical properties of the cell

- The need for an active pump that extrudes Na^+ and the role played by the pump
- The factors that contribute to the resting membrane voltage and the derivation of this voltage in terms of ionic concentrations, permeabilities, and ion pumping ratios
- The derivation and interpretation of the membrane equivalent circuit
- The general scheme for generating electric signals across the cell membrane and the upper and lower bounds of the variation of membrane voltage
- How rectification arises in a membrane system
- How a nonlinear, time-varying conductance exhibits reactance

2.1 Structure of the Cell Membrane

The basic structure of the cell membrane is that of a phospholipid bilayer. A phospholipid molecule consists of a head containing a phosphate group that is usually attached to two fatty acid chains. The head carries an electric charge, which makes it "polar" and therefore **hydrophilic**, that is, attracted to water, which is also polar. On the other hand, the fatty nature of the chains makes them **hydrophobic**, that is, they do not mix with water. Because of this, phospholipids in an aqueous medium generally assume a bilayer structure, two molecules thick, in which the polar groups face the aqueous solutions on either side, whereas the fatty acid chains face inward, away from water, as illustrated in Figure 2.1.

In living cells, the basic phospholipid bilayer structure is modified in several respects so as to serve the many important functions required of the cell membrane. Cell membranes have an abundance of cholesterol and protein molecules (Figure 2.2). Cholesterol is a small steroid-type molecule having a hydrophilic OH group, that aligns with the polar heads of the phospholipid

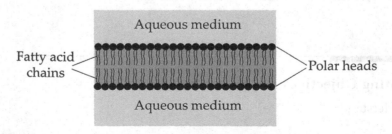

FIGURE 2.1
Phospholipid bilayer.

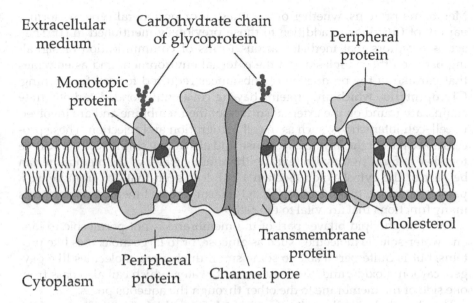

Extracellular medium

Carbohydrate chain of glycoprotein

Peripheral protein

Monotopic protein

Cholesterol

Transmembrane protein

Peripheral protein

Channel pore

Cytoplasm

FIGURE 2.2
Diagrammatic illustration of cell membrane.

molecules, and a hydrophobic portion that nestles between the fatty acid chains. On the one hand, cholesterol makes the membrane more fluid and flexible by preventing the fatty acid chains from forming a more rigid, crystal-like structure. On the other hand, the attraction to the fatty acid chains makes the membrane firmer and less permeable to small, water-soluble molecules. In addition, cholesterol plays a significant role in mechanically securing protein molecules in the membrane so that they are not adversely affected by membrane fluidity.

A wide variety of protein molecules are associated with the cell membrane. They could be **integral proteins** that are part of the membrane, or **peripheral proteins** that temporarily attach either to the membrane or to the integral proteins. Integral proteins are of two types:

1. **Transmembrane proteins** that extend across the membrane and form: (i) channels having aqueous pores, or "holes", through which small molecules and ions can flow passively down an electrochemical potential gradient; (ii) ion transporters of various types that transport ions from one side of the membrane to the other, as explained in Section 2.3.

2. **Monotopic proteins**, that is, proteins that are permanently attached to the cell membrane on only one side and do not extend across the membrane. These proteins are anchored to the membrane by having their hydrophobic regions extend into the fatty acid chain region of the membrane phospholipids.

Membrane proteins, whether of the integral or peripheral type, perform a variety of functions in addition to those previously mentioned in (1). They act as receptors that mediate various forms of communication or signaling between the cytoplasm and the external environment and as enzymes that partake in the production of substances required for cell functioning. **Glycoproteins**, which are proteins having covalently bonded carbohydrate chains, are found on the external surface of the membrane and are involved in cell–cell interactions, such as in cell recognition and rejection. These processes are part of the immune response and also allow similar cells to adhere together to form tissues. Proteins on the inner surface of the membrane can be part of the cytoskeleton (Section 1.1.4). It is seen that, far from being a passive envelope that contains the cytoplasm, the cell membrane performs many functions that are vital to the cell.

The phospholipid bilayer part of the membrane is not permeable to ions and water-soluble molecules, such as glucose, or to large molecules like proteins, but is quite permeable to some small uncharged molecules like oxygen, carbon dioxide, and, to a lesser extent, water, which can also pass from one side of the membrane to the other through the aqueous pores.

The phospholipid bilayer itself is about 4–5 nm thick. The thickness of the cell membrane, including hydration ions and attached proteins is considered to be about 7–10 nm, where 1 nm = 10 Å = 10^{-9} m = 10^{-7} cm.

2.1.1 Aqueous Pores

An **ion channel** consists of a receptor-type protein, referred to as a **channel protein**, or **aquaporin**, which surrounds an aqueous pore that forms a direct connection between the external and internal aqueous media. The pore allows passage of ions between these two media, subject to certain restrictions. The channel protein may have some carbohydrate groups attached to it on the extracellular side (Figure 2.2). Since ion channels and their aqueous pores are central to the electrical properties and behavior of the cell membrane, they will be considered here in a little more detail.

Ion channels are of various types. Some channels are in a dynamic state of spontaneous and random opening and closing and may be open for less than 1 ms at a time. Other channels are gated, that is, they are opened or closed by a variety of influences, such as the binding of some small molecules (referred to as **ligands)** to (i) the external side of the channel protein, as in the case of neurotransmitters (Section 6.1.2), or (ii) to the internal side of the protein, as in the case of second messenger systems (Section 6.3). The gating could also be due to (i) phosphorylation, (ii) the voltage across the membrane, as in the initiation and propagation of the action potential (Chapters 3 and 4), (iii) mechanical deformation, as in the case of muscle receptors (Section 9.4), or (iv) other physical stimuli such as light, as in the case of photoreceptors, or heat, as in the case of temperature receptors.

At the molecular level, all ion channels are transmembrane protein molecules that are coded by a number of genes. The aqueous pores are not of uniform diameter but usually have an aqueous cavity, or vestibule, on one or both sides of the channel. Some narrow regions of the pore, 3–8 nm in diameter, constitute a **selectivity filter** controlling channel selectivity, that is, determining which ions will pass most easily through the pore. In some K^+ channels, the selectivity filter is only about 1.2 nm in length.

The mechanisms underlying channel selectivity are rather complicated. Because water molecules are polar, ions in aqueous solutions are **hydrated**, that is, they have associated with them a number of water molecules, referred to as the **hydration number**, which increases the effective size of the ion. Na^+ have an atomic radius of 0.95 Å whereas K^+ have an atomic radius of 1.33 Å. Since both Na^+ and K^+ have the same charge, Na^+ have a larger surface charge density, so they attract more water molecules by a weak electrostatic force, and consequently have a larger hydration number than K^+. The hydrated Na^+ is therefore larger than the hydrated K^+ and is of lower mobility in solution. The hydration number for K^+ is 3–4, whereas that for Na^+ is 4–5.

The walls of the selectivity filter of the channel have charged atomic groups which could be positively charged, such as the H^+ of a hydroxyl group or an amide (NH_2) nitrogen atom in the main chain of an amino acid, or they could be negatively charged, such as oxygen-containing carbonyl ($C{=}O$), hydroxyl (OH^-), or carboxyl (COO^-) groups. Positively charged atomic groups would allow anions through the selectivity filter but not cations. Conversely, negatively charged groups would allow only cations through the filter. Selectivity between different cations, such as Na^+ and K^+, depends on how wide the selectivity filter is, the hydration number of ions in the filter, the three-dimensional geometric fit between the ion and the binding sites of the charged groups in the filter, and energy considerations at these sites. To pass through the selectivity filter having oppositely charged atomic groups, an ion must give up at least some of its hydration water molecules in exchange for bonds with charged atomic groups in the wall of the pore. To take place, such an exchange must be favorable from an energy viewpoint, that is, the energy gained through the weak electrostatic attraction with the binding site should compensate for the energy lost in shedding some of the hydration water molecules. The energy gained by binding to the charged atomic group depends on the electrostatic attraction between the ion and the charged atomic group. This depends in turn on the atomic radius of the ion and on the nature of the charged atomic group, whether it is a carboxyl or a carbonyl group, for example, as well as the spatial configuration of the binding site. The binding of ions to charged groups in the selectivity filter is transient, lasting less than about 1 μs before the ion moves under the influence of its electrochemical potential gradient. An effect of this binding is to limit the maximum current through the channel at high electrochemical potential gradients across the membrane. Thus, based on simple electrodiffusion without any binding, the current through the channel increases with the electrochemical potential gradient. Instead, the rate of binding and unbinding to the charged groups limit the channel current at high electrochemical potential gradients. Moreover, the gating of a channel from a closed to an open state, which involves a conformational change of the channel protein, may not only be due to an opening of the channel lumen, but may also be due to additional binding sites on the channel protein becoming available.

The rate flow of ions through a pore can be very high. Pore conductances and currents are typically in the ranges of 5–50 pS and 1–20 pA, respectively. A current of 1 pA is carried by about 6000 monovalent ions moving through the pore in 1 ms. Velocities of ions through pores are typically of the order of a few cm/s.

In pores having a narrow, relatively long region, ions cross single-file with repulsion between adjacent ions in the queue. In wider pores that can accommodate more than one ion species, each ion may also pass single-file, with some degree of electrical interaction between different ion species.

Problem 2.1

Calculate the conductance of a pore that is 5 nm long, 1 nm in diameter, and has a resistivity of 100 Ωcm.

ANS.: 157.1 pS.

2.2 Electrical Properties of the Cell Membrane

Artificial phospholipid bilayer membranes (Figure 2.1) have resistivities in the range of 10^{12} Ωcm to 10^{15} Ωcm because the fatty acid chains are very good insulators. On the other hand, membranes of living cells, although their structure is basically that of a phospholipid bilayer, have much lower resistivities in the range of 0.5×10^9–10^{10} Ωcm because of the presence of aqueous pores (Section 2.1.1) that allow the flow of ions under the influence of an applied voltage across the membrane.

The extracellular medium of the cell has a resistivity in the range of 15–100 Ωcm, whereas the resistivity of the cytoplasm is higher, in the range of 50–300 Ωcm because of the presence of cell organelles and a higher concentration of large molecules. Nevertheless, the cell membrane is a good insulator between two relatively good conductors. Such a structure can separate electric charges and store the electric energy expended in separating the opposite charges. It will therefore exhibit the electrical property of capacitance.

Because of the thinness of the membrane, the capacitance has a relatively high value that was estimated early on at about 1 μF/cm². More recent estimates have reduced this value to 0.7–0.8 μF/cm². A patch of unit area of a

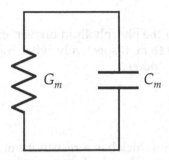

FIGURE 2.3
Basic equivalent circuit of a patch of cell membrane.

membrane having only simple, conductive channels can thus be represented by an equivalent circuit consisting of a conductance G_m of a unit area of membrane in parallel with a capacitance C_m per unit area (Figure 2.3). G_m is generally in the range of 10–100 $\mu S/cm^2$ for a neuron in the resting state. $R_m = 1/G_m$ is correspondingly in the range of 10^4–10^5 Ωcm^2. Note that C_m is in $\mu F/cm^2$ and G_m is in $\mu S/cm^2$, whereas R_m is in Ωcm^2. This is because for a given membrane voltage, increasing the membrane area increases the electric charge on each side of the membrane and hence increases the capacitance. Increasing the area increases the membrane current for a given membrane voltage because of the availability of more, parallel conducting paths, and hence increases the conductance per unit area G_m ($\mu S/cm^2$). But R_m, the reciprocal of G_m, decreases with increasing area of the membrane. The relations between R_m, G_m, **membrane resistivity** ρ_m – also referred to as the **specific membrane resistance** – and its reciprocal, **membrane conductivity** σ_m are:

$$R_m = \rho_m \delta, \text{ or } G_m = \frac{\sigma_m}{\delta} \tag{2.1}$$

where δ is the thickness of the membrane. Note that ρ_m is in $\Omega \times$(unit length) and σ_m is in S/unit length.

G_m of typical cell membranes is nonlinear (Section 2.5.1) and can be considered constant only for small variations of voltage across the membrane. The product $R_m C_m = C_m/G_m = \tau_m$ is the **membrane time constant** and is generally in the range of 10–100 ms.

Problem 2.2

Calculate the specific membrane conductance in mS/cm^2 of a membrane having 10 pores per square micrometer, assuming each pore has a conductance of 15 pS and is open, on the average, half the time, and neglecting the conductance of the membrane not having pores.

ANS.: 7.5 mS/cm^2.

Problem 2.3

Assuming resistivities of the phospholipid bilayer, membrane, and aqueous pore of 10^{12}, 10^9, and 100 Ωcm, respectively, what fraction of the area of the membrane is occupied by pores?

ANS.: $\cong 10^{-7}$.

Problem 2.4

A membrane that is 10 nm thick has a resistivity of 10^{10} Ωcm and a capacitance of 1 μF/cm^2. Determine the membrane time constant.

ANS.: 10 ms.

2.2.1 Ionic Concentrations and Permeabilities

Typically, a cubic micron of cytoplasm contains roughly: 10^{10} water molecules, 10^8 ions, 10^7 small molecules such as amino acids and nucleotides (Section 6.3.1), and 10^5 protein molecules. The ion species in the intracellular and extracellular media include Na$^+$, K$^+$, Cl$^-$, H$^+$, Ca^{2+}, Mg^{2+}, HCO$_3^-$, amino acids, proteins, and molecules with negatively charged phosphate groups (PO$_4^{2-}$). Of these ion species, Na$^+$, K$^+$ and Cl$^-$ have relatively high concentrations, so they are the only ions usually considered in the discussion of the resting membrane voltage. Generally, the intracellular medium contains an excess concentration of negative charge, which makes the inside of the neuron at a negative voltage with respect to the outside. Many of the negatively charged, relatively large molecules, such as proteins, amino acids, and energy-rich phosphates cannot diffuse outside the cell and constitute the indiffusible anions A$^-$ considered in the Gibbs–Donnan Equilibrium (Section 1.5.4).

With very few exceptions, as in the red blood cells of some sheep, Na$^+$ are more concentrated in the extracellular fluid than in the cytoplasm, whereas K$^+$ are more concentrated in the cytoplasm than in the extracellular fluid. Cl$^-$, like Na$^+$, are more concentrated extracellularly, the range of concentrations being indicated in Table 2.1 for typical mammalian cells. Note that the higher figure for [Cl$^-$]$_i$ applies to nonexcitable cells such as red blood cells.

All living cells have a voltage across the cell membrane, with the inside negative with respect to the outside. This voltage is about –10 mV in nonexcitable

TABLE 2.1

Typical Ionic Concentrations in Mammalian Cells

Ion	Cytoplasm (mM)	External Medium (mM)
Na$^+$	5–15	145
K$^+$	140–155	4–6
Cl$^-$	4–74	110–125
Ca^{++}	0.001	1.5

cells and about –70 to –90 mV in excitable cells. Assuming that Cl⁻ are at equilibrium across the cell membrane, and considering the larger concentration of $[Cl^-]_i$ typical of nonexcitable cells, the membrane voltage in these cells will be nearly $E_{Cl} = -62\log_{10}(110/74) \cong -11$ mV (Equation 1.27), whereas the lower figure for $[Cl^-]_i$ gives a membrane voltage of nearly –90 mV, which is typical of muscle cells.

The cell membrane has a high permeability to K⁺, a smaller permeability to Cl⁻, and a much smaller permeability to Na⁺, the latter being only 1/30–1/100 of the permeability to K⁺.

Bearing in mind that the inside of the cell is negative with respect to the outside, the chemical and electric potential gradients for Na⁺, K⁺, and Cl⁻ will be as in Figure 2.4, where hatched arrows denote concentration gradients, stapled arrows denote electric potential gradients, and the higher concentration of each ion species, on one side of the membrane relative to the other side, is in boldface. An important observation can be made concerning the potential gradients of Figure 2.4. *Whereas the chemical and electric potential gradients for K⁺ and Cl⁻ are in opposition, both the chemical and electric potential gradients for Na⁺ drive these ions inward.* What would then keep the concentration of

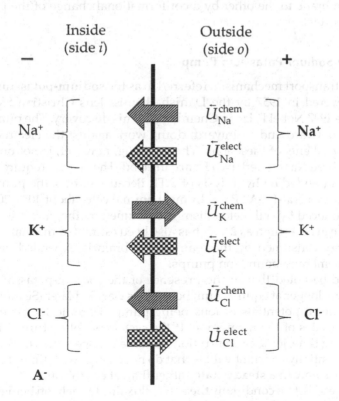

FIGURE 2.4
Chemical and electric potential gradients for Na⁺, K⁺, and Cl⁻ across a cell membrane.

Na⁺ from continuously increasing with time? The answer that seemed plausible for quite a while was to assume that the cell membrane was impermeable to Na⁺, and that K⁺ and Cl⁻ are in equilibrium, which means that the cell
is essentially in a Gibbs–Donnan equilibrium (Section 1.5.4). Measurements
of ionic concentrations and voltages were not sufficiently refined to establish
with certainty that the Nernst equation essentially applied to Cl⁻ but not
to K⁺. In the 1940s, radioactive isotope labeling of Na⁺ proved conclusively
that the cell membrane had a small, but nevertheless finite, permeability to
Na⁺ ions. So the question had to be answered as to how the concentration of
Na⁺ in the cell does not continuously increase with time despite the steady
inward flux of Na⁺ due to the electrochemical potential gradient.

2.3 Ion Transporters

The general mode of operation of ion transporters is by having special binding sites for the ions to be transported, which are then shuttled from one side
of the membrane to the other by a conformational change of the protein of
the transporter.

2.3.1 The Sodium-Potassium Pump

An active transport mechanism, referred to as the **sodium-potassium pump**,
was discovered in 1957 by the Danish chemist Jens Christian Skou, who
shared the 1997 Nobel Prize in Chemistry for his discovery. The pump transports Na⁺ outward and K⁺ inward, doing work against the electrochemical
potential gradients of these ions. The coupling, however, is not one-to-one:
for every three Na⁺ ejected, two K⁺ are injected. The energy required for this
process is provided by hydrolysis of ATP. Because of this, the pump is also
referred to as a **Na⁺-K⁺-ATPase**. In most animal cells about 10%–20% of the
energy produced by cell metabolism is consumed by the pump; in neurons,
the figure can be as high as 2/3. It has also been estimated that half the metabolic energy consumed by the mammalian brain is expended on various
types of membrane-bound ion pumps.

It should be noted that in the presence of the sodium-potassium pump,
the cell is no longer at equilibrium, because as clarified later (Section 2.4), the
electrochemical potentials of ions being pumped are no longer equalized
on the two sides of the membrane. If the system is not disturbed for a sufficiently long time, ionic concentrations and the voltage across the membrane
assume essentially constant values that do not change with time. The system
as a whole is now in a **steady state** rather than at equilibrium.

Under steady-state conditions the active flux due to each ion being pumped
is equal and opposite to the passive flux of that ion due to its electrochemical
potential gradient. Otherwise, the concentration of the given ion on either

side of the membrane will change with time, and the system will not be in a steady state. An ion that is not being pumped, and which can diffuse freely through the membrane, will be at equilibrium. That is, the net flux of that ion through the membrane is zero, and the electrochemical potential gradient of that ion is zero. The flux due to the concentration gradient of an ion at equilibrium is equal and opposite to that due to the electric potential gradient, and the Nernst equation applies to that ion.

If the cell expends a considerable amount of energy on the sodium-potassium pump, then this pump must serve some important functions. It is immediately seen that the pump results in a reduced Na^+ concentration inside the cell, which because of the resulting voltage difference, with the inside negative with respect to the outside, also results in a reduced Cl^- concentration inside the cell due to Cl^- efflux under the influence of the electric potential gradient. *The reduced concentrations of Na^+ and Cl^- compensate for the excess concentration of indiffusible anions inside the cell.* This serves to equalize the concentrations of particles in solution on both sides of the membrane in order to establish osmotic equilibrium and control cell volume (Section 1.5.1).

Moreover, as is often the case in biological systems, the same structure is made to serve more than one function, that is, different functions can be well integrated in the same structure in biological systems. The wings of a bird, for example, provide lift, propulsion, and maneuverability, all by the same structure, which is not the case with fixed-wing airplanes. In pumping Na^+ outward and K^+ inward, an electrochemical potential gradient is established for these ions. In excitable cells (Section 1.2), *the electrochemical potential gradients of Na^+ and K^+ are used to generate electric signals across the cell membrane*, as will be discussed later.

The mode of operation of the Na^+-K^+ pump is diagrammatically illustrated in Figure 2.5, which shows a pump protein having three sites that can bind Na^+ and two sites that can bind K^+. The protein can be in one of two configurations, either open to the inside of the cell, or open to the outside. When the protein is open to the inside, the Na^+ sites have a high affinity to Na^+, and the K^+ sites have a low affinity to K^+. Conversely, when the protein is open to the outside, the Na^+ sites have a low affinity to Na^+, and the K^+ sites have a high affinity to K^+. Moreover, the protein has a phosphorylation site (Section 1.1.2) that can bind an inorganic phosphate to an aspartic acid group (Figure 2.5a). The following steps are assumed to occur:

1. With the pump protein open to the inside (Figure 2.5a), the two sites that have low affinity for K^+ and the three sites that have high affinity for Na^+ are exposed to the inside, which means that the former sites will release any K^+ that were previously bound, whereas the latter sites can bind Na^+.

2. When the three Na^+ bind to their sites (Figure 2.5b), the protein is phosphorylated through hydrolysis of ATP to ADP, resulting in the binding of an inorganic phosphate P_i to the protein.

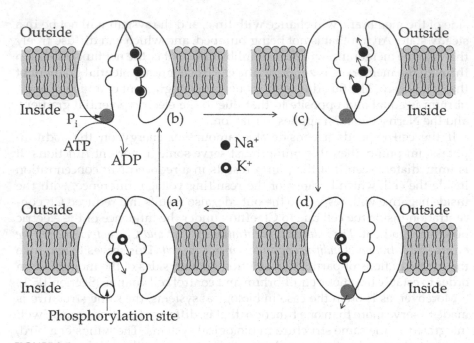

FIGURE 2.5
Diagrammatic illustration of operation of the Na$^+$-K$^+$ pump.

3. The energy released in the phosphorylation causes a conformational change of the pump protein so that it now opens to the outside (Figure 2.5c). In this configuration, the Na$^+$ sites have a low affinity to Na$^+$, which are now released, and the K$^+$ sites have a high affinity for K$^+$.

4. When the two K$^+$ bind to their sites, the protein is dephosphorylated (Figure 2.5d) and reverts to its original conformation, which opens the protein to the inside (a). The K$^+$ are now released, and the protein is ready to bind Na$^+$ and become phosphorylated, thereby repeating the cycle.

The pumping rate of the sodium-potassium pump varies considerably between various types of cells and depends on [Na$^+$]$_i$, [K$^+$]$_o$, membrane voltage, and temperature. Roughly, each pump molecule exchanges about 150 Na$^+$ for 100 K$^+$ per second, equivalent to a Na$^+$ current of about 24 aA and a K$^+$ of about 16 aA (1 aA (atto-ampere) = 10^{-18} A). The net electrogenic (see Section 2.4) pump current is then about 8 aA/pump molecule. Assuming 2000 pumping molecules/μm^2, the corresponding electrogenic current density is about 16 fA/μm^2 (1 fA (femto-ampere) = 10^{-15} A), equivalent to about 1.6 μA/cm^2. The pump is inhibited by a sufficiently high concentration of ouabain, a glycoside found in the seeds or bark of some plants, and which binds to the pump protein on the outside of the cell.

Similar to the Na$^+$-K$^+$ pump is the plasma membrane **Ca^{2+}-ATPase (PMCA)** that helps maintain a low intracellular Ca^{2+} concentration in the resting state by pumping Ca^{2+} outward against a large electrochemical potential gradient. Ion pumps are an example of **ion transporters**, which transport ions across plasma membranes and membranes of cell organelles. Other types of ion transporters are uniporters and cotransporters that will be discussed in the following section.

2.3.2 Uniporters and Cotransporters

A **uniporter** is a transmembrane protein that facilitates diffusion of a substance down a concentration gradient, without ATP hydrolysis, but at a rate that can be far higher than that of passive diffusion for that substance, the energy being derived from the concentration gradient of the transported substance. Glucose and amino acids are transported across the plasma membrane in this manner, the concentration gradient being established in these cases because these substances are used up in cell metabolism. The inner mitochondrial membrane has an efficient Ca^{2+} uniporter that allows a fast uptake of Ca^{2+} by mitochondria. This type of movement is referred to sometimes as **facilitated transport,** or **facilitated diffusion**. In a uniporter, a specific protein transports a particular substance by undergoing a contormational change, much like that illustrated in Figure 2.5 for the Na$^+$-K$^+$ pump but without ATP hydrolysis.

Facilitated transport, without directly involving an energy-consuming, or active, mechanism, could be in either direction depending on the direction of the concentration gradient of the substance being transported. If transport should always be in one direction, irrespective of the direction of the concentration gradient, then this transport should utilize an energy-consuming, active mechanism. An interesting example is the transport of glucose. In the case of most cells, where glucose is always at a higher concentration extracellularly, glucose is transported by facilitated transport. On the other hand, the transport of glucose from the lumen of the gut into the epithelial cells of the gut, for eventual movement into the blood stream, is by active transport. This is because the transport must always be in the same direction irrespective of whether the glucose concentration is higher in the lumen or in the epithelial cells. Energy-saving facilitated transport cannot be used in this case.

A **cotransporter** is a transmembrane protein that couples the transport of substances, against electrochemical potential gradients, to the transport of other substances down electrochemical potential gradients, again without directly involving ATP hydrolysis. If the substances are moved in the same direction, the cotransporter is a **symporter**. If the substances are moved in opposite directions, the cotransporter is an **antiporter**. If the substances moved by an antiporter are ions, the antiporter is an **ion exchanger**. An example of a symporter is the **K$^+$-Cl$^-$ symporter** that moves one K$^+$ and one Cl$^-$ outward. K$^+$ are moved down their electrochemical potential gradient, which provides the energy for driving Cl⁻ outward and establishing an electrochemical potential gradient for Cl⁻ that can drive them passively inward. This electrochemical potential gradient is essential for the action of some inhibitory synapses (Section 6.2.2). Note that the driving electrochemical potential of K⁺ is established in the first place by active transport that utilizes ATP hydrolysis. The **K⁺-Cl⁻ symporter** is also known as a **KCC2** (potassium chloride cotransporter 2). In some cases, a **Na⁺-K⁺-2Cl⁻ symporter** (also known as **NKCC**) transports Cl⁻ ions inward from the extracellular medium and establishes an electrochemical potential gradient for Cl⁻ that can drive them passively outward, as in presynaptic inhibition (Section 6.4). The energy for driving the NKCC symporter is the electrochemical potential for Na⁺, which again is established by active transport that utilizes ATP hydrolysis. Because of this, the action of cotransporters is often referred to as **secondary active transport**. Both the K⁺-Cl⁻ and the Na⁺-K⁺-2Cl⁻ symporters are electrically neutral, as no net transfer of charge occurs, because equal quantities of positive charge and negative charge are moved in the same direction.

An example of an ion exchanger, an antiporter, is the **Ca²⁺-Na⁺ exchanger**, which helps maintain a low $[Ca^{2+}]_i$ concentration by coupling the outflow of Ca²⁺, against an electrochemical potential gradient, to the inflow of Na⁺ down an electrochemical potential gradient. Note that in the case of the Na⁺-K⁺ pump, the pump moves both Na⁺ and K⁺ against an electrochemical potential gradient for both ions, so expenditure of energy is inevitable.

2.4 Origin of the Resting Membrane Voltage

Conceptually, it is convenient to consider the sodium-potassium pump as having two additive components: (i) an *electrically neutral* component that exchanges one Na⁺ for one K⁺, and (ii) an **electrogenic** component that only pumps Na⁺ from the inside of the cell to the outside.

Three factors can be identified as responsible for the resting voltage across membranes of living cells:

1. The neutral component of the sodium-potassium pump, in conjunction with the much lower permeability of the membrane to Na⁺ compared to K⁺.

2. The electrogenic component of the sodium-potassium pump, which transfers positive charge from the inside of the cell to the outside, thereby making the inside of the cell negative with respect to the outside.

3. The presence of an excess concentration of indiffusible anions inside the cell, as in the Gibbs–Donnan equilibrium (Section 1.5.4), which makes the inside of the cell more negative with respect to the outside.

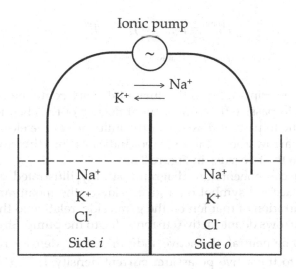

FIGURE 2.6
A model ionic system.

To appreciate how these factors contribute to a voltage across the membrane, with the inside negative with respect to the outside, consider a model system in which a membrane permeable to Na^+, K^+, and Cl^- separates two sides i and o containing NaCl and KCl (Figure 2.6), where sides i and o can be identified with the inside of the cell and its outside, respectively. In the absence of an ion pump, the concentrations of each of these ions will be the same on both sides of the membrane, under equilibrium conditions. There will be no net charge on either side of the membrane, so that the voltage across the membrane is zero.

Suppose that an ion pump is started which transports Na^+ and K^+ at equal rates in opposite directions. The effect of the pump is to increase the concentration of Na^+ on side o compared to side i, and to increase the concentration of K^+ on side i compared to side o. If the membrane has equal permeabilities to Na^+ and K^+, these two ion species will diffuse back, under the influence of their respective concentration gradients, at equal rates in opposite directions. The system will be symmetrical; the charge lost on one side because of the depletion of one cation is compensated by the charge gained because of the accumulation of the other cation. There will be no net accumulation of charge on either side of the membrane, and hence no electric voltage difference.

If the membrane has a much larger permeability to K^+ than to Na^+ ions, then Na^+ cannot diffuse back as fast as K^+ ions. There will be a net buildup of positive charge on side o. *A voltage difference is created which increases the passive flow of Na^+ from side o to side i and decreases the passive flow of K^+ from side i to side o, so that the net passive flux equals the active flux for each ion species, despite the difference in the permeabilities of the membrane to Na^+ and K^+.*

In the steady state, the following relations between current densities apply:

$$\bar{J}_{Na}^{pump} = \bar{J}_{K}^{pump} \tag{2.2}$$

$$\vec{J}_{Na}^{\text{pump}} = \vec{J}_{Na}^{\text{passive}} = \vec{J}_{Na}^{\text{chem}} + \vec{J}_{Na}^{\text{elect}} \tag{2.3}$$

$$\vec{J}_{K}^{\text{pump}} = \vec{J}_{K}^{\text{passive}} = \vec{J}_{K}^{\text{chem}} - \vec{J}_{K}^{\text{elect}} \tag{2.4}$$

where the superscripts refer to the current density components due to the pump, or due to passive flow under the influence of the chemical potential gradient, or due to passive flow under the influence of the electric potential gradient. The arrow above the J symbol indicates the direction of current with respect to the two sides in Figure 2.6.

The current components are diagrammatically illustrated in Figure 2.7, where a boldface ion symbol on a given side of the membrane denotes a higher concentration of that ion on the given side, relative to the other side. The unfilled arrows denote active currents due to the pump. Since the pump is assumed to be neutral, the active sodium current density $\vec{J}_{Na}^{\text{pump}}$ is equal and opposite to the active potassium current density $\vec{J}_{K}^{\text{pump}}$ (Equation 2.2). Under steady-state conditions, the active current of each ion is equal and opposite to the passive current of that ion (Equations 2.3 and 2.4). In the case of Na+ ions, the passive, inward current due to the voltage across the

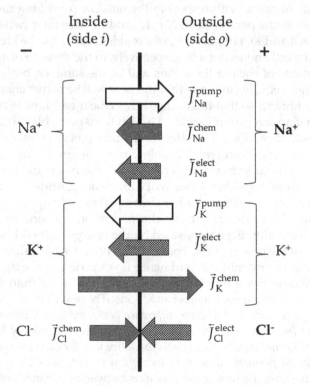

FIGURE 2.7
Components of Na+, K+, and Cl− currents through a cell membrane.

membrane $\left(\bar{J}_{Na}^{elect}\right)$ adds to the passive inward current due to the chemical potential gradient $\left(\bar{J}_{Na}^{chem}\right)$ (Figure 2.7 and Equation 2.3). In the case of K^+ ions, the passive, inward current due to the voltage across the membrane $\left(\bar{J}_K^{elect}\right)$ subtracts from the passive outward current due to the chemical potential gradient $\left(\bar{J}_K^{chem}\right)$ (Figure 2.7 and Equation 2.4). In this way, the steady-state, passive fluxes of Na^+ and K^+ are equalized, in accordance with the equality of the active fluxes due to the pump, despite the larger permeability of the membrane to K^+ compared to Na^+.

Under the influence of the voltage difference, Cl^- move from side i to side o, which increases the concentration of Cl^- on side o and reduces the net positive charge on this side. Note that because of the negative charge on Cl^-, the direction of current due to these ions is opposite to that of their flux. Despite the movement of Cl^-, side o must remain positive with respect to side i in order to maintain the required electric potential gradients for the three ions. Thus, if the movement of Cl^- were to neutralize the excess positive charge on side o, there will be no electric potential gradient to balance the chemical potential gradient for Cl^-, or equalize the passive fluxes of Na^+ and K^+. Since Cl^- are not being pumped, they will be at equilibrium in the system of Figure 2.6, when the system as a whole is in the steady-state.

The effect of indiffusible anions on the inside of the cell can be ascertained by supposing that, after a steady state has been reached, some Cl^- on side i are replaced instantaneously by an equal number of A^- ions. The charge balance will not be immediately disturbed, but Cl^- will no longer be at equilibrium, because Cl^- are now more concentrated on side o than on side i. Some Cl^- will therefore move from side o to side i under the influence of their chemical potential gradient, making side i more negatively charged still with respect to side o. Na^+ and K^+ will be attracted to side i by the negative charge. Nevertheless, the magnitude of the negative membrane voltage is larger than before A^- were introduced, because the reduced concentration of Cl^- on side i requires side i to be more negative with respect to side o than before so as to oppose the flow of Cl^- under the influence of the chemical potential gradient.

The effect of an electrogenic component of the sodium-potassium pump is to increase the voltage across the membrane because of the net transfer of positively charged Na^+ from side i to side o. This makes side i more negative still with respect to side o.

The membrane voltage in the steady state is derived quantitatively in the next section.

Problem 2.5

How does a steady state differ from a state of equilibrium? Is the state of equilibrium a special case of a steady state?

ANS.: Yes.

2.4.1 Membrane Voltage in the Steady State

Consider a cell membrane of thickness δ (Figure 2.8) through which an ion species Y flows under the influence of an electrochemical potential gradient in the resting state. U_Y, the steady-state flux of Y in the positive x direction, is the sum of the components due to the chemical potential and electric potential gradients. If the ion flows through an aqueous pore, the partition coefficient $\beta_Y = 1$, and Equation 1.22 for the permeability P_Y gives: $D_Y = \delta P_Y$, where D_Y is the diffusion constant for ion Y. Substituting for D_Y in the Nernst–Planck equation (Equation 1.17),

$$U_Y = -\delta P_Y \left(\frac{d[Y]}{dx} + \frac{z_Y q}{kT}[Y]\frac{dv}{dx} \right)$$

(2.5)

To integrate Equation 2.5, we multiply both sides by $e^{z_Y q v/kT}$, where v is the voltage at a point x inside the membrane:

$$U_Y e^{z_Y q v/kT} = -\delta P_Y \left(\frac{d[Y]}{dx} e^{z_Y q v/kT} + \frac{q z_Y}{kT}[Y]\frac{dv}{dx}e^{z_Y q v/kT} \right) = -\delta P_Y \frac{d}{dx}\left([Y] e^{z_Y q v/kT} \right)$$

(2.6)

Both sides of Equation 2.6 can be now be integrated with respect to x between the limits of 0 and δ, bearing in mind that in the steady-state: (i) U_Y is a constant that is independent of x, (ii) $v = V_{m0}$, the resting membrane voltage, at $x = 0$, and (iii) $v = 0$ at $x = \delta$ (Figure 2.8), since conventionally, the zero reference for membrane voltage is on the outside. Thus:

$$U_Y \int_0^\delta e^{z_Y q v/kT} dx = -\delta P_Y \int_0^\delta \frac{d}{dx}\left([Y] e^{z_Y q v/kT} \right) dx = -\delta P_Y \left[[Y] e^{z_Y q v/kT} \right]_0^\delta$$

$$= -\delta P_Y \left([Y]_o - [Y]_i e^{z_Y q V_{m0}/kT} \right), \text{ or } \quad U_Y = P_Y \left([Y]_i e^{z_Y q V_{m0}/kT} - [Y]_o \right) f(v)$$

(2.7)

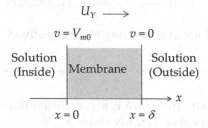

$$U_Y \longrightarrow$$

$$v = V_{m0} \qquad v = 0$$

| Solution (Inside) | Membrane | Solution (Outside) |

$x = 0 \qquad x = \delta$

FIGURE 2.8
Flux of ion Y through a cell membrane.

where,

$$f(v) = \delta \Big/ \int_0^\delta e^{z_Y q v / kT} dx \tag{2.8}$$

and $f(v)$ depends only on how v varies inside the membrane and is dimensionless.

The following should be noted concerning Equations 2.7 and 2.8:

1. When using numerical values in equations involving concentrations, rather than concentration ratios, the consistency of the units involved must be ensured. In Equation 2.7, for example, if [Y] is in particles/cm³, and P_Y is in cm/s, then U_Y is in particles/(cm²s). If [Y]* is in mM, then $[Y]^* \times 10^{-6}$ is in moles/cm³ and U_Y is in µmoles/(cm²s). The same is true of equations for current density (Equation 2.24) and for conductance (Equation 2.26) (Problem 2.15). Where concentration ratios are involved, [Y] could be in particles/cm³ or in mM. Where a distinction should be made, [Y]* is used for concentration in mM.

2. When there is no voltage across the membrane, $v_m = 0$, in Equation 2.8 and $f(v_m) = 1$. Moreover, $V_{m0} = 0$ under these conditions, and Equation 2.7 reduces to Equation 1.22 defining permeability. Hence, the permeability of an ion can be measured when there is no voltage across the membrane.

3. If $[Y]_o = 0$ in Equation 2.7, the first term on the right-hand side (RHS) of this equation represents an outward flux U_Y^{out} given by:

$$U_Y^{out} = P_Y [Y]_i \, e^{z_Y q V_{m0}/kT} f(v) \tag{2.9}$$

If $[Y]_i = 0$ in Equation 2.7, the negation of the second term in this equation represents an inward flux U_Y^{in} given by:

$$U_Y^{in} = P_Y [Y]_o \, f(v) \tag{2.10}$$

The ratio of the two fluxes is independent of $f(v)$ and is given by:

$$\frac{U_Y^{out}}{U_Y^{in}} = \frac{[Y]_i \, e^{z_Y q V_{m0}/kT}}{[Y]_o} = e^{(z_Y q/kT)(V_{m0} - E_Y)} \tag{2.11}$$

using the equilibrium voltage E_Y for ion Y (Equation 1.26). Equation 2.11 is known as the **Ussing flux ratio equation** and has been extensively applied in the study of diffusion across membranes using radioactive tracers, where initially, the concentration of the radioactively labelled ions on one side of the membrane is zero.

Problem 2.6

Determine Ussing flux ratio (Equation 2.11) assuming, $[Na^+]_i = 10$ mM, $[Na^+]_o = 145$ mM, $V_{m0} = -70$ mV and $T = 37$ °C.

ANS.: 1.06.

Consider next the case where the sodium-potassium pump actively transports γNa^+ for each K^+, with Cl^- being in equilibrium across the membrane. Then, $U_{Na}^{pump} = -\gamma U_K^{pump}$. Since the passive flux of each ion is equal and opposite to the active flux,

$$U_{Na}^{passive} + \gamma U_K^{passive} = 0 \tag{2.12}$$

$$P_{Na}\left([Na^+]_i e^{qV_{m0}/kT} - [Na^+]_o\right) + \gamma P_K\left([K^+]_i e^{qV_{m0}/kT} - [K^+]_o\right) = 0 \tag{2.13}$$

where $f(v)$ cancels out and z_Y has been set to 1. Solving for V_{m0},

$$V_{m0} = -\frac{kT}{q}\ln\frac{\gamma[K^+]_i + \dfrac{P_{Na}}{P_K}[Na^+]_i}{\gamma[K^+]_o + \dfrac{P_{Na}}{P_K}[Na^+]_o} \tag{2.14}$$

Equation 2.14 is a form of the **Goldman–Hodgkin–Katz (GHK) voltage equation**. It takes into account the electrogenic nature of the sodium-potassium pump ($\gamma > 1$), as well as the different permeabilities to Na^+ and K^+. Although the concentration of indiffusible anions does not appear explicitly in Equation 2.14, it is implicit in this equation because the presence of indiffusible anions affects $[Na^+]_i$ and $[K^+]_i$. These concentrations are in fact increased by the presence of indiffusible anions on side i, as argued previously (Section 1.5.4), which makes V_{m0} in Equation 2.14 more negative in the presence of these ions.

The following should be noted concerning Equation 2.14:

1. If P_{Na}/P_K is small, V_{m0} is much more sensitive to variations in $[K^+]_i$ and $[K^+]_o$ than to variations in $[Na^+]_i$ and $[Na^+]_o$. In the limit, if $P_{Na}/P_K = 0$, Equation 2.14 reduces to the Nernst equation for K^+ ions (Equation 1.25), bearing in mind that $V_{m0} = v_i - v_o$. On the other hand, if P_{Na}/P_K is very large, V_{m0} approaches the equilibrium voltage for Na^+. In either case, V_{m0} *approaches the equilibrium voltage of the more permeable ion.* It may be noted that some glial cells are almost selectively permeable to K^+. Their membrane voltage closely follows the Nernst equation when $[K^+]_o$ is varied. But K^+ are not at equilibrium in these cells, because a sodium-potassium pump is present, as in neurons.

2. The numerator is a function of the positive charge on the inside, whereas the denominator is a function of the positive charge on the outside.

3. If $\gamma = 1$ and $P_{Na}/P_K = 1$, then $[K^+]_i + [Na^+]_i = [K^+]_o + [Na^+]_o$, so that $V_{m0} = 0$. If $\gamma = 1$ and P_{Na}/P_K is small, the neutral pump increases $[Na^+]_o$ and decreases $[K^+]_o$, thereby increasing the magnitude of the argument of the log function and making the inside more negative than the outside. If $\gamma > 1$ and P_{Na}/P_K is small, the magnitude of the argument of the log function is further increased because $[K^+]_i$ in the numerator is larger, compared to $P_{Na}[Na^+]_i / P_K$, than is $[K^+]_o$ in the denominator, compared to $P_{Na}[Na^+]_o / P_K$. All the preceding conclusions are in accordance with the qualitative argument presented in connection with the model system of Figure 2.6.

Problem 2.7

Determine V_{m0} from Equation 2.14 assuming $P_{Na}/P_K = 1/50$, $\gamma = 1.5$, $[K^+]_i = 150$ mM, $[K^+]_o - 5$ mM, $[Na^+]_i = 10$ mM, $[Na^+]_o = 140$ mM, $k = 1.38 \times 10^{-23}$ J/K, $q = 1.6 \times 10^{-19}$ C, and $T = 37$ °C.

ANS.: −82.5 mV.

Problem 2.8

Determine V_{m0} in the preceding problem, assuming that the pump is neutral, with $[K^+]_i$ and $[Na^+]_o$ reduced by 10% and $[K^+]_o$ and $[Na^+]_i$ are increased by 10%.

ANS.: −75.6 mV.

2.5 Membrane Equivalent Circuit

In light of the preceding discussion, a more detailed and useful equivalent circuit can be developed, as shown in Figure 2.9, for a unit area of membrane. Each of the inside and outside of the cell is represented by a single node, with branches between the two nodes accounting for the flow of the various ions involved. The following assumptions are implied:

1. The inside and the outside of the cell, close to the membrane surfaces, are equipotential regions. This would apply to a small patch of membrane that is large enough to include numerous ion channels,

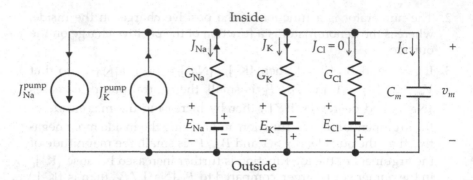

FIGURE 2.9
Equivalent circuit for a unit area of cell membrane.

or to a larger area of membrane in the absence of current flow in the direction parallel to the surface of the membrane, as in the case of a space-clamped axon (Section 3.1).

2. The ions move through the membrane independently of one another. This assumption is justified by experimental results, as explained in Section 3.2.4. It is evidently the case when each ion moves in its own channel.

3. It is assumed that Cl⁻ are in equilibrium across the cell membrane, that is, they are not transported across the membrane by other than passive electrodiffusion, which was the prevalent belief earlier on. It is now well established that this is not generally the case. It is believed that in most, but not all neurons, Cl⁻ are transported out of the cell by cotransporters (Section 2.3.1) so that there is a passive influx of Cl⁻ in the resting state. This makes E_{Cl}, the equilibrium voltage of Cl⁻ slightly more negative than the resting voltage. In muscle tissue, cotransport of Cl⁻ is relatively small, and the resting conductance for Cl⁻ is high. This means that only a small electrochemical potential difference is required to equalize the active and passive Cl⁻ fluxes through the membrane. Cl⁻ will therefore be almost at equilibrium across the membrane. For present purposes, the assumption of Cl⁻ being at equilibrium across the membrane will be retained. Active transport of Cl⁻ could be handled in the same manner as for Na⁺ and K⁺, but the assumption of Cl⁻ at equilibrium has the advantage of illustrating the behavior of an ion at equilibrium when other ions are actively transported.

The membrane capacitance is as in Figure 2.3. $J_C = C_m dv_m/dt$ is the capacitive current density in the direction of the voltage drop v_m. This is not an ionic current but is a **displacement current** due to the changes in the separated charges of the capacitor caused by electrostatic induction.

Under steady conditions, as in the resting state, $dv_m/dt = 0$, so $J_C = 0$. The sodium-potassium pump is represented by two current sources, J_{Na}^{pump} and J_K^{pump}. The passive current for each of the three ion species, Na+, K+, and Cl-, flows through a branch consisting of a conductance in series with a battery, as will be explained shortly.

Conventionally, the assigned positive polarity of voltages is a voltage drop from the inside to the outside, that is, from the intracellular medium to the extracellular medium. v_m and the battery voltages E_{Na}, E_K, and E_{Cl} are all assigned their positive polarities in this way in Figure 2.9. The batteries are drawn with their actual polarities, which makes the numerical value of E_{Na} positive, and the numerical values of E_K and E_{Cl} negative because the polarities of these batteries are opposite the assigned positive polarities of E_K and E_{Cl}. V_{m0} has a negative numerical value since the actual resting membrane voltage is inside negative with respect to outside, that is, the actual resting membrane voltage is a voltage drop from the outside to the inside.

When the assigned positive direction of voltage is a voltage drop from the inside to the outside, this makes the assigned positive direction of current outward, since such a current would flow through a positive conductance under the influence of a positive voltage. All the assigned current directions in Figure 2.9 are outward, as indicated by the line arrows associated with these currents. The actual current directions are indicated by the arrowheads drawn as part of the branch wiring. It follows that J_{Na}^{pump} and J_K have positive numerical values, whereas J_K^{pump} and J_{Na} have negative numerical values.

To interpret the ion branches, consider a general case of a branch representing the passive flow of an ionic current density J_Y of an ion Y in Figure 2.10. With the assigned positive direction of v_m a voltage drop from inside to outside, the positive direction of J_Y^{elect}, the component of the passive current density due to the electric potential gradient, is outward in the direction of the voltage drop, the same as J_Y.

The voltage v_Y across the membrane conductance per unit area G_Y represents the voltage equivalent of the "force" that drives current through the membrane. From the discussion in Section 1.4.2 on generalized Ohm's law (Equation 1.16), the driving force for current is the electrochemical potential gradient, which means that the electrochemical potential gradient for a given ion is represented by v_Y, the voltage across the membrane conductance for that ion (Figure 2.10). If the ion Y is not being pumped, then Y is in equilibrium across the membrane, the electrochemical potential gradient is zero, the net current through the membrane is zero, and the voltage across the membrane conductance is zero.

The electrochemical potential of Y on the inside is given by Equation 1.9:

$$\kappa_Y = kT \ln \frac{[Y]_i}{[Y]_o} + z_Y q(v_i - v_o) \tag{2.15}$$

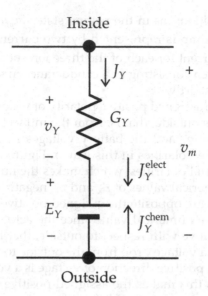

FIGURE 2.10
Equivalent circuit branch for ion Y.

where the reference state, which is quite arbitrary, is taken as the outside, and the electrochemical potential of this state, which is also arbitrary, is taken as zero.

Recall that each potential term in Equation 2.15 is potential energy per particle, and voltage is electric potential energy per unit charge. Hence, to obtain the voltage equivalent of the potential terms in Equation 2.15, both sides of the equation are divided by the charge $z_Y q$, to give:

$$\frac{\kappa_Y}{z_Y q} = v_Y = \frac{kT}{z_Y q} \ln \frac{[Y]_i}{[Y]_o} + v_m \tag{2.16}$$

where v_Y is the voltage equivalent of the electrochemical potential, as argued previously, and $v_m = v_i - v_o$ is the voltage equivalent of the electric potential, which is the membrane voltage itself. Since electrochemical potential is the algebraic sum of the chemical and electric potential gradients, it follows that the first term on the RHS of Equation 2.16 is the voltage equivalent of the chemical potential. Kirchhoff's voltage law (KVL) for the branch in Figure 2.10 is:

$$v_Y + E_Y = v_m, \text{ or } v_Y = -E_Y + v_m \tag{2.17}$$

Comparing Equations 2.16 and 2.17, it is seen that E_Y is in fact the equilibrium voltage for ion Y (Equation 1.26). In other words, for given $[Y]_i$ and $[Y]_o$, the equilibrium voltage is the voltage equivalent of the chemical potential. This is evident from the fact that if ion Y is at equilibrium, $E_Y = v_m$ in accordance with the definition of equilibrium voltage, so that $v_Y = 0$ and $J_Y = 0$. If E_Y acts alone in Figure 2.10, with v_m set to zero by connecting the two nodes together, then J_Y^{chem}, the component of the passive current density due to the chemical potential gradient is inward for $E_Y > 0$. This is in accordance with Equation 1.26, for $E_Y > 0$ means that, for a cation, $[Y]_o > [Y]_i$ so that J_Y^{chem} is inward.

The branch for ion Y in Figure 2.10 can now be identified with each of the Na+, K+, and Cl- branches in Figure 2.9. This is done in Table 2.2, which indicates actual current directions and the actual sign of E_Y. Note that the electrical component of the passive currents for Na+, K+, and Cl- is inward, that is, opposite to the assigned positive directions in Figure 2.10 because v_m is negative.

From Equation 1.26, Chapter 1, the equilibrium voltages of the three ions are:

$$E_{Na} = -\frac{kT}{q} \ln \frac{[Na^+]_i}{[Na^+]_o} > 0 \tag{2.18}$$

$$E_K = -\frac{kT}{q} \ln \frac{[K^+]_i}{[K^+]_o} < 0 \tag{2.19}$$

$$E_{Cl} = \frac{kT}{q} \ln \frac{[Cl^-]_i}{[Cl^-]_o} < 0 \tag{2.20}$$

bearing in mind that $[Na^+]_o > [Na^+]_i$, $[K^+]_i > [K^+]_o$, and $[Cl^-]_o > [Cl^-]_i$, with $E_{Cl} = V_{m0}$, since Cl- are at equilibrium. The current components due to the chemical potential gradient are in accordance with Figure 2.10, considering that E_{Na} is positive and both E_K and E_{Cl} are negative. The current directions are all in

TABLE 2.2

Ionic Currents in Equivalent Circuit

	Na+	K+	Cl-
Passive current	Inward	Outward	Zero
Current component due to electric potential gradient	Inward	Inward	Inward
Current component due to chemical potential gradient	Inward	Outward	Outward
Equilibrium voltage	Positive	Negative	Negative

Outward currents have positive numerical values. Inward currents have negative numerical values.

Positive voltages are voltage drops from inside to outside. Negative voltages are voltage drops from outside to inside.

accordance with Figure 2.7, as can be readily verified from physical reasoning bearing in mind that the current due to the negatively charged Cl⁻ is in a direction opposite to that of the flux of these ions.

Note that Kirchhoff's current law (KCL) at either node in Figure 2.9 must be satisfied for each ion individually, and not in the usual form, as the total current entering a node should be equal to the total current leaving the node. Moreover, the circuit is not a linear circuit, since the conductance in each branch is a function of v_m and of the concentrations of the given ion on the two sides of the membrane, as discussed in the following section. Moreover, as noted earlier, the rate of pumping of the sodium-potassium pump at a given temperature varies with $[Na^+]_i$, $[K^+]_o$, and v_m. Strictly speaking, the nonlinearity of the equivalent circuit of Figure 2.9 precludes, in general, the application of linear circuit theorems, such as Thevenin's and Norton's theorems.

Before ending this section, it should be pointed out that in discussing the electrical behavior of the cell membrane, voltages in the equivalent circuit are usually expressed not as absolute values but with reference to the resting membrane voltage V_{m0}. The conversion is made simply by adding the negative of the resting membrane voltage to the voltage in question. Thus, if $V_{m0} = -70$ mV, $E_{Na} = 60$ mV, and $E_K = -90$ mV, then these voltages with respect to the resting voltage V_{m0} are: $v_{mr} = -70 + 70 = 0$, $E_{Nar} = 60 + 70 = 130$ mV, and $E_{Kr} = -90 + 70 = -20$ mV, where r has been added to the subscript of a voltage symbol to indicate that the voltage in question is relative to the resting membrane voltage. E_{Clr} is of course zero if Cl⁻ are at equilibrium, since $E_{Clr} = -70 + 70 = 0$.

Problem 2.9

Assume that in Figure 2.9, $E_K = -100$ mV, G_K is 10 µS/cm², $J_K = 100$ nA/cm², $J_{Na}^{pump} = 200$ nA/cm², and $E_{Na} = 50$ mV. Determine: (a) V_{m0} and (b) G_{Na}.

ANS.: (a) $V_{m0} = -90$ mV; (b) $G_{Na} = 1.43$ µS/cm².

Problem 2.10

Assume that in Figure 2.9, $J_{Na}^{pump} = 150$ nA/cm², $J_K^{pump} = -100$ nA/cm², $V_{m0} = -70$ mV, $E_{Na} = 60$ mV, $E_K = -90$ mV, and $E_{Cl} = -70$ mV. Determine: (a) J_{Na}, (b) J_K, (c) J_{Cl}, (d) G_{Na}, and (e) G_K.

ANS.: (a) $J_{Na} = -150$ nA/cm²; (b) $J_K = 100$ nA/cm²; (c) $J_{Cl} = 0$; (d) $G_{Na} = 1.15$ µS/cm²; (e) $G_K = 5$ µS/cm².

Problem 2.11

Assume that in Figure 2.9, $J_{Na}^{pump} = 200$ nA/cm², $J_K^{pump} = -100$ nA/cm², $G_{Na} = 1$ µS/cm², $G_K = 10$ µS/cm², and $E_{Cl} = -50$ mV, determine: (a) V_{m0}, (b) E_{Na}, and (c) E_K.

ANS.: (a) $V_{m0} = -50$ mV; (b) $E_{Na} = 150$ mV; (c) $E_K = -60$ mV.

Problem 2.12

Consider the circuit of Figure 2.9 in the resting state as a linear circuit. Transform the voltage source in series with a conductance in each ion branch to its equivalent current source, then combine the current sources and conductances to show that:

$$V_{m0} = \frac{E_{Na0}G_{Na0} + E_{K0}G_{K0} + E_{Cl0}G_{Cl0} + J_{K0}^{pump} + J_{Na0}^{pump}}{G_{Na0} + G_{K0} + G_{Cl0}},$$

where zero in the subscript refers to the resting state.

2.5.1 Membrane Conductances

The conductance in the branch for a given ion represents the resistance to the flow of the ion through the membrane. In terms of the current through a given ion branch, it follows from Equation 2.17 by substituting $v_Y = J_Y/G_Y$ (Figure 2.10) that:

$$J_Y = G_Y\left(v_m - E_Y\right) \qquad (2.21)$$

To derive the expression for conductance, we must derive first the current density for a given ion when a voltage is applied across the membrane. Under these conditions, the RHS of Equation 2.13 is not zero because of a current through the membrane due to the applied voltage. $f(v)$ does not cancel out, as in Equation 2.13, so it is necessary to assume a certain voltage profile across the membrane. In the **constant-field approximation**, it is assumed that the electric field in the membrane is constant, which means that v varies linearly across the membrane. This is a good approximation for very thin membranes under steady-state conditions. Referring to Figure 2.8, the voltage variation can be expressed as:

$$v = v_m\left(1 - \frac{x}{\delta}\right) \qquad (2.22)$$

where v is the voltage at a point x inside the membrane and v_m is the voltage across the membrane. Substituting for v from Equation 2.22 in Equation 2.8 for $f(v)$,

$$f(v) = \delta \bigg/ \int_0^\delta e^{z_Y q v_m (1 - x/\delta)/kT}\,dx = 1 \bigg/ \left[\frac{kT}{z_Y q v_m}\left(e^{z_Y q v_m/kT} - 1\right)\right] \qquad (2.23)$$

The current density for a given ion Y is obtained from Equation 2.7 by substituting for $f(v)$, multiplying U_Y by $z_Y q$ to obtain current density from flux, and replacing V_{m0} with v_m. This gives:

$$J_Y = P_Y \frac{(z_Y q)^2 v_m}{kT} \frac{\left([Y]_i e^{z_Y q v_m / kT} - [Y]_o\right)}{e^{z_Y q v_m / kT} - 1}$$

$$= P_Y \frac{(z_Y q)^2 v_m}{kT} \frac{\left([Y]_i - [Y]_o e^{-z_Y q v_m / kT}\right)}{1 - e^{-z_Y q v_m / kT}} \tag{2.24}$$

where the second expression for J_Y is obtained from the first by multiplying the numerator and denominator by $e^{-z_Y q v_m / kT}$. Equation 2.24 is the **Goldman–Hodgkin–Katz (GHK) current equation**. It was originally derived assuming ions crossing a homogeneous membrane independently of one another, that is, without mutual interaction. The concept of ions crossing the membrane through individual channels and pores was not generally recognized at the time. The conductance for a given ion species is the reciprocal of the resistance presented by the membrane to the flow of current due to this ion species. It is seen from Equation 2.24 that the relation between J_Y and v_m is nonlinear, so that the membrane conductance is not constant. For a given v_m, $[Y]_i$, and $[Y]_o$, J_Y can be calculated from Equation 2.24 and E_Y from Equation 1.26. G_Y can be then determined using Equation 2.21 (Problem 2.13).

Note that the nonlinearity in Equation 2.24 is due to the exponential terms and the dependence of the current J_Y on the difference in concentrations $[Y]_i$ and $[Y]_o$ on the two sides of the membrane. If $[Y]_i = [Y]_o = [Y]$, Equation 2.24 reduces to:

$$J_Y = P_Y \frac{(z_Y q)^2 [Y]}{kT} v_m = G_Y v_m, \text{ with } \quad G_Y = P_Y \frac{(z_Y q)^2 [Y]}{kT} \tag{2.25}$$

J_Y is now directly proportional to v_m, the proportionality constant being a constant conductance G_Y.

The **incremental conductance** per unit area is $G'_Y = dJ_Y / dv_m$ and is defined as the slope of the J_Y vs. v_m curve at a particular value of J_Y or v_m. The incremental conductance per unit area can be readily derived at the equilibrium voltage of the given ion Y ($v_m = E_Y$, $J_Y = 0$), and is given by (Problem 2.14):

$$G'_Y = P_Y \frac{(z_Y q)^2}{kT} \frac{[Y]_i [Y]_o}{[Y]_o - [Y]_i} \ln \frac{[Y]_o}{[Y]_i} \tag{2.26}$$

Note that G'_Y is a positive quantity, irrespective of whether $[Y]_i$ is larger than, or smaller than $[Y]_o$. When P is in cm/s, and $[Y]$ is in particles/cm³, G'_Y is in S/cm². When $[Y]^*$ in Equation 2.26 is in mM, with q and k replaced by F and R, respectively, then G'_Y is in μS/cm² (Problem 2.15). G'_Y depends not only on permeability, as expected, but also on the concentrations of the given ion on both sides of the membrane. If these concentrations are equal, it can be shown using L'Hopital's rule that G'_Y reduces to G_Y as given in Equation 2.25, since the current-voltage relation becomes linear.

Equation 2.24 for J_K and J_{Na} are plotted in Figures 2.11 and 2.12, respectively, and membrane conductances are further discussed in Section 2.6.

Problem 2.13

The net K^+ efflux in a cell is 3.5 pmol.cm^{-2}s^{-1}, $[K^+]_i = 120$ mM, $[K^+]_o = 2$ mM, and $V_{m0} = -90$ mV. Determine G_K, assuming 37 °C.

ANS.: 18.3 μS/cm^2.

Problem 2.14

Derive Equation 2.26

Problem 2.15

Show that when $[Y]^*$ is in mM, $[Y] = N_A[Y]^* \times 10^{-6}$ particles/cm^3, where N_A is Avogadro's constant. Deduce that Equation 2.26 can be written as:

$$G_Y = P_Y \frac{(z_Y F)^2}{RT} \frac{[Y]_i^*[Y]_o^*}{[Y]_o^* - [Y]_i^*} \ln \frac{[Y]_o^*}{[Y]_i^*} \mu S/cm^2$$

where P_Y is in cm^2/s, F is Faraday's constant in C/mol (96,485 C/mol) and R is the gas constant in J/(mol.K) (8.3145 J/(mol.K)). Determine G_Y if $z_Y = 1$, $[Y]_i^* = 100$ mM, $[Y]_o^* = 10$ mM, $P_Y = 5 \times 10^{-4}$ cm/s and $T = 37$ °C.

ANS.: 4.62 mS/cm^2.

Problem 2.16

Determine the incremental conductance at the equilibrium voltage and $T = 37$ °C of: (a) K^+, assuming $[K^+]_i = 150$ mM, $[K^+]_o = 5$ mM, and $P_K = 10^{-6}$ cm/s; (b) Na^+, assuming $[Na^+]_i = 10$ mM, $[Na^+]_o = 140$ mM, and $P_{Na} = 2 \times 10^{-8}$ cm/s.

ANS.: (a) 63.5 μS/cm^2; (b) 2.05 μS/cm^2.

2.5.2 Generation of Electric Signals

Several important deductions can be made from the equivalent circuit of Figure 2.9. Generation of electric signals across the membrane means changing v_m from its resting value V_{m0} in a meaningful way. This can be accomplished by the following mechanisms:

1. Changing E_K, E_{Na}, or E_{Cl}
2. Changing the rate of pumping
3. Changing the membrane conductances

The first two mechanisms are impractical and slow. The third mechanism is the one used in practice. Changing G_{Cl} alone does not affect v_m, since Cl$^-$ are in equilibrium and no current flows in G_{Cl}. Increasing G_{Na} increases v_m in the positive direction, in opposition to the resting voltage V_{m0}. In the limit, $v_m \rightarrow E_{Na}$ as $G_{Na} \rightarrow \infty$, since the E_{Na} battery in Figure 2.9 becomes directly connected between the inside and outside nodes. Increasing v_m in the positive direction is referred to as **depolarization,** because it reduces membrane polarization. On the other hand, increasing G_K increases v_m in the negative direction, that of the resting voltage V_{m0}. In the limit, $v_m \rightarrow E_K$ as $G_K \rightarrow \infty$. Increasing v_m in the negative direction is referred to as **hyperpolarization** because it increases membrane polarization. It follows that by increasing G_{Na} and/or G_K, the variation in v_m is limited to the range:

$$E_{Na} > v_m > E_K \qquad (2.27)$$

The change in membrane conductance could be caused by:

1. Physical stimuli, such as mechanical deformation, as in touch receptors and muscle receptors, or light in the case of photoreceptors.
2. Chemical stimuli, as in taste and smell receptors as well as in nerve and muscle cells, whereby the binding of ligands to specific receptors on the outer or inner surface of the membrane change membrane conductance.
3. Membrane depolarization, which affects membrane conductances in a specific manner, as discussed in the next chapter.

Note that the previous deduction in connection with Equation 2.14 that V_{m0} approaches the equilibrium voltage of the more permeable ion follows from the equivalent circuit of Figure 2.9, since a relatively large permeability for a given ion entails a large conductance G for that ion. In the limit as $G \longrightarrow \infty$ for that ion, V_{m0} approaches the equilibrium voltage of the ion.

2.6 Membrane Rectification

Rectification is said to occur when the current through a given system depends on the polarity of the voltage applied to that system.

The simplest case of rectification in a membrane system is that due to ion movement under the influence of a difference in electrochemical potential, without any complicating factors, such as an inherent dependence of membrane conductance on membrane voltage. To illustrate this simple electrodiffusion, consider the movement of K^+ across the cell membrane when a voltage v_m is impressed across the membrane. Referring to Figure 2.9, a positive v_m causes an outward K^+ current, the current density being given by Equation 2.24, assuming the constant-field approximation. It is evident that the outward and inward currents will be unequal because of the exponential terms in v_m. From the first Equation 2.24,

$$J_K = P_K \frac{q^2 v_m}{kT} \frac{\left([K^+]_i e^{qv_m/kT} - [K^+]_o\right)}{e^{qv_m/kT} - 1}$$ (2.28)

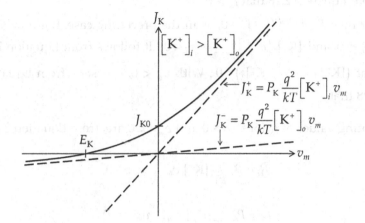

FIGURE 2.11
A plot of the variation of K^+ current density with membrane voltage.

where the numerator and denominator are multiplied by $e^{qv_m/kT}$. Equation 2.28 is plotted in Figure 2.11. It is seen that $J_K = 0$ when the bracketed term in the numerator is zero, that is, when:

$$v_m = E_K = \frac{kT}{q} \ln\left(\frac{[K^+]_o}{[K^+]_i}\right) < 0$$ (2.29)

K^+ are at equilibrium when $v_m = E_K$, so that $J_K = 0$. It also follows from Equation 2.2.7 that J_K increases in the positive direction with v_m.

Substituting for E_K from Equation 2.29, the bracketed term in the numerator of Equation 2.28 becomes:

$$\left([K^+]_i e^{qv_m/kT} - [K^+]_o\right) = [K^+]_o \left(e^{(v_m - E_K)q/kT} - 1\right)$$ (2.30)

where $q/kT \cong 37.4$ per volt at a temperature of $37°C$.

The shape of the curve of Figure 2.11 can be justified by considering three regions of the current-voltage characteristic of Equation 2.28:

1. For $v_m > 0$, and with $[K^+]_i > [K^+]_o$, $\left([K^+]_i e^{qv_m/kT} - [K^+]_o\right) > \left(e^{qv_m/kT} - 1\right)$;

    both the numerator and denominator in Equation 2.28 are positive, and $J_K > 0$.

2. For $E_K < v_m < 0$, and since v_m is negative, $e^{qv_m/kT} < 1$, and $\left(e^{qv_m/kT} - 1\right) < 0$. This makes the denominator of Equation 2.28 negative. At the same time, $(v_m - E_K) > 0$, so that $\left(e^{(v_m-E_K)q/kT} - 1\right) > 0$. It follows from Equation 2.30 that $\left([K^+]_i e^{qv_m/kT} - [K^+]_o\right) > 0$. With $v_m < 0$, it is seen from Equation 2.28 that $J_K > 0$.

3. For $v_m < E_K$, $\left(e^{qv_m/kT} - 1\right) < 0$, as in the preceding case. But now $(v_m - E_K) < 0$, and $[K^+]_o\left(e^{(v_m-E_K)q/kT} - 1\right) < 0$. It follows from Equation 2.30 that $\left([K^+]_i e^{qv_m/kT} - [K^+]_o\right) < 0$. With $v_m < 0$, it is seen from Equation 2.28 that $J_K < 0$.

The limiting values of J_K for $v_m \gg 0$ and $v_m \ll 0$ are, from Equation 2.28:

$$J_K^+ = P_K \frac{q^2}{kT}[K^+]_i v_m, \quad v_m \gg 0 \tag{2.31}$$

$$J_K^- = P_K \frac{q^2}{kT}[K^+]_o v_m, \quad v_m \ll 0 \tag{2.32}$$

Equations 2.31 and 2.32 are plotted as asymptotes in Figure 2.11. The ratio of the magnitudes of J_K^+ and J_K^- is:

$$\left|\frac{J_K^+}{J_K^-}\right| = \frac{[K^+]_i}{[K^+]_o} = e^{-qE_K/kT} > 0 \tag{2.33}$$

The current intercept ($v_m = 0$) is (Problem 2.18):

$$J_{K0} = qP_K\left([K^+]_i - [K^+]_o\right) > 0 \tag{2.34}$$

Because the outward current ($J_K > 0$) is larger than the inward current ($J_K < 0$), the current-voltage characteristic of Figure 2.11 is that of **outward rectification**. *Basically, the outward current for $v_m \gg 0$ is due to the electric field in the positive x-direction and is carried by $[K^+]_i$, whereas the inward current for $v_m \ll 0$ is due to the electric field in the negative x-direction and is carried by $[K^+]_o$. Since $[K^+]_i > [K^+]_o$, outward rectification occurs.*

If $[K^+]_i = [K^+]_o = [K^+]$, the exponential terms in Equation 2.28 cancel out and,

$$J_K = P_K \frac{q^2}{kT}[K^+]v_m \tag{2.35}$$

This is the same as Equation 2.25. No rectification occurs, and J_K becomes directly proportional to v_m. Equations 2.31 and 2.32 of the two asymptotes reduce to Equation 2.35.

If we consider Na$^+$ having $[Na^+]_o > [Na^+]_i$, the same equations derived for K$^+$ apply but give a different rectification characteristic. Equation 2.28 becomes:

$$J_{Na} = P_{Na} \frac{q^2 v_m}{kT} \frac{\left([Na^+]_i e^{qv_m/kT} - [Na^+]_o\right)}{e^{qv_m/kT} - 1} \qquad (2.36)$$

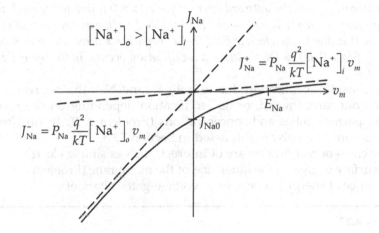

FIGURE 2.12
A plot of the variation of Na$^+$ current density with membrane voltage.

Equation 2.36 is plotted in Figure 2.12. Again, J_{Na} increases in the positive direction with v_m. The shape of the curve can be justified in the same manner as for Figure 2.11. $J_{Na} = 0$ when the bracketed term in the numerator is zero, that is, when:

$$v_M = E_{Na} = \frac{kT}{q} \ln\left(\frac{[Na^+]_o}{[Na^+]_i}\right) > 0 \qquad (2.37)$$

The limiting values of J_{Na} for $v_m \gg 0$ and $v_m \ll 0$ are:

$$J_{Na}^+ = P_{Na} \frac{q^2}{kT} [Na^+]_i v_m, \quad v_m \gg 0 \qquad (2.38)$$

$$J_{Na}^- = P_{Na} \frac{q^2}{kT} [Na^+]_o v_m, \quad v_m \ll 0 \qquad (2.39)$$

Equations 2.38 and 2.39 are plotted as asymptotes in Figure 2.12. The ratio of the magnitudes of J_{Na}^+ and J_{Na}^- is:

$$\left|\frac{J_{Na}^+}{J_{Na}^-}\right| = \frac{[Na^+]_i}{[Na^+]_o} = e^{-qE_{Na}/kT} > 0 \qquad (2.40)$$

The current density intercept ($v_m = 0$) is (Problem 2.21):

$$J_{Na0} = qP_{Na}\left([Na^+]_i - [Na^+]_o\right) < 0 \qquad (2.41)$$

Because the inward current ($J_{Na} < 0$) is larger than the outward current ($J_{Na} > 0$), the current-voltage characteristic of Figure 2.12 is that of **inward rectification**. Again, *the outward current for $v_m \gg 0$ is due to the electric field in the positive x-direction and is carried by $[Na^+]_i$, whereas the inward current for $v_m \ll 0$ is due to the electric field in the negative x-direction and is carried by $[Na^+]_o$. Since $[Na^+]_o > [Na^+]_i$, inward rectification occurs, in the opposite sense to that of K^+.*

If we consider rectification due to both K^+ and Na^+, the net rectification could be outward, inward, or no rectification depending on the relations between permeabilities and concentrations (Problem 2.20). Inward rectifying potassium channels are discussed in Section 7.3.3.

Other cases of rectification are of interest, such as simple electrodiffusion with a surface charge on the inner side of the membrane (Problem 2.22), or in the presence of energy barriers, or in voltage-gated channels.

Problem 2.17

Determine the ratio at $T = 37$ °C: (a) $\left|\dfrac{J_K^+}{J_K^-}\right|$ (Equation 2.33), assuming $E_K = -90$ mV, and (b) $\left|\dfrac{J_{Na}^+}{J_{Na}^-}\right|$ (Equation 2.41), assuming $E_{Na} = 130$ mV.

ANS.: (a) 28.96; (b) 7.73×10^{-3}.

Problem 2.18

Derive Equation 2.34 by considering that: (a) the current is a diffusion current only, (b) applying Equation 2.7. Can the intercept be obtained from Equation 2.28 by setting $v_m = 0$?

ANS.: Yes, by applying L'Hopital's rule

Problem 2.19

Equations 2.31, 2.32, and 2.35 are of the form:

$$J_K = P_K \frac{q^2}{kT}[K^+]v_m$$

Using the constant-field approximation, show that this reduces to $J_K = \sigma_K \xi$, where $\sigma_K = q^2 \mu_K [K^+]$.

Problem 2.20

Consider the K^+ and Na^+ currents due to an applied voltage, as given by Equations 2.28 and 2.36, respectively. Derive the conditions for: (a) no net rectification, (b) net outward rectification, and (c) net inward rectification.

ANS.: (a) $\left(P_K[K^+]_i + P_{Na}[Na^+]_i\right) = \left(P_K[K^+]_o + P_{Na}[Na^+]_o\right)$

(b) $\left(P_K[K^+]_i + P_{Na}[Na^+]_i\right) > \left(P_K[K^+]_o + P_{Na}[Na^+]_o\right)$

(c) $\left(P_K[K^+]_i + P_{Na}[Na^+]_i\right) < \left(P_K[K^+]_o + P_{Na}[Na^+]_o\right)$

Problem 2.21

Considering that when $[K^+]^*$ is in mM, $[K^+] = N_A[K^+]^* \times 10^{-6}$ particles/cm^3, where N_A is Avogadro's constant, deduce that Equation 2.28 can be written as:

$$J_K = P_K \frac{F^2 v_m}{RT} \frac{\left([K^+]_i^* e^{q v_m / kT} - [K^+]_o^*\right)}{e^{q v_m / kT} - 1} \mu A/cm^2$$

where P_K is in cm^2/s, F is Faraday's constant in C/mol (96,485 C/mol) and R is the gas constant in J/(mol.K) (8.3145 J/(mol.K). If $[K]_i^* = 150$ mM, $[K]_i^* = 5$ mM, $P_K = 10^{-6}$ cm/s, and $T = 310$ K, determine: (a) E_K, (b) the current intercept, (c) the asymptote for $v_m \gg 0$, and (d) the asymptote for $v_m \ll 0$.

ANS.: (a) -91 mV; (b) 14 14 $\mu A/cm^2$; (c) $542 v_m$ $\mu A/cm^2$; (d) 18.1 $\mu A/cm^2$.

Problem 2.22

Suppose that the inner side of the membrane has a surface charge. This attracts an opposite charge in the adjacent solution, resulting in an electrical double layer that causes a jump in the voltage at the inner membrane boundary. Assuming a constant electric field in the membrane, the voltage profile can be represented as in Figure 2.13. Show that Equation 2.28 is modified to:

$$J_K = P_K \frac{q^2 (v_m + V_D)}{kT} \frac{\left([K^+]_i e^{q v_m / kT} - [K^+]_o\right)}{e^{q v_m / kT} - 1}$$

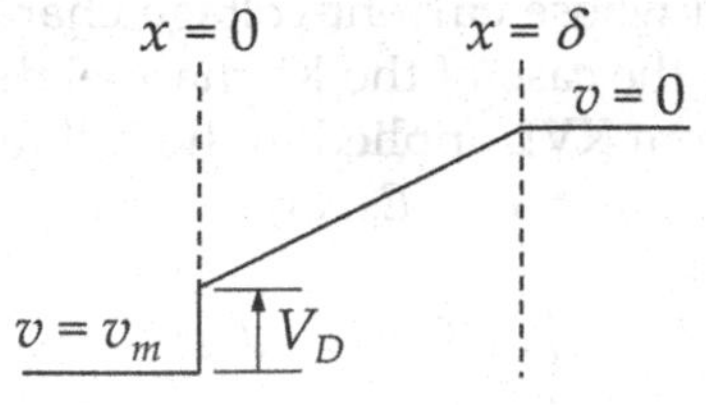

FIGURE 2.13
Figure for Problem 2.22.

By making the substitution $v'_m = v_m + V_D$, derive the ratio of the asymptotes and deduce that, irrespective of the relative values of $[K^+]_i$ and $[K^+]_o$, outward rectification, or inward rectification, or no rectification, can occur depending on whether $(E_K + V_D)$ is negative, positive, or zero, respectively. Show that if $[K^+]_i = [K^+]_o$, the rectification characteristics pass through the origin.

2.7 Membrane Reactance

When the time courses of voltage and current in a circuit are not in phase, the circuit is said to possess reactance. The membrane equivalent circuit of Figure 2.9 has reactance because of the membrane capacitance C_m. It may be thought that the conductances, although nonlinear, would not contribute to membrane reactance. This is true as long as the conductances are time-invariant. If, however, the conductances are both nonlinear and time-varying then these conductances will introduce an inductive or a capacitive reactance, as will be demonstrated in this section.

If the voltage across a time-invariant, linear or nonlinear, conductance is suddenly changed to a new value, the current through the conductance will ideally change instantly to its final value and will remain at this value if the voltage remains constant. In a membrane system, however, a change in membrane voltage to a new steady value is associated, in the general case, with a change in the distribution of ions across the membrane. The change in ion distribution cannot physically occur instantly but would take some time to assume its final state, resulting in a reactive-type behavior as explained later. Although this is true of all membrane systems, an ion channel may have some inherent characteristic time variation of conductance, as in the case of voltage-gated channels discussed in the next chapter, which will result in reactive behavior in the same manner.

2.7.1 Inductive Reactance

To illustrate qualitatively the nature of membrane reactance in a simple case, consider an ion channel whose current-voltage characteristic is that of outward rectification, as in the case of the K^+ channel discussed in the preceding section. It follows from KVL applied to the K^+ branch in Figure 2.9, as in Equation 2.21, that $J_K/G_K = v_K = v_m - E_K$, or,

$$J_K = G_K \left(v_m - E_K \right) \tag{2.42}$$

If a steady-state value of v_m, say V_{m1}, is applied across the membrane, an operating point P is defined on the current-voltage characteristic of Figure 2.11, as illustrated in Figure 2.14, together with the straight-line plot of Equation 2.42.

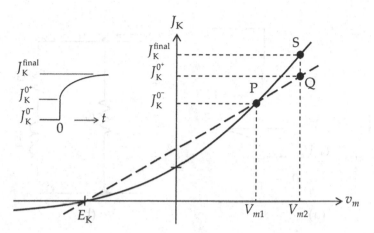

FIGURE 2.14
Variation of K⁺ current density with a step change of membrane voltage.

The current-voltage characteristic is a plot of the steady-state variation of J_K with v_m. If G_K is constant, Equation 2.42 will be a straight line passing through the voltage intercept E_K, and, if $v_m = V_{m1}$, the line will also pass through P as in Figure 2.14. The slope of this line defines a **chord conductance**, in contrast to the incremental conductance $G'_K = dJ_K/dv_m$, defined as the slope of the current-voltage characteristic at any point on this characteristic.

Let v_m be suddenly increased at $t = 0$ from V_{m1} to a new steady value V_{m2}. As explained previously, there will be no change at that instant in the ion distribution across the membrane. G_K will thus remain constant at the instant of change. J_K will instantly change from $J_K^{0^-}$, its value just before the change in v_m, to $J_K^{0^+}$ corresponding to point Q in Figure 2.13. However, the new steady value of V_{m2} will cause a redistribution of ions across the membrane, resulting in a gradual increase in current to a new steady-state value J_K^{final} corresponding to point S on the current-voltage characteristic.

The time course of J_K will be as shown in the inset of Figure 2.14. This time course is exactly analogous to that of the current i due to a step of voltage applied to the inductive circuit of Figure 2.15. Thus, if a step of voltage from 0 to V_s is applied at $t = 0$ (Figure 2.15a), the current in the inductor in Figure 2.15b will remain initially at zero. The current in G_p, and hence i, will increase stepwise from 0 to G_pV_s. As the current in the inductor builds up, i increases to a steady value $(G_p + G_L)V_s$ with a time constant LG_L (Figure 2.15a). It is seen that the behavior of the membrane system under consideration is that of an inductive circuit.

2.7.2 Capacitive Reactance

Consider next an ion channel whose current-voltage characteristic is that of inward rectification, as in the case of the Na⁺ channel discussed in

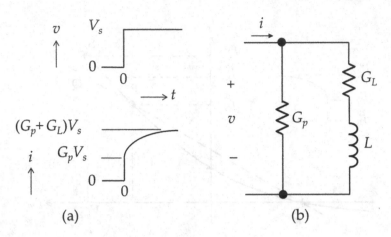

FIGURE 2.15
Variation of current with a step change of voltage (a) across an inductive circuit (b).

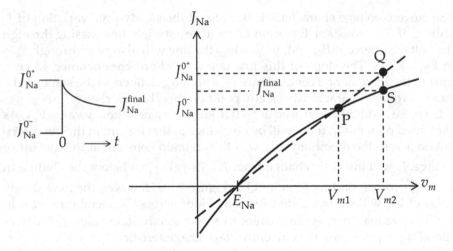

FIGURE 2.16
Variation of Na⁺ current density with a step change of membrane voltage.

the preceding section. It follows from KVL applied to the Na⁺ branch in Figure 2.9, as in Equation 2.21, that $J_{Na}/G_{Na} = v_{Na} = v_m - E_{Na}$, or,

$$J_{Na} = G_{Na}(v_m - E_{Na}) \tag{2.43}$$

Figure 2.16, analogous to Figure 2.14, shows the current-voltage characteristic and the straight line plot of Equation 2.43. Following a similar argument, it is seen that if the voltage is suddenly changed from v_{m1} to v_{m2}, the current first increases instantly from J_{Na}^{0-} to J_{Na}^{0+} along the straight line joining E_{Na} to P and then drops gradually to J_{Na}^{final}, corresponding to S, as illustrated in the inset

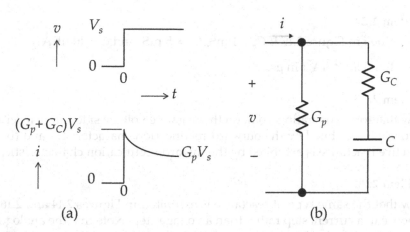

FIGURE 2.17
Variation of current with a step change of voltage (a) across a capacitive circuit (b).

of Figure 2.16. This behavior is analogous to that of the capacitive circuit in Figure 2.17. When a 0 to V_s voltage step is applied, the voltage across the capacitor remains zero, so that i jumps initially to $V_s(G_p + G_C)$. As C charges, the current falls to $G_p V_s$ with a time constant C/G_C.

In conclusion, it can be stated that, *following a sudden change in voltage magnitude, the reactance exhibited by a membrane is inductive if the incremental conductance at the operating point is larger than the chord conductance, and is capacitive if the incremental conductance at the operating point is smaller than the chord conductance.*

The following should be noted:

1. As in the case of the analogous electric circuit, the change in voltage should be fast, that is, it should occur in a time interval that is small compared to the response time of the system if the reactive effect is to be significant.

2. If the operating point is on the low-current side of the outward rectification characteristic ($v_m < 0$), then, according to the preceding conclusion, a capacitive reactance is exhibited. Similarly, if the operating point is on the high-current side of the inward rectification characteristic ($v_m < 0$), an inductive reactance is exhibited (Problem 2.25).

3. The same reactance, inductive or capacitive, is exhibited in response to a current step rather than a voltage step (Problem 2.26).

Problem 2.23

Determine i in Figure 2.15 if $V_s = 10$ mV, $G_p = 1$ mS, $G_L = 5$ mS, and $L = 10$ mH.

ANS.: $60 - 50e^{-t/50}$ µA, with t in µs.

Problem 2.24

Determine i in Figure 2.17 if $G_p = 1$ mS, $G_C = 5$ mS, and $C = 10$ nF.A

ANS.: $10 + 50e^{-t/2}$ µA, t in µs.

Problem 2.25

Show that for an operating point on the negative voltage side, (a) a capacitive reactance is exhibited by the outward rectification characteristic, and (b) an inductive reactance is exhibited by the inward rectification characteristic.

Problem 2.26

Show that the same type of reactance is exhibited in Figures 2.14 and 2.16 in response to a current step rather than a voltage step. Note that the analogous electric circuit is the dual of the circuit that gives the same time course of the current response to a voltage step.

Problem 2.27

The resistance of temperature-sensitive resistors is also nonlinear and time-varying. Deduce that: (a) a resistor having a negative-temperature coefficient exhibits an inductive reactance, whereas (b) a resistor having a positive-temperature coefficient exhibits a capacitive reactance.

Problem 2.28

Argue that a nonlinear, time-invariant resistor does not exhibit reactance.

2.8 Semiconductor Analogy

A close analogy exists between ionic systems and semiconductors because once some simplifying assumptions are made both systems are described by the same set of electrodiffusion equations. The main simplification in semiconductors is using effective masses of conduction electrons and holes to account for the interactions between these carriers and the semiconductor crystal, which allows treating these carriers as classical particles. The main simplification in ionic systems is to neglect interactions between particles in solution so that these particles can be treated as classical particles moving under the influence of electric and chemical potential gradients only. The corresponding analogous quantities between ionic systems and semiconductors are listed in Table 2.3. Note that the Fermi level in semiconductor systems is an equilibrium electrochemical potential and is not defined under nonequilibrium conditions.

TABLE 2.3

Analogy between Ionic Systems and Semiconductors

Ionic System	Semiconductor
Diffusible anions	Conduction electrons
Diffusible cations	Holes
Indiffusible anions	Acceptor impurities
Indiffusible cations	Donor impurities
Water solvent	Semiconductor matrix
Semipermeable membrane	Junction
Nernst potential	Junction potential
Equilibrium electrochemical potential	Fermi level

Summary of Main Concepts

- The basic structure of the cell membrane is that of a phospholipid bilayer having integral and peripheral proteins as well as cholesterol.
- Ion channels are essentially protein molecules of various types. Some channels open and close spontaneously, whereas other channels are gated by various influences, such as the membrane voltage, mechanical deformation, or the binding of certain substances (referred to as ligands) to the outside or the inside of the cell membrane.
- The cell membrane possesses capacitance because of the insulating properties of the phospholipid bilayer and possesses significant conductance because of the aqueous channels.
- An active pump that extrudes Na^+ must be postulated because both the electric potential gradient and the chemical potential gradient of these ions drive them inward. The sodium-potassium pump ejects three Na^+ for every two K^+ injected.
- Because of the activity of an ion pump, the membrane system is in a steady state if left undisturbed for a sufficiently long time. Under these conditions, the active flux of an ion due to the action of the pump is equal and opposite to the passive flux of that ion due to the electrochemical potential difference for that ion between the two sides of the membrane. An ion that is not being pumped and is free to diffuse through the cell membrane will be at equilibrium.
- The reduced concentrations of Na^+ and Cl^- compensate for the excess concentration of indiffusible anions inside the cell, thereby helping to equalize the concentrations of particles in solution on both sides

of the membrane in order to establish osmotic equilibrium and control cell volume.

- The resting membrane voltage is due to: (i) the neutral component of the sodium-potassium pump in conjunction with the much lower permeability of the membrane to Na^+ compared to K^+, (ii) the electrogenic component of the sodium-potassium pump, (iii) the presence of an excess concentration of indiffusible anions inside the cell.

- The electrical state of the membrane can be usefully described in terms of an equivalent circuit having, for each ion species, a branch consisting of a conductance in series with a battery. The voltage drop across the conductance is the voltage equivalent of the electrochemical potential difference for the given ion between the two sides of the membrane, and the battery voltage is the voltage equivalent of the chemical potential difference.

- Conventionally, the assigned positive direction of voltage across the membrane is a voltage drop from the inside to the outside, which makes the numerical value of membrane voltage negative and also implies that the assigned positive direction of current flow through the membrane is outward.

- The generation of electric signals involves increasing G_{Na} and/or G_K, with or without changes in G_{Cl}, the voltage variation being limited to the range between the equilibrium voltages of Na^+ and K^+.

- Rectification basically arises because of differences of ionic concentrations on the two sides of the membrane.

- Membranes exhibit reactance, which can be capacitive or inductive, because of nonlinear, time-varying membrane conductances. The membrane reactance is inductive if the incremental conductance at the operating point is larger than the chord conductance and is capacitive if the incremental conductance at the operating point is smaller than the chord conductance.

- A close analogy exists between ionic systems and semiconductors, once some simplifying assumptions are made, because both systems are then described by the same electrodiffusion equations.

3

Generation of the Action Potential

Objective and Overview

After considering in the preceding chapter the membrane at rest and deriving its equivalent circuit, the present chapter is concerned with explaining how the action potential (AP) is generated, based on the seminal work of Hodgkin and Huxley (1952) on the squid giant axon. The AP is first described under space clamp, which eliminates longitudinal variation of voltage along the axon. The essential results of voltage clamping, that is, suppressing the generation of the AP using a feedback amplifier, are then discussed and used to explain qualitatively the generation of the AP, before the mathematical equations of Hodgkin and Huxley are presented and interpreted in some detail. This is followed by a closer look at Na^+ and K^+ channels since these channels play an essential role in the generation of the AP. The chapter ends by explaining, in the light of the mechanism underlying the generation of the AP, some basic properties of the AP under space clamp, namely, an active response having a threshold, a strength-duration relationship, temperature dependence, and refractoriness.

Learning Objectives

To understand:

- The basic results of voltage-clamp experiments on the space-clamped squid giant axon
- How the AP is generated in terms of the time courses of sodium and potassium conductances following adequate stimulation
- The essential features and interpretation of the Hodgkin–Huxley equations
- Some basic features and characteristics of Na^+ and K^+ channels

- Some basic properties of the AP under space clamp, including the AP's active nature, threshold, strength-duration relationship, temperature dependence, and refractoriness

3.1 Generation of the Action Potential

Consider the experimental arrangement shown in Figure 3.1a, which is based on that used by Hodgkin and Huxley on the squid giant axon, as discussed in the next section.

A metal wire is inserted inside the axon, which is also surrounded by a metal cylinder. A current source i_{SRC} applies current pulses of short duration between the inner and outer metal electrodes. The axon is said to be **space-clamped** by the metal electrodes, which by acting as equipotential surfaces, eliminate current flow in the axial, that is, longitudinal, direction. Current flow is thereby restricted to the transverse, or radial, direction through the axonal membrane.

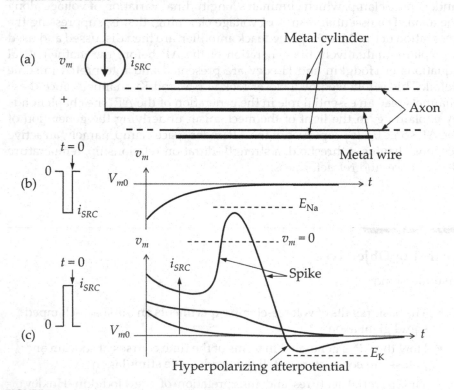

FIGURE 3.1
(a) Diagrammatic illustration of current applied to a space-clamped axon; membrane voltage responses to a hyperpolarizing pulse (b), and to depolarizing pulses of increasing amplitude (c).

Suppose that the polarity of the applied current pulse is inward through the membrane ($i_{SRC} < 0$). Effectively, the current pulse transfers positive charge from the inside of the axon to its outside, making the inside more negative with respect to the outside, that is, hyperpolarizing the membrane. In circuit terms, the membrane capacitor is charged more negatively. After the pulse is over at $t = 0$, the charge leaks through the membrane conductance, that is, the capacitance discharges, and the membrane voltage returns monotonically to the resting value (Figure 3.1b), even when the amplitude of the current pulse is relatively large. If the amplitude of the current pulse is such that the voltage excursion is small enough, the membrane conductance remains essentially constant, so that the time course of the membrane voltage is nearly exponential.

If the polarity of the current pulse is directed outward ($i_{SRC} > 0$), so as to depolarize the membrane, the time course of a small v_m for $t > 0$ is again what would be expected of a passive membrane, as in the lowest trace of Figure 3.1c. However, as the pulse amplitude is increased, the voltage response departs from what is expected of a passive membrane, as in the middle trace of Figure 3.1c. If the amplitude of the current pulse exceeds a certain value that depends on the pulse amplitude and duration, a characteristic response, known as the **action potential (AP)**, is observed. v_m initially drops somewhat, as the membrane capacitor discharges, then rapidly increases in the positive direction, turns positive, and approaches E_{Na}, the Na⁺ equilibrium voltage, before returning more slowly toward the resting voltage. It then undershoots the resting voltage toward E_K, the K⁺ equilibrium voltage, before returning more gradually to the resting voltage level. The region of rapid voltage change having $v_m > V_{m0}$ is the **spike**, whereas the undershoot having $v_m < V_{m0}$ is the **hyperpolarizing afterpotential**, or the **afterhyperpolarization**. In some nerve tissue, a "bulge" is observed on the downstroke of the spike, which is a form of **afterdepolarization**.

The following should be noted about the AP:

1. A minimum amplitude, or strength, of stimulus is required to initiate the AP, which means that the AP has a certain **threshold**. As will be explained later, the threshold is not a fixed voltage for a given membrane system, as in the case of some electronic devices and switching circuits, but depends on the magnitude and shape of the stimulus. A below-threshold stimulus is **subthreshold**, whereas an above-threshold stimulus is **suprathreshold**. Under normal conditions, and for a given stimulus, the threshold is very sharp.

2. If the stimulus strength is increased beyond threshold, the amplitude and shape of the AP are not significantly affected. However, the **latency**, or delay of the generation of the AP after the application of the stimulus, is reduced, that is, the AP occurs sooner.

3. The AP evoked in a space-clamped squid giant axon at 6°C in response to a brief current pulse that produces an initial depolarization of

20 mV has the following features: (i) a spike amplitude of about 105 mV, above a resting membrane voltage of about –60 to –70 mV, (ii) a maximum rate of rise of about 310 V/s, (iii) a spike duration of about 2.7 ms, and (iv) an afterhyperpolarization of peak amplitude of about 11 mV and duration of about 10 ms.

3.2 The Hodgkin–Huxley Model

Alan L. Hodgkin and Andrew F. Huxley, of the University of Cambridge, published in 1952 the results of a series of experiments they conducted on the squid giant axon aimed at elucidating the ionic mechanisms underlying the generation of the AP. They expressed their results in the form of a set of mathematical equations, referred to as the **Hodgkin–Huxley equations**, or **HH model**, that describe the variation of the conductances of the Na^+ and K^+ channels as a function of time and membrane voltage during the course of the AP. It should be noted that the HH model is based on fitting the experimental data obtained from voltage-clamp experiments on the squid giant axon and does not explain the molecular basis of the AP. Although derived for the squid giant axon, the basic ionic mechanisms described by the HH model have been found to apply to APs of nerve and muscle in general (Hodgkin & Huxley, 1952). In recognition of their outstanding contribution to explaining how the AP is generated and propagated, Hodgkin and Huxley shared with John C. Eccles the 1963 Nobel Prize in Physiology or Medicine.

Before describing the voltage-clamp technique and the HH model, it should be explained why the squid giant axon was used in the work of Hodgkin and Huxley. Invertebrates have "giant" axons for reasons that will be clarified later. The squid giant axon can be of up to 1 mm in diameter in some species, which allows insertion of a thin wire inside the axon to act as an inner electrode for electrical stimulation of the axonal membrane. It is also relatively easy to extrude most of the axoplasm and replace it by artificial solutions of specified ionic composition so as to investigate the effect of various ions on the behavior of the axonal membrane. The giant axon innervates the mantle muscle, which is a strong body-wall muscle of the squid. When contracted, the muscle forces a jet of sea water outward between the retracted tentacles of the squid, which rapidly propels the squid away from a source of danger.

3.2.1 Voltage Clamp Technique and Basic Results

The basic principle of the voltage-clamp technique is illustrated in Figure 3.2. The axon is space-clamped, as in Figure 3.1. The outer electrode is split into a middle section that is slightly separated from two end sections so as to eliminate "end effects" that will introduce some error into the current

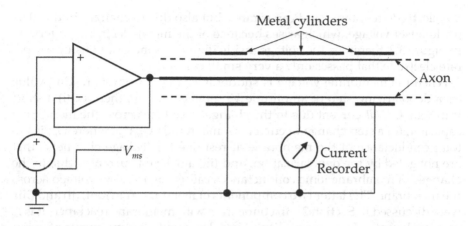

FIGURE 3.2
Diagrammatic illustration of a basic voltage clamp setup.

measurement because of non-uniformity of the transverse current at the ends of a single cylinder. Essentially, the inner electrode is connected to a high-gain, dc differential amplifier with the inverting input (the "–" terminal) connected to the output. In circuit terms, this is a **negative feedback** connection, since the output voltage is fed back to the input, such that it subtracts from the voltage applied to the noninverting input (the "+" terminal). The voltage difference is applied as a differential input voltage between the noninverting and inverting inputs of the amplifier. It is this differential input voltage that is effectively amplified by the amplifier as the output voltage. Because of the high gain of the amplifier, a very small differential input voltage is required to give a working output voltage, which means that these two inputs will be at practically the same voltage. With the output terminal connected to the inverting terminal, the amplifier will therefore provide whatever current is needed to make the output voltage follow the input voltage applied to the noninverting terminal without being physically connected to it. If a step input of magnitude V_{ms} is applied to the noninverting terminal, the output is constrained to be the same step voltage. The amplifier will therefore suppress the AP by supplying the membrane with a current equal and opposite to the current that would normally flow during an AP. Thus if during an AP, an inward current flows that would depolarize the membrane, leading to the upstroke of the spike, the amplifier will supply an equal and opposite current. This results in a net zero current through the membrane, which is what is required to keep the membrane voltage constant. The current provided by the amplifier, which can be recorded (Figure 3.2), is then equal and opposite to the membrane ionic current that would flow when the membrane is depolarized to the same voltage as that of the applied step depolarization.

It may be noted that, in practice, it is undesirable to measure voltage using the same electrode that injects current because the passage of current through an electrode affects the electrode voltage. This is not only due to the electrode resistance and capacitance but also due to changes in the electrode offset voltage, which arises because of the metal-electrolyte interface. In Figure 3.2, therefore, the voltage inside the axon is measured using a separate electrode that passes only a very small current.

When the membrane voltage is suddenly changed from its resting value by a depolarizing step of amplitude V_{ms} applied at $t=0$, there is: (i) a short transient, Cdv/dt current due to the change in voltage across the membrane capacitor, (ii) a step change in current of magnitude $G_{mr}V_{ms}$, where G_{mr} is the total conductance of the membrane at rest due to "leakage channels" that are not gated by the depolarization, and (iii) a delayed current i_m due to the changes in membrane ionic conductances caused by the new voltage across the membrane. The latter two components of membrane current, (ii) and (iii), were discussed in Section 2.7 in connection with membrane reactance. It is i_m that is relevant to our present discussion. The first two components of membrane current, (i) and (ii), are ignored in what follows.

It is to be expected that when a depolarizing voltage step is applied across the membrane, which makes the inside of the axon more positive with respect to the outside, the resulting membrane current is outward if the membrane is behaving as a conventional passive resistor, as explained in Section 2.5. However, the recorded time course of i_m in response to a depolarizing voltage step is found to have the general shape shown in the middle current trace of Figure 3.3. i_m is initially inward $(i_m<0)$ but turns outward $(i_m>0)$ soon thereafter. By changing the ionic composition of the external medium and checking their conclusions in various ways, Hodgkin and Huxley determined that i_m is the algebraic sum of two components: a fast inward component i_{Na} carried by Na^+ and a slow outward component i_K carried by K^+. The two currents i_{Na} and i_K are independent of one another, each having its own pathway, since each current can be selectively blocked by appropriate substances, as will be elaborated later.

The following should be noted:

1. It is seen from Figure 3.3 that i_K remains high with the depolarization maintained for 10 ms or so, but i_{Na} drops back to a low value. That is, i_K is sustained for the duration of the applied depolarization, but the increase in the magnitude of i_{Na} is *transient*.

2. From the current variations, Hodgkin and Huxley determined how G_{Na} and G_K vary with time and with the amplitude of the depolarization step V_{ms}. The results indicated that G_{Na} and G_K are highly nonlinear functions of both depolarization and time.

Based on these observations, it can be concluded that the AP is *essentially the result of a fast transient change in the Na^+ conductance in response to a sudden depolarization of the membrane, followed by a slower change in the K^+ conductance.*

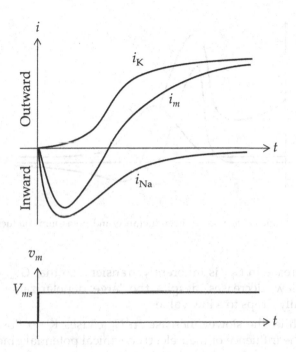

FIGURE 3.3
Membrane current and components for a step change of voltage under voltage clamp.

This concept is central to the HH model and will be used to explain, qualitatively at this stage, and with reference to Figure 3.4, how an AP is generated. Following a sudden, brief, suprathreshold stimulus, the following sequence of events occurs:

1. The large brief pulse rapidly charges the membrane capacitor. At the end of the pulse, considered to be at $t = 0$, the capacitor starts to discharge through the membrane conductance, so that v_m decreases initially.

2. After a short delay, G_{Na} increases rapidly and markedly, causing Na^+ to flow inward under the influence of their electrochemical potential gradient.

3. The inflow of Na^+ increases the membrane depolarization, which further increases G_{Na}, leading to more inflow of Na^+, more depolarization, and so on. Thus a **regenerative**, or **positive feedback**, cycle ensues:

$$\uparrow v_m \to \uparrow G_{Na} \to \uparrow Na^+ inflow \to \uparrow v_m$$

4. The positive feedback results in the rapid increase in the depolarization during the upstroke of the spike. The large increase in G_{Na} implies that $v_m \longrightarrow E_{Na}$, as explained in connection with Inequality 2.27.

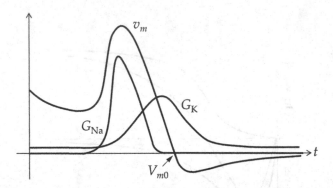

FIGURE 3.4
Variation of membrane voltage, sodium conductance, and potassium conductance during an action potential.

5. The increase in G_{Na} is inherently transient, so that G_{Na}, and hence Na⁺ inflow, decreases, despite the large depolarization, and G_{Na} eventually drops to a low value.

6. Meanwhile, the slower increase in G_K causes K⁺ to flow outward under the influence of their electrochemical potential gradient.

7. The outflow of K⁺ reduces the charge on the inner side of the membrane, which decreases the depolarization and brings the membrane voltage toward the resting value. No regenerative cycle takes place. In fact, the effect of depolarization on G_K is **degenerative**, or that of **negative feedback**. Thus,

$$\uparrow v_m \rightarrow \uparrow G_K \rightarrow \uparrow K^+\text{outflow} \rightarrow \downarrow v_m$$

A small influx of Cl⁻ into the cell, because of the depolarization, would also reduce the depolarization somewhat.

8. When $v_m = V_{m0}$, G_K is still high, as indicated in Figure 3.4, K⁺ outflow continues, which hyperpolarizes the membrane and causes the afterhyperpolarization. $v_m \longrightarrow E_K$, as explained in connection with Inequality 2.27.

9. The hyperpolarization reduces G_K, and v_m eventually returns to the resting level. The sodium-potassium pump eventually restores ionic concentrations to their resting values.

It should be emphasized that *the regenerative cycle is essential for the initiation of the AP. The regeneration, apart from causing a rapid increase in depolarization, is self-sustaining and takes over from the applied stimulus, so that the subsequent time course of v_m becomes largely independent of the stimulus, except for a reduction of the latency with increasing stimulus strength.* A larger strength of stimulus leads to a larger initial depolarization that causes an earlier and faster increase in G_{Na}, which reduces the latency of the AP.

It should be noted that using one or more integrated-circuit op amps in Figure 3.2 is nowadays relatively straightforward. At the time Hodgkin and Huxley conducted their experiments, modern transistors were not yet available, let alone integrated circuit op amps. Constructing a stable, high-gain, dc amplifier using vacuum tubes was not a trivial task. Voltage clamping has proved to be an invaluable tool for elucidating the ionic mechanisms associated with the AP. It has since been refined for application using microelectrodes in the form of patch clamping (Spotlight on Techniques 3A).

In contrast to a voltage clamp, a **current clamp** is a method of intracellular recording involving measurement of the voltage difference across the cellular membrane while injecting a constant positive or negative current (as "square" d.c. pulses) into the cell. Current clamp is used to study the response of a neuron to an inward or outward electric current, which is important, for example, for understanding how the neuron responds to neurotransmitters that act by opening membrane ion channels.

SPOTLIGHT ON TECHNIQUES 3A: PATCH CLAMP

The patch clamp is an invaluable tool developed and refined by Erwin Neher and Bert Sakmann in the 1970s and early 1980s to record from single ion channels. In recognition of the importance of their work, they shared the 1991 Nobel Prize in Physiology or Medicine.

The patch clamp allows direct measurement of the current or conductance of single ion channels using a microelectrode having a tip of about 1 μm diameter that has been smoothed by fire polishing. When the tip is brought into contact with the cell membrane and some gentle suction is applied, a **gigaseal** is formed having a high mechanical stability and a large leakage resistance of tens of gigaohms around the tip ($1\ G\Omega = 10^9\ \Omega$). The leakage resistance is in parallel with that of the channels being recorded from, which is typically around 20 pS, corresponding to $5 \times 10^{10}\ \Omega$ per channel. Unlike a gigaseal, a low leakage resistance shunting the ion channel introduces significant errors in the measurement of channel current and considerably increases the noise level. Because of its small area, the membrane patch being recorded from is essentially space-clamped, that is, the membrane voltage is practically uniform across the patch.

Figure 3.5a illustrates recording from a patch of cell membrane that has been lightly sucked into the tip of the microelectrode, resulting in **on-cell**, or **cell-attached** patch recording. The seal is so strong that rapidly pulling the microelectrode separates the patch from the rest

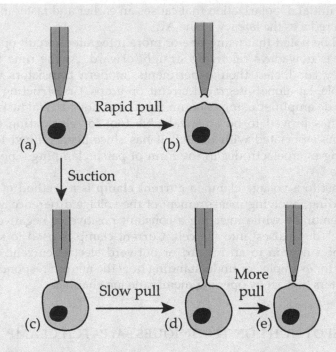

FIGURE 3.5
Patch-clamp recording; (a) membrane lightly sucked into tip microelectrode; (b) effects
of rapid pull; (c) more suction; (d) slow pull; and (e) more pull.

of the cell (Figure 3.5b), resulting in **inside-out** patch recording. The
patch now has its intracellular side exposed to the test solution surrounding the cell. Applying more suction to the patch in (a) ruptures
the patch (Figure 3.5c), thereby providing direct access to the cell's
interior through the microelectrode. The resulting **whole-cell** patch
recording allows recording currents through many channels over the
entire membrane of the cell. A disadvantage is that, over time, the cell
cytoplasm is altered by the contents of the microelectrode. The **perforated patch** technique avoids this problem. Small holes are made in
the cell-attached patch (Figure 3.5a) using pore-forming agents so that
large molecules such as proteins remain inside the cell while ions can
pass freely through the pores.

Slowly withdrawing the microelectrode in the whole-cell mode pulls
the opening in the cell membrane into an elongation stretching out of
the cell (Figure 3.5d). With further pulling, this elongation detaches
from the cell and reforms as a bulb of membrane on the end of the
microelectrode, with the extracellular side of the membrane exposed
to the test solution. This is referred to as **outside-out** patch recording

(Figure 3.5e). The inside-out and outside-out patch recordings are referred to as **excised-patch** recording.

The patch can be voltage-clamped, allowing measurement of ionic currents following a step of depolarization. Channels of ion species other than those under study can be blocked by suitable drugs. Although the patch will include, in general, many channels for the ion species of interest, the random nature of opening and closing of individual channels allows the recording of current from a single channel at some instances of time.

3.2.2 Mathematical Description

Having presented the qualitative description of the generation of the AP according to the Hodgkin and Huxley model, we consider next the equations derived by Hodgkin and Huxley to fit their voltage-clamp experimental results. Their analysis of these results led them to the following conclusions:

1. The ionic current density through the membrane J_{im} can be expressed as:

$$J_{im} = J_K + J_{Na} + J_L \tag{3.1}$$

   where J_L represents a "leakage" current due to all ions other than K^+ and Na^+. The contribution of J_L to the AP was found to be negligibly small compared to that of J_{Na} and J_K. The Cl⁻ branch of Figure 2.9 is thus replaced by a leakage branch having a constant conductance G_L. However, the battery voltage in this branch is $E_{L'} \cong 10.6$ mV with reference to the resting membrane voltage and is not zero, as in the case of $E_{Cl'}$.

2. Na^+ and K^+ flow through separate, voltage-gated channels in the membrane.

To fit the G_K and G_{Na} conductance curves, Hodgkin and Huxley chose a model based on chemical kinetics, as will be described next.

3.2.2.1 Potassium Conductance

Hodgkin and Huxley postulated that each voltage-gated K^+ channel is controlled by four charged molecular subunits, referred to as the n subunits or gates, which could be charged groups in the pores. Each of these gates can be in one of two positions that we will denote by o and c, corresponding to "open" and "closed", respectively. The open and closed positions are also referred to as **permissive** and **nonpermissive** positions, respectively. When all four gates are in the o position the channel is open to K^+, but when at least one of these gates is in the c position, the channel is closed. In the resting state, the probability is high that each of these four gates is in the c position. The effect of depolarization is to increase the probability of each gate being in the o position. The movement of each gate between the o and c positions is governed by a first-order "reversible reaction" having voltage- and temperature-dependent rate coefficients. Thus:

$$c \underset{\theta\beta_n}{\overset{\theta\alpha_n}{\rightleftharpoons}} o \tag{3.2}$$

where α_n is the rate of transition of a gate from position c to position o per unit time per gate in position c, and β_n is rate of transition from position o to position c per unit time per gate in position o. θ is a function of temperature alone, and α_n and β_n are functions of membrane voltage alone. If n is the probability that a gate is in the o position, and N is the total number of gates per unit area of membrane and is sufficiently large, then the number of gates per unit area in the o position can be considered to be nN, and the number of gates per unit area in the c position to be $(1 - n)N$. It follows that $d(nN)/dt$, the rate at which the number of gates per unit area in the o position is increasing with time, is equal to $\theta\alpha_n(1 - n)N$, the rate at which gates are moving from the c position to the o position, minus $\theta\beta_n nN$, the rate at which gates are moving from the o position to the c position. Thus, $d(nN)/dt = \theta\alpha_n(1 - n)N - \theta\beta_n nN$. Cancelling N,

$$\frac{dn}{dt} = \theta\left[\alpha_n(1-n) - \beta_n n\right] \tag{3.3}$$

In the steady state after a step of depolarization is applied ($t \longrightarrow \infty$), $dn/dt = 0$. It follows from Equation 3.3. that $n_\infty = \alpha_n/(\alpha_n + \beta_n)$. Equation 3.3 can be written in the alternative form:

$$\frac{dn}{dt} = \frac{n_\infty - n}{\tau_n} \tag{3.4}$$

where $\tau_n = 1/[\theta(\alpha_n + \beta_n)]$. At a constant membrane voltage, as under voltage clamp, n_∞ and τ_n are constant with respect to time. The solution to Equation 3.4 is (Problem 3.1):

$$n = n_\infty - \left(n_\infty - n_0\right)e^{-t/\tau_n} \tag{3.5}$$

where $n_0 = \alpha_{n0}/(\alpha_{n0} + \beta_{n0})$ is the initial value of n in the resting state ($v_{mr} = 0$) at $t = 0$.

Since the probability of a gate being in the o position is n, the probability that all four gates in a given channel are open is n^4, assuming that the movements of gates are independent of one another. If the average conductance per open K⁺ channel is γ_K, the K⁺ conductance G_K per unit area is $N\gamma_K n^4$, or,

$$G_K = G_K^o n^4 \tag{3.6}$$

where $G_K^o = N\gamma_K$ is the maximum K⁺ channel conductance per unit area, when $n = 1$, that is, all channels are in the *o* position. J_K is given by:

$$J_K = G_K(v_{mr} - E_{Kr}) \tag{3.7}$$

as explained in connection with Equation 2.21. Hodgkin and Huxley determined that the best fit to their data was obtained using the expressions for θ, α_n, and β_n given in Table 3.1. n_∞ and τ_n are plotted in Figure 3.6 as functions of v_{mr}, the membrane voltage relative to V_{m0} (Section 2.5), and G_K is plotted in Figure 3.7 as a function of time and magnitude V_{ms} of a voltage step under voltage clamp. As v_m increases with depolarization, α_n increases and β_n decreases, so that n, the new steady-state value of n increases. More of the channel gates shift to the *o* position, and G_K becomes larger. *Increasing the depolarization increases the asymptotic value of G_K as well as its maximum rate of increase, and makes this maximum rate occur earlier.*

TABLE 3.1

Parameters of the Hodgkin and Huxley Equations

$$\theta = 3^{(T-6.3)/10}$$

$\alpha_n = \dfrac{0.01(10 - v_{mr})}{e^{(10-v_{mr})/10} - 1}\,\text{ms}^{-1}$	$\beta_n = 0.125 e^{-v_{mr}/80}\,\text{ms}^{-1}$
$\alpha_m = \dfrac{0.1(25 - v_{mr})}{e^{(25-v_{mr})/10} - 1}\,\text{ms}^{-1}$	$\beta_m = 4 e^{-v_{mr}/18}\,\text{ms}^{-1}$
$\alpha_h = 0.07 e^{-v_{mr}/20}\,\text{ms}^{-1}$	$\beta_h = \dfrac{1}{e^{(30-v_{mr})/10} + 1}\,\text{ms}^{-1}$

$$\tau_n = \frac{1}{\theta(\alpha_n + \beta_n)} \qquad \tau_m = \frac{1}{\theta(\alpha_m + \beta_m)} \qquad \tau_h = \frac{1}{\theta(\alpha_h + \beta_h)}$$

In the steady state:

$$n_\infty = \frac{\alpha_n}{\alpha_n + \beta_n} \qquad m_\infty = \frac{\alpha_m}{\alpha_m + \beta_m} \qquad h_\infty = \frac{\alpha_h}{\alpha_h + \beta_h}$$

In the resting state ($v_{mr} = 0$): $n_0 = 0.318$, $m_0 = 0.053$, $h_0 = 0.596$.
$G_{Na}^o = 120\,\text{mS/cm}^2$, $G_K^o = 36\,\text{mS/cm}^2$, $G_L = 0.3\,\text{mS/cm}^2$, independently of temperature
$V_{m0} = -65\,\text{mV}$, $E_{Na} = +50\,\text{mV}$, $E_K = -77\,\text{mV}$, or with respect to the resting membrane voltage, $E_{Nar} = +115\,\text{mV}$, $E_{Kr} = -12\,\text{mV}$, and $E_{Lr} = +10.6\,\text{mV}$.

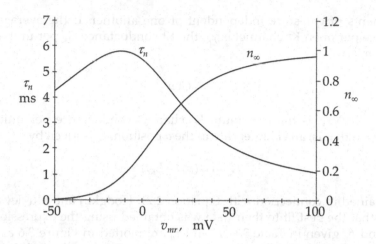

FIGURE 3.6
Variation with membrane voltage in the steady state of the potassium activation variable and its time constant.

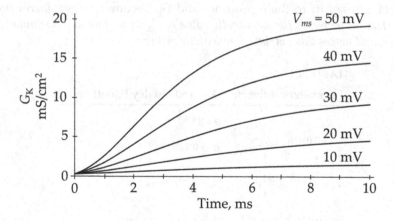

FIGURE 3.7
Time variation of potassium conductance with membrane voltage for increasing magnitude of a step change of voltage under voltage clamp.

The process by which G_K increases with depolarization is referred to as **potassium activation**, and n is referred to as the **potassium activation variable**.

The "S" shape of the variation of G_K with time in Figure 3.7 is to be expected from the first-order kinetics exemplified by Equation 3.3. The solution for n following a voltage step is Equation 3.5 with n_∞, n_0, and τ_n constant, since all these are functions of α_n and β_n, which are constant when the membrane voltage is constant during a step of depolarization. With $n_\infty > n_0$, the solution of Equation 3.5 for n is a rising, "saturating", exponential function of time.

When n is raised to the power 4, the variation of n^4 with time is essentially "S" shaped.

If the depolarization step is suddenly removed, the variation of n with time is given by Equation 3.5, with the final and initial values of n, that is n_∞ and n_0, interchanged:

$$n = n_0 - (n_0 - n_\infty)e^{-t/\tau_n} = n_0\left(1 - e^{-t/\tau_n}\right) + n_\infty e^{-t/\tau_n} \qquad (3.8)$$

If n_0 is small enough, Equation 3.8 reduces to:

$$n \cong n_\infty e^{-t/\tau_n} \qquad (3.9)$$

When raised to a power, n decreases with time exponentially at a rate that depends on τ_n and the power of n.

3.2.2.2 Sodium Conductance

Because of its transient nature, G_{Na} is a bit more complex than G_K. To account for this transient nature of G_{Na}, Hodgkin and Huxley postulated that each voltage-gated Na^+ channel is also controlled by four molecular subunits or gates. Three of these gates, referred to as the m gates, are of one type responsible for **sodium activation**, that is, the increase of G_{Na} with depolarization, and the fourth being of another type, referred to as the h gate, is responsible for **sodium inactivation**, that is, the decrease of G_{Na} with depolarization. In the resting state, the probability is high that the m gates are in the closed, or c position, and the h gate is in the open, or o position. When a sudden depolarization is applied, the m gates move *fast* with high probability to the o position, thereby opening the channel. The h gate then moves *more slowly*, to the c position, with high probability, so as to close the channel. The Na^+ channel is thus opened only transiently, even with maintained depolarization.

The kinetics of both types of gate are first-order, of the same form as that of the n gates. The equation satisfied by m, the **sodium activation variable**, is thus of the same form as Equation 3.3, namely,

$$\frac{dm}{dt} = \theta\left[\alpha_m(1 - m) - \beta_m m\right] \qquad (3.10)$$

where α_m and β_m are given in Table 3.1. Analogous to Equation 3.4, Equation 3.10 can be expressed as:

$$\frac{dm}{dt} = \frac{m_\infty - m}{\tau_m} \qquad (3.11)$$

where $m_\infty = \alpha_m/(\alpha_m + \beta_m)$ and $\tau_m = 1/[\theta(\alpha_m + \beta_m)]$. m_∞ and τ_m are plotted in Figure 3.8 as functions of v_{mr}. It is seen that m_∞, like n_∞, increases with depolarization

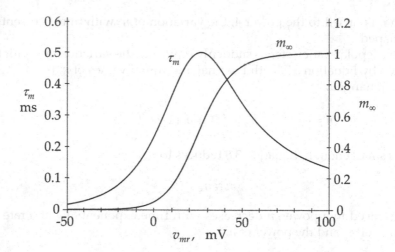

FIGURE 3.8
Variation with membrane voltage in the steady state of the sodium activation variable and its time constant.

and that τ_m is an order of magnitude smaller than τ_n, so that sodium activation is much faster than potassium activation, as discussed earlier.

The **sodium inactivation variable** h satisfies the equation:

$$\frac{dh}{dt} = \theta\left[\alpha_h(1-h) - \beta_h h\right] \qquad (3.12)$$

Analogous to Equation 3.11, Equation 3.12 can be expressed as:

$$\frac{dh}{dt} = \frac{h_\infty - h}{\tau_m} \qquad (3.13)$$

where $h_\infty = \alpha_h/(\alpha_h + \beta_h)$ and $\tau_h = 1/[\theta(\alpha_h + \beta_h)]$. h_∞ and τ_h are plotted in Figure 3.9 as functions of v_{mr}. It is seen that h_∞, unlike n_∞ and m_∞, decreases with depolarization, and that τ_h is an order of magnitude larger than τ_m but is comparable to τ_n.

G_{Na} is then given by:

$$G_{Na} = G_{Na}^o m^3 h \qquad (3.14)$$

where G_{Na}^o is the maximum Na$^+$ channel conductance per unit area. G_{Na} is plotted in Figure 3.10 as a function of time and magnitude V_{ms} of a voltage step under voltage clamp. Increasing the depolarization increases the maximum rate of increase of G_{Na} as well as its maximum value and makes this maximum rate occur earlier.

The effect of the sodium inactivation variable h is that of negative feedback, as in the case of the potassium activation variable n. Thus,

$$\uparrow v_m \rightarrow \uparrow h \rightarrow \downarrow G_{Na} \rightarrow \downarrow Na^+\text{inflow} \rightarrow \downarrow v_m$$

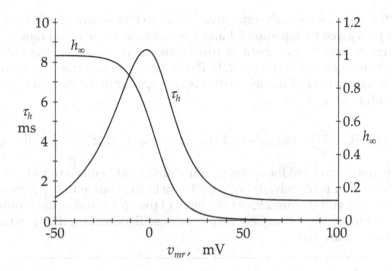

FIGURE 3.9
Variation with membrane voltage in the steady state of the sodium inactivation variable and its time constant.

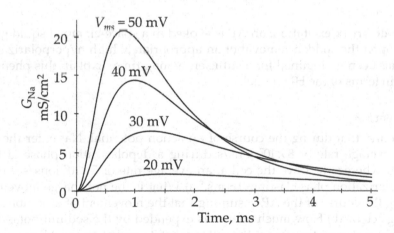

FIGURE 3.10
Time variation of sodium conductance with membrane voltage for increasing magnitude of a step change of voltage under voltage clamp.

Activation variables that involve positive feedback, such as m, are sometimes referred to as **excitation variables** or **amplification variables** because of the regenerative effect they have. Membrane variables that involve negative feedback, such as n and h, are referred to as **recovery variables** because they help restore the membrane resting voltage.

As explained in connection with Equation 2.21, J_{Na} is given by:

$$J_{Na} = G_{Na}(v_{mr} - E_{Nar}) \qquad (3.15)$$

After the stimulus ends, the total membrane current is zero, that is $J_C + J_{im} = 0$, where J_{im} is given by Equation 3.1 and $J_C = C_m dv_{mr}/dt$ is the membrane capacitor current, the assigned positive directions of all currents being outward through the membrane (Figure 2.9). The ionic current sources are assumed to be neutral in the HH model, so they do not appear in the current equation. Substituting for J_K, J_{Na}, and J_L,

$$J_C + J_{im} = C_m \frac{dv_{mr}}{dt} + G_K(v_{mr} - E_{Kr}) + G_{Na}(v_{mr} - E_{Nar}) + G_L(v_{mr} - E_{Lr}) = 0 \quad (3.16)$$

During the upstroke of the spike, J_{im} is inward ($J_{im} < 0$) and charges the membrane capacitor positively; dv_{mr}/dt and J_C are both positive in the membrane equivalent circuit (Figure 2.9). At the peak of the spike, and at the minimum or maximum of any subsequent membrane oscillations, $dv_{mr}/dt = 0$, so that J_C, and hence J_{im}, are zero.

Problem 3.1

Derive the solution to Equation 3.4 by separation of variables.

Problem 3.2

In **anode break excitation**, an AP is evoked in a space-clamped squid giant axon upon the sudden removal of an appropriately high hyperpolarization that has been maintained for a sufficiently long time. Explain this phenomenon in terms of the HH model.

Problem 3.3

It is found that during the course of an action potential, Na^+ enter the cell at an average rate of 8×10^7 ions/s during a depolarization phase lasting 0.3 ms, whereas K^+ leave the cell at an average rate of 5×10^7 ions/s during a repolarization phase lasting 0.6 ms. (a) What is the net charge movement during the course of the AP, assuming that the movement of other ions can be neglected? (b) How much energy is expended by the sodium-potassium pump in restoring the state of the cell to what it was before the AP, assuming an average membrane voltage of -70 mV?

ANS.: (a) 9.6×10^{-16} C net positive charge; (b) 6.72×10^{-17} J.

Problem 3.4

(a) A neuron has $[Na^+]_o = 150$ mM, $[Na^+]_i = 25$ mM, and $V_{m0} = -70$ mV. When a 20 mV step of depolarization is applied under voltage clamp, the peak Na^+ current is 0.8 mA/cm². What is the peak G_{Na}? (b) Determine the peak Na^+ current for the same step of depolarization but with $[Na^+]_o = 2$ mM.

ANS.: (a) 8.16 mS/cm²; (b) 0.14 mA/cm² flowing outward.

Problem 3.5

Following a step of depolarization, it is found that n in the HH model for a particular cell varies with time as $n = 0.952 - 0.463e^{-t/2.1}$ where t is in ms. Determine the final values of α_n and β_n, assuming $G_K^0 = 45$ mS/cm^2 and $\theta = 1$.

ANS.: $\alpha_n = 0.453$ ms^{-1}, $\beta_n = 0.023$ ms^{-1}.

3.2.3 Kinetic Representation of the Hodgkin–Huxley Model

In the Hodgkin–Huxley model, the K$^+$ channel is controlled by four identical and independent gates. The channel is open only when all four gates are in the open position and is closed if at least one gate is in the closed position. Hence, five states can be postulated as follows:

$$O_0 \underset{\beta_n}{\overset{4\alpha_n}{\rightleftharpoons}} O_1 \underset{2\beta_n}{\overset{3\alpha_n}{\rightleftharpoons}} O_2 \underset{3\beta_n}{\overset{2\alpha_n}{\rightleftharpoons}} O_3 \underset{4\beta_n}{\overset{\alpha_n}{\rightleftharpoons}} O \qquad (3.17)$$

O denotes open states, the numerical subscripts indicating the number of gates in the open position, with all four gates being open in state O without a subscript. α_n and β_n are as in the HH model but assumed to include the temperature parameter θ. The arrows pointing to the right denote movement of gates to the open position, whereas arrows pointing to the left denote movement to the closed position. Hence, the rate constant from O_1 to O_0 is β_n, the rate at which a single remaining gate in open position in state O_1 can move to the all-closed state O_0. On the other hand, the rate constant from O_0 to O_1 is $4\alpha_n$, since any of the four gates can move from the all-closed state to the state having one gate open. Three gates are in the closed position in state O_1 and two gates are in the open position in state O_2. Hence the rate constant from state O_1 to state O_2 is $3\alpha_n$, whereas the rate constant from state O_2 to state O_1 is $2\beta_n$. The other rate constants are similarly interpreted in terms of the number of gates in the closed and open positions.

In an analogous manner, the kinetic model for the Na$^+$ can be postulated as follows:

$$(3.18)$$

where O_0 to O are the open states for the activation gates m, the numerical subscripts indicating the number of m gates in the open position, with all three gates being open in state O without a subscript. In the O states, the inactivation gate h is assumed to be in the open state. But since the inactivation gate can also be in the closed state, there corresponds to each of the O states a state where the inactivation gate is closed. These states, denoted as

In_0 to In_3, are shown in the bottom row in Reactions 3.18, where In denotes inactivation and the numerical subscript indicates the number of m gates in the open position. The rate constants between each O state and the corresponding In state are α_h in the direction of closed to open h gate, and β_n in the opposite direction. The rate constants between the In states are the same as those between the corresponding O states, since the transitions between the In states involve movement of the m gates, while the h gate is closed.

3.2.4 Na⁺ and K⁺ Channels

The independence of the Na⁺ and K⁺ voltage-gated channels is strongly supported by the fact that different drugs can selectively block one channel and not the other. Thus, the voltage-gated Na⁺ channel of neurons is selectively blocked by tetrodotoxin (TTX), a potent toxin produced by symbiotic bacteria and found in many animal species, particularly aquatic animals such as the puffer fish. Local anesthetics, such as procaine, are synthetic drugs that, by blocking the voltage-gated Na⁺ channels of axons carrying pain signals, prevent pain-induced APs in the periphery from reaching the central nervous system and hence the brain centers responsible for the sensation of pain. The voltage-gated K⁺ channel in axons is selectively blocked by the synthetic drug tetraethylammonium (TEA).

The density of voltage-gated Na⁺ channels has been estimated at about $330/\mu m^2$ in the squid giant axon and up to about $2,000/\mu m^2$ in the node of Ranvier (Section 4.2) of a frog nerve. The corresponding figures for voltage-gated K⁺ channels are about 30 and 1,000, respectively. The size of the selectivity filter of the Na⁺ channel has been estimated at 3 Å×5 Å. In frog muscle, voltage-gated Na⁺ channels have a mean conductance of 10.5 pS, the current varying with the membrane voltage in accordance with Equation 3.15, between about 0.7 pA to 1.5 pA, over a membrane voltage range of -15 to -50 mV.

It should not be expected that all voltage-gated Na⁺ or K⁺ channels have the same properties in all types of excitable cells irrespective of the function of the cell or part of the cell. The axon is required to propagate the AP efficiently (Chapter 4), whereas the AP of the cardiac muscle, for example, is of a long duration that enables the muscle fiber to develop maximal tension (Section 10.5.3). Cardiac pacemaker cells generate APs continuously at variable heart rates. Hence, the voltage-gated K⁺ channels, for example, would not be expected to have the same properties in all these cases. In fact, different voltage-gated K⁺ channels in various types of neurons differ in their speed of operation, and some inactivate partially or completely, like the voltage-gated Na⁺ channels discussed previously, and at different rates. They also respond differently to different drugs. Some K⁺ channels in neurons, described as inward rectifiers, are normally blocked and are opened by hyperpolarization, which removes the blockage. Even in the same type of cell, such as the squid giant axon, three types of K⁺ channels have been identified having conductances of 10 pS, 20 pS, and 40 pS and different gating properties, the 20 pS channel being the main contributor to the behavior of K^+ conductance of the axon (Equation 3.6). Voltage-gated Na^+ channels show less variability between different excitable cells compared to voltage-gated K^+ channels.

The different types of channels also differ in their selectivity to various ions. No channels have perfect selectivity to only one ion species. Some Na^+ channels, for example, have a permeability to K^+ that is about one tenth of that to Na^+. Some K^+ channels pass Na^+ at a rate that is only 1/10,000 of the rate for K^+.

It should be kept in mind that individual voltage-gated channels open and close at random with probabilities that depend on the voltage. The observed behavior at the macroscopic level is the aggregate behavior of many channels and will be independent of time, under steady-state conditions, but with some statistical variation. This is diagrammatically illustrated in Figure 3.11 for Na^+ channels following a step of depolarization (a). The individual channels are shown in (b) opening with some variability and closing at random, after some interval, with more variability. The ensemble sum of conductances (c) will have the shape observed for the axon (Figure 3.11c). The individual channels are shown in (b) as being either fully open or fully closed,

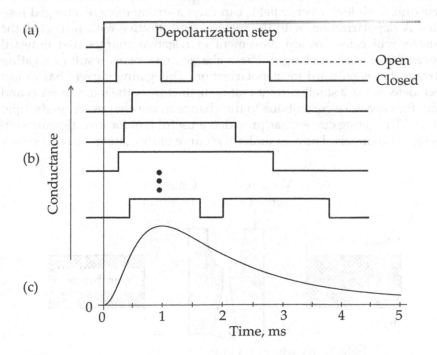

FIGURE 3.11
(a) Step depolarization applied to a voltage-gated channel; (b) changes in conductances of individual channels opening with some variability in time and closing at random; (c) ensemble of conductance changes of all channels.

the channel conductance in the open state being nominally fixed for a given type of channel. This is assumed to be generally true, although some types of channels have partially open states and are described as having **subconductance states**. The kinetics of channel opening and closing are discussed in Section 5.4.2.

Figure 3.12 is a diagrammatic illustration of a voltage-gated channel comprising three main functional units whose arrangement is for functional purposes only and is not intended to reflect the actual architecture of the channel. These units are: (i) an aqueous pore of non-uniform diameter having an enlargement or vestibule on one or both sides, as well as a constriction that acts as a **selectivity filter** (Section 2.1.1), (ii) a **voltage sensor** that is responsive to the transmembrane voltage, and (iii) a gate that opens or closes according to the transmembrane voltage and the dynamics of the channel. The voltage sensor generally includes **gating charges** in the form of charged residues that are constituent molecular units of the channel protein. These are acted on by the electric field due to the depolarization, resulting in a conformational change that opens or closes the gate. It should be noted that a transmembrane voltage of 100 mV across a structure 10 nm thick and of uniform dielectric constant establishes an electric field of 100 kV/cm. The maximum electric field could be even larger across parts of a nonuniform structure. Such high electric fields can exert a strong force on charged residues. A depolarization, with the inside going positive with respect to the outside, will cause inward movement of negative charges and outward movement of positive charges. These charge movements result in a **gating current** that is outward for depolarization. This gating current has in fact been detected as a small current that is distinct from the ionic currents and from the capacitive current due to the change in the charge across the lipid bilayer. The gating current has provided a useful tool for investigating voltage-gated channels. The area under the curve of the gating current density

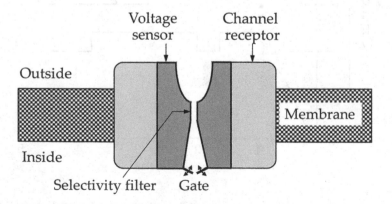

FIGURE 3.12
Diagrammatic illustration of voltage-gated channel.

as a function of time represents charge movement equal to qzN per unit area, where z is the number of charges per gating residue and N is the number of gating residues contributing to the gating current per unit area. Moreover, studies of the gating current have revealed that Na^+ activation and inactivation are coupled processes in that inactivation may be due to some protein units blocking the open channel sometime after the channel is opened due to the movement of the charged residues that activate the channel. But while the channel is inactivated, these inactivating protein units prevent the activating charged residues from returning to their positions in the resting state.

In some large-conductance K^+ channels, charged residues in the vestibules store K^+ through electrostatic attraction, thereby making more of these ions available for transiting through the channel under appropriate conditions.

Voltage-gated Na^+ and K^+ channels play an essential role in the generation and propagation in practically all excitable cells, the upstroke of the spike being due to the regenerative effect of a large Na^+ influx in almost all cases, with few exceptions. In pacemaker cells of the heart (Section 10.5.3.2), the upstroke of the spike is due to influx of Ca^{2+}, whereas in some freshwater algae, the upstroke of the spike is due to efflux of Cl^- through voltage-gated Cl^- channels. In neurons, Cl^- channels may play a role in stabilizing the membrane voltage by providing a shunt conductive path that reduces the effects of permeability changes on the membrane voltage (Figure 2.9).

Voltage-gated Ca^{2+} channels are also ubiquitous and play a critical role in neurotransmission, modulation of neuronal firing, and muscular contraction, as discussed in Chapters 5, 6, 7, and 9. Again, several types of Ca^{2+} channels have been identified that differ in their voltage dependence, inactivation rate, selectivity to various ions, and response to drugs, the properties of these channels being adapted to the functions they serve. Ion channels in neurons are discussed in more detail in Section 7.3.

It should be mentioned that the study of the properties of ion channels has involved several disciplines, such as: (i) electrophysiology, aimed at experimental investigation of the electrical properties of ion channels utilizing techniques such as patch clamping (Spotlight on Techniques 3A) applied to cells in vivo or in vitro using neuron cultures or slices of nerve tissue, (ii) biophysics, concerned with the mathematical description of the electrical behavior of membranes in terms of electrodiffusion, and chemical kinetics, as in the discussions of this and preceding chapters, (iii) X-ray crystallography, electron microscopy and electron diffraction, which seek to obtain the structural details of channels, (iv) pharmacology, concerned with the effects of drugs on channel behavior, and (v) molecular biology, which aims at determining the molecular structure of the channel proteins and correlating this structure with the results of electron microscopy, pharmacology, and biophysical studies. Patch-clamping has proved to be an invaluable tool, as has **mutagenesis**, in which genes for channel proteins are modified so as to change the structure of the channel protein in a known manner and determine the resulting effect on channel behavior (Spotlight on Techniques 5A).

Finally, a note concerning Equation 2.21 when applied to voltage-gated channels:

$$J_Y = G_Y (v_m - E_Y) \tag{3.19}$$

G_Y in this equation is a chord conductance (Section 2.7). If v_m is suddenly changed, then G_Y will not be expected to change instantaneously and would therefore remain constant at the instant of change, as explained in Section 2.7. With E_Y constant, J_Y is said to obey Ohm's law "instantaneously". This was verified by Hodgkin and Huxley in their classical work on the squid giant axon in normal seawater. However, Equation 3.19 is not obeyed by all channels. In the case of Ca^{2+} channels, for example, where $[Ca^{2+}]_i$ is about 100 nM and $[Ca^{2+}]_o$ is 1–5 mM (Table 2.1), which means the ratio $[Ca^{2+}]_o/[Ca^{2+}]_i$ could be as large as 50,000, an influx of Ca^{2+} would be expected to change $[Ca^{2+}]_i$ significantly, at least in the immediate vicinity of the channel on the inside, and virtually instantaneously. E_Y in Equation 3.19 cannot be considered a constant under these conditions, independently of J_Y. In such cases, the current is expressed by the GHK current equation (Equation 2.24) rather than Equation 3.19 (see Section 8.2.2.2). Another example is the myelinated axon of the toad, in which the instantaneous current-voltage relation is nonlinear and is consistent with assuming a constant permeability to Na^+, rather than a constant conductance, as in the case of the squid giant axon in normal seawater.

3.3 Properties of the Action Potential under Space Clamp

We will consider in this section some of the more important properties of the AP under space clamp, which is sometimes referred to as the **membrane AP** to distinguish it from the **propagating AP**, considered in Chapter 4.

3.3.1 Active Response

We begin by defining an **active response**, in contrast to a passive response, as a response in which the energy delivered by the system under consideration is larger than that supplied by the applied source, or stimulus, the balance of energy being provided by some additional source of energy. This is true of a transistor circuit, for example, where the energy delivered to a load is larger than that supplied by the signal source at the input of the transistor, the balance of energy being provided by the one or more dc bias supplies. In the case of the AP, if we consider a suprathreshold, brief current stimulus that depolarizes the membrane by about 20 mV, the energy delivered by the stimulus is that stored in a capacitor of about

1 $\mu F/cm^2$, which is $(1/2)(1 \ \mu F/cm^2)(20 \ mV)^2 \equiv 200 \times 10^{-12} \ J/cm^2$. The "bias" sources in the membrane equivalent circuit (Figure 2.9) are the batteries and the ionic pump current sources. Since the ionic pumps in the HH model are assumed neutral, no net energy is expended by these pumps during the AP. As for the energy dissipated in the conductances during the course of an AP, we can make a very rough estimate assuming an AP of 100 mV amplitude, 1 ms duration, and an average membrane conductance of 10 mS/cm^2 during the time course of the spike. This gives 10^{-7} J/cm^2, which far exceeds that supplied by the stimulus. The response is therefore active in nature.

3.3.2 Threshold

At the end of a narrow, depolarizing current pulse, the membrane capacitor is less charged than in the resting state, with the inside less negative with respect to the outside. The added positive charge on the inside is mainly carried by K^+, these being the cations of highest intracellular concentration. The ensuing depolarization sets off three processes involving the voltage-gated Na^+ and K^+ channels: (i) a sodium activation process that increases G_{Na} and hence the Na^+ inward current, (ii) a sodium inactivation process that curtails the increase in G_{Na} and the Na^+ inward current, and (iii) a potassium activation process that increases G_K and hence the K^+ outward current. Being the fastest of the three processes, sodium activation takes hold first. But when the stimulus ends, and before G_{Na} increases significantly, the membrane capacitor starts to discharge, causing an initial drop in v_m (Figure 3.1). The capacitor discharge current, which is outward (Figure 2.9), is in fact accounted for by a reduction in the inward current, due to the reduction in membrane voltage, that is carried by K^+ and other ions in the leakage channel referred to in the preceding section. If the depolarization is not large enough, the inward Na^+ current due to the opening of the Na^+ channels remains less than the outward current, and v_m continues its drop to the resting value. If, however, the increase in inward current exceeds the outward current, there will be a net inward current that triggers the regenerative cycle, thus initiating the AP. It follows that *the threshold of the AP under space clamp is when the inward Na^+ current equals the outward current through the membrane. Just beyond threshold, the net membrane current is inward, thereby triggering the regenerative cycle that results in the upstroke of the spike.* It is seen that the threshold of the AP is not a fixed voltage, but depends on the time course of the stimulus.

It would be expected that if the depolarization is slow enough, as when the stimulating current is a slow ramp function of time, for example, the speed advantage of sodium activation is lost, thereby allowing sodium inactivation and potassium activation to exercise their effects concomitantly. If the effects of sodium inactivation and potassium activation are strong enough, sodium activation will not be able to turn the net current inward and initiate an AP, irrespective of the depolarization level of the membrane. The threshold voltage is then virtually infinite. The failure to generate an AP under these conditions is referred to as **accommodation** of the membrane to the applied stimulus.

A threshold of the AP is advantageous in that a finite, nonzero, threshold protects against unwanted APs due to random fluctuations of membrane voltage around the resting level.

3.3.3 Strength-Duration Relationship

Consider a depolarizing current pulse of a given amplitude and duration applied to the membrane. If the duration of the pulse is very small compared to the membrane time constant, practically all the charge of the current pulse, which is the product of the amplitude of the current pulse and its duration, is deposited on the membrane capacitor, with negligible charge flowing through the membrane conductance over the duration of the pulse. The change in membrane voltage would be fast, favoring sodium activation, which means that the threshold voltage will be low, as explained in the preceding section.

If the duration of the current pulse is T_D, a certain minimum amplitude of the current pulse, i_{mp}, is required to generate an AP. It is found in practice that if the amplitude of the pulse is reduced, a longer duration of the pulse is needed to bring the membrane to threshold and generate an AP. A plot of i_{mp} vs. T_D is the **strength-duration relationship** and has the general shape shown in Figure 3.13. The horizontal asymptote is the **rheobase** and represents the smallest amplitude of a current pulse of very long (theoretically infinite) duration that will generate an AP. The **chronaxie** is the value of T_D for a current pulse whose amplitude is twice the rheobase.

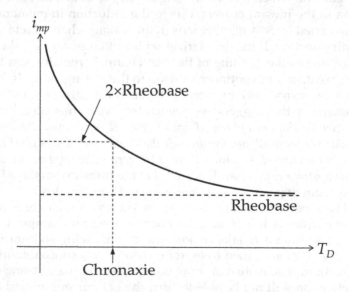

FIGURE 3.13
Strength-duration relationship.

The general shape of Figure 3.13 can be readily interpreted by considering a current pulse applied to an RC circuit (Figure 3.14), where R_m and C_m are the resistance and capacitance of a patch of space-clamped axon, or the total membrane resistance and capacitance of a neuron whose inside can be considered equipotential, as well as its outside. Assume for simplicity to begin with, that the threshold for generating an AP is a constant V_{thr}. If a current step of amplitude i_{mp} is applied, it can be readily shown (Problem 3.6) that the voltage v across the parallel combination will increase exponentially with time as:

$$v = R_m i_{mp}\left(1-e^{-t/\tau_m}\right) \tag{3.20}$$

where $\tau_m = R_m C_m$ is the membrane time constant. As $t \longrightarrow \infty$, v approaches a steady value $R_m i_{mp}$ when all of i_{mp} flows through R_m and no current flows C_m. If $R_m i_{mp} = V_{thr}$, $i_{mp} = V_{thr}/R_m$, corresponds to the asymptote, or rheobase, in Figure 3.13. If $i_{mp} > V_{thr}/R_m$, then V_{thr} is reached at a time T_D, which from Equation 3.20 is such that:

$$V_{thr} = R_m i_{mp}\left(1-e^{-T_D/\tau_m}\right) \tag{3.21}$$

Once V_{thr} is reached and an AP is generated at $t = T_D$, the current step can be terminated and becomes a current pulse of amplitude i_{mp} and duration T_D. Equation 3.21 can be expressed as:

$$i_{mp} = \frac{V_{thr}}{R_m\left(1-e^{-T_D/\tau_m}\right)} \tag{3.22}$$

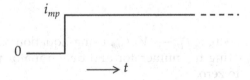

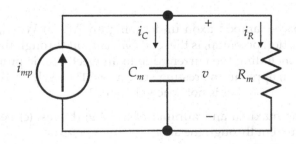

FIGURE 3.14
Current step applied to a parallel RC circuit.

The plot of i_{mp} vs. T_D has the general shape of Figure 3.13. As T_D increases, e^{-T_D/τ_m} decreases, $\left(1-e^{-T_D/\tau_m}\right)$ increases, and i_{mp} decreases. As $T_D \longrightarrow \infty$, i_{mp} becomes a step of amplitude V_{thr}/R_m, as argued previously. If $T_D \longrightarrow 0$, $i_{mp} \longrightarrow \infty$, but $i_{mp}T_D \longrightarrow V_{thr}C_m$ (Problem 3.7). $i_{mp}T_D$ becomes a current impulse, or **Dirac delta function** of area $i_{mp}T_D$. This impulse instantly deposits a charge $i_{mp}T_D = V_{thr}C_m$ on C_m, resulting in a voltage V_{thr} that is just sufficient to generate an AP.

In practice, R_m is not constant, as the membrane voltage rises toward threshold because of the nonlinear behavior of an excitable membrane noted earlier. Moreover, V_{thr} is not a constant voltage, as explained previously. As T_D increases, potassium activation and sodium inactivation would have more time to exercise their effects, which increases V_{thr}. This would require a larger i_{mp} to generate an AP than if V_{thr} remained constant (Equation 3.22). The actual strength-duration relationship would then be "flatter" than is given by Equation 3.22.

Although the strength-duration relationship was explained with reference to a space-clamped axon, the same form of graph of Figure 3.13 applies to excitable tissues in general, including muscle, and not necessarily under space clamp.

Problem 3.6

Derive Equation 3.20 by writing Kirchhoff's current law in Figure 3.14 as: $C_m \dfrac{dv}{dt} + \dfrac{v}{R_m} = i_{mp}$. Show that the solution to this equation is of the form: $v = Ae^{-t/\tau_m} + R_m i_{mp}$, where A is an arbitrary constant and $\tau_m = R_m C_m$. Determine A from the initial condition that $v = 0$ at $t = 0$.

Problem 3.7

Show that as $T_D \longrightarrow 0$, $i_{mp}T_D \longrightarrow V_{thr}C_m$, using Equation 3.22. Use L'Hopital's rule by differentiating the numerator and denominator with respect to T_D before setting T_D to zero.

Problem 3.8

Consider a space-clamped axon undergoing an AP. (a) When, during the course of the action potential, is the *total ionic current* through the membrane equal to zero, including the current due to an electrogenic pump? (b) Does the answer in (a) apply in the resting state as well? (c) Why is the answer in (a) not true of an axon that is not space-clamped?

ANS.: (a) at the maxima and minima of the AP; (b) yes; (c) because of the longitudinal current through the axon.

3.3.4 Effect of Temperature

Membrane conductance for a given ion Y is proportional to the ratio of P_Y/T, where P_Y is the permeability of the membrane to ion Y and T is the absolute temperature (Equations 2.25 and 2.26). P_Y is proportional to the product of T and the mobility of a given ion (Equation 1.22). Theoretically, the temperature dependence of conductance is therefore that of mobility. Experimentally, mobility increases with temperature because a rise in temperature weakens the bond between a hydrated ion and its hydrating water molecules, which decreases the effective size and mass of the hydrated ion, leading to a higher mobility of ions in solution. The variation of permeability with temperature depends not only on the temperature variation of mobility of ions in solution but also on factors affecting the movement of ions through the pores, such as the temporary binding of ions to the charged groups in a pore. In practice, membrane conductance generally increases with temperature by about 2%–4% per °C. The conductances G_K^o and G_{Na}^o (Equations 3.6 and 3.14) are affected in this manner.

The main effect of temperature in the HH model is on the time constants τ_n, τ_m, and τ_h, which vary inversely with the temperature (Table 3.1). The corresponding rates of change, dn/dt, dm/dt, and dh/dt, will all increase with temperature (Equations 3.4, 3.11, and 3.13). It follows that sodium activation, sodium inactivation, and potassium activation are all speeded up by an increase in temperature. Effectively, the time scale is contracted, so that a rise in temperature increases the rates of rise and fall of the spike and reduces its duration. However, the maximum rate of change of v_m is limited by the membrane time constant C_m/G_m, with the following consequences: (i) As the temperature is raised, the effect of faster sodium inactivation and potassium activation is felt sooner, while the ability of sodium activation to cause a more rapid increase in v_m is limited by the membrane time constant. The peak of the spike is therefore reduced and the threshold increased. (ii) The downstroke of the spike, being slower than the upstroke, is less limited by the membrane time constant and is therefore more affected by temperature than the upstroke.

A rise in temperature has a profound effect on the propagating AP, as discussed in Section 4.3.2.

3.3.5 Refractoriness

An important consideration is how soon after the generation of an AP can a new AP be generated. Clearly, no new AP can be generated during the regenerative cycle. Nor can a new AP be generated during the early part of the recovery phase when sodium inactivation and potassium activation are still strong. The inability to generate a new, normal AP soon after an AP has been initiated is referred to as **refractoriness**. The period, after the initiation of an AP, during which it is not possible to generate a new AP, no matter how strong the second stimulus, is the **absolute refractory period,** and the axon is said to be absolutely refractory during this period. It is analogous to the "dead time" of an electronic monostable circuit, when a second output pulse cannot be generated.

The effects of sodium inactivation and potassium activation decline during the course of the recovery phase of the AP. While they are significant, (i) it would take a stronger depolarization to turn the net current inward to initiate an AP, which implies that the threshold level is higher than normal, and (ii) the larger G_K means that, for a given increase in G_{Na}, the amplitude of the AP is reduced, as follows from voltage division in the membrane equivalent circuit (Figure 2.9). As recovery continues, the threshold and the amplitude of the AP approach their normal values (Figure 3.15). The time interval following the absolutely refractory period, and until the membrane conductances virtually assume their steady-state values, is the **relative refractory period**. Typically, the relative refractory period ranges from being about equal to the absolute refractory period to being about twice this period. In the space-clamped squid giant axon at 6.3°C, both the absolute and relative refractory periods are about 5 ms each.

Refractoriness limits the highest frequency at which an AP can be generated to about the reciprocal of twice the absolute refractory period. It also has an important bearing on the propagating AP, as will be discussed in the following chapter.

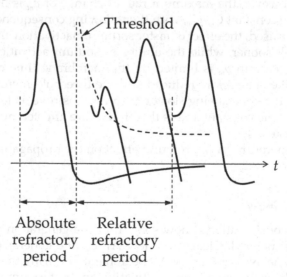

FIGURE 3.15
Membrane refractoriness.

Summary of Main Concepts

- Space-clamping eliminates the longitudinal variation of voltage and current along an axon.

- Voltage clamping is an invaluable tool for elucidating the ionic mechanisms associated with the AP.

- The AP is the result of a fast, transient increase in the Na^+ conductance in response to a sudden depolarization of the membrane, followed by a slower increase in the K^+ conductance.

- The regenerative cycle is essential for initiation of the AP. The regeneration, apart from causing a rapid increase in depolarization, is self-sustaining and takes over from the applied stimulus, so that the subsequent time course of membrane voltage becomes largely independent of the stimulus, except for a reduction of the latency with increasing stimulus strength.

- Voltage-gated Na^+ and K^+ channels are independent channels that can be selectively blocked by various substances.

- A voltage-gated channel has three functional components: (i) an aqueous pore for the passage of the ion, including a constriction that acts as a selectivity filter, (ii) a voltage sensor that is responsive to the voltage across the membrane, and (iii) a gate that opens and closes according to the transmembrane voltage and channel dynamics.

- Voltage-gated channels open and close at random with probabilities that depend on the voltage. The aggregate behavior of a large number of channels for a given ion is what is observed at the macroscopic level as variation of channel conductance for that ion as a function of time, in response to an applied voltage.

- The AP is an active response that depends on the electrochemical potential gradients for Na^+ and K^+ that are established by the sodium-potassium pump.

- The threshold of the AP under space clamp is when the inward Na^+ current equals the outward current through the membrane. Just beyond threshold, the net membrane current is inward, thereby triggering the regenerative cycle that results in the upstroke of the spike.

- The threshold level depends on the time course of the stimulus. If this time course is slow enough, no AP will be generated, even for depolarizations that exceed the threshold for a fast-changing stimulus. This is referred to as accommodation of the membrane.

- The strength-duration relationship is a plot of the amplitude of the current pulse that just evokes an AP as a function of the duration of the pulse.

- Membrane ionic conductances increase with temperature by about 2%–4% per °C.

- A rise in temperature speeds up sodium activation, sodium inactivation, and potassium activation, resulting in a reduction in spike amplitude and duration as well as an increase in the threshold.

- Following the initiation of an AP, there is an absolute refractory period during which it is not possible to generate another AP followed by a relative refractory period during which an AP of reduced amplitude and higher threshold can be generated.

4

Propagation of the Action Potential

Objective and Overview

After having considered the generation of the action potential (AP) in the preceding chapter, the present chapter is concerned with the propagation of the AP. Propagation along an unmyelinated axon is considered first, qualitatively to begin with, and then quantitatively in terms of RC cable theory. The cable equation is derived in detail and solved for a semi-infinite cable subjected to different types of stimuli that illustrate some important features of cable behavior, including the variation of the speed of conduction of the AP as the square root of axon diameter.

Myelinated axons are considered next, explaining qualitatively how they conduct the AP at a faster speed and the maximization of the speed of conduction with respect to the ratio of axon-to-fiber diameter and the ratio of internodal length to fiber diameter. Some quantitative aspects are then discussed, subject to certain assumptions, and leading to some important conclusions concerning propagation of the AP along myelinated axons, including the variation of the speed of conduction with fiber diameter.

Some basic properties of the propagating action potential are then examined, namely, the threshold and the effect of temperature. The chapter ends by comparing active propagation with passive propagation.

Learning Objectives

To understand:

- How the AP is propagated along unmyelinated and myelinated axons and the factors affecting the speed of conduction
- The basics of the theory of RC cables and some important features of cable behavior

- Some quantitative aspects of the propagation of APs along a myelinated axon
- The threshold behavior of nerve fibers when stimulated by external electrodes
- The effect of temperature on the propagating AP
- How active propagation compares with passive propagation

4.1 Propagation of the Action Potential along an Unmyelinated Axon

Axons in invertebrates are unmyelinated, that is, they lack the relatively thick myelin sheath of myelinated axons of vertebrates. Propagation of the AP along an unmyelinated axon is simpler to analyze because, ideally, the axon can be modeled as an *RC* cable of uniform diameter.

Suppose that an AP has been generated by the neuron and is travelling along the axon away from the cell body, toward the axon terminations. How does the AP propagate? Consider a part of the axon, or a patch of the axonal membrane, that is in the spike part of an AP. This patch, shown in gray in Figure 4.1, is depolarized and at a more positive voltage on the inside of the axon with respect to neighboring regions, causing current flow from the depolarized patch to neighboring regions, as illustrated in Figure 4.1. These currents are referred to as **local-circuit currents,** and they form closed loops as required by conservation of charge. The current intensity is highest immediately adjacent to the depolarized patch and decreases away from this region. The basis for an orderly propagation of the AP is that *the intensity of the local-circuit current is sufficient to initiate an AP in the region immediately*

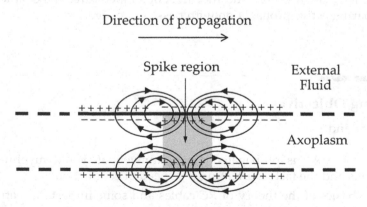

FIGURE 4.1
Propagation of action potential along an unmyelinated axon.

ahead of the depolarized patch of membrane undergoing an AP. The region in the
wake of the AP has just undergone an AP and is not excited again because of refrac-
toriness. By the time this region has recovered, the AP would have moved
further along the axon, so that the intensity of the local-circuit current will
not be sufficient for re-excitation.

In some experimental situations, an axon may be excited by a voltage pulse
applied between two electrodes placed near the outer surface of the axon, as
illustrated in Figure 4.2. The stimulating current flows from anode to cathode and through the axonal membrane at these locations. The membrane is
hyperpolarized at the anode and depolarized at the cathode. If the stimulus
is strong enough, an AP is generated in a patch of axonal membrane at the
cathode. The local-circuit current from this patch will then excite the two
regions on either side, causing two APs to propagate in opposite directions
away from the cathode, as indicated in Figure 4.2. If the membrane hyperpolarization at the anode is large enough, the circulating currents may not
be able to depolarize the membrane regions in the vicinity of the anode to
threshold, so that the AP propagating toward the anode will not be able to
continue beyond the anode, a condition known as **anodal block.**

Considering Figure 4.1, it stands to reason that the speed of conduction
of the AP depends on the intensity of the local-circuit currents and on their
rate of increase. The larger the intensity of the local circuit currents and the
more rapid their buildup, the faster is the depolarization of neighboring
regions to threshold and the sooner is an AP initiated in these regions. It is
to be expected, therefore, that the speed of conduction of the AP increases
with a decrease in: membrane capacitance, membrane conductance, internal resistance of the axon, or resistance of the external medium. For a given
membrane depolarization due to the action potential, all these changes are
conducive to a larger and faster depolarization of neighboring regions. The
variation of the speed of conduction of the AP with axon diameter will be
investigated quantitatively in the following sections after presenting the
cable model of the axon.

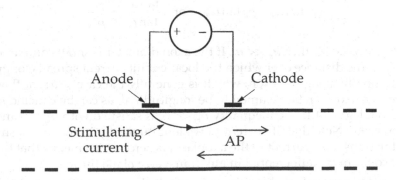

FIGURE 4.2
Axon stimulated by surface electrodes.

4.1.1 Cable Model

The following assumptions are made in deriving the classical equations for an *RC* cable representing an axon:

1. The structure is tubular and uniform; in practice, the axon diameter varies by about 5% over most of the length of the axon.

2. The axoplasm and external fluid behave as purely ohmic resistances. Inductance is neglected for two reasons: (i) the full magnetic field, due to the current in the axon, is confined to the membrane between the inner and outer cylinders, which makes the inductance relatively small in the case of a thin membrane, and (ii) the effect of the inductance, which appears as a series element, is swamped by the much larger effect of the series resistance.

3. Current flows transversely through the membrane and longitudinally through the axoplasm and external medium. This can be justified by considering current flow between two media, such as the membrane and the axoplasm (Figure 4.3). Conservation of charge requires that the component of current density normal to the interface be the same, that is:

$$J_a \cos\alpha_a = J_m \cos\alpha_m \qquad (4.1)$$

Conservation of energy requires that the component of the electric field parallel to the interface be the same, that is $\xi_a = \xi_m$. Substituting $\xi = \rho J$,

$$\rho_a J_a \sin\alpha_a = \rho_m J_m \sin\alpha_m \qquad (4.2)$$

where ρ_i and ρ_m are the resistivities of the axoplasm and membrane, respectively. Dividing equation 4.2 by Equation 4.1:

$$\rho_a \tan\alpha_a = \rho_m \tan\alpha_m, \quad \text{or} \quad \frac{\tan\alpha_m}{\tan\alpha_a} = \frac{\rho_a}{\rho_m} \qquad (4.3)$$

Since $\rho_a \ll \rho_m$, then $\alpha_m \ll \alpha_a$. If the axon diameter is small compared with the distance over which the local-circuit current spreads longitudinally along the axon, which is generally the case, current flow in the axon can be assumed to be longitudinal, as can be concluded from Figure 4.1. The inequality $\alpha_m \ll \alpha_a$ is satisfied with $\alpha_m \cong 0$ and $\alpha_a \cong 90°$. Note that, if current flows longitudinally in the axoplasm, the transverse current in the axoplasm is zero, which means that the axoplasm is equipotential in any transverse plane through it.

The external medium is assumed to be a thin film (Problem 4.7b), in which case the current through this medium is effectively

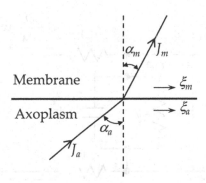

FIGURE 4.3
Current flow through two media of different resistivities.

longitudinal. If the axon is surrounded by a relatively large conducting medium, as is usually the case, the assumption of longitudinal
current in the external medium is not justified, but the resistance of
the external medium will be negligibly small in this case.

4. The cable extends from $x = 0$ to $x \longrightarrow \infty$. Such a cable is said to be
semi-infinite, in contrast to an **infinite cable,** which extends from
$x = -\infty$ to $x \longrightarrow \infty$.

Subject to the aforementioned assumptions, an element of the axon of length
Δx can be represented as in Figure 4.4, where,

v_{mr}: membrane voltage with respect to the resting value V_{m0}
v_a: voltage of the axoplasm with respect to an arbitrary zero reference
v_e: voltage of the external medium with respect to the same zero reference
r_a: resistance of the axoplasm per unit length
r_e: resistance of the external medium per unit length
i_a: longitudinal current in the axoplasm
i_e: longitudinal current in the external medium
i_m: membrane current per unit length
g_m: membrane conductance per unit length
c_m: membrane capacitance per unit length

We begin by applying Kirchhoff's current law (KCL), that is conser-
vation of charge, to an element of length Δx in Figure 4.4. At location
x, KCL requires that $i_a = -i_e$, and at location $x + \Delta x$, KCL requires that
$i_a + (\partial i_a / \partial x)\Delta x = -i_e + (\partial i_e / \partial x)\Delta x$. Hence,

$$i_a = -i_e \quad \text{and} \quad \frac{\partial i_a}{\partial x} = -\frac{\partial i_e}{\partial x} \tag{4.4}$$

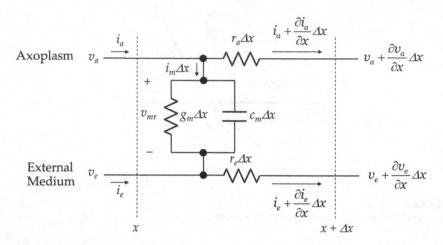

FIGURE 4.4
Voltages and currents in an infinitesimal element of *RC* cable.

From KCL at the upper node, $i_a = i_m \Delta x + i_a + (\partial i_a / \partial x) \Delta x$, and KCL at the lower node gives: $i_e = i_m \Delta x + i_e + (\partial i_e / \partial x) \Delta x$. It follows that:

$$i_m = -\frac{\partial i_a}{\partial x} = \frac{\partial i_e}{\partial x} \tag{4.5}$$

From Kirchhoff's voltage law (KVL) on the LHS of the element Δx:

$$v_{mr} = v_a - v_e \tag{4.6}$$

Applying Kirchhoff's voltage law (KVL) to the upper series path in Figure 4.4, $v_a - r_a \Delta x (i_a + (\partial i_a / \partial x) \Delta x) = v_a + (\partial v_a / \partial x) \Delta x$. Neglecting the second-order infinitesimal gives:

$$\frac{\partial v_a}{\partial x} = -r_a i_a \tag{4.7}$$

Similarly, applying Kirchhoff's voltage law (KVL) to the lower series path in Figure 4.4, $v_e - r_e \Delta x (i_e + (\partial i_e / \partial x) \Delta x) = v_e + (\partial v_e / \partial x) \Delta x$. Neglecting the second-order infinitesimals gives:

$$\frac{\partial v_e}{\partial x} = -r_e i_e \tag{4.8}$$

The current $i_m \Delta x$ is the sum of the current through the conductance $g_m v_{mr} \Delta x$, and the current through the capacitor $c_m \Delta x \partial v_{mr} / \partial t$. Cancelling out Δx,

$$i_m = g_m v_{mr} + c_m \frac{\partial v_{mr}}{\partial t} \tag{4.9}$$

We will next derive from the preceding equations the standard equation for an RC cable in normalized form. Differentiating Equation 4.6 partially with respect to x, and substituting for $\partial v_a/\partial x$ and $\partial v_e/\partial x$ from Equations 4.7 and 4.8, respectively,

$$\frac{\partial v_{mr}}{\partial x} = -r_a i_a + r_e i_e \tag{4.10}$$

Differentiating Equation 4.10 again partially with respect to x and substituting from Equation 4.5,

$$\frac{\partial^2 v_{mr}}{\partial x^2} = (r_a + r_e) i_m \tag{4.11}$$

Substituting for i_m from Equation 4.9 and rearranging.

$$\frac{1}{g_m(r_a + r_e)} \frac{\partial^2 v_{mr}}{\partial x^2} = \frac{c_m}{g_m} \frac{\partial v_{mr}}{\partial t} + v_{mr} \tag{4.12}$$

Next, we introduce dimensionless variables, $X = x/\lambda$ and $\varphi = t/\tau_m$, where $\lambda = 1/\sqrt{g_m(r_a + r_e)}$ and $\tau_m = c_m/g_m$ is the **membrane time constant**. Equation 4.12 becomes:

$$\frac{\partial^2 v_{mr}}{\partial X^2} = \frac{\partial v_{mr}}{\partial \varphi} + v_{mr} \tag{4.13}$$

Equation 4.13 is the standard equation for an RC cable in normalized form. It is a partial differential equation involving both the normalized space variable X and the normalized time variable φ.

In order to interpret λ, the solution of Equation 4.13 will be derived for a step of depolarization of amplitude V_{ms} applied at $X = 0$ and $t = 0$. It is assumed that V_{ms} is sufficiently small so that the membrane voltage remains well below threshold, which means that the membrane conductance can be considered constant at substantially its value under resting conditions. In the steady state, as $t \longrightarrow \infty$, v_{mr} does not vary with time, so that $\partial V_{mr}/\partial \varphi = 0$ and Equation 4.13 reduces to:

$$\frac{\partial^2 v_{mr}}{\partial X^2} = v_{mr} \tag{4.14}$$

The general solution to Equation 4.14 is:

$$v_{mr} = Ae^{-X} + Be^{+X} \tag{4.15}$$

where A and B are arbitrary constants, as can be readily verified by substituting for v_{mr} from Equation 4.15 in Equation 4.14. Physically, v_{mr} decreases with X because of the voltage division between the series resistances and the shunt resistance. As $X \longrightarrow \infty$, $v_{mr} \longrightarrow 0$ and cannot increase without limit. The second term on the RHS of Equation 4.15 must therefore be zero, that is, $B = 0$. At $x = 0$, $v_{mr} = V_{ms}$, which gives $A = V_{ms}$. Equation 4.15 becomes:

$$v_{mr} = V_{ms}e^{-X} = V_{ms}e^{-x/\lambda} \tag{4.16}$$

According to Equation 4.16, v_{mr} decreases exponentially with x at a rate determined by λ, in the same manner as the variation in time of the response of a first-order electric circuit, such as that representing the voltage across an initially charged capacitor C that is discharging through a resistance R, the rate of discharge depending in this case on the time constant RC. For this reason, λ is referred to as the **space constant**. At $x = \lambda$, $v_{mr} = V_{ms}/e$, and at $x = 3\lambda$, $v_{mr} \cong 0.05V_{ms}$. For a squid giant axon of 0.5 mm diameter in the resting state at 20°C, assuming $g_{mr} \cong 0.1$ mS/cm and $r_i \cong 20$ kohm/cm, gives $\lambda \cong 0.7$ cm, neglecting r_e. It is seen that the voltage is rapidly attenuated along the axon and becomes negligibly small for x larger than about 2 cm. It follows that passive propagation is not feasible except over distances that are short compared to λ.

4.1.2 Solution of the Cable Equation

Equation 4.13 is conveniently solved using the Laplace transform, which transforms the partial differential equation to an ordinary one. Taking the Laplace transform of both sides of Equation 4.13 with respect to the time variable φ,

$$\frac{d^2V_{mr}(s)}{dX^2} = (s+1)V_{mr}(s) \tag{4.17}$$

where $V_{mr}(s)$ is the Laplace transform of v_{mr} with respect to time, and $v_{mr} = 0$ in the resting state, just before the stimulus is applied, so that the Laplace transform of $\partial v_{mr}/\varphi$ is $sV_{mr}(s)$. As in the case of Equation 4.13, the general solution to Equation 4.17 is:

$$V_{mr}(s) = Ae^{-X\sqrt{s+1}} + Be^{X\sqrt{s+1}} \tag{4.18}$$

where A and B are independent of X but depend, in general, on the boundary conditions and on s. For the type of stimuli of interest, $v_{mr} \longrightarrow 0$ as $X \longrightarrow \infty$. This means that $B = 0$ so that v_{mr} does not increase without limit as $X \longrightarrow \infty$. Equation 4.18 reduces to:

$$V_{mr}(s) = Ae^{-X\sqrt{s+1}} \tag{4.19}$$

The responses of a semi-infinite cable to two types of stimulus applied at $X = 0$ and $t = 0$ will be considered next: a current step and a current impulse, as the deductions from these solutions are needed for the discussion of neuro-nal responses in Chapter 7. For simplicity, and in accordance with common practice, it will henceforth be assumed that $r_e = 0$. If the external medium is a thin film, r_a is replaced by $(r_e + r_a)$ in all expressions involving r_a. If the external medium is a volume conductor, as is usually the case, then r_e can generally be neglected compared to r_a.

4.1.2.1 Response to a Current Step

To determine A in Equation 4.19 when a current step of magnitude I_0 is applied to the cable at $t = 0$ and $X = 0$, Equation 4.10 is used with $r_e = 0$ and $X = x/\lambda$. The equation becomes:

$$\frac{dv_{mr}}{dX} = -\lambda r_a i_a \tag{4.20}$$

Taking the Laplace transform of both sides with respect to φ,

$$\frac{dV_{mr}(s)}{dX} = -\lambda r_a I_a(s) \tag{4.21}$$

where $I_a(s)$ is the Laplace transform of i_a with respect to φ. Substituting for $dV_{mr}(s)/dX$ from Equation 4.19,

$$A\sqrt{s+1}e^{-X\sqrt{s+1}} = \lambda r_a I_a(s) \tag{4.22}$$

At $X = 0$, $i_a = I_0 u(\varphi)$, where $u(\varphi)$ is the unit step function at the origin. It follows that $I_a(s) = I_0/s$. Substituting in Equation 4.22 with $X = 0$ gives:

$$A = \frac{\lambda r_a I_0}{s\sqrt{s+1}} \tag{4.23}$$

Substituting in Equation 4.19,

$$V_{mr}(s) = \frac{\lambda r_a I_0}{s\sqrt{s+1}}e^{-X\sqrt{s+1}} \tag{4.24}$$

From a table of inverse Laplace transform pairs,

$$v_{mr}(X,\varphi) = \frac{\lambda r_a I_0}{2}\left\{e^{-X}\text{erfc}\left(\frac{X}{2\sqrt{\varphi}} - \sqrt{\varphi}\right) - e^{X}\text{erfc}\left(\frac{X}{2\sqrt{\varphi}} + \sqrt{\varphi}\right)\right\} \tag{4.25}$$

where erfc$(y) = 1 - erf(y)$ is the complementary error function, and erf(y) is the error function defined as the area under a Gaussian-shaped curve, with some scaling, as follows:

$$\text{erf}(y) = \frac{2}{\sqrt{\pi}}\int_0^y e^{-\sigma^2}d\sigma \tag{4.26}$$

It is of interest to determine some steady-state conditions. As $\varphi \longrightarrow \infty$, $\text{erf}(\infty) = 1$, and $\text{erf}(-\infty) = -1$, since the error function is odd, and $1 - \text{erfc}(\infty) = 0$, whereas $1 - \text{erfc}(-\infty) = 2$. Equation 4.25 reduces to:

$$v_{mr}(X,\infty) = \lambda r_a I_0 e^{-X} \tag{4.27}$$

The voltage in the steady state decreases exponentially with distance with a space constant λ, in agreement with Equation 4.16.

At $X = 0$, Equation 4.27 reduces to:

$$v_{mr}(0,\infty) = \lambda r_a I_0 \tag{4.28}$$

The input resistance seen by the source is:

$$\frac{v_{mr}(0,\infty)}{I_0} = R_{in} = \lambda r_a \tag{4.29}$$

Substituting $\lambda = 1/\sqrt{g_m r_a}$,

$$R_{in} = \frac{r_a}{\sqrt{g_m r_a}} = \sqrt{r_m r_a} \tag{4.30}$$

where $r_m = 1/g_m$. Substituting $r_m = r_a \lambda^2$, or $r_a = r_m / \lambda^2$,

$$R_{in} = \lambda r_a = \frac{r_m}{\lambda} \tag{4.31}$$

It follows from Equations 4.29–4.31 that R_{in} can be interpreted as: (i) the geometric mean of the series resistance per unit length r_a in Ω/cm and the membrane resistance per unit length r_m in Ωcm, (ii) the series resistance of a section of the cable of length λ, or (iii) the membrane resistance of a section of the cable of length λ.

Returning to the general case, $v_{mr}(X,\varphi)$ can be normalized with respect to its steady-state value at $X = 0$ by dividing Equation 4.25 by Equation 4.28 to give:

$$\frac{v_{mr}(X,\varphi)}{v_{mr}(0,\infty)} = \frac{1}{2}\left\{e^{-X}\text{erfc}\left(\frac{X}{2\sqrt{\varphi}} - \sqrt{\varphi}\right) - e^{X}\text{erfc}\left(\frac{X}{2\sqrt{\varphi}} + \sqrt{\varphi}\right)\right\} \tag{4.32}$$

Equation 4.32 is plotted in this normalized form in Figure 4.5 for $X = 0$, $X = 0.6$, and $X = 1.2$.

It follows from Equation 4.32 that at $X = 0$,

$$\frac{v_{mr}(0,\varphi)}{v_{mr}(0,\infty)} = \text{erf}\sqrt{\varphi} \tag{4.33}$$

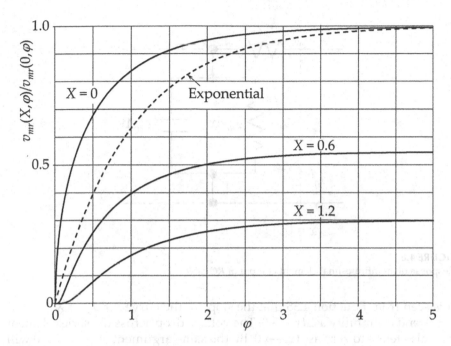

FIGURE 4.5
Normalized voltage vs. time at various distances along a semi-infinite cable in response to a current step applied at $X = 0$. The dashed plot is an exponential-type response.

The slope of the curve in Figure 4.5 for $X = 0$ is the derivative of $\text{erf}\sqrt{\varphi}$:

$$\frac{d}{d\varphi}\left[\frac{v_{mr}(0,\varphi)}{v_{mr}(0,\infty)}\right] = \frac{d\left(\text{erf}\sqrt{\varphi}\right)}{d\varphi} = \frac{e^{-\varphi}}{\sqrt{\pi\varphi}} \qquad (4.34)$$

This slope tends to infinity at $\varphi = 0$ (Figure 4.5). In interpreting this result, it should be kept in mind that voltage and current signals propagate along the cable at a finite speed. Hence, at very small values of time, the source sees only a very short segment of the cable, which in the limit is infinitesimally small, as illustrated in Figure 4.6 for the hypothetically first infinitesimal element connected to the current source. Because I_0 is suddenly applied, and since the stored energy in the capacitor element $c_m\Delta x$ cannot change instantaneously, the voltage across the parallel element $g_m\Delta x$ does not change instantaneously. Hence, I_0 from the current source flows at $\varphi = 0$ through the capacitive element irrespective of g_m, which, in effect, is being short-circuited. It follows from the relation $i = Cdv/dt$ for a capacitor that:

$$\frac{dv_{mr}(0,0)}{d\varphi} = \frac{I_0}{c_m\Delta x} \qquad (4.35)$$

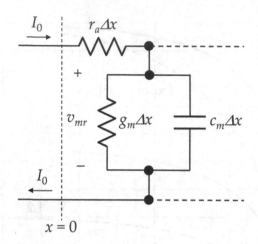

FIGURE 4.6
Response of an infinitesimal element at input of *RC* cable.

It is seen from Equation 4.35 that the slope of the curve for v_{mr} at $X = 0$ and $\varphi = 0$ tends to infinity as $\Delta x \longrightarrow 0$. The voltage drop across the series element $r_a \Delta x$ also tends to zero as $\Delta x \longrightarrow 0$. By the same argument, v_{mr} for $X > 0$ will not change until a later time because of the finite speed of conduction, and the larger X, the longer it will take v_{mr} to rise significantly above zero, as illustrated in Figure 4.5 for $X = 0.6$, and $X = 1.2$ (see Problem 4.1).

It should be noted that if the step current I_0 is applied to a patch of membrane represented by C_m in parallel with G_m in per unit area units, the voltage across the parallel combination, as a fraction of the final voltage, increases with time as $\left(1 - e^{-t/\tau_m}\right)$, where $\tau_m = C_m/G_m$. The voltage reaches 63% of its final value at $t = \tau_m$ (Figure 4.5, $\varphi = 1$). On the other hand, if the step current I_0 is applied to a cable having the same per unit membrane parameters so that $c_m/g_m = \tau_m$, the voltage at the cable input as a fraction of the final voltage increases with time as $\mathrm{erf}\sqrt{t/\tau_m}$ (Equation 4.33), reaching 84% ($\mathrm{erf}(1) = 0.84$) of the final value $t = \tau_m$ (Figure 4.5). That the voltage increases faster at the input of a cable, in response to an applied current, than across an equipotential patch of membrane having the same per unit membrane capacitance and inductance, is counterintuitive at first sight but has an important bearing on synaptic integration (Section 7.4.1).

A measure of the speed at which the voltage spreads along the cable upon application of the current step of magnitude I_0 can be obtained by considering the function $W = v_{mr}(X,\varphi)/v_{mr}(X,\infty)$, where W is the ratio of the voltage variation $v_{mr}(X,\varphi)$ with φ, for any X, to the steady-state value $v_{mr}(X,\infty)$ for the same X. It can be shown analytically that a plot of X vs. φ for the half-steady-state value point ($W = 0.5$) has a slope of almost exactly 2, so that $x/t = 2\lambda/\tau_m$ (Jack et al., 1983). This speed varies as the square root of the diameter of the axon (Problem 4.2) and is a useful indicator of the speed at which signals spread passively and the delay involved.

Problem 4.1

Considering that erf(0.5) $\cong$ 0.5 and erf(1.5) $\cong$ 0.97, argue from Equation 4.32 that at one space constant, $v_{mr}(X,\varphi)/v_{mr}(0,\varphi)$ remains near zero for about a quarter time constant, and that at three space constants, $v_{mr}(X,\varphi)//v_{mr}(0,\varphi)$ remains near zero for about one time constant.

Problem 4.2

Show that $\dfrac{2\lambda}{\tau_m} = \dfrac{1}{C_m}\sqrt{\dfrac{G_m}{\rho_a}}\,a$, where C_m is the membrane capacitance per unit area, G_m is the membrane conductance per unit area, ρ_a is the resistivity of the axoplasm, and a is the axon diameter.

4.1.2.2 Response to a Current Impulse

Consider a current impulse $i_a = Q_0\delta(t)$ applied at $X = 0$, where $\delta(t)$ is the unit impulse function at the origin and Q_0 is the charge represented by the area under the impulse. Equation 4.21 applies but with a different $I_a(s)$, the Laplace transform of i_a with respect to φ. By definition of the Laplace transform with respect to φ:

$$I_a(s) = \int_0^\infty Q_0\delta(t)e^{-s\varphi}d\varphi = \int_0^\infty Q_0\delta(t)e^{-st/\tau_m}d(t/\tau_m) = \frac{Q_0}{\tau_m}\int_0^\infty \delta(t)e^{-st/\tau_m}dt$$

$$(4.36)$$

$$= \frac{Q_0}{\tau_m}\int_0^\infty \delta(t)dt = \frac{Q_0}{\tau_m}$$

since the exponential is unity at $t = 0$, with $\delta(t)$ being zero everywhere except at $t = 0$, by definition of $\delta(t)$. Note that $u(t)$ is dimensionless, so that its Laplace transform is not affected by time scaling from t to φ. On the other hand, $\delta(t)$ has the dimensions of t^{-1}, and its Laplace transform is affected by time scaling. Substituting in Equation 4.22,

$$A = \frac{\lambda r_a Q_0}{\tau_m\sqrt{s+1}}$$

$$(4.37)$$

It follows from Equation 4.19 that:

$$V_{mr}(s) = \frac{\lambda r_a Q_0}{\tau_m\sqrt{s+1}}e^{-X\sqrt{s+1}}$$

$$(4.38)$$

From a table of inverse Laplace transform pairs, and substituting $r_a = 1/(\lambda^2 g_m)$ and $\tau_m g_m = c_m$,

$$v_{mr}(X, \varphi) = \frac{Q_0}{c_m \lambda \sqrt{\pi \varphi}} e^{-\frac{x^2}{4\varphi} - \varphi} \tag{4.39}$$

$$\text{and, } v_{mr}(0, \varphi) = \frac{Q_0}{c_m \lambda \sqrt{\pi \varphi}} e^{-\varphi} \tag{4.40}$$

At $t = 0$, $v_{mr}(0, 0) \longrightarrow \infty$, as to be expected from the charge being applied to an infinitesimally small capacitance. For $\varphi > 0$, $v_{mr}(0, \varphi)$ decreases at a faster rate than an exponential because of φ in the denominator of Equation 4.40.

4.1.2.3 Wave Propagation

We consider, next, the case of a wave propagating in the steady state, such as an AP. Such a propagating wave can be expressed as:

$$v_{mr} = f(x - \theta t) \tag{4.41}$$

where θ is the speed of the wave and f depends on the shape of the wave. If we consider a time $(t + t')$, the argument of the function becomes $(x - \theta t - \theta t')$. This equals $(x - \theta t)$ if x is replaced by $(x + x')$ where $x' = \theta t'$. In other words, the function has the same value at (x, t) and $(x + x', t + t')$, so that in time t' the wave has traveled a distance x' in the positive x direction, the speed of conduction being $x'/t' = \theta$.

Equation 4.41 satisfies the wave differential equation:

$$\frac{\partial^2 v_{mr}}{\partial x^2} = \frac{1}{\theta^2} \frac{\partial^2 v_{mr}}{\partial t^2} \tag{4.42}$$

as can be verified by direct substitution of Equation 4.41 in Equation 4.42. If the RHS of Equation 4.42 is equated to the RHS of Equation 4.11, with $r_e = 0$, and the substitution $i_m = \pi a J_m$ is made, where a is the axon diameter and J_m is the membrane current density, the resulting equation is:

$$J_m = \frac{1}{\pi a r_a \theta^2} \frac{\partial^2 v_{mr}}{\partial t^2} \tag{4.43}$$

Dividing both sides Equation 4.9 by the perimeter of the axon and rearranging, the equation becomes:

$$C_m \frac{\partial v_{mr}}{\partial t} + G_{mp} v_{mr} = J_m \tag{4.44}$$

where the current per unit length i_m becomes the current density J_m, the capacitance per unit length c_m becomes the capacitance per unit area C_m, and the conductance per unit length g_m becomes the conductance per unit area G_{mp}, the effective conductance per unit area of the *active* membrane undergoing an action potential. Equation 4.44 is a differential equation that defines the time course of v_{mr} at a particular x as a function of J_m. It is noteworthy in that it indicates that the time course of v_{mr} depends only on the per unit area quantities J_m, G_{mp} and C_m. It is independent of other parameters including axon diameter.

It also follows from Equation 4.44 that the functional relation between the time course of the AP and J_m depends only on C_m and G_{mp}, independently of any other parameter of the axon. This means that the relation between J_m and $\partial^2 V_{mr}/\partial t^2$ in Equation 4.43 must also depend only on C_m and G_{mp} independently of the term multiplying $\partial^2 V_{mr} / \partial t^2$ in this equation. In other words, for given C_m and G_{mp},

$$\frac{1}{\pi a r_a \theta^2} = \text{constant} \tag{4.45}$$

Two deductions concerning the speed of conduction can be made from Equation 4.45:

1. For a given axon, a is a constant, so that

$$\theta \propto \frac{1}{\sqrt{r_a}} \tag{4.46}$$

or $\theta \propto 1/\sqrt{r_a + r_e}$, if r_e is included.

2. If $r_e = 0$, then substituting $r_a = 4\rho_a/\pi a^2$, where ρ_a is the resistivity of the axoplasm, Equation 4.45 becomes:

$$\frac{a}{\theta^2} = \text{constant} \tag{4.47}$$

This means that *the speed of conduction of the AP along an unmyelinated axon varies as the square root of the axon diameter, neglecting the resistance of the external medium and assuming the same membrane properties and the same resistivity of the axoplasm in axons of different diameter.*

This is an important result that explains the need for "giant" axons in invertebrates. In the case of the squid, for example, it was mentioned that the giant axon innervates the mantle muscle whose contraction allows the squid to escape danger. To be effective, such an escape mechanism must be fast. Since contraction of the mantle muscle requires APs to propagate along the axon to the mantle muscle, the speed of conduction of APs must be sufficiently high, which in turn requires an axon of relatively large diameter. The speed of conduction for a typical 0.5 mm diameter squid axon is about 20 m/s at 20°C.

As mentioned earlier (Section 4.1.2.1), passive signals also spread along an RC cable structure at a speed that is nominally proportional to the square root of the diameter (Problem 4.2).

It should be mentioned that, in practice, the conditions that axons of different diameters have the same membrane properties and the same resistivity of axoplasm may not be strictly valid, nor can the resistance of the external medium be neglected. It is found in practice that the conduction speed along unmyelinated axons can vary as the diameter raised to a power between 0.7 and 0.8, rather than 0.5. The deviation is most marked in axons of small diameter.

Problem 4.3

Show that ideally, the time constant of an axonal membrane is $\rho_m \varepsilon_m$ independently of the length of the axon, its diameter, or the thickness of the membrane, where ε_m is the permittivity, or dielectric constant, of the membrane.

Problem 4.4

Show that if the resistance of the external fluid is neglected, $\lambda = \dfrac{\sqrt{a}}{2\sqrt{\rho_a / (\rho_m \delta)}}$,

where a is the diameter of the axon, ρ_a is the resistivity of the axoplasm, ρ_m is the resistivity of the membrane, and δ is the membrane thickness.

Problem 4.5

Using Equation 4.16, show that the steady-state input resistance of a cable of infinite length is λr_a, as in Equation 4.31.

Problem 4.6

Using Equations 4.8 and 4.10, with $i_a = -i_e$, show that: $\dfrac{\partial v_e}{\partial x} = -\dfrac{r_e}{r_a + r_e} \dfrac{\partial v_{mr}}{\partial x}$. This equation is usefully applied to the case where a nerve fiber is surrounded by a thin conducting fluid and immersed in an insulating medium. The voltage recorded externally is then a fraction of the AP in accordance with simple voltage division.

Problem 4.7

An axon of diameter a is surrounded by a cylindrical fluid of diameter b. If ρ_m, ρ_a, and ρ_e are the resistivities of the membrane, axoplasm, and external fluid, respectively, show that:

(a) The longitudinal conductance of a length l of the external fluid is:

$$g_e = \dfrac{\pi \left(b^2 - a^2\right)}{4 \rho_e l} \text{ S.}$$

(b) The transverse conductance of a length l of the external fluid per unit length is: $g_{et} = \dfrac{2\pi l}{\rho_e \ln(b/a)}$ S. Note that if the transverse conductance is large compared to the longitudinal conductance, the transverse voltage drop is small compared to the longitudinal drop, so that the extracellular medium can be considered thin, or "confined". Show that this is equivalent to having: $(b^2 - a^2)\ln(b/a) \ll 8l^2$.

(c) The space constant is: $\lambda = \sqrt{\dfrac{\rho_m a(b^2 - a^2)}{4\left(\rho_a(b^2 - a^2) + \rho_e a^2\right)}}$.

Problem 4.8

Argue that if an axon of 10 μm diameter conducts an AP having a rise time of 200 μs at a speed of 5 m/s, the rising phase of the AP is spread over a distance of 1 mm.

Problem 4.9

Show that if one electrode is placed at the surface of an axon and another electrode at the cut end of the axon, the recorded action potential is of the same shape as AP of the axon, but its magnitude is reduced by a factor $\dfrac{g_a}{g_e + g_a}$, where g_a is the conductance of the axoplasm per unit length, and g_e is the conductance per unit length of a thin film surrounding the axon.

Problem 4.10

Consider two unmyelinated axons having the same specific electrical properties and diameters a and αa, where $\alpha > 1$. (a) Show that the space constant $\lambda = 1/\sqrt{g_m(r_a + r_e)}$ varies as $\sqrt{\alpha}$, neglecting r_e. (b) Deduce that the speed of conduction must vary as the square root of the diameter.

Problem 4.11

Consider the two unmyelinated axons of the preceding problem. (a) Assume that a small patch of membrane is being stimulated to threshold by direct application of current across the membrane. Based on the same current density required to excite a patch of membrane, deduce that the threshold will vary as $\alpha\sqrt{\alpha}$, or as $\alpha^{1.5}$. (b) Deduce that if the stimulus is applied externally between two distant electrodes, the threshold will vary inversely with the square root of the diameter.

Problem 4.12

Consider a cable of finite length l. Its normalized length with respect to the space constant λ is the **electrotonic length**. Thus, $L = l/\lambda$. Let the cable be terminated at $X = L$ by an open circuit, so that the longitudinal current i_a at the termination is zero. In the case of axons or dendrites, this is the physiological condition at the termination, where a membrane seal closes off the intracellular medium. Electrically, the membrane seal presents a very high impedance at the end of the RC cable, so that the cable is effectively terminated with an open circuit. Using this boundary condition and assuming that the voltage at $X = 0$ is V_{m0} in the steady state, show that the voltage at any value X, where $0 \leq X \leq L$, is given by $v_{mr}(X) = V_{m0} \dfrac{\cosh(L-X)}{\cosh(L)}$. Argue that the voltage along the cable is less attenuated than in an infinite cable, by comparing the voltages at $X = L$ in both cases.

Problem 4.13

Consider a cable of electrotonic length L, as in the preceding problem, that is short-circuited at $X = L$, so that the voltage at this point is zero. This is the case when an axon is cut, for example, so that the intracellular and extracellular voltages are equal. Show that the voltage at any value X, where $0 \leq X \leq L$, is given by $v_{mr}(X) = V_{m0} \dfrac{\sinh(L-X)}{\sinh(L)}$. Evidently, the voltage along the cable is more attenuated than in an infinite cable because of the current sink termination.

Problem 4.14

Consider a cable of electrotonic length L, as in the preceding problems, that is terminated by a resistance R_T, the voltage across which is V_T. Show that the voltage at any value X, where $0 \leq X \leq L$, is given by $v_{mr}(X) = \dfrac{V_{m0} \sinh(L-X) + V_T \sinh(X)}{\sinh(L)}$. Note that this reduces to the result of Problem 4.13 when $V_T = 0$.

Problem 4.15

Use the result of Problem 4.14 and Equation 4.20 in the form $\left. \dfrac{dv_{mr}(X)}{dX} \right|_{X=L} = -\lambda r_a i_T = -R_{in\infty} \dfrac{V_T}{R_T}$, where $R_{in} = \lambda r_a$ (Equation 4.31), to eliminate V_T between the two equations and obtain, after simplification using hyperbolic functions identities: $v_{mr}(X) = V_{m0} \dfrac{\cosh(L-X) + (R_{in\infty}/R_T)\sinh(L-X)}{\cosh(L) + (R_{in\infty}/R_T)\sinh(L)}$.
This gives the voltage in the steady state at any point X along a cable of electrotonic length L that is terminated by a resistance R_T.

4.2 Propagation of the Action Potential along a Myelinated Axon

In vertebrates, the maintenance of posture and the execution of fast reflexes and rapid voluntary movements requires AP conduction speeds of 100 m/s or more along millions of axons in the periphery and in the spinal cord. To scale up diameters of unmyelinated fibers to achieve such speeds is clearly not feasible. A solution had to be found for achieving higher speeds of conduction of the AP along axons of smaller diameter.

The solution, in the form of myelinated axons, is ingeniously simple and highly effective. The axon is surrounded by a myelin sheath consisting of up to 200 layers or so of passive cell membrane interrupted at regular intervals in what are referred to as the **nodes of Ranvier** (Figure 4.7). The region between adjacent nodes is the **internode**, whose length is roughly 100–150 times the axon diameter and ranges in length between about 200 μm and 2.5 mm, depending on axon diameter. The sheath is wrapped around the axon during embryonic development by specialized satellite cells of the nervous system – the glial cells (Section 1.2.3). In the central nervous system, the glial cells that form the myelin sheath are referred to as **oligodendrocytes**, with each oligodendrocyte forming one internode of myelin for up to about 50 adjacent axons. In the peripheral nervous system, a glial cell referred to as a Schwann cell forms one internode of only a single axon.

The electrical behavior of the axon in the internode is modified in two ways:

1. The axonal membrane is inexcitable, which means that APs cannot normally be generated in the internode, only at the nodes of Ranvier, because the internode regions lack voltage-gated Na^+ channels. Moreover, both central and peripheral myelinated axons have K^+ channels, under the myelin sheath, near the two ends of each internode, whose function is to suppress any action potential that may be generated by the axon membrane under the myelin sheath.

2. Having 200 additional layers of cell membrane around the axon in the internode means that g_m and c_m, the radial conductance and capacitance of the membrane per unit length, become nearly $g_m/200$ and $c_m/200$, respectively.

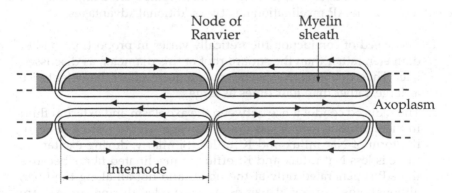

FIGURE 4.7
Propagation of action potential along a myelinated axon.

Consequently, when an AP is generated at a given node, only a small fraction of the local-circuit current crosses the membrane in the internode because of the reduced conductance and capacitance; the bulk of the current flows from the excited node to the adjacent nodes on either side (Figure 4.7). The strength of the current is sufficient to evoke an AP in the node ahead, whereas the node behind had just undergone an AP and is incapable of generating a new AP because of refractoriness. The AP effectively jumps from one node to the next in the direction of propagation, the mode of conduction being referred to as **saltatory conduction**. The reason for the faster speed of conduction along a myelinated axon compared to an unmyelinated axon is not difficult to see. *Assuming that the time required to initiate a new AP is roughly the same in both cases, the new AP is initiated in the immediate neighborhood in an unmyelinated axon but at the much longer internodal distance in a myelinated axon. Effectively, a much longer distance is covered by the AP during the same time, which means a higher speed of conduction or a much smaller diameter of axon for the same speed of conduction.*

It should be noted that the preceding discussion about the AP jumping from one node to the next is oversimplistic in reality and only serves to explain the basis for the higher speed of conduction in a myelinated axon. If one considers a 20 μm fiber conducting an AP at a speed of 100 m/s and having a spike duration of 0.5 ms, then the AP travels a distance of 50 mm in 0.5 ms. Assuming an internodal distance of 2.5 mm, this means that the spike is being regenerated over almost 20 nodes simultaneously.

In discussing myelinated axons, the axon diameter is considered as that of the axon proper, not including the myelin sheath, and is denoted by a. The external diameter, including the myelin sheath, is denoted by d and is referred to as the **fiber diameter**. The fiber and axon diameters would then be the same in unmyelinated axons.

As an indication of the effectiveness of myelination, consider the frog, a vertebrate that is cold-blooded like the squid. A 14 μm fiber diameter of the frog has the same speed of conduction of the AP as a squid giant axon of 0.5 mm diameter but only 1/1200 of the area. Apart from a higher speed of conduction of the AP, myelination has two additional advantages:

1. The speed of conduction theoretically varies in proportion to fiber diameter, rather than the square root of this diameter, as discussed later. Thus, doubling the speed would require a fiber diameter twice as large, rather than four times as large.

2. The reduced conductance between the axoplasm and external fluid in the internode means less ionic "leakage" in the resting state, in the form of Na^+ influx and K^+ efflux. Similarly, during excitation, there is less Na^+ influx and K^+ efflux in myelinated fibers because the AP is generated only at the nodes and not all along the fiber, although the current densities at the nodes during an AP are

about an order of magnitude larger than in unmyelinated fibers. As a result, ionic movements in a frog myelinated fiber of 14 μm diameter are about 1/5000 of those in a squid unmyelinated fiber of 0.5 mm diameter. This means that much less energy has to be expended in myelinated axons by the sodium-potassium pump in maintaining steady-state ionic concentrations and in restoring them after excitation.

If the conduction speed of the AP is proportional to the fiber diameter of a myelinated axon and to the square root of the diameter of an unmyelinated axon, and if the proportionality constant has the same numerical value when diameters are expressed in micrometers, then the conduction speed will be the same if both types of axon had a fiber diameter of 1 μm, since $\sqrt{1} = 1$. For fibers of smaller diameter, say 0.8 μm, the unmyelinated fiber has a faster speed of conduction, since $\sqrt{0.8} > 0.8$. Conversely, a myelinated fiber of diameter larger than 1 μm will have a faster speed of conduction. However, there is experimental evidence that the critical diameter at which the speed of conduction is the same for myelinated and unmyelinated fibers is about 0.2 μm. In practice, the largest unmyelinated fibers in vertebrates have diameters that are only slightly larger than 1 μm. The smallest fiber diameters of myelinated fibers are about 1 μm in the periphery and about 0.2 μm in the central nervous system. It should be noted that in the peripheral nervous system, axons are primarily required to conduct APs reliably and at maximum speed for a given diameter. On the other hand, axons in the central nervous system may be subject to other considerations such as timing relationships and minimization of energy expenditure. The area of the node of Ranvier, which decreases with fiber diameter, sets a lower limit on the diameter of myelinated fibers, so if the diameter of a myelinated fiber is too small, the node will contain only a small number of voltage-gated channels, which detracts from the reliability of generation of the AP.

It can be argued qualitatively that for a given fiber diameter, there is an optimum thickness of the myelin sheath that maximizes the speed of conduction along the axon. If the sheath is too thick, the axon diameter is reduced, and the internal resistance of the axon, r_a, is increased, which slows down the rate of change of voltage along the internode and reduces its amplitude. This also delays the generation of the AP at the node. Both of these effects reduce the speed of conduction. But a sheath that is too thin also decreases the speed of conduction by having larger capacitance and inductance per unit length, which increases the current leakage in the internode region. The effect of an increase in r_a in reducing the speed of conduction dominates when the sheath is too thick, whereas the effect of an increase in c_m and g_m in reducing the speed dominates when the sheath is too thin. The result is an optimum thickness of the myelin sheath, for a given fiber diameter, that maximizes the speed of conduction.

Another optimization condition applies to the ratio l/d, or l/a, assuming that the ratio d/a has been optimized, as discussed in the preceding paragraph. An l that is too small sacrifices some of the advantage of myelination and reduces the speed of conduction by reducing the distance over which the AP "jumps". If l is too large, the axial resistance of the internode increases together with the total internodal capacitance and conductance of the membrane, which reduces the speed of conduction, as discussed in the preceding paragraph. Moreover, if l is too large and the generation of the AP at a given node is blocked for some reason, the strength of the current at the next node further along in the direction of propagation may not be sufficient to generate an AP at this node. This reduces the safety margin for propagation of the AP. In practice, blocking the generation of the AP even at two adjacent nodes does not stop the propagation of the AP. It may be noted that the effect of local anesthetics is to inhibit the activation of the voltage-gated sodium channels, thereby blocking the propagation of APs along fibers that conduct pain signals to the central nervous system.

In practice, the aforementioned optimization conditions apply reasonably closely to real myelinated fibers. The ratio a/d ranges between 0.55 and 0.74, and the ratio l/d ranges between 75 and 150, around a broad maximum in both cases, as indicated by computer simulations. The ratios a/d and l/d are 0.71 and 143, respectively, for the frog myelinated fiber having $a = 10$ μm, $d = 14$ μm, and $l = 2$ mm. The broadness of the maxima allows for considerable variations in the a/d and l/d ratios between species, and between nerve fibers of different diameters in any given species, without appreciable effect on the maximum speed of conduction. The l/d ratio for fibers in the central nervous system is generally smaller than that for peripheral fibers.

An interesting consequence of the constancy of l/d is that as the animal grows, both l and d increase. But if the speed of conduction is proportional to d and l/d remains constant, the conduction time of the AP from the periphery to the central nervous system remains the same. This means that the basic timing relationships for the conduction of APs remain the same and need not be modified with growth, both for automatic activities such as breathing and maintenance of posture as well as learned voluntary movements.

It should be mentioned that an unmyelinated axon is not "bare" but is usually surrounded by non-myelinating Schwann cells. The squid giant axon, for example, is enveloped by several Schwann cells, with narrow channels between adjacent Schwann cells connecting the immediate exterior of the axon to the extracellular medium. In some cases, several small, unmyelinated fibers may be embedded in one Schwann cell.

There is evidence of glial cell-axon interactions in the process of myelination. Before myelination occurs during embryonic development, the voltage-gated Na^+ channels are uniformly distributed along the axon. The oligodendrocytes induce clustering of the Na^+ channels at the nodes of Ranvier, leaving the internodal regions inexcitable. The locations of the nodes of Ranvier and the thickness of the myelin sheath seem to be regulated by the axon. This is not surprising, since an oligodendrocyte myelinates several axons of different diameters with internodes having different lengths and thickness of the myelin sheath depending on axon diameter. Moreover, following a cut of a peripheral axon, the regenerating axonal stump follows a path that is guided by surviving Schwann cells.

4.2.1 Quantitative Considerations

The myelin sheath can be considered as layers of membrane that are tightly packed on top of one another, with no axoplasm in between, so that the myelin sheath is effectively a single membrane that extends from a radius of $a/2$ to a radius of $d/2$. If the membrane resistivity is ρ_m, the transverse resistance of a cylindrical shell of internodal length l, radius r, and thickness dr is $\dfrac{\rho_m dr}{2\pi rl}$. The total transverse resistance is $\displaystyle\int_{a/2}^{d/2} \dfrac{\rho_m dr}{2\pi rl} = \dfrac{\rho_m}{2\pi l}\ln\left(\dfrac{d}{a}\right)$ and the conductance g_m of a unit length of the membrane is:

$$g_m = \frac{2\pi}{\rho_m}\Big/\ln\left(\frac{d}{a}\right) \tag{4.48}$$

The resistance per unit length of the axoplasm, r_a, equals $4\rho_a/\pi a^2$. Substituting in the expression for the space constant λ, assuming $r_e = 0$:

$$\lambda^2 = \frac{\rho_m}{8\rho_a}a^2\ln\left(\frac{d}{a}\right) \tag{4.49}$$

For a given d, it is of interest to determine a that maximizes λ. Setting $d\lambda/da = 0$, gives:

$$\ln\left(\frac{a}{d}\right) = -\frac{1}{2}, \text{ or } a = d/\sqrt{e} = 0.61d \tag{4.50}$$

It can be shown (Problem 4.3) that the time constant of the membrane is $\rho_m \varepsilon_m$, where ε_m is the permittivity of the membrane and is independent of l, a, and d. It follows that for the same time constant, maximizing λ maximizes the spread of voltage along the internode and hence the conduction speed. According to Equation 4.50, conduction speed is maximized if $a = 0.61d$, assuming an ideal cable, negligible r_e, and ρ_m and ρ_a that are independent of a. The preceding analysis also neglects the effect of the capacitance at the node, which loads the internode cable. If the effect of this capacitive loading is taken into account assuming that the capacitance of the node is 0.4 times that of the internode, the optimal value of a/d becomes about 0.7.

Computer simulation of propagation along myelinated fibers indicates that the speed of conduction has a broad maximum at $a/d \cong 0.7$ and decreases to about 98% of maximum at $a/d = 0.6$ and 0.75, and to 95% of maximum at

$a/d = 0.55$. Computer simulation also indicates that the speed of conduction has a broad maximum at $l/d \cong 110$, and decreases to about 95% of maximum at $l/d = 60$ and 180.

It will be argued next that the speed of conduction of the AP for myelinated axons is directly proportional to fiber diameter, based on the following assumptions:

1. The specific properties of myelinated axons are the same, irrespective of fiber diameter.
2. The ratio d/a is constant, irrespective of fiber diameter.
3. The ratio l/d is constant, irrespective of fiber diameter.

As mentioned previously, the last two assumptions are in accordance with experimental observations, and are optimized with a broad maximum, as indicated by computer simulations.

Equation 4.12 applies over the internode region bearing in mind that, in the case of the unmyelinated axon, C_m and G_m are independent of axon diameter assuming the same properties and thickness of the axonal membrane. In the case of the internode region of a myelinated axon, the membrane conductance and capacitance per unit area depend on the thickness of the myelin sheath and hence the fiber diameter. With c_m and g_m both being inversely proportional to $\ln(d/a)$ (Equation 4.48 and Problem 4.16), and in view of Assumption 2, c_m and g_m are independent of fiber diameter. Hence, the coefficient c_m/g_m in Equation 4.12 is a constant independent of fiber diameter. It equals, in fact, the membrane time constant $\rho_m \varepsilon_m$ (Problem 4.16). With $r_e = 0$, the coefficient $1/g_m r_a$ in Equation 4.12 equals $\pi a^2/4g_m \rho_a$ and is a constant multiplied by a^2, or a constant multiplied by l^2, in accordance with Assumptions 2 and 3. Equation 4.12 for the internode region becomes:

$$K_1 l^2 \frac{\partial^2 v_{mr}}{\partial x^2} = K_2 \frac{\partial v_{mr}}{\partial t} + v_{mr} \qquad (4.51)$$

where K_1 and K_2 are constants independent of fiber diameter. If a dimensionless space variable is defined as $X_i = x/l$, with respect to the internode length l, Equation 4.51 becomes:

$$K_1 \frac{\partial^2 v_{mr}}{\partial X_i^2} = K_2 \frac{\partial v_{mr}}{\partial t} + v_{mr} \qquad (4.52)$$

According to Equation 4.52,

$$v_{mr} = f(X_i, t), \quad 0 \le X_i \le 1, \qquad (4.53)$$

where $X_i = 0$ at the beginning of an internode in the direction of propagation, and $X_i = 1$ at the end of the internode. Under conditions of steady propagation, the time course of v_{mr} is the same at all points having the same X_i along the internodes of a given fiber, that is, v_{mr} is repetitive over successive internodes. According to Equation 4.53, v_{mr} also has the same time course at the same X_i along internodes of fibers having different diameters, since v_{mr} is a function of only X_i and t independently of fiber or axon diameter. This means that, for all myelinated fibers, irrespective of fiber or axon diameter: (i) the voltage drop across the internode region is the same, and (ii) the propagation delay τ_{int} of the membrane voltage over the internode region is the same. Moreover, if the time course of v_{mr} is the same over internodes, irrespective of fiber and axon diameter, then the time course of the AP at the nodes must also be the same. Otherwise, the voltages at the beginning of internodes will not be the same. The same time course of the AP at the nodes implies that the current density J_m at the nodes is independent of fiber or axon diameter. It follows that the delay τ_n in the generation of the AP at the nodes is the same, irrespective of fiber or axon diameter. Several deductions follow from these considerations:

1. The speed of conduction of the AP along a myelinated axon can be expressed as:

$$\theta = \frac{l+w}{\tau_{int} + \tau_n} \qquad (4.54)$$

   where w is the width of the node. With $w \ll l$, and $(\tau_{int} + \tau_n)$ constant, θ is proportional to l and hence to d because of Assumption 3. It follows that *the speed of conduction of the AP along a myelinated fiber is proportional to fiber diameter, assuming the same specific membrane properties and constant ratios of d/a and l/d.*

2. If the voltage drop across the internode is the same independently of a, then the longitudinal current i_l in the axoplasm is proportional to the conductance of the axoplasm in the internode region, neglecting the small leakage current. As the conductance is proportional to a^2/l, it follows that:

$$i_l \propto a^2 / l, \quad \text{or} \quad i_l \propto a \qquad (4.55)$$

   since l/a is independent of a.

3. The node current i_n equals the difference between the longitudinal currents in the axoplasm at each end of the node. Hence, i_n must also be proportional to a:

$$i_n \propto a \qquad (4.56)$$

4. With the current density at the node being the same, as argued previously,

$$i_n \propto aw \qquad (4.57)$$

   where w is the width of the node.

5. If i_n is proportional to a (Equation 4.56), it follows from Equation 4.57 that w is constant. w is, in fact, about 1–2 µm for peripheral fibers of different diameters but could be wider for fibers in the central nervous system.

As in the case of unmyelinated axons, the assumptions made in the preceding analysis do not hold exactly in practice. The spike duration is found to vary nearly inversely with the conduction speed over a range of approximately 0.32–0.60 ms at 37.1°C. However, the spike duration of most fibers lies between 0.4 and 0.5 ms at 37°C. The spike rise time varies inversely with the conduction speed for speeds below about 40 m/s and is almost independent of the speed at higher speeds. The spike fall time decreases almost linearly with conduction speed. There is also evidence that the conduction time across the internode region is longer for smaller-diameter fibers than for larger-diameter fibers, as would be predicted from the variation of spike rise and fall times with conduction speed and hence with fiber diameter. This variation in spike rise and fall times may reflect differences in the specific properties of myelinated fibers of different diameters. It is found in practice that conduction speed of the AP is nearly proportional to fiber diameter over a wide range of diameters, although in some cases the speed of conduction was found to increase nearly as the diameter raised to the power 1.5.

Problem 4.16

Show that the capacitance per unit length in the internode region is: $c_m = 2\pi\varepsilon_m \Big/ \ln\left(\dfrac{d}{a}\right)$, where ε_m is the permittivity of the myelin sheath. Note that $c_m/g_m = \rho_m\varepsilon_m$, as in Problem 4.3.

4.3 Properties of the Propagating Action Potential

4.3.1 Threshold

We have seen in Section 3.3.2 that at the threshold of the AP under space clamp, the inward current equals the outward current. Although the membrane is depolarized, there is no current flow in the longitudinal direction because of the space clamp. In the propagating AP, there is no space clamp, which means that as soon as a patch of membrane is depolarized current would flow to neighboring regions of the axon.

Consider two identical patches of axonal membrane that are excited just beyond threshold; one patch is in a space-clamped axon, the other patch in an identical axon that is not space-clamped. The inward current in the latter case is not only required to just exceed the outward current in order to trigger the regenerative cycle, as in the former case, but should also supply longitudinal current to neighboring regions of the axon. This means that the stimulus should be larger, so that the threshold of the propagating AP is higher than that of the AP under space clamp.

A peripheral nerve in a limb may include several thousand fibers of different diameters. If an external stimulus is applied to such a nerve, and the stimulus strength is increased from a low value, which fibers will be excited first, the larger-diameter fibers or the smaller-diameter fibers? Assume that all fibers have the same specific electrical properties and that the stimulating electrodes are sufficiently far apart so that the stimulating current density is uniform across the nerve in the region of stimulation. Consider two fibers, the larger fiber having an axon diameter that is α times that of the smaller fiber, where $\alpha > 1$. If both fibers are to be excited at the same time, the current density at the nodes must be the same. Assuming the width of the node is roughly the same, the current through the node will therefore have to be α times in the larger fiber, which requires that the longitudinal current varies in proportion to α. But the longitudinal current in the larger fiber varies as α^2 of that in the smaller fiber, assuming the same current density. Hence, as the stimulus strength is increased from a low value, the internal current will reach the value required for excitation in the larger fiber before it does in the smaller fiber, and the larger fiber will be excited first. In practice, the proximity to the stimulating electrode affects the current through the fibers, but it remains generally true that, in response to an *external stimulus*, larger-diameter fibers are excited before smaller diameter ones, in both myelinated and unmyelinated fibers. However, the opposite is true in the case of a stimulus applied directly across the membrane, because a larger current is then required for a larger-diameter fiber to give the same current density in the node (Problem 4.17).

An aspect of propagation of the AP related to threshold is that if the stimulus applied to an axon is subthreshold, the change in membrane voltage in the vicinity of the stimulating electrode decays rapidly with distance along the axon. A recording electrode that is at least several space constants away from the point of stimulation will not record any significant change in membrane voltage. On the other hand, if the stimulus is suprathreshold, a full-fledged AP is recorded. The recorded response is therefore described as all-or-none.

4.3.2 Effect of Temperature

It was argued in Section 3.3.4 that the main effect of a rise in temperature on the membrane AP is to increase the rate of depolarization of the spike voltage and decrease its amplitude and duration as well as increasing the threshold. In the case of the propagating AP, a faster rate of depolarization of the spike will shorten the time required to initiate an AP. In opposition, a smaller spike amplitude reduces the magnitude of the current at the node, and a shorter duration reduces the depolarization caused by a given magnitude of current at the node. These opposing effects are such that the speed of conduction of the AP increases with a rise of temperature from an initially low value, reaches a maximum, and then decreases with temperature. At a high enough temperature, the current at the node will not be sufficient to generate an AP and conduction ceases. In the case of the squid giant axon of 0.5 mm diameter, the speed of conduction increases from about 9 m/s at 6.3°C, to 20 m/sec at 20°C, to a maximum of approximately 30 m/s at about 33°C. Conduction ceases at about 38°C to 40°C. The amplitude of the AP falls slowly at first from about 85 mV at 5°C to about 80 mV at to 20°C more rapidly thereafter to 63 mV at 30°C. The duration of the AP decreases from about 2.7 ms at 6°C to about 1 ms at 18.5°C.

In mammalian myelinated nerve fibers, experimental data indicate a Q_{10} of approximately 1.6 for the speed of conduction in the temperature range 27°–37°C, that is, the speed of conduction is multiplied by this factor over the specified 10°C increase in temperature.

Problem 4.17

Consider two *myelinated* axons having the same specific electrical properties and diameters a and αa, where $\alpha > 1$. Deduce that the threshold varies as: (a) α, for direct application of a stimulus across the membrane; (b) $1/\alpha$ for stimulation with distant external electrodes.

4.3.3 Active vs. Passive Propagation

It was argued in Section 3.3 that the AP is an active response. Because of the relatively large series resistance of the axoplasm and the relatively large shunt conductance and capacitance of the membrane, passive propagation along axons is not feasible except over short lengths as in dendrites (Section 1.2 Chapter 1) or internodes where the shunt conductance and capacitance are much reduced. Table 4.1 compares the cable properties of a squid giant axon at 20°C, myelinated fiber of the frog at 20°C, and a typical electrical coaxial cable. In a coaxial cable, passive propagation is feasible because of a

TABLE 4.1

Comparison of Cable Properties ($T = 20°C$)

Cable Property	Squid Giant Axon	Frog Myelinated Fiber (internode)	Electrical Coaxial cable
Outer diameter, mm	0.5	0.014	0.9
Shunt conductance, g_m, mS/cm	0.1	2.3×10^{-5} to 3.7×10^{-5}	5×10^{-13}
Shunt capacitance, c_m, µF/cm	0.15	10×10^{-6} to 16×10^{-6}	10^{-6}
Series resistance, r_a, Ω/cm	2×10^4	140×10^6	2.5×10^{-4}
Space constant, $\lambda = 1/\sqrt{g_m r_a}$, cm	0.7	0.4–0.6	2.8×10^9
Time constant c_m/g_m, ms	1.5	0.27–0.7	2×10^6
Conduction speed, m/s	20	20	2×10^8

large space constant of about 2800 km compared to a fraction of 1 cm for the squid giant axon and the internode region of the frog nerve, which necessitates an active mode of propagation along axons. In this mode of propagation, the AP is regenerated all along an unmyelinated axon but only at the nodes of myelinated axons.

Active propagation differs from passive propagation in several important respects:

1. In passive propagation, the signal is attenuated as it propagates because of power losses in the cable, since these losses can only be supplied by the power associated with the signal. In contrast, an actively propagating signal is not attenuated because it is regenerated as it propagates along a myelinated or an unmyelinated axon.

2. The shape of an actively propagating signal is predetermined by the properties of the signal-generating mechanism. In passive propagation, the shape of the propagating signal is that of the applied signal modified by attenuation and frequency response of the medium of propagation. *Since the actively propagating AP is of fixed shape, information can only be coded in the spatiotemporal patterns of excitation, that is, in the spatial distribution of fibers carrying the APs and in the timing of successive APs along these fibers.*

3. The speed of an actively propagating AP is limited by the time it takes an AP to generate a new AP. The speed in a passive cable is determined by the dielectric constant and magnetic permeability of the insulating medium. Thus, passive propagation is generally much faster than active propagation (Table 4.1). The relative slowness of active propagation along an axon, entails delays that can be in the tens of milliseconds, and which the nervous system must cope with in its control and communications functions.

4. The highest signal frequency in passive propagation is determined by the passive electrical elements of the cable, that is, the series impedance and the shunt admittance, whereas in the case of the AP, the highest frequency is limited by refractoriness. The maximum frequency of APs along an axon is nearly the reciprocal of twice the absolute refractory period. This gives approximately 1000 APs/s for nerve fibers having an absolute refractory period of about 0.5 ms.

5. An important advantage of active propagation over passive propagation is the absence of reflections at terminations and branch points. The AP simply dies out at an axon termination because refractoriness does not allow back propagation in the wake of an AP, as discussed earlier. Similarly, when an AP reaches a branch point of an axon, the local-circuit currents flow into both branches and evoke APs in each branch as long as the branches are not too large compared to the parent trunk. Again, no AP is possible in the wake of the AP because of refractoriness, so no reflection takes place at the branch point.

However, some propagation delay is introduced by the branching. In some experimental situations, as in the case of the H-Reflex (Section 11.4.1), two APs may travel in opposite directions along a given axon. The APs annihilate one another when they collide because neither AP can propagate through the refractory region of the other AP.

Summary of Main Concepts

- The AP propagates as a result of excitation due to local-circuit currents. The new excitation occurs in the adjacent region in the direction of propagation in unmyelinated axons and in the next node of Ranvier in myelinated axons. Considering that the time to generate an AP is roughly the same in both cases, the speed of conduction is much larger in myelinated axons.

- Theoretically, the speed of conduction of the AP is proportional to the square root of the diameter of unmyelinated axons, assuming the same membrane properties, the same resistivity of the axoplasm in axons of different diameter, and neglecting the resistance of the external medium.

- It is found that the ratios d/a, l/d, and the width of the node of Ranvier are roughly constant in myelinated fibers of different diameters.

- Assuming the same specific properties of myelinated axons, irrespective of fiber diameter, and constant ratios d/a and l/d, the speed of conduction of the AP along myelinated axons is proportional to fiber diameter.

- Myelination reduces the ionic fluxes per unit length of the fiber in both the resting state and during the course of the AP.

- Compared to passive propagation, active propagation is slow, allows only a signal of fixed size and shape (but without attenuation), requires a distributed source of energy, and causes no reflections at terminations or branch points.

- When a current step is applied to a semi-infinite cable, the rate of change of voltage with respect to time, at the point of application of the current stimulus, tends to infinity, and the voltage increases toward its steady-state value at a faster rate than an exponential. The rise of voltage, at points away from the point of stimulation, is progressively delayed because of propagation time.

- When a current impulse is applied to a semi-infinite cable, the voltage, at the point of application of the current stimulus, decreases with time at a faster rate than an exponential because of charge leakage along the cable.

5

The Neuromuscular Junction

Objective and Overview

Having discussed the generation and propagation of the action potential (AP), the next step is to consider what the AP does when it reaches the terminals of the axon. In most cases, the AP causes the release of neurotransmitters at chemical synapses – these being specialized structures between neurons and their target cells, such as other neurons or skeletal muscle cells. The neuromuscular junction (NMJ), which is a special type of chemical synapse between the axon terminations of α-motoneurons and skeletal muscle fibers, has provided much of the basic understanding of operation of chemical synapses, as a result of extensive investigations on the frog neuromuscular junction and the crayfish neuromuscular junction, as well as the squid giant synapse that gives rise to the squid giant axon discussed in Chapter 3. The neuromuscular junction is therefore discussed in detail including its morphology, its operation, the statistics of ACh release, the structure and kinetics of the ACh channel, the generation of the endplate current and endplate voltage, and the effects of various chemical substances. This provides much of the foundation for the discussion of synapses in the nervous system in the following chapter.

Learning Objectives

To understand:

- The structure and basic operation of the neuromuscular junction
- The statistics of neurotransmitter release
- The basic structure of the acetylcholine receptor of the neuromuscular junction

- Channel kinetics, both at the macroscopic level and at the level of single channels, including channel desensitization
- How a muscle action potential is generated by endplate currents and by the endplate voltage
- How the normal operation of the neuromuscular junction can be interfered with by various chemical agents

5.1 Structure

As the axon of an α-motoneuron approaches the muscle it innervates, it divides into short branches, each of which terminates on a muscle fiber in a number of bulbous terminals that in mammals look like a bunch of grapes (Figure 5.1). Each bulbous terminal, termed a **synaptic bouton**, forms a synaptic structure known as the **neuromuscular junction**, or **myoneural**

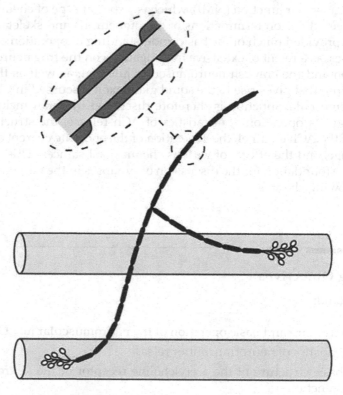

FIGURE 5.1
Terminations of an α-motoneuron axon on muscle fibers.

junction. The synaptic bouton loses the myelin sheath of the axonal branch but remains covered by a Schwann cell.

Although the structural details of the NMJ vary between different verte-brate species or even between different muscles in the same species, a "typi-cal" ultrastructure of a mammalian NMJ is illustrated in Figure 5.2.

The synaptic bouton becomes embedded in a small, shallow depression in the muscle, where it is closely apposed to the muscle membrane but sepa-rated from it by a **synaptic cleft** (Figure 5.2). The muscle membrane in the synaptic region is termed the **endplate**, or the **motor endplate**. The synap-tic bouton contains an abundance of mitochondria and **synaptic vesicles** of 40–50 nm diameter that are filled with the neurotransmitter acetylcholine (ACh). The endplate is folded into many troughs about 500 nm deep and about 100 nm wide separated by crests of about the same width. The width of the synaptic cleft at the crests of the endplate is about 20–60 nm. Facing the opening of the troughs are thickened regions of the synaptic bouton referred to as **active zones**. These zones have a high concentration of vesicles, with many of these vesicles touching the inner side of the membrane of the syn-aptic bouton.

The cleft contains a **basement membrane** shown in Figure 5.2 as a line parallel to the membrane of the endplate. The basement membrane is com-posed of collagen and other proteins but does not constitute a barrier to the diffusion of small molecules and ions in the cleft. At the edges of the NMJ, the basement membrane merges with both the basement membrane

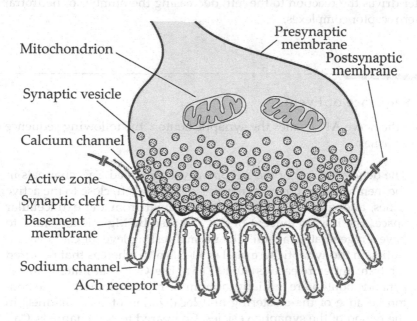

FIGURE 5.2
Ultrastructure of neuromuscular junction.

that surrounds the muscle fiber and the basement membrane covering of the Schwann cell.

There are about 10^7–10^8 ACh receptors on the postsynaptic membrane in the mammalian NMJ, each receptor being essentially a protein molecule embedded in the membrane of the endplate. The concentration of ACh receptors is at a maximum of 15,000–30,000/μm^2 in the crest region and tapers off sharply into the folds. The depths of the folds contain voltage-gated Na^+ channels required for the generation of action potentials. The folds of the postsynaptic membrane greatly increase the surface area, allowing for a large number of ionic channels.

In understanding the action of a neurotransmitter at a receptor site, it is important to bear in mind that, in general, the binding of a neurotransmitter to a receptor is reversible:

$$\text{Neurotransmitter} + \text{Receptor} \rightleftharpoons \text{Neurotransmitter} - \text{Receptor Complex} \quad (5.1)$$

The neurotransmitter thus binds to the receptor, and the neurotransmitter-receptor complex dissociates by the unbinding of the neurotransmitter from the receptor due to random thermal motion of the neurotransmitter and the receptor atoms at the binding site. A higher concentration of neurotransmitter in the medium surrounding the receptor increases the number of neurotransmitter-receptor complexes, that is, drives the process described by Reaction 5.1 more to the right, whereas a lower concentration of neurotransmitter drives the reaction to the left, decreasing the number of neurotransmitter-receptor complexes.

5.2 Sequence of Events

When the nerve AP invades the synaptic bouton, the following sequence of events ensues:

1. The depolarization of the AP opens voltage-gated Ca^{2+} channels in the membrane on the presynaptic side of the cleft, close to the active zones, allowing Ca^{2+} to enter the terminal from the extracellular space. The inward current of Ca^{2+}, blocked by Mg^{2+}, is too small to have a regenerative action. The normal resting level of Ca^{2+} concentration in the synaptic terminal is 50–100 nM, whereas that required for transmitter release is 10–100 μM. The Ca^{2+} that enter into the synaptic bouton are, initially, not uniformly distributed in the bouton because of the clustering and localization of Ca^{2+} channels in the region of the synaptic vesicles. Compared to Na^+ channels, Ca^{2+} channels open more slowly and inactivate more slowly as well.

2. The influx of Ca^{2+} causes the synaptic vesicles, whose membrane is identical to that of the axonal membrane, to fuse with the axonal membrane facing the synaptic cleft, thereby releasing ACh into the cleft by exocytosis (Section 1.1.4). The contents of 30–300 vesicles are released by the AP in a synaptic bouton, depending on the concentration of Ca^{2+} and the NMJ involved, each vesicle containing between several thousand and 10,000 ACh molecules. The contents of a vesicle are considered to be the **quantal unit**, or quantum, of transmitter release. The number of vesicles released per AP is termed the **quantal content**. It can be shown that about 6000 ACh molecules per vesicle would be nearly isotonic with the intracellular medium (Problem 5.1). The time between the entry of Ca^{2+} and neurotransmitter release is 0.2 ms or less. The relationship between depolarization and transmitter release is S shaped. Below a certain threshold depolarization, no transmitter is released. Above threshold, transmitter release increases very rapidly with depolarization, then saturates at a level beyond which neurotransmitter release does not increase with depolarization. Exocytosis first involves the formation of a channel, or fusion pore, about 5 nm wide that spans the membrane of the vesicle and the presynaptic membrane. This channel then rapidly expands in diameter to about 50 nm when fusion is complete.

3. The ACh molecules diffuse through the synaptic cleft toward ACh receptors on the muscle membrane. Not all the ACh molecules released into the synaptic cleft will reach the receptors. Some will diffuse sideways away from the endplate region, and a good number will be hydrolyzed by the enzyme ACh-esterase (AChE) in the synaptic cleft, as described later. An estimated 1000–2000 ACh molecules per vesicle arrive at the endplate. Diffusion time across the narrow cleft is typically in microseconds.

4. The binding of two ACh molecules to the receptor causes a conformational change in the receptor protein, which opens an ion channel that is impermeable to anions but is wide enough to allow several cationic species to pass through. However, because of the relatively high concentrations of Na^+ and K^+, the ion channel current is mainly due to these ions, which pass through the channel with almost equal permeability.

5. Although the Na^+ current is inward, whereas the K^+ is outward, the net effect is an inward **endplate current** (epc) because of the much larger electrochemical potential difference driving Na^+ inward. The epc depolarizes the endplate resulting in a voltage commonly referred to as the **endplate potential** (epp). The time courses of the epc and the epp are considered in more detail in Sections 5.5.1 and 5.5.2, respectively.

6. The depolarization produced by the epp activates the voltage-gated Na^+ channels well beyond threshold resulting in a muscle AP that

invades the whole muscle fiber in vertebrates, producing a single contraction, or twitch, as described in Section 9.2.1.

7. The synaptic cleft contains the enzyme ACh-esterase (AChE), which breaks down ACh into choline and acetic acid. The AChE molecules, synthesized mainly by the muscle cell, are tethered to the muscle side of the basement membrane at a relatively high concentration of about 0.2 mM. Note that the active site of an enzyme such as AChE can successively hydrolyze a large number of ACh molecules. Following the release of ACh into the synaptic cleft, the concentration of free ACh rises and then falls rapidly, within a fraction of a millisecond, because of binding of ACh to the receptors and breakdown of unbound ACh by AChE.

The NMJ is an example of an **obligatory synapse**, or a **non-integrating synapse**, since a single AP invading the synaptic bouton is sufficient to elicit a muscle AP. Because of this, it may be thought that the nerve terminal could have been connected directly to the muscle fiber without the need for a NMJ. However, the big mismatch between an axon terminal of diameter not exceeding few micrometers and a muscle fiber of 30–100 μ diameter having about 5 times the capacitance per unit area (Section 9.1.2) implies that, if all the charge due to the AP in the axon terminal were deposited directly on the muscle fiber, the resulting depolarization would be quite small, well below threshold (Problem 5.2). The effect of the nerve AP is greatly amplified by the combined action at the endplate of the large number of ACh molecules released. However, the NMJ introduces a "weak link" in that it can be adversely affected by many factors, as described later.

A number of "housekeeping operations" takes place in the aftermath of the AP. Following exocytosis of ACh and the fusing of the vesicular membrane with the presynaptic membrane, parts of the presynaptic membrane close to the release site pinch off by endocytosis (Section 1.1.4) and reform vesicles that are subsequently filled by active transport with newly synthesized ACh. Choline resulting from breakdown of ACh by AChE in the synaptic cleft is reabsorbed into the presynaptic terminal and used in the synthesis of new ACh, whereas the acetyl portion is derived from acetyl coenzyme A (acetyl-coA) that is found between the inner and outer membranes of mitochondria. The mitochondria, which are abundant in the presynaptic terminal, also play an essential role in the supply of ATP and in the uptake of Ca^{2+}, as mentioned previously (Section 1.1.2). ATP is believed to be required to "prime" the vesicles to make them ready to accept Ca^{2+} and fuse rapidly with the presynaptic membrane. It may be noted that at low rates of transmitter release, exocytosis may not be complete. That is, the neurotransmitter is released while the fusion pore described under item 2 above has not dilated fully. The pore then recloses, rapidly reforming the vesicle, without endocytosis or bulk retrieval of vesicles, which predominate at higher rates of transmitter release.

The events just described at the NMJ are entirely under local control over the short term. However, over the long term, two-way communication with the "center", that is, the cell body, through axonal transport (Section 1.2.2), is essential for many operations, such as the degradation of "worn out" vesicles or subsynaptic membrane and cell organelles, including mitochondria, and their replacement by new parts as well as the supply of enzymes and various other substances required for cell function.

Following the depolarization-induced influx of Ca^{2+} into the presynaptic terminal, the intracellular concentration of Ca^{2+} is eventually restored to its low resting level by a Ca^{2+} pump and a Ca^{2+}-Na^+ exchanger (Section 2.3) in the presynaptic terminal.

The quantal content decreases with repetitive activity, the decrease being more marked for fast muscle fibers compared to slow muscle fibers (Section 9.3.2). Voluntary endurance exercise increases quantal content and decreases the reduction of quantal content with repetitive activity.

Problem 5.1

Show that a concentration of about 6000 ACh molecules in a spherical vesicle of 50 nm diameter will result in an osmolarity of about 0.3 osmoles, which is almost isotonic with the intracellular medium.

Problem 5.2

Assume that a large, depolarized axon terminal at the NMJ can be represented as a cylinder of 5 μm diameter, 10 μm in length, having a capacitance of 1 μF/cm², and that the muscle fiber can be represented as a cylinder of 30 μm diameter, 100 μm in length, having a capacitance of 5 μF/cm². Calculate the resulting depolarization of the muscle membrane if the axon cylinder is depolarized to 70 mV and instantly connected to the muscle fiber.

ANS.: Approx. 0.124 mV.

SPOTLIGHT ON TECHNIQUE: FREEZE-FRACTURE AND CRYO-ELECTRON MICROSCOPY

Freeze-fracture electron microscopy was used in the late 1960s to reveal structural details of the cell membrane and cell organelles. It is less commonly used nowadays, but is of interest, nevertheless, because of the informative intricacies involved. The main steps of freeze-fracture are:

1. Rapid freezing of a suitably mounted specimen by immersion in liquid nitrogen (–196°C), which instantly immobilizes cell components. Unless ultra-rapid freezing is used, the standard

method of plunge freezing produces, in most biological specimens, ice crystals that damage ultrastructure. This can be
avoided by treating the specimen with a cryoprotectant, such
as glycerol. Because the cryoprotectant may produce structure
artifacts, the specimen is first fixated with glutaraldehyde.

2. Fracturing the frozen specimen under vacuum using a liquid-
nitrogen-cooled microtome blade or some hinged device. An
essential feature of freeze fracturing is that the fracture tends
to split the lipid bilayers of frozen cell membranes, leaving
one layer attached to the cytoplasmic side and another layer
attached to the external side of the membrane.

3. Making a replica of the frozen-fractured surface, by first evapo-
rating, under vacuum, a fine layer of platinum-carbon onto the
specimen, followed by a backing carbon layer. The topographi-
cal features of the frozen-fractured surface are thus replicated
by variations in thickness of the shadowing layer of deposited
platinum.

4. Removal of biological material from the replica by applying
some cleaning agent, after bringing the replica to atmospheric
pressure and room temperature.

5. Examining the replica in a transmission electron microscope.

The resulting electron micrograph reveals three-dimensional details
of membrane structure at macromolecular resolution. When applied to
the neuromuscular junction, cleavage can occur along the lipid bilayers of the presynaptic membrane or the postsynaptic membrane, which
reveals some interior ultrastructures involved in neuromuscular transmission. Figure 5.3 illustrates a reconstruction of the presynaptic and
postsynaptic membranes of the frog neuromuscular junction using the
freeze-fracture technique. The two halves of the presynaptic membrane
show protruding particles on the cytoplasmic half and corresponding
pits on the outer half, the protrusions arising from vesicles fusing with
the membrane. The particles on the cytoplasmic half of the postsynaptic membrane are believed to be ACh receptors.

Ideally, it is desirable to freeze cells directly from the living state
without resorting to pretreatment involving fixating agents and cryo-
protectants, as these may distort the ultrastructure under investigation.
Ultra-rapid freezing techniques have been developed for this purpose,
involving jet freezing, spray freezing, or freezing by impact against a
cold metal block. The ultra-rapid freezing allows temporal resolution
of cellular events, that is, examining these events at various stages of
their evolvement.

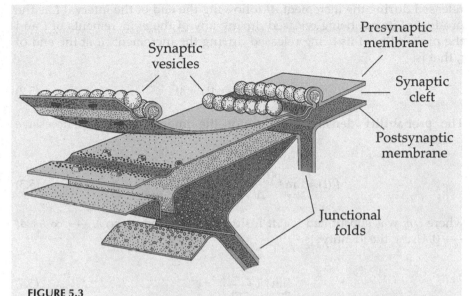

Synaptic vesicles

Presynaptic membrane

Synaptic cleft

Postsynaptic membrane

Junctional folds

FIGURE 5.3
Ultrastructure of the frog neuromuscular junction as revealed by the freeze-fracture technique.

5.3 Statistics of Neurotransmitter Release

5.3.1 Spontaneous Release

A relatively small number of vesicles continuously leak from the presynaptic bouton, in the absence of presynaptic APs, giving rise to **miniature end-plate potentials (mepps)**. The amplitude of the mepps varies in accordance with a unimodal, approximately normal distribution around a mean value that is generally between 0.4 and 0.5 mV. The variance in the distribution is due to variation in the number of neurotransmitter molecules per vesicle, in the number of neurotransmitter molecules that bind to receptors, and in the kinetics of channel opening after the binding of a neurotransmitter.

We will first derive the probability density function for the interval between successive releases. A basic assumption is that the release of a vesicle does not affect the release of other vesicles, that is, the release of a vesicle does not depend on the past history of vesicle release. If vesicles are released at an average rate of r per second, the probability of a vesicle released in a small increment of time Δt is $r\Delta t$ and the probability of its not being released is $(1 - r\Delta t)$. Assuming that a vesicle is released at $t = 0$ and that the interval t until the next release is divided into m increments of Δt, that is, $m\Delta t = t$, then $p(t + \Delta t)$, the probability of the next vesicle being released during the increment Δt following the end of the interval t, is the product of its not being released during any of the m increments of t and the probability of its being released during the increment Δt at the end of t, that is,

$$p(t + \Delta t) = (1 - r\Delta t)^m r\Delta t \tag{5.2}$$

The **probability density function** of the interval between successive releases is:

$$f_1(t) = \lim_{\Delta t \to 0} \frac{p(t + \Delta t)}{\Delta t} = \lim_{m \to \infty} r\left(1 - \frac{rt}{m}\right)^m \tag{5.3}$$

where t/m was substituted for Δt inside the parentheses, so $m \longrightarrow \infty$ as $\Delta t \longrightarrow 0$. Using the identity:

$$\lim_{m \to \infty}\left(1 + \frac{x}{m}\right)^m = e^x \tag{5.4}$$

it follows that:

$$\lim_{m \to \infty}\left(1 - \frac{rt}{m}\right)^m = e^{-rt} \tag{5.5}$$

Equation 5.3 then gives:

$$f_1(t) = re^{-rt} \tag{5.6}$$

$f_1(t)$ is an **exponential distribution**. From the definition of the probability density function, $f_1(t)dt$ is the probability that the interval X between successive releases will have a value between t and $t + dt$. This means that in a large number N of intervals between releases, the number of occurrences dn of any interval between t and $t + dt$ is:

$$dn = Nf_1(t)dt = Nre^{-rt}dt \tag{5.7}$$

In experimental work, dt is approximated by a small increment Δt, and the number of intervals Δn within each increment Δt is:

$$\Delta n = Nf_1(t)\Delta t = Nre^{-rt}\Delta t \tag{5.8}$$

A plot of Δn vs. t gives a histogram of bin width Δt, the number of counts Δn in each bin falling exponentially with time (Figure 5.4) in accordance with Equation 5.8. The total number of counts in the histogram is N.

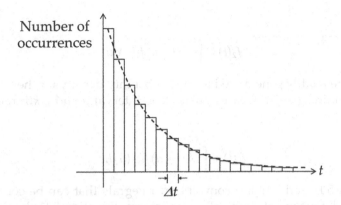

FIGURE 5.4
Histogram illustrating an exponential distribution.

The number of intervals n with a duration less than or equal to t is the integral of dn in Equation 5.7 from 0 to t, or:

$$n = \int_0^t Nre^{-rt}dt = N\left(1 - e^{-rt}\right)$$ (5.9)

The ratio n/N is the probability distribution function $F_1(t)$ obtained by integrating Equation 5.6 between 0 and t. The mean interval is of course $1/r$ (see also Equation 5.25).

Experimental recordings of intervals between successive mepps show an exponential distribution for short intervals and a power-law distribution over intervals of 1 s or more. The exponential distribution reflects a random, memoryless process, as assumed in the preceding discussion. The power law is due to a nonrandom process, such as that of recycling and repackaging of synaptic vesicles.

The probability density for the time to release the second, third, or kth vesicle, following the release of a vesicle at $t = 0$, will be determined next. To have a second release at time t, there must be a first release at an intermediate time $u < t$. The probability of a first release between u and $(u + du)$ is $f_1(u)du$, and the probability of a second release between t and $t + dt$ is $f_1(t - u)dt$. The product of the two probabilities $f_1(u)du$ and $f_1(t - u)dt$ is the probability of a second release between t and $t + dt$ for only one value of u out of an infinite number of possible values. Hence this product is the infinitesimal $df_2(u)dt$, where $f_2(t)$ is the probability density for the time to release the second vesicle. Thus:

$$df_2(u)dt = f_1(u)du \times f_1(t - u)dt$$ (5.10)

or,

$$f_2(t) = \int_0^t f_1(t-u)f_1(u)du \tag{5.11}$$

This can be readily generalized to the probability density for the kth release, as this requires $(k - 1)$ releases during the interval u and a kth release at t, which gives:

$$f_k(t) = \int_0^t f_1(t-u)f_{k-1}(u)du \tag{5.12}$$

Equations 5.11 and 5.12 are convolution integrals that can be conveniently evaluated by means of the Laplace transform. Equation 5.11 then takes the form:

$$\mathcal{L}\left[f_2(t)\right] = \mathcal{L}\left[f_1(t)\right] \times \mathcal{L}\left[f_1(t)\right] = \left\{\mathcal{L}\left[f_1(t)\right]\right\}^2 \tag{5.13}$$

Bearing in mind that $\mathcal{L}\left[f_1(t)\right] = r/(s+r)$, Equation 5.12 becomes:

$$\mathcal{L}\left[f_k(t)\right] = \mathcal{L}\left[f_1(t)\right] \times \left\{\mathcal{L}\left[f_1(t)\right]\right\}^{(k-1)} = \left\{\mathcal{L}\left[f_1(t)\right]\right\}^k = \frac{r^k}{(s+r)^k} \tag{5.14}$$

where $f_k(t)$, is the probability density function for the kth release. Taking the inverse Laplace transform in Equation 5.14 gives:

$$f_k(t) = \frac{r^k t^{k-1} e^{-rt}}{(k-1)!} \tag{5.15}$$

$f_k(t)$ is a **gamma distribution**, plotted in Figure 5.5 for $r = 1.0$ and different values of k. The following should be noted:

1. For $k = 1$, $f_k(t) = f_1(t)$, that is, the probability density function for a single release is the same as the probability density function of the interval between successive releases, which is an exponential distribution (Equation 5.6). At $t = 0$, $f_1(t) = r$, in accordance with the interpretation that $f_1(t)\Delta t$ is the probability of a single release between $t = 0$ and $t = \Delta t$, which is $r\Delta t$.

2. The probability of more than one release in the time interval between $t = 0$ and $t = \Delta t$ is small and decreases with increasing k.

3. For any $f_k(t)$, $\int_0^\infty f_k(t)dt = 1$, since k releases are certain to occur as $t \longrightarrow \infty$.

4. The gamma distribution approaches a Gaussian, or normal, distribution for large k.

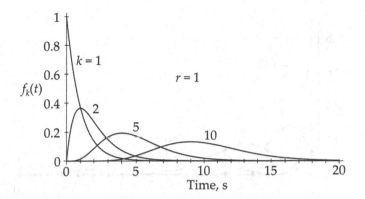

FIGURE 5.5
Family of curves for a gamma distribution.

5. Consider $k = 10$. The maximum occurs at nearly 9 s. The probability of ten releases by this time is nearly 0.5, the area under the curve from 0 to 9s. By 20 s, the probability of ten releases is very high, as the area under the curve becomes nearly one. More than ten releases can also occur by 20 s because curves for $k > 10$ have a finite area at this time. Hence one can say that there is a high probability that at least ten releases will occur by 20 s.

5.3.2 Evoked Release

Following an action potential, quanta are released at a high rate r per second within a short period of time. In this case, the probability of a certain number of quanta being released is of interest, rather than the intervals between released quanta. According to the preceding discussion, the probability p_k of release of k quanta can be obtained as the time integral of Equation 5.15. Referring to a table of integrals, the integral is:

$$p_k = \int_0^t \frac{r^k t^{k-1} e^{-rt}}{(k-1)!} dt = \frac{r^k t^k e^{-rt}}{k!} \tag{5.16}$$

Let $m = rt$ be the mean number of vesicles released in time t. Substituting for rt in Equation 5.16,

$$p_k = \frac{m^k e^{-m}}{k!} \tag{5.17}$$

Equation 5.17 is a **Poisson distribution,** plotted in Figure 5.6, where a smooth curve is drawn through the discrete values of k. m is both the mean and the variance of the Poisson distribution.

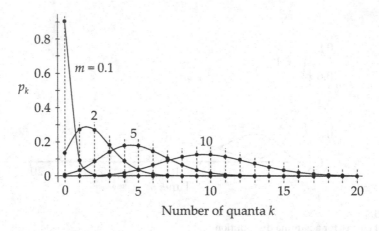

FIGURE 5.6
Family of curves for a Poisson distribution.

It is implicit in the derivation of Equation 5.17 that the number of vesicles available for release is virtually limitless, which means that if m is not too large the probability of vesicle release at each site is small. In practice, p is small under conditions of low Ca^{2+} or high Mg^{2+}, in which case the Poisson distribution applies.

To account for the finite number of vesicles n available for release, it is assumed that p, the probability of vesicle release, is the same for all sites, that vesicles are released independently of one another, and that both p and n do not vary with time, that is, the release process is a stationary stochastic process. The probability of release of k quanta requires k successes, of probability p^k, and $(n - k)$ failures, of probability of failure $(1 - p)^{n - k}$. However, the k vesicles could be released in various combinations. Thus, the first vesicle could be released in n different ways, that is out of the n vesicles available for release, the second in $(n - 1)$ different ways, and the kth vesicle in $[n - (k - 1)]$ different ways. But the k vesicles are indistinguishable from one another, and can be arranged in $k!$ number of ways. Hence, the total number of ways of selecting the k vesicles out of n vesicles is:

$$C_k^n = \frac{k(n-1)\ldots[n-(k-1)]}{k!} = \frac{n!}{k!(n-k)!} \tag{5.18}$$

and,

$$p_n(k) = \frac{n!}{k!(n-k)!} p^k (1-p)^{n-k} \tag{5.19}$$

$p_n(k)$ is a **binomial distribution**, plotted in Figure 5.7 for various values of n. The mean of the distribution is $\mu = np$, the mean number of vesicles released per site. The variance of the distribution is $\sigma_n = np(1 - p)$.

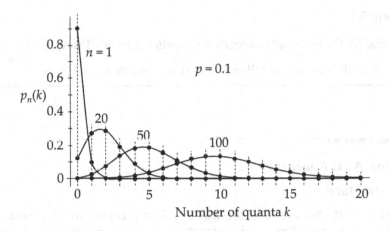

FIGURE 5.7
Family of curves for a binomial distribution.

If $n \gg 1$, the binomial distribution can be approximated by a normal distribution. Equation 5.19 becomes:

$$P_n(k) = \frac{n!}{k!(n-k)!} p^k (1-p)^{n-k} = \frac{1}{\sigma_n \sqrt{2\pi}} e^{-(k-\mu)^2 / 2\sigma_n^2} \tag{5.20}$$

which is a normal distribution centered around μ.

If $n \longrightarrow \infty$, with np finite, $p \longrightarrow 0$ and the binomial distribution reduces to a Poisson distribution. Thus, Equation 5.19 can be written as:

$$P_n(k) = \frac{n(n-1)...(n-k+1)}{k!} \left(\frac{\mu}{n}\right)^k \left(1-\frac{\mu}{n}\right)^{n-k} \tag{5.21}$$

As $n \longrightarrow \infty$, and using Equation 5.4, Equation 5.21 becomes:

$$P_n(k) = \frac{n^k \mu^k}{k! n^k} \left(1-\frac{\mu}{n}\right)^n = \frac{\mu^k}{k!} \left(1-\frac{\mu}{n}\right)^n = \frac{\mu^k e^{-\mu}}{k!} \tag{5.22}$$

Equation 5.22 is a Poisson distribution of mean μ.

Problem 5.3

Assume that vesicles are released from a NMJ according to Poisson statistics with a mean quantal content of 5. In 100 observations, (a) what is the expected number of observations in which 2 vesicles are released? (b) what is the expected number of failures, that is when no vesicles are released?

ANS.: (a) $N_2 = 100 \dfrac{4^2 e^{-4}}{2!} = 14.7 \cong 15$; (b) $N_0 = 100 \dfrac{4^0 e^{-4}}{0!} = 1.8 \cong 2$.

Problem 5.4

Show that for the binomial distribution (Equation 5.19) $\mu = \ln\left(N_0 / N\right)\dfrac{p}{\ln(1-p)}$, where N_0 is the number of failures, when no vesicles are released.

5.4 The ACh Receptor

5.4.1 Structure

The ACh NMJ channel receptor was the first receptor to be investigated in detail, in terms of molecular structure, electron microscopic imaging, and unitary channel current measurement. It is an **ACh nicotinic receptor** because it responds to nicotine in the same way as ACh.

The structure of the ACh receptor has been determined using cryo-electron microscopic images of many receptors isolated from the muscle-derived, electric organ of the Torpedo ray. Figure 5.8a illustrates a longitudinal section through the channel. The ACh receptor is a large transmembrane protein having a molecular mass of about 290 kDa (kilodaltons, where the dalton is a unit of atomic mass equal to 1/12 of the atomic mass of ^{12}C, or nearly 1.67×10^{-27} kg, the mass of a hydrogen atom). The receptor is about 7–8 nm in diameter and 16 nm in length. It protrudes about 7 nm into the external medium

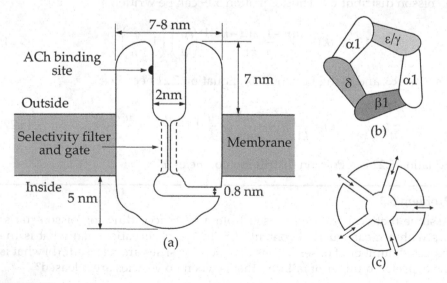

FIGURE 5.8

(a) Structure of an ACh receptor; (b) arrangement of the subunits of the receptor; (c) lateral openings on the cytoplasmic side of the receptor.

and about 5 nm into the cytoplasm. The receptor consists of five longitudinal strips arranged in a cylindrical configuration like the staves of a barrel (Figure 5.8b). The strips are subunits of the protein, identified clockwise, as α1, β1, δ, α1, and γ in fetal ACh receptors and ε in adult ACh receptors. Because it consists of five subunits, the receptor is a **pentamer**. Since the subunits are not all the same, the receptor is a **heteromer**, in contrast to a **homomer** having identical subunits. The receptor is therefore a pentameric heteromer.

The pore that is enclosed by the five subunits has a long vestibule of about 2 nm diameter on the extracellular side, and narrows along the region spanning the membrane, where it includes the gate. The gate is shown closed in Figure 5.8a but opens, as indicated by the dashed lines, as a result of binding of ACh at the binding sites The diameter of the narrowest part of the pore when closed is less than 0.6 nm. On the cytoplasmic side, the pore has a shorter vestibule, also about 2 nm diameter, with lateral openings into the cytoplasm (Figure 5.8c). These openings are narrow, approximately 0.8 nm wide, and occur between the subunits. Although the vestibules and openings into the cytoplasm are lined with negative charges of amino acids that block anions and facilitate the movement of cations, it is believed that the charge selectivity of the pore is mainly due to the negative charges in the inner vestibule. The configuration of the five openings and the relatively wide regions of the rest of the pore allow bidirectional movement of all the alkaline metal cations, such as Li^{+}, Na^{+}, and K^{+}, as well as alkaline earth metal cations, such as Ca^{2+} and Mg^{2+}, and several small organic cations. Large, charged drug molecules can block the channel by binding to charged groups on the external side of the receptor. The ACh receptor is characterized by fast opening and closing of the channel.

The two ACh binding sites are mainly in the α subunits, on nearly opposite sides of the pore, at the α–γ and α–δ subunit interfaces. The binding sites are about 4 nm from the membrane surface. It is believed that the binding of ACh causes a twisting movement in parts of the α subunits, which opens the gate by breaking weak hydrophobic interactions that keep the gate closed. The fast action of ACh is believed to be due to the intimate association of the ACh binding sites with the pore.

Details about the subunits are shown in Figure 5.9. Each of the five subunits contains four segments of approximately 20 amino acids each, referred to as M1–M4, that span the membrane, coursing back and forth across the membrane four times in the form of an α-helix. Note that it is customary to show segments of a subunit of a channel protein side-by-side longitudinally along the membrane (Figure 5.9a) although these segments may not in fact be configured in this manner, as illustrated diagrammatically in Figure 5.9b. Each subunit has, on the extracellular side, a large NH_2 terminus and a short carboxylic acid (COOH) terminus. The ACh binding sites are in the NH_2 terminus regions of the α subunits. The walls of the pore are formed by the M2 segments of each subunit and by the loop connecting this segment to the M3 segment. Pore-lining protein segments are hydrophylic.

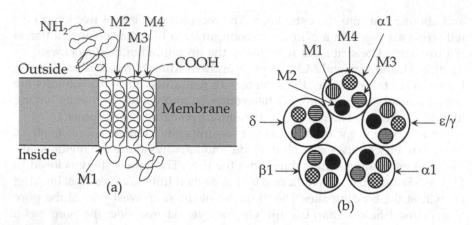

FIGURE 5.9
(a) Segments of a subunit of an ACh receptor; (b) segments and subunits around the pore.

In human adults, the average half-life of ACh receptors is about 10 days, the synthesis of these receptors being regulated by muscle fiber nuclei close to the NMJ. When a muscle is denervated (Section 9.3.3), ACh receptors will appear all over the denervated muscle fibers. The newly formed ACh receptors are of the fast turnover type (referred to as fAChR) having a shorter half-life of about 1 day and differ from the normal receptors in having their ε subunit replaced by a γ subunit.

Other forms of nicotinic ACh receptors are found in other parts of the nervous system. Parasympathetic postganglionic neurons have ACh receptors that consist of two α3 and three β4 subunits. There are also nicotinic ACh receptors in the brain consisting of different types of α and β subunits, or of α subunits only (pentameric homomers), the latter having an appreciable permeability to Ca^{2+} in addition to Na^+ and K^+. The other type of ACh receptor is the **muscarinic** type, because the alkaloid **muscarine**, a toxin derived from a species of poisonous mushrooms, acts as an ACh agonist at these receptors. Found in the CNS, these receptors are coupled to G proteins (Section 6.3) and differ in structure from nicotinic ACh receptors.

5.4.2 Channel Kinetics

All channels open and close randomly because of thermal agitation of constituent atoms and groups of atoms of the channel protein, as well as thermal agitation of water molecules and ions in the channel. Thus, even a ligand-gated channel, such as an ACh channel, has a very small, but finite, probability of opening, even in the absence of any ACh. However, the probability is many orders of magnitude larger in the presence of two ACh molecules. Similar considerations apply to voltage-gated channels.

The durations of channel opening and closing are exponentially distributed (Equation 5.6), as discussed later, in accordance with a memoryless random process. In other words, the probability of transition between the open and closed state does not depend on the past history of opening or closing. For a given membrane voltage, it depends simply on the present state of the channel, that is, being open or closed. Thus, an open channel may close at any time t, with zero time considered to be anywhere during opening – similarly for opening of a closed channel.

Let the probability of closing an open channel during a time Δt be $\sigma \Delta t$. Assume that the channel opened at $t = 0$ and remains open at time t. Let t be divided into m increments of Δt, so that $m\Delta t = t$. The probability of the channel closing during the interval $(t + \Delta t)$ is the product of the probability of its remaining open during the interval 0 to t and the probability of closing during Δt, that is:

$$p(t + \Delta t) = (1 - \sigma \Delta t)^m \times (\sigma \Delta t) \qquad (5.23)$$

Proceeding as in Equations 5.2–5.6 gives a probability density function:

$$f_1(t) = \frac{1}{\tau_o} e^{t/\tau_o} \qquad (5.24)$$

where $\tau_o = 1/\sigma$ is the mean open time, or mean **open lifetime** of the channel. This follows from the definition of the mean of an exponential distribution as:

$$\text{mean} = \int_0^\infty \frac{t}{\tau_o} e^{-t/\tau_o} dt = \left[-te^{-t/\tau_o} \right]_0^\infty - \int_0^\infty (-e^{-t/\tau_o}) dt = \tau_o \qquad (5.25)$$

According to the interpretation of probability density, $f_1(t)dt$ is the probability that the open channel closes between t and $t + dt$.

The exponential variation is also encountered in simple kinetics, as was considered in Section 3.2. Thus, consider the "reversible reaction" of Equation 3.2, expressed as:

$$N_c \underset{\alpha}{\overset{\beta}{\rightleftarrows}} N_0 \qquad (5.26)$$

where N_c and N_o denote the number of closed channels and open channels, respectively, and α and β are the corresponding rate constants. Note that, in accordance with current usage, α and β are interchanged compared to Equation 3.2 of the Hodgkin–Huxley model. The rate dN_o/dt at which N_o increases is the rate βN_c at which closed channels open, minus the rate αN_o at which open channels close:

$$\frac{dN_o}{dt} = \beta N_c - \alpha N_o \qquad (5.27)$$

If N is the total number of channels and is a constant, then $N_c = N - N_o$. Substituting for N_c in equation 5.27:

$$\frac{dN_o}{dt} = \beta N - (\alpha + \beta) N_o \qquad (5.28)$$

Equation 5.28 can be integrated to give:

$$N_o(t) = N_{o\infty} - (N_{o\infty} - N_{o0}) e^{-(\alpha + \beta)t} \qquad (5.29)$$

where N_{o0} is the initial value of N_o, and $N_{o\infty}$ is the final equilibrium value of N_o, which from Equation 5.28, with $dN_o/dt = 0$, equals $\beta N/(\alpha + \beta)$. It is seen that $N_o(t)$ approaches equilibrium exponentially with a time constant $1/(\alpha + \beta)$.

Equations 5.27–5.29 represent chemical kinetics on a *macroscopic* scale, considering a large number of channels. How are they to be interpreted for single channels, as observed under patch clamp? If the rate at which N_o open channels close is αN_o, then α is the mean rate at which a single open channel will close. This means that $1/\alpha$ is the **mean open lifetime** of a channel. Although the open lifetimes of individual channels will be randomly distributed, according to an exponential distribution, the mean open lifetime of a large number of channels is $1/\alpha$. Identifying $1/\alpha$ as τ_o (Equation 5.25), it follows that in a system such as that of Equation 5.26 having a single open state, *the mean open lifetime is equal to the reciprocal of the rate constant for channel closing*. It should be noted that rate β in Equation 5.26 increases the number of open channels but cannot affect the time a channel remains open. In fact, α can be estimated by taking the mean of observed open lifetimes of a fairly large number of channels. Similar considerations apply to the mean closed lifetime of channels.

Returning to the binding of ACh to the channel receptor, channel opening can be described by the following kinetic diagram:

$$A + R \underset{k_{-1}}{\overset{k_1}{\rightleftharpoons}} A_1 R \underset{k_{-2}}{\overset{\overset{\displaystyle A}{\overset{+}{k_2}}}{\rightleftharpoons}} A_2 R^c \underset{\alpha}{\overset{\beta}{\rightleftharpoons}} A_2 R^o \qquad (5.30)$$

The first reaction represents the binding of one ACh molecule (A) to the receptor (R), resulting in the complex $A_1 R$, followed by the binding of a second ACh molecule to produce the complex $A_2 R^c$, where the c superscript denotes a closed channel. In this state, the channel opens with a fairly high rate constant β to produce the complex $A_2 R^o$, where the o superscript denotes an open channel. Although, in accordance with random opening or closing of channels, the channel may open with no ACh bound, or with only one ACh molecule bound, the rates involved are negligibly small.

The following numerical values for the mouse ACh receptor at 22°C and a membrane voltage of –100 mV give some idea about the kinetics involved:

$$k_1 = 4 \times 10^8 \text{ M}^{-1}\text{s}^{-1} \qquad k_{-1} = 2.5 \times 10^4 \text{ s}^{-1}$$
$$k_2 = 2 \times 10^8 \text{ M}^{-1}\text{s}^{-1} \qquad k_{-2} = 5 \times 10^4 \text{ s}^{-1}$$
$$\beta = 5 \times 10^4 \text{ s}^{-1} \qquad \alpha = 1.5 \times 10^3 \text{ s}^{-1}$$

The dimensions of the various rate constants are explained in Section 1.6.3. The following should be noted concerning the kinetics expressed by Reaction 5.30:

1. The binding of ACh to a receptor is very rapid, as indicated by large k_1 and k_2.
2. The mean open lifetime of the channel is $1/\alpha$, or about 0.67 ms.
3. When an open channel closes and returns to the state A_2R^c, the probability of reopening is $\beta/(\beta + k_{-2}) = 0.5$, which means that a channel may close and reopen once or twice, resulting in a burst of openings interrupted by closures of mean duration $1/(\beta + k_{-2}) = 10$ µs. The mean duration of a burst is about 1.4 ms, after which the channel closes, and the neurotransmitter leaves the receptor.
4. After the initial binding of ACh to the receptor, the concentration of ACh in the cleft falls rapidly because of hydrolysis by AChE. It is highly unlikely, therefore, that ACh will bind to the receptor again before the arrival of the next AP at the presynaptic terminal.

5.4.3 Channel Desensitization

If the endplate is subjected to prolonged exposure to ACh, it **desensitizes**, that is the endplate conductance drops to a low value within a few seconds due to closure of channels in the presence of ACh. Single channels may stay desensitized, that is in a closed state, for a second or so or for many tens of seconds. They may take a few seconds or minutes to recover. Thus, with continuous application of ACh, a channel may go into cycles of bursts of openings, interrupted by short closures, followed by desensitization for short or longer periods, then recovery in a burst of openings, and so on.

Desensitization is analogous to inactivation of voltage-gated channels and is caused by conformational changes in the receptor. It effectively curtails the action of any neurotransmitter that remains bound to the receptor for a duration that is longer than that required by the function of the synapse in question. Desensitization is manifested as a reduced responsivity induced by previous activation.

Channel desensitization is a distinct state from channel closure, although in both states the channel is closed in the presence of the neurotransmitter. Kinetically, the rate constants of the transitions between an open channel and a closed channel are quite different from the rate constants of the transitions between an open channel and a desensitized channel.

5.5 Generation of the Muscle Action Potential

5.5.1 The Endplate Current

Following the binding of two ACh molecules to a channel receptor, the channel opens and current begins to flow through the channel. The **endplate current (epc)** is the aggregate current through all the activated channels and starts to increase in about 0.5–0.8 ms after depolarization of the presynaptic terminal, at 20–25°C, and in less time at 37°C. An open channel has a conductance of 20–40 pS, over a temperature range of 8°–27°C, with a temperature dependence having a Q_{10} of 1.2–1.5, that is, the conductance is multiplied by this factor for a 10°C temperature rise. The conductance of an open channel is hardly affected by the membrane voltage, that is, it behaves like an ohmic resistor that obeys Ohm's law.

The time course of the epc is shown in Figure 5.10, as a negative, that is inward, current. The epc increases very rapidly, reaching its peak in a few hundred microseconds. This rise time is due to the dispersion in the timing of the release of vesicles, their arrival at the receptor sites, and their fast action on the channel. The peak of the epc is given by: (the sum of the conductances of the largest number of open channels)×(the difference between the membrane voltage and the equilibrium voltage of the epc). The equilibrium voltage of the epc is about 0 mV, intermediate between the equilibrium voltages of Na$^+$ and K$^+$ (Problem 5.5). Because channel opening is a stochastic process, the number of open channels at any given time equals the total number of channels multiplied by the probability of channel opening.

The epc decays exponentially in a few ms in the frog sartorius muscle at 25°C with a time constant τ_{epc} due to channel kinetics, while ACh is still bound to the receptors, as discussed earlier. The exponential decay is in accordance with Equation 5.29. If all channels eventually close, as $t \longrightarrow \infty$, $N_{0\infty} \longrightarrow 0$ and $N_o(t)$ together with the overall conductance and epc decays

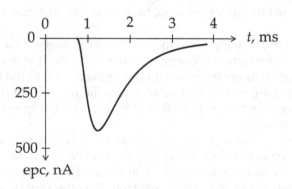

FIGURE 5.10
Time course of the endplate current following depolarization of the presynaptic terminal.

exponentially with a time constant $\tau_{epc} = 1/(\alpha + \beta)$, for the two-state transition of Reaction 5.26.

τ_{epc} decreases somewhat with increased depolarization and with a rise in temperature. The rate of decay of the epc increases e-fold per 110 mV, which is weak compared to the voltage-gated Na$^+$ and K$^+$ channels of the squid giant axon where the voltage dependence is an e-fold increase per 4–6 mV. The weak voltage dependence of τ_{epc} is attributed to a small gating charge that is involved in the conformational change that closes the channel.

The requirement of two ACh molecules for opening the channel, rather than a single ACh molecule, reduces "noise", or undesired opening of the channel, due to the continuous spontaneous release, or "leakage", of single vesicles of ACh, as well as individual ACh molecules in some cases. Moreover, if only a single ACh molecule were required to activate a channel, the peak of the epc, as a measure of the epc response, would increase linearly with concentration of ACh and will eventually saturate when ACh is bound to all the receptors. If two ACh molecules are required to activate a channel, it follows from the simple "reaction":

$$[ACh][ACh] \rightarrow epc \tag{5.31}$$

that the peak of the epc will increase as the square of ACh concentration. Below saturation, the variation will be sigmoidal instead of linear; that is, it increases slowly at low ACh concentration and more rapidly at higher ACh concentration before saturation.

Problem 5.5

Assume that the ionic channel of the NMJ has conductances G_{SK} and G_{SNa} to K$^+$ and Na$^+$, respectively, the equilibrium voltages of these ions being E_K and E_{Na}. Show that the effective equilibrium voltage is $E_S = \dfrac{G_{SNa}E_{Na} + G_{SK}E_K}{G_{SK} + G_{SNa}}$.
Note that if $G_{SK} = G_{SNa}$, $E_S = \dfrac{E_{Na} + E_K}{2}$. It should be kept in mind that the foregoing analysis applies assuming a *linear time-invariant* circuit, which is not strictly justified.

SPOTLIGHT ON TECHNIQUES: POWER SPECTRAL ANALYSIS

If a constant ACh concentration is applied externally to the NMJ, the individual channels will still open and close randomly, causing noise-like fluctuations in the epc whose magnitude depends on the membrane voltage. The exponential decay of the epc referred to in connection with Figure 5.10 will have a low-pass filtering effect that is reflected in the power spectrum of the current fluctuations.

A simple derivation of the power spectrum of the fluctuations is to assume the epc to be of the form ke^{-at}, $t > 0$, and 0 for $t < 0$. The Fourier transform of the epc is then:

$$\frac{k}{a+j\omega} = \frac{ka}{a^2+\omega^2} - j\frac{\omega}{a^2+\omega^2}$$

The power spectrum $S(\omega)$ is the square of the magnitude of this Fourier transform. That is:

$$S(\omega) = \frac{k^2}{a^2+\omega^2} = \frac{k^2/a^2}{1+\left(2\pi f/a\right)^2}$$

The plot of $S(\omega)$ against f on a log-log scale has two asymptotes: a horizontal low-frequency asymptote and a high-frequency asymptote of slope -40 dB/decade. The two asymptotes intersect at a corner frequency $a/2\pi$.

Figure 5.11 illustrates the power spectrum of current fluctuations recorded by Anderson and Stevens (1973) from a frog NMJ that is voltage clamped at -60 mV at 8°C and subjected to a constant ACh iontophorectic application. The experimental points are fitted by a Lorentzian function having a corner frequency of 21 Hz, which agreed with the decay time constant of the epc. The current through individual channels, when open, is typically about 2 pA at a membrane voltage of -60 mV.

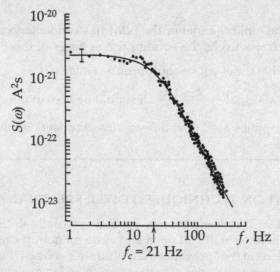

FIGURE 5.11
Power spectrum of current fluctuations in frog NMJ.

5.5.2 The Endplate Voltage

The **endplate voltage**, conventionally referred to as the **endplate potential (epp)**, recorded following a nerve AP is the depolarization caused by the epc acting on the parallel combination of resistance and capacitance of the muscle membrane. The general shape of the epp is illustrated in Figure 5.12. The amplitude of the epp is 50–70 mV above a resting membrane voltage of about –90 mV for the muscle. Since the threshold of the muscle AP is 15–25 mV above the resting voltage, a muscle AP is generated well before the epp reaches its peak. Hence, the full time course of the epp can only be recorded with the muscle AP suppressed. The presence of voltage-gated Na^+ channels in the depths of the junctional folds serves to reduce the threshold for the muscle AP. The size of the epp rapidly declines with high-frequency repetitive stimulation of 20–40 APs because of a decrease in the number of vesicles released, but levels off thereafter at about 65% of its size for a single AP. As may be expected, this decline in the size of the epp with repetitive stimulation is more marked at the NMJs of fast muscle fibers compared to those of slow muscle fibers (Section 9.3.2).

The general shape of the epp can be better understood if it is assumed that the epc is ideally a current pulse. Since channels are open during this pulse, membrane conductance G_{mo} is relatively high. The membrane time constant C_m/G_{mo} is relatively small, causing a rapid rise of membrane voltage toward the equilibrium voltage of the epp which is the same as that of the epc since, at this voltage, the epc current is zero. When the epc pulse ends, the channels are closed and the membrane conductance is considerably reduced. The epp decays toward the resting membrane voltage with a considerably larger time constant.

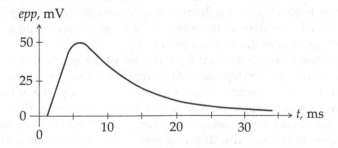

FIGURE 5.12
Time course of the endplate potential following depolarization of the presynaptic terminal.

It should be emphasized that, unlike the AP, the epp is not an active response that involves a regenerative cycle. This means that it has neither a threshold nor refractoriness, and it propagates passively in accordance with the cable properties of the membrane. It becomes progressively smaller in amplitude and slower in time course as it spreads from the inexcitable region of endplate to neighboring regions of the muscle membrane, where it generates a muscle AP. Because the epp is a passive response, epps that occur in rapid succession summate; in fact, the epp can be considered as the sum of a large number of miniature epps (mepps), each of which is due to the effect of a single vesicle out of the total number of vesicles released when the nerve terminal is depolarized by the nerve AP.

5.6 Interference with Normal Operation

Normal operation of the NMJ can be interfered with at several levels, including synthesis and release of ACh at the presynaptic terminal, blocking of ACh channels, desensitization of ACh receptors, or their damage. Moreover, a slow, sustained depolarization of the endplate can inhibit generation of the muscle AP in neighboring muscle membrane through accommodation of the AP mechanism (Section 3.3.2). **Myasthenia gravis**, derived from the Greek for severe muscle weakness, is a disease in which neuromuscular transmission is impaired. Some forms of the disease are due to genetic mutations, but the most prevalent form of the disease is autoimmune in which the body's immune system produces antibodies that block, alter, or destroy ACh receptors. Moreover, the normal infolding at the endplate is reduced and the synaptic cleft is enlarged. The overall effect is a drastic reduction of the epp to around the threshold for the muscle AP or even less. The disease usually affects the eyelid muscles first, causing drooping eyelids. It can also affect the eye muscles, resulting in inability to move the eyes. The effect of depression of neuromuscular transmission in other parts of the body due to the aforementioned factors ranges from muscle weakness to muscle paralysis, including muscle fatigue, that is, progressive weakness of a muscle during activity. Symptoms are partially relieved by drugs that inhibit AChE, which enhances the action of ACh. Respiratory failure can occur if the breathing muscles are severely affected.

The drug hemicholinium-3 inhibits the reuptake of choline at the presynaptic terminal, which depresses ACh synthesis because most of the choline needed for ACh synthesis is provided through reuptake of the choline resulting from hydrolysis of ACh in the synaptic cleft, and only a relatively small amount of choline is transported from the cell body. α-latrotoxin is an extremely potent neurotoxin, that is a poison of the nervous system, contained in the venom of the black widow spider. It causes massive exocytosis of ACh from presynaptic terminals either by forming open pores in the presynaptic membrane that allow the influx of Ca^{2+} and Na^+ or by binding to special receptors, thereby initiating processes that lead to exocytosis. The depletion of ACh eventually leads to muscle paralysis. Botulinum toxin, produced by a bacterium found in poisoned foods, is one of the most toxic substances known. Only 2 ng of a form of this toxin, when injected intravenously, can kill a human adult by preventing ACh vesicles from fusing with the presynaptic membrane and releasing ACh into the synaptic cleft. Extremely small doses of other forms of this toxin, commercially known as **Botox**, are injected into the skin to relax muscles causing wrinkles. Botox is also used in the treatment of disorders caused by overactive muscle movement or conditions arising from hyperactivity of some nerves.

Many substances act as reversible or irreversible **anticholinesterases**, that is substances that prevent the enzyme AChE from hydrolyzing ACh. This increases both the level and duration of action of ACh and results in an initial strong contraction followed by desensitization, accommodation, and eventual muscle paralysis. The effects depend on the dose and vary with different muscles. Examples of reversible anticholinesterases are eserine and neostigmine, whereas some constituents of snake venoms and organophosphates found in insecticides and nerve gases, such as sarin, act as essentially irreversible anticholinesterases.

It should be noted that ACh is also a neurotransmitter of the postganglionic and preganglionic neurons of the parasympathetic system, so that anticholinesterases would stimulate the parasympathetic system, which reduces the heart rate, lowers blood pressure, increases secretion from some glands, constricts air passageways of the lungs, increases motility of the gastrointestinal tract, and reduces intraocular pressure.

5.6.1 ACh Agonists and Antagonists

Agonists of a given neurotransmitter are substances that bind to the receptor of the neurotransmitter and mimic its action. There are many ACh agonists that bind to ACh receptors and depolarize the endplate. Generally, however, they do not have the same kinetics as ACh, do not desensitize the channel in the same manner, and are not similarly affected by AChE. In low doses, they mimic the action of ACh, but in high doses, or with prolonged application, they initially cause muscular contraction followed by desensitization, accommodation, and eventual neuromuscular block. Because of this, they are termed **depolarizing blocking agents**. Examples of ACh agonists are nicotine, succinylcholine, and decamethonium. Succinlycholine is applied intravenously during surgery as a short-acting muscle relaxant. It has a rapid onset of about 30 s and a duration of action of 5–10 minutes. It is not hydrolyzed by AChE but by other cholinesterases normally found in the blood.

Antagonists of a given neurotransmitter, also referred to sometimes as **blockers**, are substances that bind to the receptor of the neurotransmitter but reduce or block its action. Antagonists could be **competitive**, **noncompetitive**, or **uncompetitive**. As its name implies, a competitive antagonist competes with the neurotransmitter or agonist for the receptor sites. The binding of the competitive antagonist to the receptor site could be reversible (or surmountable), or it could be irreversible (or insurmountable). In the case of a reversible competitive antagonist, the bond to the receptor site is chemically reversible, so that the blocking action depends on the concentration of the antagonist and is reduced by a higher concentration of the neurotransmitter. On the other hand, increasing the concentration of the neurotransmitter does not reduce the blocking effect of an irreversible competitive antagonist that has bound to the site because the bond of the antagonist to the receptor site is chemically irreversible.

A noncompetitive antagonist blocks the action of a neurotransmitter, or agonist, by binding to a site on the receptor other than the binding site of the neurotransmitter, or agonist, that is, it is not competing for the same binding site. The site to which the noncompetitive antagonist binds is referred to as an **allosteric** site, which in general is a site other than the active site of an enzyme that allows the binding of a substance that regulates the enzyme's activity. An uncompetitive antagonist differs from a noncompetitive neurotransmitter in that it requires receptor activation by the neurotransmitter, or agonist, before it can bind to a separate allosteric binding site. Noncompetitive and uncompetitive antagonists could also be reversible or irreversible.

ACh competitive antagonists do not depolarize the membrane; hence, they are termed **non-depolarizing blocking agents**. d-tubocurarine, a well-known example of a reversible competitive ACh antagonist, is a constituent of curare, a substance that was used by South American Indians to poison their arrows. Its onset is relatively slow, 5 minutes or more, and its action lasts for 1–2 hours. α-bungarotoxin, found in snake venom, is an irreversible competitive ACh antagonist.

Summary of Main Concepts

- Synapses are specialized structures that constitute the main channel of communication by means of electric signals between a neuron and its target cells.

- The neuromuscular junction is a type of chemical synapse in which the depolarization due to an AP in the presynaptic neuron triggers the release of a neurotransmitter from the presynaptic membrane. The neurotransmitter is ACh in the case of the neuromuscular junction.

- The release of ACh is mediated by the influx of Ca^{2+} through voltage-gated Ca^{2+} channels that are opened by the depolarization due to the AP.

- ACh rapidly diffuses across the synaptic cleft and binds to specialized receptors in the postsynaptic membrane.

- The binding of two ACh molecules to a receptor allows the passage of Na^+ and K^+ through the channel, which causes a net inward epc that depolarizes the endplate.

- The action of ACh is terminated as a result of channel kinetics, desensitization, and the action of the enzyme AChE.

- Leakage of vesicles at the NMJ causes mepps in the postsynaptic membrane, whose amplitude varies according to an approximately normal distribution around a mean value that is generally between 0.4 and 0.5 mV.

- For the spontaneous release of vesicles leading to mepps, the probability density function of the interval between successive releases is an exponential distribution, whereas the probability density function for the release of the kth vesicle is a gamma distribution. For evoked release, the probability density function of a certain number of quanta being released is a binomial distribution.

- The ACh receptor is described as a pentameric heteromer, and its selectivity filter is wide enough to allow bidirectional movement of small cations, those making a significant contribution to the epc being Na^+ and K^+. An open channel has a conductance of 20–40 pS, over a temperature range of 8°–27°C, with a temperature dependence having a Q_{10} of 1.2–1.5.

- All channels open and close randomly because of thermal agitation, the durations of channel opening and closing being exponentially distributed, in accordance with a memoryless random process. The mean open lifetime of the channel is equal to the reciprocal of the macroscopic rate constant for channel closing.

- Because of channel kinetics, an open channel may close for short intervals and reopen once or twice after which the channel closes and ACh dissociates from the receptor.

- Channels desensitize and close but with kinetics that are different from normal closure in the presence of ACh.

- The epc is the aggregate current through all the activated channels and starts to increase in about 0.5–0.8 ms after depolarization of the presynaptic terminal, at 20°–25°C. It reaches its peak in a few hundred microseconds. It has an equilibrium voltage of about 0 mV.

- The endplate voltage results from the action of the epc on the membrane conductance in parallel with the membrane capacitance. The endplate membrane is inexcitable, and the epp is a suprathreshold stimulus that generates a muscle AP in the muscle membrane adjoining the endplate.

- The normal operation of the NMJ can be interfered with through agonists, antagonists, drugs, and diseases, such as myasthenia gravis.

Appendix 5A Chapman–Kolmogorov Equation

The exponential distribution of the time a channel remains open or closed indicates that channel-gating is a memoryless Markov process. This enables a mathematical description of channel kinetics in terms of the theory of continuous-time Markov chains. According to this theory, the transition probabilities for channel-gating obey the Chapman–Kolmogorov equation which will be derived next for the simplest, two-state case and the result generalized to the n-state case.

Equation 5.26 is reproduced here for convenience:

$$N_c \underset{\alpha}{\overset{\beta}{\rightleftharpoons}} N_o \tag{5A.1}$$

where N_c and N_o denote the number of closed channels and open channels, respectively, and α and β are the corresponding rate constants. It follows from Equations 5.27 and 5.28 that in the steady state ($dN_c/dt = 0 = dN_o/dt$),

$$\frac{N_c}{N} = \frac{\alpha}{\alpha + \beta}, \quad \frac{N_o}{N} = \frac{\beta}{\alpha + \beta} \tag{5A.2}$$

N_c/N is the probability of a closed state, whereas N_o/N is the probability of an open state. α and β can therefore be interpreted in terms of probabilities. Formally,

Probability that a channel that is open at t closes at $t + \Delta t = \alpha \Delta t + \delta(\Delta t)$ (5A.3)

where $\delta(\Delta t)$ is a higher-order infinitesimal, remainder term that represents the probability of more than one transition occurring during Δt and which vanishes as $\Delta t \longrightarrow 0$. Equation 5A.3 can be expressed in terms of conditional probabilities as:

$\alpha \Delta t$ = probability(channel closes at $t = t + \Delta t$ | channel is open at t)   (5A.4)

That is, $\alpha \Delta t$ is the probability of the channel closing at $t = t + \Delta t$ given that the channel was open at t. Similarly,

$\beta \Delta t$ = probability(channel opens at $t = t + \Delta t$ | channel is closed at t)   (5A.5)

It follows that:

$1 - \alpha \Delta t$ = probability(channel is open at $t = t + \Delta t$ | channel is open at t)   (5A.6)

$1 - \beta \Delta t$ = probability(channel is closed at t and $t + \Delta t$ | channel is closed at t)

(5A.7)

We next define the transition probability P_{ij} as:

$p_{ij}(\Delta t)$ = probability(channel in state j at $t = \Delta t$ | channel in state i at $t = 0$)

(5A.8)

If the open state is denoted by a subscript 1 and the closed state by a subscript 2, then,

$$p_{11}(\Delta t) = 1 - \alpha \Delta t \tag{5A.9}$$

$$p_{12}(\Delta t) = \alpha \Delta t \tag{5A.10}$$

$$p_{21}(\Delta t) = \beta \Delta t \tag{5A.11}$$

$$p_{22}(\Delta t) = 1 - \beta \Delta t \tag{5A.12}$$

According to these definitions,

$p_{11}(t)$ = probability(channel is open at t | channel is open at $t = 0$)   (5A.13)

$p_{11}(t + \Delta t)$ = probability(channel is open at $t + \Delta t$ | channel is open at $t = 0$)

(5A.14)

But $p_{11}(t + \Delta t)$ equals the probability that the channel is open at $t = t + \Delta t$, is open at t, and is open at $t = 0$, plus the probability that the channel is open at $t = t + \Delta t$, is closed at t, and is open at $t = 0$. Thus:

$$p_{11}(t + \Delta t) = [p_{11}(\Delta t)][p_{11}(t)] + [p_{21}(\Delta t)][p_{12}(t)] \tag{5A.15}$$

$$p_{11}(t + \Delta t) = (1 - \alpha \Delta t)p_{11}(t) + \beta p_{12}(t) \tag{5A.16}$$

$$= p_{11}(t) - [\alpha p_{11}(t) - \beta p_{12}(t)]\Delta t \tag{5A.17}$$

$$p_{11}(t + \Delta t) - p_{11}(t) = -\alpha[p_{11}(t) - \beta p_{12}(t)]\Delta t \tag{5A.18}$$

$$\lim_{\Delta t \to 0} = \frac{p_{11}(t + \Delta t) - p_{11}(t)}{\Delta t} = \frac{dp_{11}(t)}{dt} = -\alpha p_{11}(t) + \beta p_{12}(t) \qquad (5A.19)$$

Similarly,

$$\frac{dp_{12}(t)}{dt} = \alpha p_{11}(t) - \beta p_{12}(t) \qquad (5A.20)$$

$$\frac{dp_{21}(t)}{dt} = -\alpha p_{21}(t) + \beta p_{22}(t) \qquad (5A.21)$$

$$\frac{dp_{22}(t)}{dt} = \alpha p_{21}(t) - \beta p_{22}(t) \qquad (5A.22)$$

Equations 5A.19–5A.22 can be written in matrix form as:

$$\frac{d}{dt}\begin{bmatrix} p_{11}(t) & p_{12}(t) \\ p_{21}(t) & p_{12}(t) \end{bmatrix} = \begin{bmatrix} p_{11}(t) & p_{12}(t) \\ p_{21}(t) & p_{12}(t) \end{bmatrix}\begin{bmatrix} -\alpha & \alpha \\ \beta & -\beta \end{bmatrix} \qquad (5A.23)$$

or,

$$\frac{dP(t)}{dt} = P(t)Q \qquad (5A.24)$$

where $P(t)$ is **the transition matrix** $[P_{ij}(t)]$ and Q is known as the infinitesimal matrix, defined as:

$$Q = [q_{ij}] = \lim_{\Delta t \to 0} \frac{P(\Delta t) - I}{\Delta t} \qquad (5A.25)$$

and I is the identity matrix,

$$I = [\delta_{ij}] = \begin{matrix} 1 & i = j \\ 0 & i \neq j \end{matrix} \qquad (5A.26)$$

Equation 5A.25 can be readily verified for the two-state case. The matrix $P(\Delta t)$ has the elements: $p_{11}(\Delta t)$, $p_{12}(\Delta t)$, $p_{21}(\Delta t)$, and $p_{22}(\Delta t)$ given by Equations 5A.9–5A.12. When 1 is subtracted from the diagonal elements the matrix becomes:

$$\begin{bmatrix} -\alpha \Delta t & \alpha \Delta t \\ \beta \Delta t & -\beta \Delta t \end{bmatrix}$$

Dividing by Δt gives the $\mathbf{Q}$ matrix.

Equation 5A.24 is the **Chapman–Kolmogorov equation**, derived for the two-state case. However, the form expressed by Equations 5A.24 and 5A.25 is quite general and applies to a single channel that can exist in one of n states, where $p_{ij}(\Delta t)$ is defined as in Equation 5A.8 and q_{ij} in equation 5A.25 is given by:

$$q_{ij} = -(\text{sum of transition rates leading away from the state}), \text{ for } i = j \quad (5A.27)$$

$$= +(\text{transition rate from state } i \text{ to state } j), \text{ for } i \neq j \quad (5A.28)$$

$\mathbf{Q}$ is in general a singular matrix, which means that the determinant of $\mathbf{Q}$ is zero, that is, $\Delta\mathbf{Q} = 0$, as can be readily verified for the two-state case. Because it is a singular matrix, $\mathbf{Q}$ has only $(n - 1)$ nonzero eigenvalues (λ) defined by the equation: $\Delta(\mathbf{Q} - \lambda\mathbf{I}) = 0$, where Δ denotes the determinant.

The following conclusions can be derived from the Chapman–Kolmogorov equations:

Conclusion 1: Once a state is entered, the time interval spent in the state is exponentially distributed with a mean lifetime equal to $-1/q_{ii}$. For the two-state model, the probability density function for the open lifetime is $\alpha e^{-\alpha t}$ and the mean open lifetime is $1/\alpha$ (Equations 5.24 and 5.25). Similarly, the probability density function for the closed lifetime is $\beta e^{-\beta t}$ and the mean closed lifetime is $1/\beta$.

Conclusion 2: The general solution of the Chapman–Kolmogorov equation is:

$$\mathbf{P}(t) = e^{\mathbf{Q}t} = \mathbf{I} + \mathbf{Q}t + (\mathbf{Q}t)^2 / 2! + \ldots \quad (5A.29)$$

and,

$$p_{ij}(t) = p_j(\infty) + w_1 e^{\lambda_1 t} + w_2 e^{\lambda_2 t} + \ldots \quad (5A.30)$$

where $p_j(\infty)$ is the equilibrium probability that the channel is in state j, λ_1, λ_2, ..., are the nonzero eigenvalues of the matrix $\mathbf{Q}$, and w_1, w_2, ..., are functions of the rate constants.

For the two-state case, there is only one eigenvalue, $\lambda = -(\alpha + \beta)$. Equation 5A.30 becomes:

$$p_{11}(t) = p_1(\infty) + w_1 e^{-(\alpha+\beta)t} \quad (5A.31)$$

Setting $t = 0$ in Equation 5A.31,

$$p_{11}(0) = p_1(0) = p_1(\infty) + W_1 \quad (5A.32)$$

where $p_1(0)$ is the initial probability of the channel being open at $t = 0$. From equation 5A.32,

$$w_1 = p_1(0) - p_1(\infty) \tag{5A.33}$$

Substituting in Equation 5A.31,

$$p_{11}(t) = p_1(\infty) - [p_1(\infty) - p_1(0)]e^{-(\alpha+\beta)t} \tag{5A.34}$$

This is the same as Equation 5.29 when divided by the number of channels N and the ratios of channel numbers are interpreted as probabilities.

From Equation 5A.19 in the steady state, as $t \longrightarrow \infty$,

$$\frac{dp_{11}(t)}{dt} = 0 = -\alpha p_{11}(\infty) + \beta p_{12}(\infty) \tag{5A.35}$$

$p_{11}(\infty) = p_1(\infty)$ is steady-state probability of the channel being open, and $p_{12}(\infty)$ is the steady-state probability of the channel being closed, which equals $1 - p_1(\infty)$. Substituting in Equation 5A.35 gives:

$$p_1(\infty) = \frac{\beta}{\alpha + \beta} \tag{5A.36}$$

Equation 5A.34 can be written as:

$$p_{11}(t) = \frac{\beta}{\alpha + \beta} + \left(p_1(0) - \frac{\beta}{\alpha + \beta} \right) e^{-(\alpha+\beta)t} \tag{5A.37}$$

Similarly,

$$p_{12}(t) = \frac{\alpha}{\alpha + \beta} + \left(p_{12}(0) - \frac{\alpha}{\alpha + \beta} \right) e^{-(\alpha+\beta)t} \tag{5A.38}$$

$$p_{21}(t) = \frac{\beta}{\alpha + \beta} + \left(p_{21}(0) - \frac{\beta}{\alpha + \beta} \right) e^{-(\alpha+\beta)t} \tag{5A.39}$$

$$p_{22}(t) = \frac{\alpha}{\alpha + \beta} + \left(p_2(0) - \frac{\alpha}{\alpha + \beta} \right) e^{-(\alpha+\beta)t} \tag{5A.40}$$

$(\alpha + \beta)$ is termed the **relaxation rate constant**; $1/(\alpha + \beta)$ is termed the **relaxation time constant** and is also the time constant for the macroscopic decay.

Problem 5.6

A neuron contains 50,000 Na⁺ channels of 10 pS conductance and which open and close in accordance with a two-state transition (Reaction 5.26) having $\alpha = 120e^{-v_M/40}$/s and $\beta = 30e^{v_M/20}$/s, where v_m is in mV. Derive the Q matrix for the neuron at $v_m = -80$ mV and determine the mean Na⁺ current in the steady state.

ANS.: $Q = \begin{vmatrix} -887 & 887 \\ 0.53 & -0.53 \end{vmatrix}$; -40.3 pA.

Problem 5.7

A ligand-gated channel follows the two-state transition scheme:

$$A + R \underset{\alpha}{\overset{\beta}{\rightleftharpoons}} AR^o$$

where $\beta = 10^5$/s, $\alpha = 50$/s, and $u_A = 10^{-5}$ M. (a) Derive the Q matrix; (b) determine the mean open and closed lifetimes; (c) calculate the percentage of the time the channel is open.

ANS.: (a) $Q = \begin{vmatrix} -100 & 100 \\ 1 & -1 \end{vmatrix}$; (b) 0.02 s and 1 s, respectively; (c) about 1%.

Problem 5.8

A channel is described by the following transition scheme:

$$C_1 \underset{k_{-1}}{\overset{k_1}{\rightleftharpoons}} C_2 \underset{k_{-2}}{\overset{k_2}{\rightleftharpoons}} C_3 \underset{\alpha}{\overset{\beta}{\rightleftharpoons}} O$$

(a) Derive the Q matrix; (b) determine the mean lifetimes in each of the four states; (c) if the rate constants k_1, k_{-1}, k_2, and k_{-2} are all much larger than α and β, what is the approximate relaxation time constant from the open state?

ANS.: (a) $Q = \begin{vmatrix} -\alpha & \alpha & 0 & 0 \\ \beta & -(\beta + k_{-2}) & k_{-2} & 0 \\ 0 & k_2 & -(k_{-1} + k_2) & k_{-1} \\ 0 & 0 & k_1 & -k_1 \end{vmatrix}$; (b) state 1: $1/\alpha$, state 2: $1/(\beta + k_{-2})$,

state 3: $1/(k_{-1} + k_2)$, state 4: $1/k_1$; (c) approximately $1/(\alpha + \beta)$.

6

Synapses

Objective and Overview

Having discussed the neuromuscular junction (NMJ) in the previous chapter, the main focus in the present chapter is on synapses in the central nervous system. The chapter begins with a general overview of the various types of synapses, including an equivalent electric circuit and a careful definition of the criteria for considering a synapse to be excitatory or inhibitory. Neurotransmitter types are discussed in some detail. Fast chemical synapses, both excitatory and inhibitory, are considered next, including the highly important NMDA and non-NMDA types of receptors. The basic principles underlying the operation of second-messenger systems are explained. These systems are slower in speed than the fast chemical synapses but are more varied and have a profound effect on many aspects of cell function. The main neuromodulatory systems, which act mostly through second-messenger systems are presented, followed by presynaptic inhibition and facilitation as a means for selectively affecting certain inputs to a neuron. Synaptic plasticity, manifested as short-term and long-term modification of synaptic efficacy, is then discussed in terms of its operational effects and basic molecular mechanisms. Synaptic plasticity is of great importance, as it affects neuronal responses to stimulation and underlies learning and memory. The Chapter ends with a discussion of the less common electrical synapses, which have some important properties that distinguish them from chemical synapses.

Learning Objectives

To understand:

- The overall properties and types of synapses
- The main features of operation of fast chemical synapses compared to the neuromuscular junction

- The basic molecular mechanisms underlying second-messenger systems and the effects of these systems on ionic phenomena
- The main neuromodulatory systems of the CNS
- The mechanisms underlying presynaptic inhibition and facilitation
- The types and basic mechanisms of synaptic plasticity
- The structure and distinguishing features of electrical synapses

6.1 Overview of Synapses

6.1.1 General

Neurons communicate over the short term, that is, over time spans not exceeding a few hundred milliseconds, by means of electric signals that mostly involve synapses. But synapses are far from being just a means of transmitting electric signals between neurons and their target cells. By virtue of their properties, number, and spatial distribution, they play a major role in the processing of electric signals in the nervous system (Chapter 7). Moreover, synapses are considered by some to be the basic unit of storage of information in the brain. It is estimated that the brain has roughly 10^4 times as many synapses as neurons, that is, of the order of 10^{15} synapses.

Synapses can be divided into two broad categories: electrical and chemical. **Electrical synapses** have conductive pathways between the presynaptic and postsynaptic cells so that electric charge is transferred directly through the synapse, as discussed in Section 6.6, thereby changing the membrane voltage of the postsynaptic cell. The majority of synapses, however, are **chemical synapses** in which the presynaptic and postsynaptic cell membranes are separated by a narrow cleft, 20–40 nm wide. The operation of chemical synapses is similar to that of the neuromuscular junction (NMJ) but differs from it in many important respects, as will be clarified in this and the following sections. Overall, an action potential (AP) in the transmitting, or **presynaptic** neuron, produces in the **postsynaptic** neuron a change in membrane voltage from the resting value, termed the **postsynaptic potential (psp)**, as illustrated diagrammatically in Figure 6.1. Operationally, the depolarization due to the AP in the presynaptic neuron triggers the release of a chemical substance, the neurotransmitter, from the presynaptic membrane. The neurotransmitter diffuses across the synaptic cleft and binds to specialized receptors in the postsynaptic membrane. In the case of **ionotropic receptors**, the binding of the neurotransmitter to the receptor opens an ion channel that allows ions to flow between the cytoplasm of the postsynaptic cell and the extracellular medium, thereby changing the membrane voltage of the postsynaptic cell. In **metabotropic receptors**, the binding of the neurotransmitter to the receptor may trigger changes in cell metabolism via intracellular second-messengers,

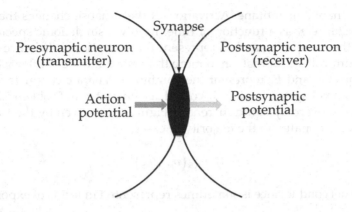

FIGURE 6.1
Overall operation of a chemical synapse.

eventually leading to the gating of ion channels, as discussed in Section 6.3. As to be expected, the direct channel-gating by ionotropic receptors is faster than the indirect gating by metabotropic receptors, producing a change in the voltage of the postsynaptic membrane within a few milliseconds or less of the depolarization of the presynaptic terminal. Consequently, synapses having ionotropic receptors are termed **fast chemical synapses**. However, second-messenger systems are of great diversity and can be highly complex, with far-reaching effects on many aspects of cell function.

A simplified electrical equivalent circuit for a fast chemical synapse and a surrounding patch of postsynaptic membrane is illustrated in Figure 6.2, where v_m represents the instantaneous membrane voltage, and G_{m0} and V_{m0} represent the resting membrane conductance and voltage, respectively, of

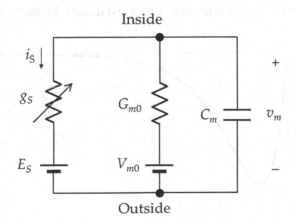

FIGURE 6.2
Simplified electrical equivalent circuit for a fast chemical synapse and a surrounding patch of postsynaptic membrane.

the postsynaptic membrane. Activation of the synapse changes the synaptic conductance g_S as a function of time. If only a single ionic species flows through the synapse, g_S and E_S represent the synaptic conductance and the equilibrium voltage for that ion. If more than one ionic species flows through the synapse, g_S and E_S represent the combined synaptic conductances and equilibrium voltages of the ions involved, as illustrated in Problem 5.5 for the NMJ. The outward synaptic current in Figure 6.2 is given by the following direct proportionality in the majority of cases:

$$i_S = g_S(v_m - E_S) \tag{6.1}$$

The channel conductance is sometimes represented in terms of exponentials as:

$$g_S = G_S\left(1 - e^{-t/\tau_r}\right)\left(ae^{-t/\tau_{ff}} + (1-a)e^{-t/\tau_{fs}}\right), \quad t \geq 0 \tag{6.2}$$

where G_S and $a \leq 1$ are constants, τ_r is the time constant for the rising phase. τ_{ff} is the time constant for a fast component of the falling phase and τ_{fs} is the time constant for a slow component. The channel conductance, or the synaptic current, can also be represented more simply by an **alpha function** $f(t)$ of the form:

$$f(t) = A_{\max}(t/\alpha)e^{(1-t/\alpha)}, \quad t \geq 0 \tag{6.3}$$

where $A_{\max}$ is the maximum value of $f(t)$ and occurs at $t = \alpha$. Figure 6.3 illustrates a synaptic current i_S in the form of an alpha function having $A_{\max} = -I_m$, where the negative sign denotes an inward current.

Chemical synapses can be excitatory or inhibitory. The basic distinction between the two is that activation of an **excitatory synapse** enhances the

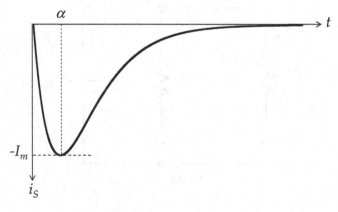

FIGURE 6.3
Alpha function representation of synaptic current.

likelihood of generation of an AP in the postsynaptic cell, whereas the activation of an **inhibitory synapse** reduces this likelihood. If the threshold for the generation of an AP in the postsynaptic cell is $V_{thr} > V_{m0}$, then the synapse is *excitatory* if:

$$E_S > V_{thr} > V_{m0} \tag{6.4}$$

This is because the activation of the G_S–E_S branch depolarizes the membrane and moves v_m toward E_S, closer to V_{thr}, as illustrated by the vertical arrow and the dashed line in Figure 6.4a, where it is assumed that $V_{m0} = -60$ mV, $V_{thr} = -50$ mV, and $E_S = E_{S1} = -45$ mV. Inequality 6.4 implies that $i_S < 0$ in Equation 6.1. This means that cations will move inward, or anions outward, thereby depolarizing the membrane. The synapse is excitatory since depolarization brings the membrane voltage closer to threshold (Figure 6.4a), which enhances the likelihood of generating an AP by other psps. The psp is termed in this case the **excitatory post-synaptic potential (epsp)**.

In an **obligatory synapse** or a **non-integrating synapse**, a single epsp depolarizes the postsynaptic membrane beyond threshold and generates an AP in the postsynaptic cell, as in the NMJ. Most synapses, however, are **integrating synapses**, in which 50–100 epsps are typically required to generate an AP, so that a single epsp only enhances the likelihood that additional epsps will depolarize the membrane to threshold.

On the other hand, the synapse is inhibitory if:

$$E_S < V_{m0} < V_{thr} \tag{6.5}$$

In Figure 6.4b, $V_{m0} = -60$ mV, $V_{thr} = -50$ mV, and $E_S = E_{S2} = -65$ mV, so $E_{S2} < V_{thr}$ and is also less than V_{m0}; $i_S > 0$ in Equation 6.1, which means that cations will move outward, or anions inward, thereby hyperpolarizing the membrane.

$E_{S1} = -45$ mV

$V_{thr} = -50$ mV

$E_{S3} = -55$ mV

$V_{m0} = -60$ mV

$E_{S2} = -65$ mV

(a)　　　　　(b)　　　　　(c)

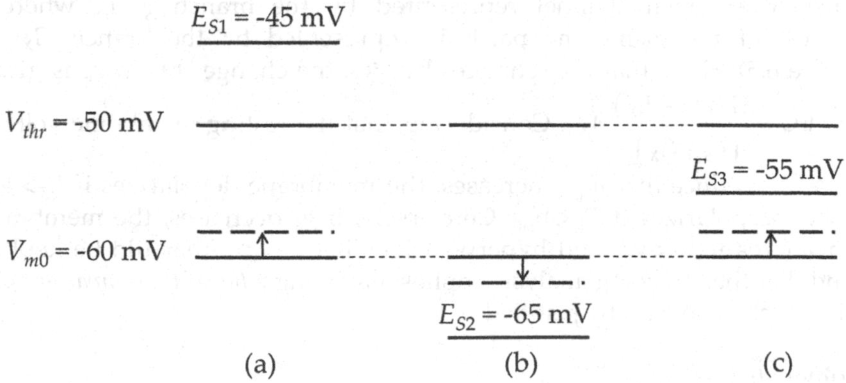

FIGURE 6.4
Voltage relations at an excitatory synapse (a) and at an inhibitory synapse (b); (c) synaptic activation is depolarizing but the effect is inhibitory because of the shunting effect.

The subtractive effect of the hyperpolarization makes it more difficult for epsps to depolarize the membrane to threshold, which reduces the likelihood of generating an AP and makes the synapse inhibitory. The psp in this case is termed the **inhibitory post-synaptic potential** (ipsp).

It should be noted, however, that inhibitory action can also be due to effectively clamping the membrane voltage at a subthreshold level. In Figure 6.4c, $V_{m0} = -60$ mV, $V_{thr} = -50$ mV, as in the preceding cases, but $E_S = E_{S3} = -55$ mV, so $E_{S3} < V_{thr}$ still, but $E_{S3} > V_{m0}$. Synaptic activation moves v_m closer to E_{S3}, which means that the membrane is depolarized and v_m is brought closer to threshold. However, a large g_S produces a "shunting" effect which, in the extreme case of $g_S \longrightarrow \infty$, rigidly clamps v_m at -55 mV, making it impossible for epsps to depolarize the membrane to threshold. The effect of synaptic activation is therefore inhibitory in this case. The psp, although depolarizing, has an inhibitory effect, like that of an ipsp. Thus, whereas *an epsp always depolarizes the membrane from the resting state, an ipsp may either depolarize or hyperpolarize the membrane from the resting state.* An epsp will also tend to clamp v_m at E_S in Figure 6.4a, but since $E_S > V_{thr}$, the threshold is reached before v_m equals E_S. Similarly, a hyperpolarizing ipsp will also have a shunting effect because of increased membrane conductance, which makes it more difficult still for an epsp to depolarize the membrane to threshold. Note that the shunting effect would also prevent linear summation of psps from adjoining synapses, even if membrane conductances were constant (see Problem 6.2).

Although most chemical synapses are axo-dendritic or axo-somatic, synapses can also be axo-axonic (Section 6.4) as well as dendro-dendritic (Section 6.2.1). Moreover, electrical communication between neurons is not restricted to synapses but can take place through so-called field potentials caused by electrical activity in neighboring cells (Section 7.2.4.2).

Problem 6.1

Consider an open channel represented by the branch g_S–E_S whereas the rest of the membrane patch is represented by the branch G_X–E_X (Figure 6.5). Show that if g_S changes by Δg_S, the change Δv_m in v_m is given by: $\Delta v_m = \dfrac{G_X(E_S - E_X)}{(g_S + G_X)^2} \Delta g_S$. Considering that the resting membrane voltage $V_{m0} \cong E_X$, deduce that if g_S increases, the membrane depolarizes if $E_S > V_{m0}$ and hyperpolarizes if $E_S < V_{m0}$. Conversely, if g_S decreases, the membrane depolarizes if $E_S < V_{m0}$ and hyperpolarizes if $E_S > V_{m0}$. It should be kept in mind that the foregoing analysis applies assuming a *linear time-invariant* circuit, which is not strictly justified.

Problem 6.2

Consider two synapses represented as in Figure 6.5 by two branches, g_{s1}–E_{s1} and g_{s2}–E_{s2}. Show that if synapse 1 acts alone ($g_{s2} = 0$), $v_{m1} = \dfrac{E_{S1}g_{S1} + E_X G_X}{g_{S1} + G_X}$,

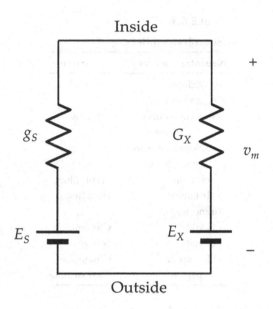

Inside

FIGURE 6.5
Figure for Problem 6.1.

if synapse 2 acts alone ($g_{S1} = 0$), $v_{m2} = \dfrac{E_{S2}g_{S2} + E_X G_X}{g_{S2} + G_X}$, and if both synapses act together, $v_{m12} = \dfrac{E_{S1}g_{S1} + E_{S2}g_{S2} + F_X G_X}{g_{S1} + g_{S2} + G_X}$. Note that $v_{m12} < v_{m1} + v_{m2}$. Superposition in linear circuits applies to voltages and currents only.

6.1.2 Neurotransmitters

6.1.2.1 Types of Neurotransmitters

Neurotransmitters are predominantly small molecules that are packaged in vesicles in presynaptic terminals, released into the synaptic cleft, and typically reabsorbed into the presynaptic terminal. The most prevalent neurotransmitters are listed in Table 6.1 and are briefly discussed in what follows. All the neurotransmitters listed in Table 6.1, with the exception of ACh, are amino acids or related compounds.

ACh was discussed in Chapter 5. It is the neurotransmitter not only of the neuromuscular junction but also all preganglionic neurons of the autonomic nervous system – both sympathetic and parasympathetic – as well as postganglionic neurons of the parasympathetic system. However, the ACh receptors of autonomic neurons are distinct from, though similar to, the nicotinic ACh receptors (nAChRs) of the NMJ. ACh is also found throughout the brain and is a major neuromodulator (Section 6.3.2). It is synthesized from choline and acetyl-coA. Choline is mostly derived from food, but only a small

TABLE 6.1

Neurotransmitters

Neurotransmitter	Precursor
Acetylcholine	Choline
Biogenic amines	
Catecholamines	Tyrosine
Dopamine	
Norepinephrine	
Epinephrine	
Serotonin	Tryptophan
Histamine	Histidine
Amino acids	
GABA	Glutamine
Glycine	Serine
Glutamate	Glutamine
Aspartate	Oxaloacetate

amount is made in the human liver. Acetyl-coA participates in many biochemical reactions involved in protein, carbohydrate, and lipid metabolism and is synthesized in mitochondria.

Biogenic amines, also referred to as **monoamines** because they have a single amino group (NH_2, formally derived from ammonia, NH_3), are derived from aromatic amino acids by the action of aromatic amino-acid-decarboxylase enzymes and have their amino group connected to an aromatic ring by a two-carbon chain ($-CH_2-CH_2-$ or derivatives). The biogenic amines include: (i) the **catecholamines:** dopamine, norepinephrine, and epinephrine, all derived from the essential amino acid tyrosine; (ii) serotonin, derived from the essential amino acid tryptophan; and (iii) histamine, derived from the essential amino acid histidine. Dopamine is a neurotransmitter in the brain and also a major neuromodulator, discussed in Section 6.3.2. Norepinephrine, also referred to as noradrenalin, is the neurotransmitter of the postganglionic neurons of the sympathetic system. Its effect on the heart is to increase the heart rate as well as the force of contraction of the heart. It causes contraction of the smooth muscles of veins and arterioles, leading to vasoconstriction, except in the arterioles of skeletal muscle, which are not affected by norepinephrine. Norepinephrine is a neurotransmitter in the brain and also a major neuromodulator, discussed in Section 6.3.2. Epinephrine, also referred to as adrenalin, is derived from norepinephrine and is secreted as a hormone into the blood stream by the adrenal medulla. Epinephrine is vasodilatory on the arterioles of skeletal muscle in order to cater to the requirements of the "fight-or-flight" response. Its effect on the heart is similar to that of norepinephrine. In the brain, only a small number of neurons use epinephrine as a neurotransmitter. Serotonin, also known as 5-Hydroxytryptamine (5-HT), is an important neurotransmitter in the brain and also a major neuromodulator, discussed in Section 6.3.2. Histamine is a neurotransmitter that is mainly concentrated in the hypothalamus, one of whose main functions is to link the nervous system to the endocrine system through its action on the pituitary gland.

Because of their widespread effects in the brain, defects in biogenic amine functions are implicated in many psychiatric disorders. Many substances of abuse also exercise their effects through their actions on biogenic amine pathways.

GABA (γ-aminobutyric acid), derived from the nonessential amino acid glutamine, is a major neurotransmitter of inhibitory neurons, such as cerebellar Purkinje cells, and inhibitory interneurons in the brain, as well as some inhibitory interneurons in the spinal cord. Glycine, a nonessential amino acid that is derived from the nonessential amino acid serine, is the major neurotransmitter of inhibitory interneurons in the spinal cord. The amino acids glutamine, glycine, and serine are abundant in living cells and partake in many important cellular functions and structures. The phospholipid phosphatidylserine is a major constituent of cell membranes and the myelin sheath.

Glutamate is the major excitatory neurotransmitter of the central nervous system. It is the monovalent cation that arises from the loss of an H^+ from the terminal carboxyl group of glutamic acid – a non-essential amino acid. Glutamate is chemically related to glutamine. Aspartate, derived mainly from oxaloacetate, is also a major excitatory neurotransmitter.

Synaptic receptors, as well as synapses, and even neurons, are sometimes characterized by the neurotransmitter associated with them. Thus, receptors activated by glutamic acid, for example, are referred to as **glutamatergic receptors**, and those activated by γ-aminobutyric acid (GABA) are referred to as **GABAergic receptors**. Some receptors may be activated by more than one neurotransmitter, and some neurons may release more than one neurotransmitter from the same site or from different sites, as mentioned in the next section.

The amino acid β-alanine acts as an inhibitory neurotransmitter in parts of the brain. The amino sulfonic acid taurine is also an inhibitory neurotransmitter released by cerebellar stellate interneurons.

Other substances can also act as neurotransmitters. ATP acts as an excitatory neurotransmitter for the trimeric (that is, consisting of three subunits) P2X purinergic receptor family that is expressed in motoneurons, as well as neurons in some regions of the brain, and is widely expressed in autonomic neurons, where it is involved in the contraction of cardiac cells and some smooth muscle cells as well as in mediating pain. ATP is released by tissue damage and serves to transmit pain signals to the spinal cord via the axon terminals of nociceptive dorsal root ganglion cells.

Adenosine (the nucleoside of adenine, Section 6.3.1) is derived extracellularly from ATP by enzymatic action and has an inhibitory action in the brain. The stimulatory effect of caffeine is due to its antagonistic action on the adenosine receptors.

Neuroactive peptides, also referred to as **neuropeptides**, can act as neurotransmitters when released by neurons, with some neuropeptides released by glia. Neuropeptides can act as hormones when secreted by glands into the blood stream. More than 100 neuropeptides have been identified and are involved in a wide range of brain functions including sensory perception, feeding behavior, emotions, social behavior, learning, and memory. Examples are oxytocin, vasopressin, substance P, and opioids, which include endorphins, these being endogenous compounds that mimic the action of morphine. However, the site of synthesis of neuropeptides, the type of vesicles in which they are stored, and the mechanism and site of exocytosis are quite different from those of the aforementioned neurotransmitters, which are often referred to as **small-molecule neurotransmitters** to distinguish them from neuropeptides.

Neuropeptides are synthesized in the rough endoplasmic reticulum (Section 1.1.1), initially as part of larger precursor molecules, the **prepropeptides**. They then pass through the Golgi apparatus, which is a cell organelle consisting of vesicles and folded membranes (Figure 1.1) that is involved in secretion and intracellular transport. The neuropeptides undergo some processing in the Golgi apparatus and are eventually packaged into **large dense-core vesicles** (**LDCVs**). Because of the way they are synthesized and packaged, more than one neuropeptide can be found in the same vesicle. The vesicles are translocated to their release sites, which could be in the soma, axon, or dendrites. Neuropeptides act almost exclusively via activation of G protein-coupled receptors (Section 6.3).

It should be noted that an abundance of neurotransmitters is found in the **enteric nervous system** (**ENS**), a massive network of over 100 million neurons located in the gastrointestinal tract and which coordinate the contractions in this tract. The ENS accounts for about 95% of the body's serotonin and about 50% of the body's dopamine, in addition to a variety of other neurotransmitters, including ACh and substance P. There seems to be an intimate, two-way relationship between the brain and the gut, including the ENS and the gut's **microbiome**, that is, the microorganisms that inhabit the gut.

6.1.2.2 Neurotransmitter Cycle

Small-molecule neurotransmitters, or simply neurotransmitters, are synthesized mainly in axon terminals and are actively packaged into small electron-lucent vesicles, about 40 nm in diameter, by transporters driven by a H^+ electrochemical potential gradient. This gradient is established by a vacuolar-type **H⁺-ATPase** (**V-ATPase**) which uses the energy from ATP hydrolysis to pump H^+ (protons) into vesicles. This results in a lower pH inside the vesicle and in a positive voltage of tens of millivolts or more with respect to the cytoplasm. The resulting electrochemical potential gradient for H^+ is used to accumulate various substances inside the vesicle. The vesicles are released into the synaptic cleft at active zones through exocytosis (Section 1.1.3) mediated by the inflow of Ca^{2+} resulting from depolarization of the presynaptic terminal, as in the NMJ. However, there is evidence that in some cases neurotransmitters may be released by exocytosis from the axoplasm directly into the synaptic cleft. Vesicles may be round or flat in shape. It is believed that the former are found in excitatory synapses, whereas the latter are found in inhibitory synapses.

It is essential for proper synaptic action that the neurotransmitter binds to the postsynaptic receptor for a relatively short interval. Otherwise, the synapse becomes inoperative, as the neurotransmitter remains bound to the receptor. ACh is hydrolyzed by the enzyme AChE, as in the NMJ (Section 5.2). In the case of other neurotransmitters, the neurotransmitter concentration in the cleft rapidly decreases, which unbinds the neurotransmitter from the receptor. This decrease is mainly due to uptake by transporters that are specific for each neurotransmitter and which are driven by the electrochemical potential gradient of Na^+. In the case of ACh, only choline is transported back into the presynaptic terminal, as in the NMJ (Section 5.2). Dopamine, norepinephrine, serotonin, most of GABA and glycine, and some of the glutamate are transported back directly into the presynaptic terminal. Some of GABA and glycine and most of the glutamate are transported into glial cells, where GABA and glutamate are converted to glutamine that is transported back to the presynaptic terminal. Glial cells can also synthesize, store, and release neurotransmitters and can also have receptors for neurotransmitters that modulate their function. The action of neuropeptides is terminated by peptidases, which are extracellular enzymes that break the neuropeptides into inactive amino acids.

It was once believed that a neuron released only one neurotransmitter, as embodied in Dale's Principle. It is now well-established that a neuron may release more than one neurotransmitter, a phenomenon referred to as **neurotransmitter corelease**. It was previously mentioned that multiple neuropeptides may be released from the same vesicle. Moreover, a neuron may corelease neuropeptides and small-molecule neurotransmitters or more than one small-molecule neurotransmitter. The vesicles for coreleased neurotransmitters can be found in the same synaptic bouton or in different locations. For example, GABA is coreleased with the neuropeptide somatostatin by neurons in the hippocampus and with neuropeptide Y in the arcuate nucleus. Dopamine is released with the neuropeptide glucagon-like peptide-1 in the nucleus accumbens, located in the basal forebrain, and is anatomically part of the basal ganglia (Section 12.2.3). Neurons in the tuberomammillary nucleus in the posterior hypothalamus corelease histamine, GABA, and the neuropeptides galanin, enkephalin (an opioid), and substance P. Retinal amacrine cells corelease GABA and ACh from different vesicle populations, subserving different physiological functions. ATP and GABA are coreleased in dorsal horn and lateral hypothalamic neurons. Glycine and GABA are coreleased by spinal interneurons and by cerebellar Golgi and Lugaro cells (Section 12.2.4.3). Dopamine and glutamate are coreleased by neurons in the ventral tegmental area (VTA) of the midbrain, and serotonin and glutamate are coreleased by neurons of the dorsal raphe nucleus in the brainstem (Section 12.2.5.2).

Problem 6.3

Consider a vesicle undergoing exocytosis, beginning with the formation of a "fusion pore" of conductance g_p, as diagrammatically illustrated in Figure 6.6. Assume that the vesicle has a capacitance C_V and a voltage v_V across its membrane of the polarity shown. Determine: (a) the capacitance added to that of the presynaptic membrane when the vesicle fuses with the membrane; (b) the outward current i_p through the pore at the instant that this pore is formed; (c) the total outward charge through the pore.

ANS.: (a) C_V; (b) $(v_V + v_m)g_p$; (c) $(v_V + v_m)C_V$

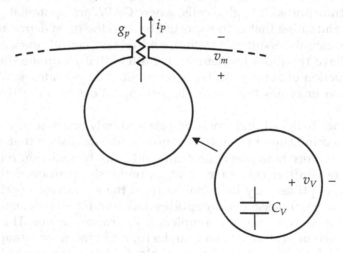

FIGURE 6.6
Figure for Problem 6.2.

6.2 Fast Chemical Synapses

6.2.1 General

The same general principles discussed for the NMJ in the preceding chapter apply to fast chemical synapses in the rest of the nervous system. The depolarization of the presynaptic AP causes Ca^{2+} influx, which triggers the release of a neurotransmitter from vesicles by exocytosis. The neurotransmitter diffuses across the synaptic cleft and binds to ionotropic receptors on the postsynaptic membrane, thereby opening an ion channel and generating a postsynaptic potential. However, some notable features of these synapses, compared to the NMJ, should be emphasized:

1. A fast chemical synapse usually contains only one or, at most, a few active zones for neurotransmitter release. In contrast, an NMJ contains several hundred active zones. Since release of neurotransmitter vesicles from an active zone is probabilistic, vesicle release due to a presynaptic AP at a fast chemical synapse typically occurs only 10%–40% of the time, with a significant probability of failure. On the other hand, because of the large number of active zones at the NMJ, the probability of an AP failing to evoke an epp is almost nil.

2. Because an AP in one of the presynaptic axons of a fast chemical synapse typically does not release more than one vesicle of neurotransmitter, the peak of the postsynaptic potential (psp) at a fast chemical synapse does not exceed a few hundred microvolts. Whereas a NMJ is obligatory, or non-integrating, fast chemical synapses are integrating, as discussed in Section 6.1.1.

3. At 37°C, the delay between a presynaptic spike and the beginning of the psp in a fast chemical synapse is 0.2–0.5 ms, and the time to peak of the psp is typically less than 1.5 ms. The decay time constant of the postsynaptic current (psc) is typically a fraction of a millisecond, which is also the mean open time of the channel (Equation 5.24).

4. Unlike the NMJ, which is excitatory, fast chemical synapses can be excitatory or inhibitory, as explained in Section 6.1.1. Moreover, whereas the ion channel of the NMJ is relatively wide and allows practically all small cations to flow in both directions, ion channels of fast chemical synapses are generally more selective.

5. Neuronal nicotinic ACh receptors differ from the nicotinic ACh receptors of the neuromuscular junction (Section 5.4) in that they comprise only α and β subunits in the ratio of $3\alpha{:}2\beta$ and are not sensitive to α-bungarotoxin.

6. As to be expected, neuronal synapses are of varied structure and organization that suit their function. For example, ribbon synapses are commonly associated with receptors of complex sensory systems, as in retinal photoreceptors, vestibular-organ receptors, and cochlear hair cells. In these synapses, 100 or more vesicles are tethered to a ribbon, there being 10–100 such ribbons located several nanometers away from the presynaptic membrane. Graded voltages of the presynaptic membrane cause a sustained release of vesicles at a high rate for long periods, as required for high rates of signal transmission. Sensory systems also have dendro-dendritic reciprocal synapses in which each of the two dendrites involved is both presynaptic and postsynaptic, the two synapses being often adjacent to one another (Figure 6.7). Both synapses may be excitatory or one synapse is excitatory, the other inhibitory. A presynaptic membrane may form synapses on two different postsynaptic cells.

Some particular characteristics of inhibitory and excitatory, fast chemical synapses are considered next.

6.2.2 Fast Inhibitory Synapses

The most common inhibitory neurotransmitters in vertebrates are γ-aminobutyric acid (GABA) and the amino acid glycine, the latter being a widespread neurotransmitter of inhibitory interneurons of the brainstem and spinal cord. The GABA receptor of fast synapses is denoted as $GABA_A$ to distinguish it from a different receptor $GABA_B$ encountered in second-messenger systems (Section 6.3). $GABA_A$ and glycine channels are pentameric, like nicotinic ACh receptors (Section 5.4.1). When opened by the binding of the inhibitory neurotransmitter, they are permeable to small anions, but because Cl^- have a higher concentration than other anions, the ipsp is caused mainly by inflow of Cl^- down their electrochemical potential gradient, resulting in a hyperpolarization. In spinal motoneurons, the equilibrium voltage of Cl^- is generally in the range −70 to −80 mV, close to a less negative, resting voltage. Figure 6.8 illustrates an ipsp in a motoneuron elicited by an inhibitory interneuron. In addition, the increased Cl^- conductance has an added inhibitory effect due to shunting, as explained in Section 6.1.1.

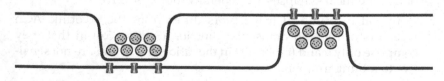

FIGURE 6.7
A dendro-dendritic reciprocal synapse.

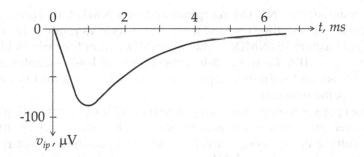

FIGURE 6.8
Illustration of an ipsp in a motoneuron.

In some central neurons, the equilibrium voltage of Cl⁻ is nearly equal to the resting voltage. In such cells an increase in Cl⁻ conductance does not hyperpolarize the membrane, but the effect is still inhibitory because of the shunting effect. In other neurons, such as those with metabotropic GABA$_B$ receptors, inhibition is caused by the opening of K⁺ channels, which results in both hyperpolarization and shunting.

Inhibitory synapses are found on dendrites, cell bodies, and axonal initial segments of neurons, those on cell bodies and axonal initial segments being particularly effective in controlling the generation of the AP outputted by the neuron. Not all GABA$_A$ receptors of fast chemical are the same, there being many subtypes having different subunits which results in distinct functional characteristics.

Both GABA$_A$ and glycine channels have two or three conductance levels in the range of 10–30 pS. Glycine channels tend to have a somewhat larger conductance than GABA$_A$ channels due to a larger diameter of the pore. The intermediate conductances could be due to conformational changes in only some of the subunits of the receptor, rather than in all the subunits that can open the channel. Bicuculline, found in some plant alkaloids, is a competitive GABA$_A$ antagonist (Section 5.6.1), whereas picrotoxin, also found in some plants, is believed to block the GABA$_A$ channel rather than act on the receptor site. By reducing inhibition in the nervous system these toxins cause convulsions. On the other hand, substances such as barbiturates, some tranquilizers, and alcohol enhance the size and duration of GABA ipsps, thereby inducing a state of calm and sedation.

6.2.3 Fast Excitatory Synapses

In the central nervous system, L-glutamate is the dominant excitatory neurotransmitter, although neurons also release several other amino acids, such as L-cysteine, L-homocysteine, and L-aspartate, that can act as excitatory neurotransmitters on some glutamate receptors (GluRs). These ionotropic glutamate receptors, believed to be tetramers, can be divided into two broad classes designated as **NMDA receptors** and **non-NMDA receptors** according to whether or not they respond to the amino acid derivative dicarboxylic N-methyl-D-aspartate (NMDA). The non-NMDA receptors are divided into two groups: **AMPA (α-amino-3-hydroxyl-5-methyl-4-isoxazolepropionic acid)** receptors and **kainate** receptors, kainate being a natural marine acid present in some seaweed.

The fast epsps of motoneurons are due to non-NMDA receptors (Figure 6.9). The channels are permeable to Na⁺, K⁺, and, to a lesser extent, Ca²⁺, the equilibrium voltage being about 0 mV. Non-NMDA epscs have a fast rise time of less than 1 ms, and non-NMDA channels rapidly desensitize, typically within a few milliseconds. AMPA receptors are more prevalent than kainate receptors and their i_{epsc} is considerably larger, faster rising, and faster decaying. The conductance of non-NMDA channels is usually represented by an exponential function of instantaneous rise at $t = 0$ and an exponential decay with a time constant of a few milliseconds.

The NMDA receptors have some unusual properties. In addition to a binding site for glutamate, the NDMA receptor has binding sites for glycine, Zn²⁺, and Mg²⁺. At normal resting levels, the channel has a low conductance due to blockage by extracellular Mg²⁺ under the influence of the inwardly directed electric field associated with the resting membrane voltage. But when the membrane is depolarized, the blockage is relieved and the channels conduct. In the conducting state, the channels are 5–10 times more permeable to Ca²⁺ than to Na⁺ and K⁺. Compared to non-NMDA receptors, their sensitivity to glutamate is higher, and they require the presence of glycine or D-serine in the external medium to become responsive. Under normal conditions, the concentration of glycine in the extracellular medium is sufficient to allow glutamate to activate the channel.

In most central synapses having glutamate as a neurotransmitter, both NMDA and non-NMDA receptors are present together in the same synapse, particularly NMDA and AMPA receptors, so that the depolarization required to activate the NMDA receptors is provided by the activation of the non-MDA receptors. The synaptic currents due to both types of receptors can be separated by using pharmacological agents and by holding the cell

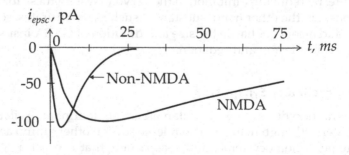

FIGURE 6.9
epsps due to NMDA and non-NMDA receptors.

at different voltages. Whereas the time course of the epsc of the non-NMDA channel is typical of that of an ionotropic channel, the epsc of the NMDA channel is not. Its rise and decay times are much slower and its decay time is slower than an exponential (Figure 6.9), presumably due to the slow opening and closing of the NMDA channel. It appears that the primary function of NMDA receptors is not to generate a psp but rather to allow the influx of Ca^{2+} at the time the postsynaptic membrane is depolarized. This can activate various Ca^{2+}-dependent processes, including potentiation of the synapse, as discussed in Section 6.5.2.1. Current flow through NMDA receptor channels is enhanced by some antipsychotic drugs and is inhibited by the hallucinogenic drug phencyclidine (PCP).

The conductance of the NMDA channel has been modeled as (see Bower and Beeman, 1998):

$$g_S(v_m, t) = G_{S0} \frac{e^{-t/\tau_1} - e^{-t/\tau_2}}{1 + \eta \left[Mg^{2+} \right]_o e^{-\gamma v_m}} \tag{6.6}$$

where $G_{S0} = 0.2$ nS, $\tau_1 = 80$ ms, $\tau_2 = 0.67$ ms, $\eta = 0.33$/mM, and $\gamma = 0.06$/mV. A typical value of extracellular $[Mg^{2+}]_o$ is 2 mM.

In some fast excitatory synapses, serotonin (5-HT) is the neurotransmitter. The receptor, designated as a 5-HT$_3$ receptor, is similar to nACh receptors. It is the only ionotropic serotonin receptor. All of the other dozen or so serotonin receptors are metabotropic.

6.3 Second-Messenger Systems

6.3.1 General Description

Second messengers are molecules that relay signals from postsynaptic receptors activated by extracellular substances, termed first messengers, to molecules inside the cell, which in turn affect cell activity in some way. Second messengers are involved in a variety of cell functions, including hormone activity and gating of ion channels, where the first messengers are hormones and neurotransmitters, respectively. Some substances can act both as neurotransmitters, when released from presynaptic terminals, and as hormones, when secreted by specialized organs into the blood stream. Examples are epinephrine, norepinephrine, serotonin, and dopamine.

To gain a better understanding of second-messenger systems, it is necessary to review first some biochemistry in order to explain the structure of an important, typical second-messenger cyclic AMP (cAMP) and the guanine-derived phosphates. It should be recalled that two of the basic constituents of nucleic acids are the purine compounds adenine and guanine.

When attached to the 1' carbon atom of a ribose sugar molecule (Figure 6.10), they become the nucleosides adenosine and guanosine, respectively. When a single phosphate group is attached to the 5' carbon atom of a ribose sugar molecule, these nucleosides become the nucleotides adenosine monophosphate (AMP) and guanosine monophosphate (GMP), respectively. However, another phosphate group can attach to the first phosphate group to give adenosine diphosphate (ADP) and guanosine diphosphate (GDP), respectively. The attachment of a third phosphate group to the second phosphate group, gives adenosine triphosphate (ATP) and guanosine triphosphate (GTP), respectively.

We have encountered ATP as the energy-rich compound that fuels energy-consuming cell activities (Section 1.1) by breaking down into the lower energy form ADP. AMP is a lower-energy form still and can exist as cyclic AMP (cAMP), which is synthesized from ATP by the membrane-bound enzyme adenylate cyclase. This enzyme liberates two phosphate groups from ATP and loops back the remaining phosphate group to the 3' carbon atom of the ribose ring, as illustrated in Figure 6.10.

We next consider some enzymes and proteins that play important roles in second-messenger systems. A second messenger can activate a **protein kinase (PKA)**, which is an enzyme that has catalytic subunits and regulatory subunits. In the inactive state the catalytic subunits are inhibited by the regulatory subunits, but this inhibition is removed by the binding of a second messenger, which allows the catalytic subunits to phosphorylate target proteins (Section 1.1.2). In the phosphorylation of a target protein, a phosphate group from an ATP molecule is substituted for an OH^- group in one of

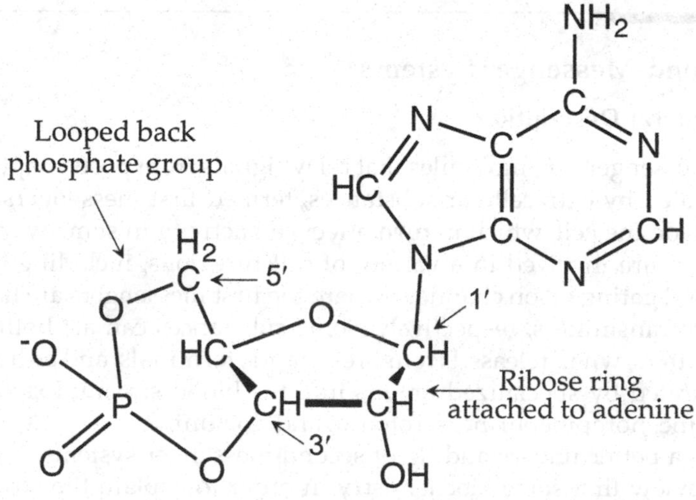

FIGURE 6.10
cAMP molecule.

the amino acid side chains of the target protein. Phosphorylation alters the structural conformation of a protein, thereby activating it or modifying its function. The phosphorylated, or target, proteins can perform many functions, including gating of ion channels, regulation of metabolism of glycogen, sugar, and lipids; or expression of specific genes. Protein kinases can also be activated not by second messengers but by other protein kinases in a cascade of protein kinases. The amino acids that are commonly phosphorylated are serine, threonine, and tyrosine.

The phosphate group is hydrolyzed back to an OH^- group by enzymes referred to as **phosphatases**, and the process is known as dephosphorylation. Protein phosphatase 1 (PP1) dephosphorylates a variety of proteins as well as K^+ and Ca^{2+} channels, NMDA, and AMPA glutamate receptors. Protein phosphatase 2A (PP2A) also dephosphorylates a range of proteins that overlap with those of PP1, in addition to tau protein that stabilizes microtubules of the cytoskeleton. Excessive phosphorylation of tau protein is associated with Alzheimer's disease. Protein phosphatase 2B (PP2B), also known as calcineurin, is abundant in neurons and is activated by Ca^{2+}. It activates T cells of the immune system and dephosphorylates AMPA receptors. Protein phosphorylation and dephosphorylation are of fundamental importance in cell functioning as it is the major molecular mechanism through which protein activity in a cell is regulated both in and outside the nervous system.

G proteins are a group of proteins anchored to the lipid part of the phospholipid molecules of the membrane on the cytoplasmic side. They are composed of three subunits, Gα, Gβ, and Gγ, together with a GDP molecule that is bound to the membrane by the α subunit. A specific G protein, the G_s protein, stimulates cAMP synthesis when activated.

Second-messenger systems can be quite complicated and of great diversity, so only some of the basic features will be presented here. Figure 6.11 illustrates a "typical" second-messenger system, the sequence of events being as follows:

1. A first messenger, usually a hormone or a neurotransmitter, binds to a special receptor (1), the G protein-coupled receptor which exposes a binding site on the receptor to an inactive, membrane-bound G_s protein. The G protein-coupled receptor family is quite large and includes receptors for norepinephrine (α- and β-adrenergic receptors), ACh (muscarinic receptors), GABA ($GABA_B$ receptors), dopamine, glutamate, serotonin, and neuropeptides, as well as receptors for hormones such as the thyroid-stimulating hormone (TSH) and the parathyroid hormone (PTH).

2. The G_s protein binds to the receptor (2) and exchanges a GDP molecule for a GTP molecule. This activates the G_s protein and breaks it up into a Gα-GTP subunit (3) and a Gβγ subunit (not shown in the figure), both of which can move along the inner part of the membrane.

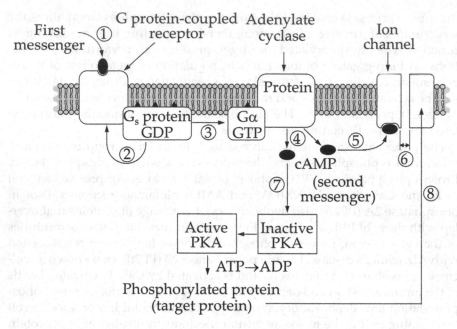

FIGURE 6.11
Diagrammatic illustration of a second-messenger system.

The Gα-GTP subunit eventually contacts another membrane-bound protein, which, in the illustration, is the enzyme **adenylate cyclase**. Adenylate cyclase is stimulated by the Gα-GTP subunit in some cases and by the Gβγ subunit in other cases. When stimulated, adenylate cyclase converts many molecules of ATP from the cytoplasm to the second messenger, cAMP, which diffuses freely in the cytoplasm (4). Adenylate cyclase is referred to as the **first effector**.

3. Figure 6.11 illustrates two possible pathways for cAMP:

   (a) It can bind to a receptor (5), thereby gating an ion channel (6), or

   (b) It can activate a protein kinase (PKA) (7), which phosphorylates a target protein. The protein kinase is referred to as the **second effector**. The phosphorylated protein is shown in Figure 6.10 gating an ion channel (8).

4. As in the case of fast synapses, it is essential for proper operation that the system is restored to its initial state in a timely manner, in readiness for renewed activity. cAMP is converted back to AMP by a **phosphodiesterase** (**PDE**) enzyme. The phosphorylated proteins are dephosphorylated by phosphatases. The Gα-GTP subunit has an intrinsic **GTPase** activity that converts GTP to GDP, which stops cAMP synthesis. The Gα-GDP subunit then recombines with a Gβγ subunit to reform an inactive G$_s$ protein. The restoration of the initial

state is affected by many factors, such as the concentration levels of the substances involved and the effect of some proteins on the GTPase rate. Typically, the Gα-GTP subunit remains active for a few seconds before its bound GTP is hydrolyzed to GDP.

5. The process is repeated as long as the first messenger remains bound to the G protein-coupled receptor.

The following points should be noted concerning second-messenger systems:

1. When the stimulus is a neurotransmitter released by a nerve terminal and affects membrane voltage, this effect is often referred to as **slow synaptic action**. Apart from the speed difference, slow synaptic action differs from fast synaptic action, previously discussed, in the following respects:

   (a) The metabotropic neurotransmitters of slow synaptic action (Section 6.1) include most of the ionotropic neurotransmitters of fast synaptic action, such as dopamine, norepinephrine, serotonin, histamine, as well as many more. Where the same neurotransmitter is both metabotropic and ionotropic, as in the case of ACh, glutamate, serotonin, and GABA, the receptors are different. Thus, the metabotropic receptor for GABA is $GABA_B$ to distinguish it from the ionotropic receptor $GABA_A$. The metabotropic receptor for ACh is described as a **muscarinic ACh receptor** (mAChR) because it is particularly responsive to muscarine (found in some poisonous mushrooms), whereas the ionotropic nicotinic ACh receptor (nAChR) is much more responsive to nicotine. Moreover, the effect of a metabotropic transmitter may be excitatory or inhibitory – again, depending on the receptor. Thus, whereas the action of ionotropic glutamate receptors is always excitatory or depolarizing, as the reversal voltage of their ionic current is near zero, the metabotropic glutamate receptors (mGluRs) can produce either excitation or inhibition depending on the reversal voltage of the ionic currents that they regulate. Opioids, which are used as narcotics and are also prescribed to manage pain, act as metabotropic neurotransmitters. Melatonin, the hormone that synchronizes biological rhythm with the 24 hr day/night (circadian) cycle acts on G-coupled protein receptors.

   (b) The channels affected by slow synaptic action are mainly Ca^{2+} and K^+ channels, but the effect can be much more far-reaching than merely channel opening or increased conductance. In some cells, such as those with metabotropic $GABA_B$ receptors, inhibition is caused by the opening of K^+ channels. Because the K^+ equilibrium voltage of neurons ($E_K = -80$ mV) is always negative to the resting voltage, opening K^+ channels inhibits the cell even more strongly than opening Cl⁻ channels, assuming a synaptic conductance of similar value.

The action of second messengers on the presynaptic membrane can enhance or depress the entry of Ca^{2+} on the presynaptic side. This regulates the efficacy of transmitter release and, hence, the size of the psp mediated by ionotropic receptors of the postsynaptic membrane. The action of these receptors can also be regulated directly by second messengers on the postsynaptic side. The conductance of K^+ channels may be increased or decreased. Moreover, second messengers can also affect resting channels in the soma and dendrites and could alter the voltage dependence and the speed of gating of voltage-gated channels of the cell, thus altering many electrical properties of the cell including resting voltage, input resistance, space and time constants, threshold, as well as the shape and duration of the AP. These diverse effects of metabotropic receptors are referred to as **modulatory synaptic actions**.

(c) Ionotropic channels are located within a few tens of nanometers of the presynaptic active zones that release the neurotransmitter, so synaptic action is fast and localized. In slow synaptic action, on the other hand, there is no tight coupling between the presynaptic and postsynaptic sides, which allows the neurotransmitter to diffuse over a considerable distance and affect several neighboring cells, as in the case of the heart (Section 10.5).

2. The effect of a stimulus is greatly amplified. Thus, a single G-protein-coupled receptor for a first messenger may activate many G_s proteins in succession, assuming that the subunits of the activated G_s protein break away from the receptor. An activated adenylate cyclase may produce many cAMP molecules, and an activated PKA can successively phosphorylate many protein molecules.

3. Many second messengers other than cAMP have been identified, with interactions between the pathways of some of them. cGMP can act as a second messenger where, much like cAMP, its synthesis is stimulated by nitric oxide (NO). Also like cAMP, cGMP can activate a protein kinase, resulting in a cGMP-dependent protein kinase (PKG) enzyme, which, in turn, phosphorylates a variety of proteins. NO is sometimes considered a second messenger because of its diverse effects within its cellular targets. NO is synthesized in neurons and can readily permeate the plasma membrane to act on neighboring cells. NO can also modify target proteins by **nitrosylation**, that is, adding a nitryl (NO_2) group to some amino acids. It can regulate neurotransmitter release and is a powerful vasodilator. It has been implicated in the regulation of excitability and firing, in long-term potentiation and long-term depression, as well as in memory processes.

4. Different subunits of G proteins can have different effects and give rise to other second messengers. Thus, the Gα subunit previously mentioned is more properly termed a $G_{\alpha s}$ subunit because of its association with the G_s protein and its stimulation of adenylate cyclase to produce cAMP. Another subunit, $G_{\alpha i}$ of the $G_{i/o}$ protein family inhibits the production of cAMP by binding to a different site on adenylate cyclase. In this case, the Gβγ subunit of the $G_{i/o}$ protein directly opens a G protein-activated inwardly rectifying K^+ (GIRK) channel (Section 7.3.3), leading to membrane hyperpolarization without the involvement of a second messenger. Other subunits of G proteins mediate different pathways that produce other second messengers. Thus, in the **phosphoinositol second-messenger system**, subunit $G_{\alpha q}$ can activate a membrane enzyme, phospholipase Cβ (PLC), which cleaves a membrane phospholipid PIP_2 (phosphatidylinositol-4,5-bisphosphate) to yield two second messengers: inotisol-1,4,5-triphosphate (IP_3) and diacylglycerol (DAG). IP_3 releases Ca^{2+} into the cytoplasm from intracellular stores of the sarcoplasmic reticulum. DAG remains within the plasma membrane and activates protein kinase C (PKC). PKC and DAG can phosphorylate many protein substrates in the cell. The cleavage and hence reduction of PIP_2 lead to the closure of K_v7 potassium channels (Section 7.3.3). The actions of IP_3 and DAG are terminated by enzymes that convert them into inactive forms that can be recycled to produce PIP_2.

5. Ca^{2+} function as second messengers in many respects, including neurotransmitter release, synaptic plasticity (Section 6.5), and muscle contraction (Section 9.2). This is because normal intracellular concentration of Ca^{2+} is very small (100 nM or less), so that even a relatively small influx of these ions can lead to a large percentage of change in intracellular Ca^{2+} concentration that can trigger various biochemical reactions. One of the targets of Ca^{2+} is the *calcium-modula*ted protein, calmodulin (CaM), which, when activated by the binding of Ca^{2+}, can in turn interact with a variety of target proteins including kinases and phosphatases. In neurons, the dominant calmodulin-dependent protein kinase is CaMKII. Being similar in structure to troponin C (Section 9.1.2) and found abundantly in the cytosol of all cells, CaM is involved in many important processes such as smooth muscle contraction, intracellular movement, metabolism, inflammation, apoptosis, short-term and long-term memory, and the immune response. CaMKII's role in synaptic plasticity is discussed in Section 6.5.2.1.

6. As explained in connection with Figure 6.11, the chain of events triggered by the stimulus, that is, the binding of a first messenger to receptors, is not limited to ion channels but can affect many aspects of cell function including ion pumps and transporters, metabolism,

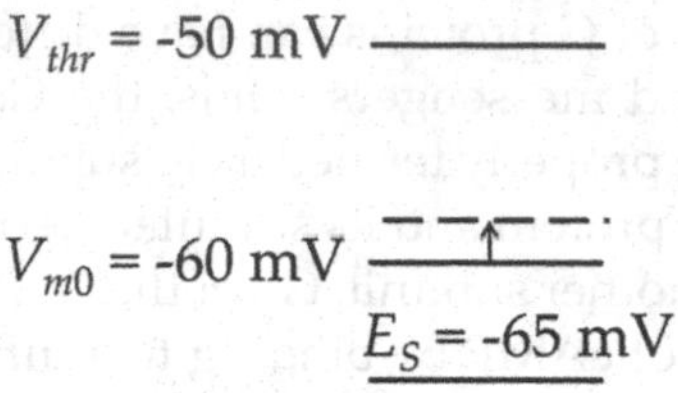

$$V_{thr} = -50 \text{ mV} \quad \underline{\hspace{2cm}}$$

$$V_{m0} = -60 \text{ mV} \quad \underline{\uparrow}$$
$$\underline{E_S = -65 \text{ mV}}$$

FIGURE 6.12
Voltage relations used to illustrate the excitatory effect of a reduction in K⁺ conductance.

various enzymes and proteins, growth factors, and even gene expression, with generally more than a single pathway between a given receptor and a target effector. Because of the time taken by the chain of processes involved, the end effects are delayed with respect to the stimulus and can last from several seconds to several minutes, and much longer if gene expression is included, as discussed at the end of Section 6.5.3. This requires at least 30–60 minutes.

Before ending this section, and in view of the fact that slow synaptic action may reduce channel conductance, it is instructive to consider the effect of this reduction on the psp and on the excitability of the cell. Consider the case illustrated in Figure 6.4b and reproduced in Figure 6.12, where the ion involved is K⁺, as is, in fact, the case of the postsynaptic neuron in autonomic ganglia having muscarinic ACh receptors. In the resting state, K⁺ efflux is balanced by Na⁺ influx. If the slow synaptic action reduces g_S, K⁺ efflux is reduced so that the membrane depolarizes, as can also be argued from the equivalent electric circuit of Figure 6.2. The effect is therefore excitatory, since v_m is moved closer to threshold. This is opposite to what is shown in Figure 6.4b. The depolarization reduces the electrochemical potential difference driving Na⁺ inward and increases the electrochemical potential difference driving K⁺ outward, so that K⁺ efflux is balanced by Na⁺ influx. The effect of the slow synaptic action on the repetitive firing of the ganglionic neuron is discussed in Section 7.3.3 in connection with K_v7 channels and I_M current.

Similarly, it can be argued that in a case like that of Figure 6.4a, a decrease in g_S results in an inhibitory effect, opposite that of an increase in g_S (Problem 6.1).

SPOTLIGHT ON TECHNIQUES 5A: METHODS OF STUDYING ION CHANNELS

Following are some of the main methods used to investigate the structure and functioning of ion channels.

Patch Clamp. A valuable tool for studying ion channels is the patch-clamp technique (Spotlight on Techniques 3A). This technique has been continually refined, and is now sensitive enough to measure the change in capacitance due to the fusion of a single vesicle with the presynaptic membrane. Patch clamping can be used to investigate ion channel kinetics, that is, the timing and duration of channel opening and closing, channel current and conductance, ions carrying the channel current, channel selectivity to various ions, and the effects of various pharmacological agents and gene mutations (**mutagenesis**). Structural and functional details of channels can be inferred from the results of these patch-clamp investigations.

Molecular Biology. Molecular biology techniques are used in many ways in studying ionic channels. To begin with, these techniques allow the identification of genes that express classes of ion channels. Once identified, genes can be cloned to obtain the large number of channels required for structural analysis using X-ray crystallography, for example. The identified genes can be used to derive the amino acid sequence of the channel protein. Once this is known, a structural model of the channel protein can be constructed, based on the amino acid sequences of known protein structures. This allows the prediction of the three-dimensional structure of the channel protein as well as the identification of regions that are likely to form the hydrophobic membrane-spanning segments and the hydrophilic segments that line the pore. For example, voltage-gated channels that are selective for different ions have membrane-spanning segments that contain specific, charged amino acids. The proposed structure can be explored and tested in many ways using molecular biology techniques. Antibodies can be produced against specific regions in the protein sequence and then attached to the channel protein in order to locate these regions. Mutagenesis can be used to delete or substitute specific amino acid residues, and naturally occurring mutations can be utilized to determine the contribution of these residues to channel functioning and hence assess the importance of their locations. **Chimeric channels** can be produced by combining parts that are derived from the genes of different species, where differences in channel properties in different species are known. The functions of specific regions of the channel can then be assessed by comparing the properties of the chimeric channel to those of the two original channels.

X-Ray Crystallography. Information on the structure of various biological entities can be obtained using various spectroscopic techniques based on irradiating the specimen with light, electrons, or X rays, where the spatial resolution depends on the wavelength of the radiation. Visible light, whose shortest wavelength is about 300 nm, can be used to visualize cell organelles. Electrons of wavelength less than 10 nm can be used to derive the shape and overall structure of large protein molecules. A particularly useful form of electron microscopy (EM) is cryo-EM, in which cooling to –160 °C or less avoids the complications of staining or having the specimen in a vacuum. X rays, on the other hand, having a wavelength of about 0.1 nm, have the potential of providing information on the location of nearly all the atoms of a protein. The usefulness of this technique, however, depends on having a high-quality three-dimensional crystal of the protein. The first step in X-ray crystallography is to obtain a sufficiently large and pure sample of the target protein. This is done by either isolating the protein from a natural, plentiful source or by cloning its gene into a high expression system. The protein sample is tested by various means to ensure that it is pure, homogenous, soluble, active, and stable. A somewhat elaborate procedure is then followed to form a supersaturated solution from which the protein precipitates in crystalline form. The crystals are flash-cooled to 100 K and accurately positioned using a special apparatus in order to expose it to a focused X-ray beam. The resulting diffraction pattern is recorded on an X-ray detector. The crystal is rotated through 1° or less and another diffraction pattern is recorded. The process is repeated for a total rotation of up to 180°. The recorded diffraction patterns are computer-processed using the Fourier transform to obtain an electron density map. The map is then interpreted to derive a three-dimensional model of the protein, which can be refined against the electron diffraction data.

6.3.2 Neuromodulators

Neuromodulation is broadly defined as the alteration of neuronal activity through application of a stimulating agent, which could be electrical or chemical. In the present context, the term **neuromodulator** refers to a chemical agent that affects neurons over a relatively large area of the brain or the spinal cord, mostly through G protein-coupled receptors. Neuromodulators could be neurotransmitters, such as norepinephrine, acetylcholine, dopamine, serotonin, and histamine released by certain groups of neurons but not reabsorbed by the presynaptic neuron – nor broken down. They may be released at nerve endings, as in slow chemical synapses, or they may be carried by the cerebrospinal fluid. Neuromodulators could also be hormones circulating through the blood, such as thyroid hormones, steroid hormones (such as androgen and estrogen), metabolic hormones (such as insulin), stress hormones (such as cortisol), sex hormones (such as testosterone), or neuropeptides such as adenosine or oxytocin.

Generally speaking, a neuromodulatory input does not directly activate a neuron to produce an output but can completely transform the behavior of the neuron, changing everything from its level of excitability to the pattern of firing it produces for a given input. The neuromodulatory effects of serotonin and norepinephrine on motoneurons are discussed in Section 11.2.3.

The following are the major neuromodulators of the central nervous system and their effects on this system, excluding their effects in the rest of the body:

1. Norepinephrine – originating in the CNS, mainly in the locus coeruleus (in the pons), and, to a lesser extent, in the lateral tegmentum (in the brainstem). This system essentially mobilizes the brain for action in the same manner as the sympathetic system mobilizes the body. It regulates the sleep–wake cycle and arousal. Thus, it enhances processing of sensory inputs, attention, and formation and retrieval of both long-term and working memory. It is involved in a broad range of behavioral, cognitive, and emotional responses associated with enhanced vigilance.

2. Dopamine – originating in the CNS in the ventral tegmentum (in the midbrain), the substantia nigra (part of the basal ganglia), and the arcuate nucleus of the hypothalamus. It is involved in motor function, particularly in the initiation of movement, and in the learning of new motor skills. Damage to the substantia nigra resulting in a deficiency of dopamine in this area is associated with Parkinson's disease. Dopamine in the CNS plays a major role in reward-motivated behavior, that is, behavior based on predicting rewards or pleasurable experiences, such as drug addiction, gambling, cravings, and sex-related activity. It is also involved in some hormonal functions because of its influence on the pituitary gland.

3. Serotonin (5-HT) – originating in the CNS from about 165,000 neurons in the raphe nuclei in the brainstem. Serotonin generally promotes a feeling of well-being. It is involved in regulation of mood, emotions, appetite, sleep, learning and memory, pain modulation, and sexual arousal. It is also a modulator of stereotyped motor functions mediated in the spinal cord and in the brainstem, such as locomotion and urination.

4. Acetylcholine – originating in the CNS in the pedunculopontine nucleus and the dorsolateral tegmental nuclei of the brainstem, in the nucleus basalis in the basal forebrain, and in the medial septal nucleus below the rostrum of the corpus callosum. The cholinergic system is involved in rewards, enhancing signal detection and sensory attention, regulating homeostasis, mediating the stress response, motor learning and control, and the formation of memories.

6.4 Presynaptic Inhibition and Facilitation

In vertebrates, presynaptic inhibition was first discovered and studied most intensively in the spinal cord. It is manifested at axo-axonic synapses, as illustrated in Figure 6.13. The primary afferent fiber is the axon of a sensory neuron, whose cell body is in the dorsal root ganglion (Section 11.1.3) and which typically is a Ia fiber making an excitatory monosynaptic connection with an α-motoneuron (Section 11.2.1). The axo-axonic synapse is made on the presynaptic terminal of the primary afferent by the axon of a GABAergic interneuron in the spinal cord. The membrane of the primary afferent that is postsynaptic on the axo-axonic synapse contains both GABA$_A$ and GABA$_B$ receptors, which means that presynaptic inhibition can be mediated by either ionotropic or metabotropic actions on the primary afferent terminals. When GABA is released by the axon of the interneuron and binds to GABA$_A$ receptors, Cl$^-$ channels are opened, allowing Cl$^-$ to flow outward from the presynaptic terminal under the influence of their electrochemical potential gradient. This electrochemical potential gradient is maintained across the membrane of the presynaptic terminal not by an active Cl$^-$ pump but by an electrically neutral Na$^+$–K$^+$–2Cl$^-$ symporter that transports these ions inward from the extracellular space (Figure 6.13). The outflow of Cl$^-$ due to the activation of GABA$_A$ receptors in the presynaptic terminal results in a depolarization of this terminal by up to 25 mV and lasting for more than 100

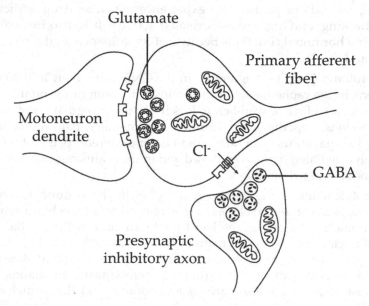

FIGURE 6.13
Presynaptic inhibition.

ms, referred to as **primary afferent depolarization (PAD)**. The depolarization, as well as the attendant increase in membrane conductance, reduces the amplitude of the AP in the presynaptic terminal so that less neurotransmitter is released from the terminal. This, in turn, decreases the amplitude of the epsp induced in the postsynaptic membrane of the motoneuron. The overall effect is a depression of the transmission through the primary afferent-motoneuron synapse, an effect referred to as **presynaptic inhibition**. It is seen that the presynaptic inhibition in Figure 6.13 selectively reduces only the effectiveness of the input from the primary afferent fiber on the motoneuron. In contrast, if the inhibition from the inhibitory interneuron is applied directly to the motoneuron, it would reduce the effectiveness of inputs other than those from the primary afferent fiber.

The binding of GABA to GABA$_B$ receptors in the presynaptic terminal can also cause the G$\beta\gamma$ subunit of a G protein to close voltage-gated Ca^{2+} channels and open K$^+$ channels. This reduces the influx of Ca^{2+} and enhances repolarization of the presynaptic terminal, thereby reducing the width of the presynaptic AP. Both of these effects decrease transmitter release. Another effect of the G$\beta\gamma$ subunit is believed to be a reduction in efficacy of Ca^{2+} to cause fusion of the vesicle membrane with the membrane of the presynaptic terminal. These, however, are slower processes compared to the ionotropic effect on GABA$_A$ receptors.

In presynaptic facilitation, the release of Ca^{2+} in the presynaptic terminal is enhanced. One mechanism is believed to involve closure of K$^+$ channels by second-messenger systems in the presynaptic terminal, which increases the duration of the presynaptic AP, thereby enhancing Ca^{+2} influx and neurotransmitter release, resulting in a larger epsp. Presynaptic facilitation can also occur by the activation of Ca^{2+} receptors in the presynaptic terminals either directly by a neurotransmitter or through depolarization of voltage-gated Ca^{2+} channels.

Presynaptic inhibition and facilitation are not confined to the spinal cord but are fairly widespread at synapses in the CNS, as in the hippocampus and retina, for example.

6.5 Synaptic Plasticity

Plasticity of neuronal circuits is the alteration of the behavior of these circuits in response to experience or injury. Plasticity could result from: modification of the response of neurons to synaptic inputs (Chapter 7), such as a change in the intrinsic excitability of neurons; the addition or removal of synaptic pathways; or modification of the strength or efficacy of synapses, the latter being referred to as **synaptic plasticity**. In view of the complexity of the subject, however, only some basics of synaptic plasticity will be presented in this section.

It is convenient to distinguish between short-term synaptic plasticity, generally lasting for up to a few minutes, and long-term plasticity, lasting from minutes to hours and even longer if synaptic growth is included. The following should be noted concerning synaptic plasticity:

1. Both short-term and long-term synaptic plasticity can be an enhancement or a depression of synaptic activity.

2. Several processes, and not a single process, are involved in synaptic plasticity, the time courses of these processes, and their relative contributions depending on the particular synapses.

3. The underlying molecular mechanisms of synaptic plasticity may differ in detail between various parts of the brain, but some general principles apply in almost all cases.

4. As will become clear from the following discussion, short-term plasticity is mainly presynaptic, whereas long-term plasticity is mainly postsynaptic.

5. Whereas long-term synaptic plasticity is involved in learning and memory, short-term plasticity affects the encoding of neuronal responses (Chapter 7).

6.5.1 Short-Term Synaptic Plasticity

Virtually all synapses, including the NMJ, exhibit some form of short-term plasticity. In general, short-term plasticity is manifested under two conditions:

1. Paired-pulse stimulation, resulting in **paired-pulse facilitation** or **paired-pulse depression,** according to whether the response to the second pulse is larger or smaller than the response to the first pulse, respectively.

2. Stimulation by a train of high-frequency stimuli, referred to as **tetanic stimulation**.

In paired-pulse facilitation, the response to a second stimulus that follows the first stimulus by about 10 ms, or so, can be five times the response to the first stimulus. The ratio of the response to the second stimulus to the response to the first stimulus decreases as the interval between the pair of stimuli is increased, falling to unity for intervals between a few hundred milliseconds and about one second. In many cases, the decline in facilitation as a function of the separation between the two stimuli can be divided into a more rapid phase lasting for several tens of milliseconds and a slower phase lasting for several hundreds of milliseconds.

Enhancement or depression of synaptic function can also be observed with tetanic stimulation at frequencies of 10–200 Hz, lasting from 200 ms to 5s.

Augmentation is enhancement that grows and decays with a time constant of 5–10 s, whereas **posttetanic potentiation** (**PTP**) can last from about 30 seconds to several minutes and up to an hour or more at some synapses. Because all these effects are generally due to different processes, their time courses and relative prominence depends on the synapses under consideration.

In discussing short-term plasticity, it is helpful to divide the pool of neurotransmitter vesicles in the presynaptic terminal into two groups:

1. The "readily releasable pool" of vesicles in contact with the membrane, ready for exocytosis, or are at least available for immediate replenishment of release sites vacated by exocytosis. The size of this pool varies from less than 10 to about 30 in most synapses. In most synapses, only one or two vesicles are released per AP, per active zone or release site, although multiple releases seem to occur in some cases.

2. The "reserve pool" of more distant vesicles that respond more slowly to the need for replenishment of release sites. The restoration of this pool depends on the formation of new vesicles from membrane recovered after exocytosis and the filling of these vesicles with neurotransmitter.

Ca^{2+} are involved in both the replenishment of the readily releasable pool from the reserve pool and in the restoration of this latter pool.

Operationally, the important factors for short-term plasticity are:

1. the kinetics, and hence the probability, of vesicle release,
2. the number of active zones,
3. the sizes of the aforementioned pools of vesicles, and
4. the rate of replenishment of the readily releasable pool.

It stands to reason, that an increase in the probability of vesicle release in response to a stimulus can deplete the readily releasable pool and will result in synaptic depression for a second stimulus, the depression lasting until the pool is replenished. Similarly, prolonged tetanic stimulation can deplete the readily releasable pool of vesicles, resulting in synaptic depression. Conversely, a reduction in the probability of vesicle release by the first stimulus can result in synaptic enhancement as more vesicles may be released by the next AP.

The major factor underlying short-term plasticity is the Ca^{2+} concentration $[Ca^{2+}]_i$; specifically, the extent and time course of $[Ca^{2+}]_i$ build up in the presynaptic terminal following excitation and the time course of the return of $[Ca^{2+}]_i$ to its resting level. As noted previously, depolarization of the presynaptic terminal opens Ca^{2+} channels, causing an immediate high elevation in the local Ca^{2+} concentration that triggers the release of neurotransmitter vesicles. This release is highly sensitive to the concentration of Ca^{2+}, varying as the fourth power of the concentration.

Ca^{2+} influx is influenced by the inherent properties of Ca^{2+} channels, and by the extracellular Ca^{2+} concentration – the larger this concentration the larger the influx, and conversely. Ca^{2+} influx is also influenced by the number of metabotropic receptors in the presynaptic terminal. Activation of these receptors affects, through second messengers, Ca^{2+} channels and both voltage-dependent and voltage-independent K^+ channels in the presynaptic terminal. Inactivation of some K^+ channels causes broadening of the AP, which increases Ca^{2+} influx, whereas gradual activation of a Ca^{2+}-dependent K^+ current reduces the amplitude of the AP, or its duration, or causes its failure altogether, which reduces Ca^{2+} influx. Some of these metabotropic receptors are **autoreceptors**, that is, they are receptors of the presynaptic membrane that are activated by neurotransmitters released from the same presynaptic terminal. Examples of such neurotransmitters are GABA and ATP. The activation of autoreceptors is a negative feedback mechanism that reduces Ca^{2+} influx and hence the release of neurotransmitters. The depression of transmission through autoreceptors is termed **homosynaptic inhibition**. The presynaptic metabotropic receptors could also be activated by neurotransmitters released into the extracellular space by other neurons in which case the inhibition is termed **heterosynaptic inhibition**.

Following the initial, localized, and non-uniform concentration of Ca^{2+} as a result of presynaptic depolarization, these ions are subjected to:

1. diffusion to regions of lower concentration,
2. uptake and subsequent release by mitochondria, sarcoplasmic reticulum, and Ca^{2+}-binding proteins, and
3. extrusion by a Ca^{2+} pump and a Ca^{2+}–Na^+ exchanger.

Diffusion in the presence of uptake and binding, termed **buffered diffusion**, facilitates diffusion of Ca^{2+} by increasing the concentration gradient.

Until the Ca^{2+} concentration decays to the resting level, there is a residual concentration of Ca^{2+} above the resting level. The total Ca^{2+} concentration due to a second, closely spaced stimulus, will therefore be increased, leading to net facilitation if there is a sufficient pool of readily releasable vesicles. Augmentation is influenced by the activity of the Ca^{2+} pump and the Ca^{2+}–Na^+ exchanger, whereas post-tetanic potentiation is affected by the uptake of Ca^{2+} by intracellular entities and their subsequent release.

All the preceding mechanisms of synaptic plasticity act on the presynaptic terminal. There are, in addition, mechanisms that involve the postsynaptic cell. Synaptic activity is depressed, for example, by desensitization of $GABA_A$, glutamate NDMA, and AMPA receptors. In some cells, AMPA channels are blocked by endogenous intracellular polyamines. The block is relieved by depolarization, resulting in enhancement of synaptic activity. Postsynaptic cell activity may also release substances, such as ATP, dopamine, GABA, glutamate, and the hormone neurotransmitter oxytocin, all of which can influence vesicle release from the presynaptic terminal through their action on metabotropic receptors in this terminal, as mentioned earlier. This mechanism is an example of **retrograde control** of neurotransmitter release.

To complicate matters further, glia that are intimately associated with synapses have been implicated in short-term synaptic plasticity by controlling the extent and speed of their uptake of neurotransmitters. Glia can also respond to the presence of neurotransmitters in the extracellular space by releasing substances that can enhance or depress vesicle release from the presynaptic terminal.

6.5.2 Long-Term Synaptic Plasticity

6.5.2.1 Long-Term Potentiation

Long-term potentiation (LTP), was first studied at excitatory synapses on CA1 pyramidal cells of the hippocampus which plays an important role in spatial navigation and in the formation and retrieval of **declarative memory**. This is the type of memory associated with consciousness and can be expressed in language, such as remembering a quote or a phone number. LTP has been observed at other excitatory synapses in the mammalian brain, including the motor cortex, amygdala, and cerebellum (Chapter 12). It is induced by stimulating a sufficiently large number of presynaptic fibers at a frequency of about 100 Hz for 1–3 seconds and is manifested as a several-fold increase in synaptic strength long after the high-frequency stimulation is over. Stimulation of a large number of fibers ensures that when a given synapse is activated, its postsynaptic membrane is depolarized by the activity of other synapses. Thus, a requirement of LTP is that strong postsynaptic depolarization should be paired with transmitter release from the presynaptic terminal, that is, it should occur within 100 ms or so after presynaptic depolarization.

LTP is triggered by activation of NMDA glutamate receptors (NMDARs) concomitant with depolarization of the postsynaptic membrane. As mentioned earlier, the NMDA channel is blocked at the resting voltage by extracellular Mg^{2+}. High-frequency stimulation causes prolonged depolarization of the postsynaptic membrane that removes the blockage, so that the glutamate-activated channel of the NMDA receptor conducts and allows the influx of Ca^{2+} and Na^+ (Figure 6.14). The ensuing rise in $[Ca^{2+}]_i$ triggers LTP through several mechanisms; the dominant one seems to be through CaMKII and, to a lesser extent, through PKC (Section 6.3.1). An interesting property of CaMKII is that its 12 subunits not only phosphorylate other proteins but can also phosphorylate each other. Because of this **autophosphorylation**, the activity of CaMKII can persist long after $[Ca^{2+}]_i$ returns to its normal level. CaMKII phosphorylates AMPA receptors (AMPARs), thereby increasing the conductance of this receptor channel.

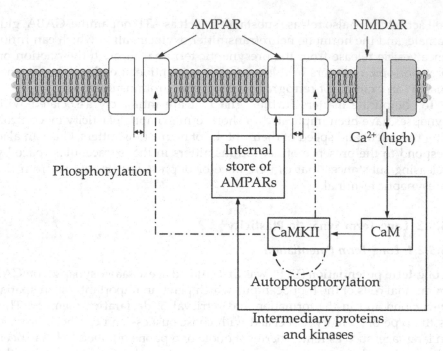

FIGURE 6.14
Diagrammatic illustration of long-term potentiation.

The increase in conductance of the AMPAR channel induced by CaMKII cannot by itself explain LTP that persists long after the cessation of tetanic stimulation. LTP has been explained by the presence of **silent synapses** having NMDARs but not AMPARs. When LTP is triggered as described in the preceding paragraph, AMPARs move to the silent synapses partly through lateral diffusion along the plasma membrane but mainly from intracellular stores of these receptors in the form of recycling endosomes (Section 1.1.4). The receptors are thus inserted, or expressed in the plasma membrane, leading to a larger synaptic response. This expression of AMPARs is regulated by Ca^{+2} activating some intermediary proteins and protein kinases (Figure 6.14). In some synapses, at least in lower animals, non-conducting AMPARs may be always present at silent synapses but become functional in LTP because of the aforementioned effect of phosphorylation. *LTP is thus considered to be due to the recruitment of new AMPARs in both synapses that had been silent as well as synapses that already possess functional AMPARs* (Figure 6.14).

6.5.2.2 Long-Term Depression

Long-term depression (LTD) is manifested following low-frequency stimulation at 1–5 Hz for 3–15 minutes. Like LTP, it can be induced by activation of NMDARs leading to an increase in $[Ca]_i$. The difference, however, is that

LTD is induced by a modest, slow rise in $[Ca]_i$, whereas LTP requires a fast increase in $[Ca]_i$ beyond a fairly high, critical threshold. The threshold, however, is not a fixed level but depends on the history of the synapse. In at least some cases, LTP raises the threshold, thereby increasing the probability that a calcium influx will yield LTD. This "negative feedback" exercises some regulatory effect on synaptic plasticity. LTD can completely reverse LTP, a process termed **depotentiation**, and, conversely, LTP can erase LTD.

Although the molecular mechanisms underlying LTD may differ somewhat in different parts of the brain, the general principle is that since LTP is caused mainly by phosphorylation and insertion of AMPARs in the plasma membrane, LTD involves the reversal of both of these processes. Thus, a modest rise in $[Ca]_i$ favors protein phosphatases rather than protein kinases, particularly PP1 and PP2B (Section 6.3.1). PP1 can dephosphorylate both CaMKII and AMPARs. However, since PP1 is not directly influenced by Ca^{2+}, a phosphatase cascade is initiated with the activation of PP2B, which then dephosphorylates a phosphoprotein inhibitor 1 (I1) (Figure 6.15). In its phosphorylated state, I1 is a potent inhibitor of PP1, so that its dephosphorylation by PP2B removes this inhibition, thereby activating PP1. The removal of inhibition is termed **disinhibition**.

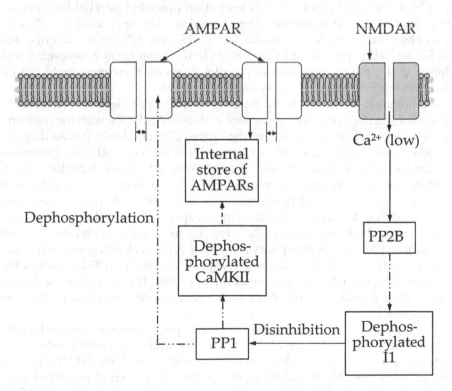

FIGURE 6.15
Diagrammatic illustration of long-term depression.

In addition, LTD involves removal of synaptic AMPARs from the plasma membrane back into recycling endosomes, a process that is also described as **receptor internalization**. The recycling of AMPARs between their internalized form and their membrane-expressed form is also termed **AMPA receptor trafficking**. The trafficking requires elevations in postsynaptic Ca^{2+}, mainly through activation of NMDARs.

Internalization of AMPARs occurs by a different mechanism in cerebellar Purkinje cells but leads to LTD just the same. The circuitry of the cerebellar cortex is described in detail in Section 12.2.4. Purkinje cells have two types of excitatory inputs – from climbing fibers and from parallel fibers. A climbing fiber makes 500–1000 synapses on the main dendritic branches of the Purkinje cell, so that a climbing fiber AP results in intense depolarization of the dendrites. The synapses of parallel fibers, on the other hand, are found on fine dendrites and on dendritic spines. Paired activation of the two inputs at a low frequency of about 1 Hz gave some mixed results in cerebellar slice experiments. When climbing fiber activation followed parallel fiber activation by 250 ms, LTD of the parallel fiber synapses was observed with 100 paired stimuli. The depression was still about 25%, 30 minutes after stimulation. There was no LTD when the interval between stimulations was reduced to 125 ms or when climbing fiber activation preceded parallel fiber activation by 250 ms. When the number of pairings was increased to 600, LTD was observed at the four intervals tested (250 ms, 125 ms, 0, −250 ms, that is, CF activation preceding PF activation in the latter case). Moreover, it was found that a strong activation of parallel fibers on their own could induce LTD through elevation of postsynaptic Ca^{2+} levels.

Glutamate is released at both climbing fiber and parallel fiber synapses. At the climbing fiber synapses, glutamate activates AMPARs, with the resulting intense depolarization opening voltage-gated Ca^{2+} and leading to an increase in $[Ca^{2+}]_i$. At the parallel fiber synapses, there are no NMDARs; glutamate receptors activate AMPARs and metabotropic glutamate receptors which initiate the phosphoinositol second-messenger system (Section 6.3.1) that produces the two second messengers IP3 and DAG. IP3 and Ca^{2+} resulting from climbing depolarization act on the receptors of the endoplasmic reticulum causing the release of more Ca^{2+}. These ions, together with DAG, activate the kinase PKC which phosphorylates other proteins leading to internalization of AMPARs. The loss of these receptors results in LDP because of the reduced epsps produced by parallel-fiber activity. It is seen that the loss of AMPARs is due to the action of protein kinases rather than phosphatases as in Figure 6.15.

Whether LTP or LTD occur at a synapse appears in some cases to be governed by the relative timing of the presynaptic and postsynaptic APs, a phenomenon referred to as **spike timing-dependent plasticity (STDP)**. Thus, if the postsynaptic AP occurs after presynaptic AP, the resulting depolarization will relieve the Mg^{2+} block of NDMARs, resulting in a large Ca^{2+} influx, leading to LTP. On the other hand, if the postsynaptic AP occurs before presynaptic activity, then the depolarization due to the postsynaptic AP will have subsided by the time presynaptic AP occurs, thereby reducing the Ca^{2+} influx and leading to LTD.

6.5.3 Structural Changes in Dendritic Spines

The LTP and LTD mechanisms of the preceding discussion are believed to be a prelude to activity-dependent, structural changes in synapses that can last for years. The synapses mainly affected are those on character-istic, knob-shaped, outgrowths on dendrites known as **dendritic spines** (Figure 6.16). As can be seen from this figure, the spine shape can vary from stubby, to mushroom-like, to spindly with a thin and relatively long neck. The spine length varies between a fraction of a μm and few μms, the neck diameter being generally less than 0.1 μm, and the head volume ranging from 0.01 μm³ to 0.8 μm³. Spines are commonly found on the dendrites of most principal neurons in the brain. Neocortical and hippocampal pyrami-dal cells (Section 7.1) have tens of thousands of spines that may constitute up to about 40% of the total dendritic membrane area. Cerebellar Purkinje cells are believed to have more than 100,000 spines comprising about 75% of the total membrane area. It is estimated that more than 90% of excitatory synapses in the brain terminate on spines, with generally one synapse per spine on the spine head.

Spines are very plastic in that they can change significantly in shape, size, and number. The change in spine shape, referred to as **spine motility**, occurs in a matter of seconds to minutes under the influence of synaptic

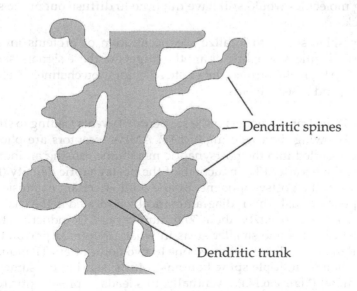

Dendritic spines

Dendritic trunk

FIGURE 6.16
Dendritic spines.

transmission, intracellular calcium, and many other ions and molecules. The motility is mediated by a cytoskeleton of microfilaments (Section 1.1.3) of the protein actin (Section 9.1.2), which dynamically changes between its monomer and polymer forms. Moreover, spine number is variable and spines can appear and disappear in a matter of hours, although the larger mushroom-shaped spines are the most stable.

Spine heads contain AMPA and NMDA receptors, as to be expected from the preceding discussion. The spine shape serves several important functions:

1. The long, thin neck introduces a high internal resistance between the head and the dendritic shaft or trunk. A given synaptic current injected into the spine head will therefore produce a larger depolarization at the head than if the synapse were located on the dendritic shaft. The larger depolarization is more effective in activating voltage-gated channels in the spine. However, the larger depolarization in the spine head, while it lasts, results in a smaller driving force and hence less current flow caused by a subsequent change in the conductance due to synaptic activity at the head of the spine.

2. The long, thin neck constrains the outflow of signaling molecules, most notably Ca^{2+}, resulting in a large, transient increase in the concentration of these molecules that is restricted to the spine. This allows them to exercise their functions more effectively, particularly those related to plasticity. However, long-lasting signaling molecules would still have the time to diffuse out of the spine neck.

3. The spine serves to localize a concentration of proteins involved in biochemical signaling or are the targets of these signals, such as NMDARs, metabotropic glutamate receptors, ion channels, CaMKII, actin, and other proteins.

Figure 6.17 illustrates a model for the sequence of events leading to structural changes following the triggering of LTP. AMPA receptors are phosphorylated and inserted into the postsynaptic membrane causing an increase, in about 10 minutes after LTP induction, of the **postsynaptic density** (**PSD**) or active area of the postsynaptic membrane with attendant expansion of the presynaptic terminal. This is diagrammatically indicated in the change from Figure 6.17a to Figure 6.17b. About 30 minutes after LTP induction, the postsynaptic area splits into smaller areas in what are termed **perforated synapses** (Figure 6.17c). This is followed one to two hours after LTP induction by the formation of **multiple spine boutons** (**MSBs**) sharing the same presynaptic terminal (Figure 6.17d). Eventually, this leads to presynaptic remodeling and new synapses, with each of the newly formed spines having its own

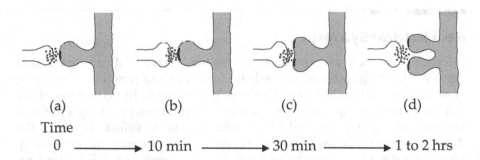

Time

0 ⟶ 10 min ⟶ 30 min ⟶ 1 to 2 hrs

FIGURE 6.17
Synaptic growth. (a) Time of LTP induction; (b) increase of postsynaptic density; (c) formation of perforated synapses; (d) formation of multiple spine boutons.

presynaptic terminal. It is believed that with disuse the reverse process of reduction in the number of synapses takes place.

Structural changes in dendrites and synapses require synthesis of new proteins, which implies that second messengers and some of their targets are involved in activating **transcription factors**, that is, proteins that control the transcription of genetic information from DNA to messenger RNA. One such ubiquitous transcription factor is the **cAMP response element-binding protein (CREB)** that is activated through phosphorylation by PKA, CaMKIV, a relative of CaMKII (Section 6.3.1), and other kinases that translocate to the nucleus. Dendrites have the metabolic machinery for local protein synthesis, as required for fast changes in structure and function.

6.5.4 Hebbian Synapses

The idea that synaptic efficacy increases with simultaneous activity in the presynaptic and postsynaptic sides of a synapse was formulated by Donald Hebb in 1949 based on theoretical considerations. The requirement for simultaneous activity has since been shown to be activity in the presynaptic terminal concomitant with depolarization on the postsynaptic side of the synapse, as served by NMDA receptors. This depolarization may be due to repeated activity of the presynaptic terminal or to activity in neighboring synapses. The increase in synaptic efficacy was assumed to result from metabolic changes or from synaptic growth. Hebb's proposal has had a major impact on theories of learning and was included in some form or other in practically all models of learning, as in the form of synaptic weights in neural networks and other computer and mathematical models of learning. As is clear from the preceding discussion on long-term plasticity, synaptic efficacy is indeed influenced by activity. Synapses whose efficacy depends on activity in both the presynaptic and postsynaptic sides are termed **Hebbian synapses**, whereas those synapses whose efficacy is influenced by activity on either side alone are termed **non-Hebbian synapses**.

6.6 Electrical Synapses

Electrical synapses, as commonly understood, have a direct connection between the cytoplasms of two cells by means of **gap junctions** consisting of at least several hundreds of channels (Figure 6.18). In vertebrates, these channels are made up of isoforms of the protein connexin having a molecular mass in the range of 26–57 kD, where a protein **isoform** is one of the different forms of the same protein. Six connexin molecules, not necessarily identical, form a hexameric hemichannel, or **connexon**, about 5–7.5 nm long and with an external diameter of about 7 nm. A connexon spans across the cell membrane and extends for 1–1.5 nm into the extracellular space. Each connexon has six protrusions, one from each of the connexin molecules, that fit into the depressions between the protrusions of the other connexon of the channel. A channel having a pore of 1.2–2 nm diameter is thus formed, with tight interlocking in the extracellular gap of 2–3.5 nm separating the membranes of the two cells. The tight interlocking is necessary to prevent leakage of ions between the channel and the extracellular space. The two connexons are held together noncovalently by hydrogen, hydrophobic, and ionic bonds between the extracellular loops of the connexin molecules.

The pore is wide enough to allow the passage of ions and small molecules of molecular mass up to about 1 kD, which includes ATP and second messenger molecules. Conductances of individual channels of gap junctions are in the range of 10–300 pS. Assuming a cylindrical pore 15 nm long and 2 nm

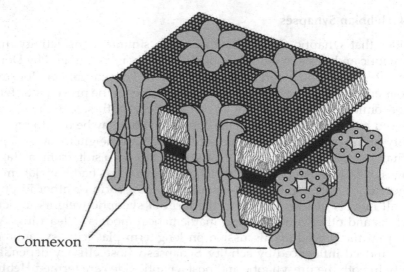

Connexon

FIGURE 6.18
Gap junctions between two cells.

in diameter, with a cytoplasm resistivity of 200 Ωcm, the conductance of a single pore is about 105 pS, corresponding to a resistance of about 10,000 MΩ.

A connexon is **homomeric** if it consists of one form of connexin, and **heteromeric** if it consists of more than one form. A gap junction is **homotypic** if its two connexons are of the same kind, and **heterotypic** if its connexons are not of the same kind, due to the neurons forming the gap junction not being of the same type.

Most types of cells other than nerve cells are connected by gap junctions, including glial cells and cardiac muscle cells (Section 10.5). Electrical synapses were first discovered in invertebrates and, later, in some parts of the mammalian brain. However, they were considered a rarity at the time compared to chemical synapses. Since then, electrical synapses have been shown to be widespread in almost all parts of the brain including the cerebral cortex, hippocampus, hypothalamus, and cerebellum, particularly between inhibitory interneurons of the same subtype and, in some cases, between neurons and glia. Gap junctions between nerve cells are commonly observed during early embryonic development but mostly disappear later.

Gap junctions are not simple, conductive channels that are identical in all cells. Depending on the connexins forming the gap junctions, the channels may differ in conductance, gating properties, and permeability to ions and small molecules. In most electrical synapses, the gap junctions are bidirectional and sign-preserving, that is, depolarization and hyperpolarization are transmitted as such, and in both directions, and have a linear voltage-current relation over a considerable range of voltage. In other cases, the voltage-current relation is highly nonlinear, whereas in still other cases the conductance drops to a lower value, at a voltage of about 30 mV in either direction between the two cells of the gap junction. The giant motor synapse of the crayfish is rectifying, that is, it allows transmission in only one direction, due to a conformational change that causes twisting of one end of the connexons and results in the six subunits pinching together and closing the channel. The conformational change is triggered by the polarity of the transjunction voltage, that is, the voltage between the two cells of the gap junction. At least in invertebrates, rectifying properties of gap junctions vary over a wide range from weakly rectifying to moderate to strongly rectifying.

Gap junction channels may not always be open. Their conductance can be regulated by various protein kinases (Section 6.4), by ions such as H^+, Ca^{2+}, Mg^{2+} and Na^+, and ATP. The probability of channel-opening is reduced by long-chain alcohols, such as heptanol and octanol, and by the gaseous anesthetic halothane. They manifest activity-dependent plasticity, at least in lower animals, that involves Ca^{2+}, as in chemical synapses.

The principal functional property of electrical synapses is transmission of electrical signals, even at subthreshold levels, and bidirectionally in most cases, with virtually no delay between the presynaptic voltage and the postsynaptic current. This is in contrast to chemical synapses where a minimum depolarization is required to release the neurotransmitter and where transmission is unidirectional. The fast transmission through electrical synapses has two important consequences:

1. Fast response, which is important in the defense against danger in some cold-blooded invertebrates, where the low temperature accentuates the difference in response times between electrical and chemical synapses. It is also important in the tail-flip response of fish.
2. Near-synchrony of response of a group of cells, as in the case of cardiac cells (Section 10.5) and inhibitory interneurons in the cerebral cortex. In the latter case, the electrical coupling subserves synchronization of rhythmic, subthreshold, as well as firing activity among groups of neurons.

Finally, the following properties of electrical synapses should be noted:

1. In contrast to chemical synapses, there is no amplification at electrical synapses. In fact, the presynaptic cell sees what is essentially the resistance of the channel in series with the parallel resistance and capacitance on the postsynaptic side (Problem 6.5). This causes some attenuation as well as slowing of the voltage response because of the time it takes to charge the membrane capacitor on the postsynaptic side. In other words, the voltage response is low-pass. The attenuation and delay are smaller, the larger the presynaptic cell is compared to the postsynaptic cell, and the more numerous are the gap junctions connecting the two cells. The delay is typically a fraction of a millisecond but can be comparable in some cases to that at the fastest chemical synapses.
2. Because of the electrical coupling between the presynaptic and postsynaptic cells, the size and shape of the voltage signal at the postsynaptic cell are directly related to the size and shape of the voltage signal at the presynaptic cell.
3. A current flowing between a more depolarized cell A to a less depolarized cell B tends to excite cell B and inhibit cell A because of the reduced depolarization in cell A due to charge movement to cell B. This is discussed in Section 7.2.4.1.
4. Repetitive stimulation of electrical synapses does not lead to facilitation or depression as in chemical synapses.
5. The gap junctions allow chemical continuity for small molecules. The chemical continuity appears to be particularly important during development, as many gap junctions between cells disappear after maturation.
6. Mutations in connexin genes can cause a variety of diseases, including peripheral neuropathy associated with abnormal Schwann cells that lack normal connexins.

Problem 6.4

It has been suggested that the larger the number of subunits of a channel protein the less selective is the resulting pore. Compare the ion selectivity of: (i) a voltage-gated channel (a tetramer), (ii) ACh receptor (a pentamer), and (iii) a gap junction (a hexamer).

Problem 6.5

Consider that the gap junction is represented by a conductance G_j and that each of the presynaptic and postsynaptic cells is represented by a parallel combination of capacitance and conductance (Figure 6.19). Assuming very slow changes, show that if the changes in the voltages across the membranes of the two cells are ΔV_1 and ΔV_2, respectively, the **coupling coefficients** k_{12} and k_{21} are given by: (a) $k_{12} = \Delta V_2/\Delta V_1 = G_j/(G_j + G_{m2})$ and (b) $k_{21} = \Delta V_1/\Delta V_2 = G_j/(G_j + G_{m1})$. Note that if cell 2 is larger than cell 1 ($G_{m2} > G_{m1}$), there is larger attenuation in going from cell 1 to cell 2 than in going from cell 2 to cell 1 ($k_{12} < k_{21}$).

Problem 6.6

If a current i_1 is injected into cell 1 in Figure 6.19, show that in the steady state:

(a) the input resistance seen by the source is $\dfrac{v_1}{i_1} = R_{11} = \dfrac{G_{m2} + G_j}{G_{m1}G_j + G_jG_{m2} + G_{m2}G_{m1}}$

(b) $\dfrac{v_2}{i_1} = R_{12} = \dfrac{G_j}{G_{m1}G_j + G_jG_{m2} + G_{m2}G_{m1}}$. Note that if the subscripts 1 and 2 are interchanged, the expression remains the same, that is, $R_{12} = R_{21}$. This resistance is known as the **transfer resistance** between the two cells. In linear circuit analysis, $R_{12} = R_{21}$ is an expression of the reciprocity theorem.

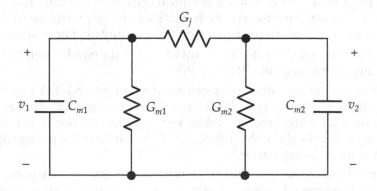

FIGURE 6.19
Figure for Problem 6.5.

Summary of Main Concepts

- Synapses are specialized structures that constitute the main channel of communication by means of electric signals between a neuron and its target cells.

- In chemical synapses, including the neuromuscular junction, the depolarization due to the AP in the presynaptic neuron triggers the release of neurotransmitter from the presynaptic membrane. The neurotransmitter diffuses across the synaptic cleft and binds to specialized receptors in the postsynaptic membrane.

- In the case of ionotropic receptors, the binding of the neurotransmitter to the receptor opens an ion channel that allows ions to flow between the cytoplasm of the postsynaptic cell and the extracellular medium, thereby changing the membrane voltage of the postsynaptic cell. In metabotropic receptors, the binding of the neurotransmitter to the receptor may trigger changes in cell metabolism via intracellular second messengers in addition to the gating of ion channels.

- The activation of an excitatory synapse enhances the possibility of generation of an AP in the postsynaptic cell, whereas the activation of an inhibitory synapse reduces this possibility.

- Proper synaptic operation requires termination of the action of the neurotransmitter in a timely manner. In the NMJ, this is accomplished by channel kinetics, desensitization, and the action of the enzyme AChE. In other chemical synapses, hydrolysis by AChE is replaced by uptake of neurotransmitter by neurons and glia.

- The peak of the psp at a fast chemical synapse is typically less than a few hundred microvolts. Such synapses are integrative, so that the generation of an AP requires activation of many excitatory synapses.

- ipsps are caused mainly by the inflow of Cl^-, the equilibrium voltage being in the range of -70 to -80 mV.

- Receptors of excitatory synapses are divided into NMDA and non-NMDA receptors. The NMDA receptors are normally blocked by intracellular Mg^{2+} and are unblocked by depolarization. Their main function is to allow the influx of Ca^{2+} at the time the postsynaptic membrane is depolarized.

- In second-messenger systems a neurotransmitter or hormone, the first messenger, binds to a metabotropic receptor, thereby activating a G protein and causing the production of second-messenger molecules. These can act directly on ion channels or indirectly by activating particular protein intermediaries. Other proteins activated by second messengers can regulate a variety of cell functions, including metabolism and gene expression.

- Synaptic plasticity could be short-term plasticity, generally lasting for up to a few minutes, and long-term plasticity, lasting from minutes to hours. In both cases, plasticity could be enhancement or depression. Short-term plasticity is mainly presynaptic, whereas long-term plasticity is mainly postsynaptic.

- Short-term plasticity is manifested under conditions of paired-pulse stimulation and stimulation by a train of high-frequency stimuli. In paired pulse stimulation, facilitation or depression depends on the probability of vesicle release by the first stimulus. Facilitation is mainly due to a residual Ca^{2+} at the time of the second stimulus due to the first stimulus. Augmentation is influenced by the activity of the Ca^{2+} pump and the Ca^{2+}–Na^+ exchanger, whereas PTP is affected by the uptake of Ca^{2+} by intracellular entities and their subsequent release.

- LTP is triggered by activation of NMDA glutamate receptors (NMDARs) concomitant with depolarization of the postsynaptic membrane. It is due to the recruitment of new AMPARs in both synapses that had been silent as well as synapses that already possess functional AMPARs.

- LTD is due to reversal of the processes leading to LTP, namely dephosphorylation and internalization of AMPARs. It is triggered by a low Ca^{2+} concentration, unlike LTP, which is triggered by a high Ca^{2+} concentration.

- Activity-dependent structural changes in synapses are initiated by LTP and LTD.

- As commonly understood, electrical synapses have a direct connection between the cytoplasms of two cells by means of gap junctions, which are made up of the six molecules of connexin forming a connexon on each of the presynaptic and postsynaptic sides. The individual channels of gap junctions have a pore of 1.2–2 nm diameter.

7

Neurons

Objective and Overview

After having discussed the action potential and synapses, it is time to consider neurons, the most important constituents of the nervous system, where synaptic activity eventually leads to the generation of action potential (AP) outputs. The chapter begins with a brief overview of the diverse types and forms of neurons and their classifications. This is followed by a discussion of the basic synaptic mechanisms of temporal summation and spatial summation of excitatory and inhibitory inputs, the basic configurations of inhibitory interneurons and their target cells, and various nonsynaptic mechanisms that influence the generation of neuronal action potentials, namely, extrasynaptic receptors and electrical mechanisms acting through gap junctions or through extracellular field currents.

The classification and properties of various Na^+, Ca^{2+}, K^+, and Cl^- ion channels are considered next, together with the most important ionic currents in these channels, their main characteristics, and their effects on neuronal firing and other aspects of synaptic and neuronal functions. The channels and currents contributing to afterhyperpolarization and afterdepolarization are discussed.

The chapter ends by considering various types of dendritic responses, including those resulting from cable properties and from the presence of ion channels, leading to: (i) modulation of synaptic inputs and of backpropagation of the generated action potential, (ii) Na^+, Ca^{2+}, and NMDA spikes, and (iii) bistable behavior.

Learning Objectives

To understand:

- The basic synaptic mechanisms of temporal summation and spatial summation of excitatory and inhibitory synaptic inputs

- How neuronal firing is influenced by various extrasynaptic mechanisms, both chemical and electrical

- How the interplay of many ionic current components flowing through channels, whose behavior can be influenced by a multitude of factors, leads to an incredible variety of neuronal characteristics and responses

- How dendritic responses arise from the cable and active properties of dendrites, how the latter modulate synaptic activity and back-propagation of the action potential, and how they can lead to dendritic Na^+, Ca^{2+}, and NMDA spikes as well as dendritic bistability

7.1 Overview of Neurons

Neurons are the core of the nervous system, their primary functions being the reception, processing, and transmission of information. Estimates of the total number of neurons in the whole of the nervous system vary widely, but the number is generally believed to be well in excess of 100 billion neurons, of which about 86 billion are in the brain. Neurons fall into a large number of distinct morphological classes, reflecting a great variety of sizes and shapes that serves their particular functions, as illustrated by representative examples in Figure 7.1. It is seen that the structure of various types of neurons can depart very significantly from that of the "typical" neuron, the motoneuron that was presented in Section 1.2.1.

Neurons can be classified in a number of ways. One of the earliest classifications is anatomical, according to which neurons can be unipolar, bipolar, or multipolar. **Unipolar** neurons have only one process emanating from the cell body. In the case of sensory neurons of the dorsal root ganglia (Figure 11.2), this single process divides at some distance from the cell body into two main branches that conduct APs from peripheral sensory receptors, past this branch point and into the spinal cord. On the other hand, most amacrine cells of the retina (Figure 7.1a) have dendrites and no axons. They neither generate nor conduct action potentials; instead they actively conduct synaptic signals through their dendrites to their target cells. Figure 7.1b illustrates a **bipolar** cell of the retina having an axon and a single dendrite on opposite ends of the soma. The axon of the dopamine-releasing neuron of the substantia nigra, which is a nucleus of the basal ganglia (Section 12.2.3), emerges from a dendrite up to 240 μm from the soma.

Most neurons are **multipolar** having one axon and one or more dendritic trees, or main dendritic trunks, leaving the cell body. Neurons are divided into **Golgi I** and **Golgi II** neurons. Golgi I neurons, also known as **principal neurons**, usually have axons that project outside the region of gray matter in which their cell bodies and dendrites are located, the axons providing

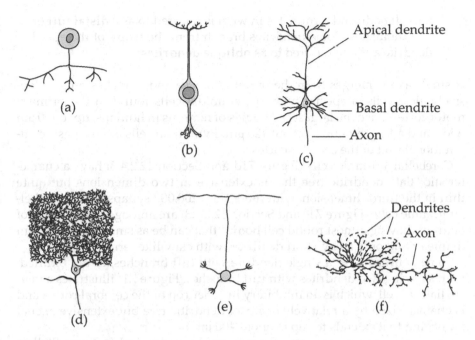

FIGURE 7.1

(a) Amacrine neuron; (b) bipolar cell; (c) pyramidal cell; (d) Purkinje cell; (e) granule cell; (f) neurogliaform cell.

the major neuronal output from the given region. Whereas most principal neurons are excitatory, such as motoneurons (Figure 1.4) and pyramidal cells (Figure 7.1c), some are inhibitory, like cerebellar Purkinje cells (Figure 7.1d), which have the most elaborate dendritic tree of all neurons. On the other hand, multipolar **Golgi II** neurons, also referred to as **interneurons** or **association neurons**, have axons that are restricted to the gray matter in which their cell bodies and dendrites are located. Some interneurons are excitatory, like cerebellar granule cells (Figure 7.1e), but most are inhibitory, like the Renshaw cells (Figure 1.5).

Neurons also differ greatly in the shapes and sizes of their cell bodies and in the extent of their dendritic and axonal arborizations. Neuronal cell bodies may be round, oblong, fusiform, or pyramidal in shape. The structure of pyramidal cells (Figure 7.1c), found in the cerebral cortex, hippocampus, and amygdala, differs somewhat between different regions but is characterized by two sets of dendrites:

1. **Basal dendrites** that arise from the base of the pyramidal cell body and then branch more extensively, and

2. A single thick apical dendrite of a few microns diameter that stems from the apex of the pyramid and branches with distance away from

the cell body and terminates in what is referred to as a **distal tuft** or **tuft dendrites**. Some dendrites branch from the trunk of the apical dendrite and are referred to as **oblique dendrites**.

A single axon emerges from the pyramidal base and branches quite extensively. Betz cells, a special type of pyramidal cells found in the primary motor cortex, have the largest cell bodies of neurons in humans, up to 100 μm wide, and a length of up to about 120 μm. Pyramidal cells are the basic computational unit of the cerebral cortex.

Cerebellar Purkinje cells (Figure 7.1d and Section 12.2.4.3) have a characteristic "flat" dendritic tree that is extensive in two dimensions but quite thin in the third dimension, with more than 200,000 synapses. The cerebellar granule cells (Figure 7.1e and Section 12.2.4.3) are amongst the smallest of neurons, having almost round cell bodies that can be as small as about 5 μm diameter and only a few, short dendrites with claw-like terminations.

Some neurons have a single dendrite with tuft branches, or are bitufted, that is having two dendrites with tuft branches. Figure 7.1f illustrates a neurogliaform cell, which is an inhibitory interneuron of the cerebral cortex and is characterized by a relatively compact dendritic tree but extensive axonal branching that extends for up to about 400 μm.

Some interneurons are named according to anatomical features. Stellate cells are interneurons having several dendrites radiating from the cell body in a star-like manner. They are found in the cerebral and cerebellar cortices and could be excitatory or inhibitory. Basket cells are inhibitory interneurons having highly branched axonal arborizations that are like baskets surrounding part of the soma of the target cell. Basket cells are found throughout the brain. Chandelier cells are inhibitory interneurons so called because their axons make highly specialized candle-shaped synaptic "cartridges" on initial segments of axons of pyramidal neurons. Double-bouquet cells are numerous cortical inhibitory interneurons characterized by tightly interwoven bundles of vertically oriented axonal arborizations resembling a horsetail. Martinotti cells are small inhibitory interneurons found in various layers of the cerebral cortex. Their axons project vertically to layer I and terminate on the dendritic tufts of pyramidal neurons.

The adjectives afferent and efferent (Section 1.2.1) are sometimes applied to neurons. An **afferent neuron** to a given region inputs APs to that region, whereas an **efferent neuron** outputs APs from that region.

Neurons can also be classified according to the neurotransmitter they release at their axon terminals. The neuron is described by adding the suffix "ergic" to part of the name of the neurotransmitter. Thus, neurons that release ACh are **cholinergic** neurons. Similarly, neurons can be **glutamatergic, GABAergic, glycinergic, dopaminergic, serotonergic**, etc. if they release the respective neurotransmitter concerned (Section 6.1.2.1).

Neurons can be further classified according to their excitability and their firing patterns, as discussed in Section 8.1.2.

7.2 Triggering of Neuronal Spikes

Typically, a neuron has several thousand synaptic inputs from other neurons and in turn makes synapses on tens or hundreds of other neurons. Bearing in mind the very large number of neurons in the brain, it is seen that neurons form complicated, interconnected networks in which the generation of electric signals by neurons depends on the activities of fast and slow chemical synapses and their spatial distribution, on electrical synapses, and on changes in the chemical or electrical extracellular environment of neurons. The resulting effects on generation of APs in neurons are in general very complex, as will become apparent from the discussion that follows. Before broaching these complexities, however, we will first consider some basic mechanisms by which synaptic inputs can lead to the generation of APs in neurons, a process commonly referred to as neuronal firing or spiking, since neuronal APs are often referred to as spikes. We will then present factors that can influence the basic mechanisms.

7.2.1 Basic Synaptic Mechanisms

The simplest case to consider is that illustrated in Figure 7.2a, which shows a neuron having some representative synapses A, B, and C, the former two being excitatory, the latter inhibitory. Conventionally, an excitatory synapse is shown as an unfilled triangular shape, whereas an inhibitory synapse is shown as a filled one. The part of the cell body that connects to the axon is the **axon hillock**. At the time of early investigations on the electrophysiology of neurons, it was assumed that the dendrites and cell bodies were passive structures, so that generated excitatory postsynaptic potentials (epsps) and inhibitory postsynaptic potentials (ipsps) propagate passively as **electrotonic spread** governed by the cable (*RC*) properties of the membrane. These postsynaptic voltages summate in the region of the axon hillock, and when the net sum exceeds threshold, a neuronal AP, or spike, is generated in the initial, unmyelinated segment of the axon as the output AP of the neuron. There is evidence that in some cases, the output AP may be initiated at the first node of Ranvier. It is well established that the density of voltage-gated Na^+ channels at the nodes of Ranvier is high, which lowers the threshold for the AP. Once initiated, the AP propagates in the forward direction along the axon and its depolarization spreads backwards into the cell body and dendrites.

The summation of postsynaptic voltages can take place temporally as well as spatially. As mentioned in Section 5.2, the amplitude of a single epsp is normally too small to generate an AP, particularly after it is attenuated as it propagates passively from the synapse to the axon hillock. This is illustrated in the upper trace of Figure 7.2b, where the epsp, due to the excitatory synapse A, is below the threshold for the generation of the AP. However, if synapse A is excited repetitively by presynaptic APs that occur sufficiently

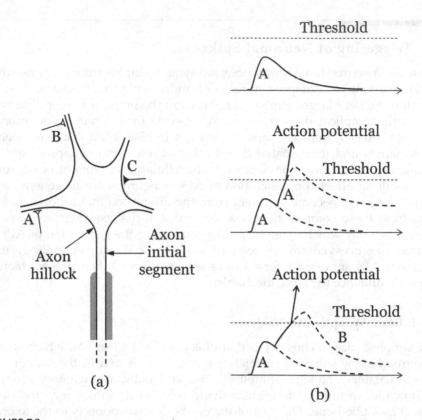

(a)

(b)

FIGURE 7.2
(a) Site of generation of action potential; (b) epsp (top trace), temporal summation (middle trace), and spatial summation (bottom trace).

close together in time, **temporal summation** occurs, which, if the threshold is exceeded, results in a neuronal AP. This is illustrated in the middle trace of Figure 7.2b by the summation of two successive epsps from synapse A, bearing in mind that, in practice, the summation of a large number of epsps is generally required to reach threshold. It should be noted that the summation of epsps is not linear because later epsps encounter a different membrane conductance due to the earlier epsps, as well as a smaller net depolarization for driving the synaptic current, compared to earlier epsps.

Alternatively, an epsp due to synapse A can summate with an epsp due to synapse B at a different location from synapse A. If the two epsps occur sufficiently close in time, their sum may exceed threshold and result in an AP. This is described as **spatial summation** and is illustrated in the lower trace of Figure 7.2b. Spatial summation can also occur at dendritic branches, where smaller-diameter dendrites combine to a form a larger diameter dendrite.

Even in the presence of temporal and spatial summation, the simultaneous activation of inhibitory synapse C can prevent the generation of an AP in two ways:

1. The hyperpolarization of the ipsp can subtract directly from the depolarization due to the epsps, thereby preventing the total depolarization from reaching threshold.

2. The increased membrane conductance due to a fast inhibitory synapse increases the shunt conductance in the path of passively propagating synaptic voltages, which increases the attenuation of these voltages and hence reduces the total depolarization due to epsps.

It is seen that, in general, neuronal firing is the result of a dynamic interplay of a large number of excitatory and inhibitory inputs to a given neuron, a process described as **integration** *of these inputs – not in the mathematical sense but rather in the sense of combining these inputs in complex, unspecified ways to generate neuronal APs.* Moreover, the process of generation of neuronal APs is subject to many **modulating** influences that alter or adjust this process in some way, as described later. Synaptic integration is further elaborated in Section 7.4.1.

Before ending this subsection, it should be noted that the aforementioned synaptic responses, whether excitatory or inhibitory, are described as **phasic**, in the sense that they are transient in nature, in contrast to tonic effects that will be discussed in Section 7.2.3. Moreover, the direct, short-term responses to activation of chemical synapses are mediated through ion channels restricted to the postsynaptic membrane of the synapse itself in the case of fast synaptic action and generally restricted to the target cell in question in the case of slow synaptic action. When well-defined signal pathways are involved in signal transmission, the signal transmission is described as **wiring transmission**, in contrast to volume transmission discussed in Section 7.2.3.

7.2.2 Synaptic Connections between Neurons

In general, a neuron receives synaptic inputs from many neurons and, in turn, makes synapses with many other neurons. Consider two neuronal populations A and B; if every A neuron makes synaptic connection with a number X of B neurons, on the average, X is the **divergence number from A to B**; if every B neuron receives synaptic input from a number Y of A neurons, on the average, Y is the **convergence number of A on B**. An example of a relatively large convergence number of more than 100 is that from the neurons of the dorsal striatum on the neurons of the globus pallidus in the basal ganglia (Section 12.2.3). An example of a relatively large divergence number of roughly 300 is that from granule cells to Purkinje cells in the cerebellum (Section 12.2.4.3).

The inhibitory synaptic inputs referred to in the preceding section are generally provided through two basic configurations of inhibitory interneurons with respect to their target neurons and the direction of signal propagation. In **feedforward inhibition** the inhibitory interneurons are excited by the same excitatory input as the neurons they inhibit. The inhibition, from the

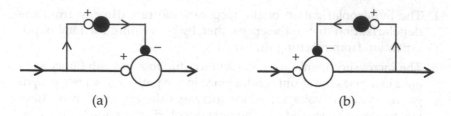

FIGURE 7.3
(a) Feedforward inhibition; (b) feedback inhibition.

interneuron to its target neuron, is in the direction of signal propagation. This is diagrammatically illustrated in Figure 7.3a by a single-line diagram in which *small unfilled* circles denote excitatory synapses, *small filled* circles denote inhibitory synapses, a *large unfilled* circle denotes an excitatory neuron, and a filled circle of intermediate size denotes the inhibitory interneuron. The direction of signal propagation in Figure 7.3 is that through the principal, excitatory neuron, from left to right. In contrast, Figure 7.3b illustrates **feedback inhibition** in which the interneuron is excited by the neuron it inhibits.

It has been estimated that approximately 20% of the synaptic contacts in the brain are inhibitory. Note that a single-line diagram of neuronal connections as in Figure 7.3 should be interpreted in terms of neuron types rather than literally. Thus, in Figure 7.3a, the incoming excitation may excite a given excitatory neuron, but the inhibitory neuron excited by this incoming excitation may inhibit other excitatory neurons of the same type as the given excitatory neuron. Similarly, in Figure 7.3b, the output of the excitatory neuron excites an inhibitory interneuron, which may then inhibit other excitatory neurons of the same type. In **center-surround-inhibition**, an excited neuron is surrounded by inhibited neurons of the same type. In **lateral inhibition**, the inhibited neurons are located mainly laterally with respect to the excited neuron of the same type. Examples of these types of inhibition are given in connection with Figure 12.14.

Problem 7.1

Given two neuron populations A and B. Each A neuron receives inputs from a number Y of B neurons. On the other hand, each A neuron makes synaptic connection with a number X of B neurons in such a manner that every B neuron receives an input from just one A neuron. Argue that the ratio of the number of B neurons to that of A neurons is Y/X.

7.2.3 Nonsynaptic Mechanisms

Many receptors on the dendrites and cell bodies of neurons are **extrasynaptic receptors** that do not belong to any synapse, that is, they are not part of the postsynaptic side of any synapse. Such receptors have received considerable attention, particularly GABAergic extrasynaptic receptors, that is, receptors responsive to GABA.

GABA$_A$ receptors are composed of α or β subunits in combination with δ, γ, or ϵ subunits. Different combinations of these subunits, particularly those involving the δ subunit, have high GABA sensitivity and little desensitization to continuous GABA exposure. This makes extrasynaptic receptors of this type particularly responsive to low, ambient, extracellular concentration of GABA, resulting in sustained inhibition, described as **tonic inhibition**, in contrast to the phasic inhibition mentioned earlier. The ambient concentration of GABA can come from different sources:

1. Spillover from synapses. It is generally true that when a neurotransmitter is released into the synaptic cleft, it may spill over, by diffusing sideways away from the activated synapse to the extracellular space, where it may activate receptors in the same cell or in neighboring cells. However, transmitter reuptake involving GABA transporters in neurons and glia, normally limits the spatial extent of this spillover, unless the presynaptic stimulation is especially intense or involves sustained firing of many neurons synapsing close to one another. The time course of the response to neurotransmitter spillover can be quite prolonged, depending on the neurotransmitter reuptake mechanisms.

2. Reversal of the GABA transporter in neurons. The transporter normally removes GABA from the extracellular space at high GABA levels, but when the extracellular GABA level falls too low, the transponder reverses and maintains a base, background level.

3. Release from glial cells, particularly astrocytes (Section 1.2). These cells not only remove GABA at high levels from the extracellular space, following synaptic activity, but can also release GABA into the extracellular space.

It should be emphasized that tonic inhibition, as commonly referred to in the literature, may not only be due to excitation of extrasynaptic receptors but could also be due to sustained activity of inhibitory neurons affecting their target neurons through chemical synapses in the usual manner. Irrespective of how it is generated, tonic inhibition influences neuronal firing in several ways:

1. By reducing the probability of firing of neurons, tonic inhibition causes an overall decrease in the excitability of whole regions of the brain, such as the cerebellum, the hippocampus, and the neocortex. Reduced excitability generally decreases the mean firing rates of neurons.

2. Tonic inhibition can alter the pattern of neuronal firing in response to excitatory inputs. Moreover, many neurons show regular,

spontaneous, pacemaker-like spiking in the absence of excitatory inputs (Section 8.1.5). Tonic inhibition can make this spiking pattern irregular. Modeling studies show that irregularly firing neurons in a network can increase the speed and sensitivity of the response of the network to external inputs and make the response more linear.

3. Tonic inhibition affects cable properties. By increasing membrane conductance, the attenuation by electrotonic spread is increased and the time constant is reduced, so that changes in membrane voltage are speeded up, as is most evident in the accelerated rates of decay of synaptic voltages. This reduces the time integral of synaptic responses and hence the transfer of charge.

The degree of tonic inhibition can be controlled by various factors, such as neuronal activity, chemical agents, and the activity of the GABA transporter, which depends in turn on membrane voltage as well as intracellular concentrations of GABA and Na^+.

A starker example of nonsynaptic transmission is encountered in neurogliaform cells (Figure 7.1f), which, as mentioned before, are inhibitory, cortical interneurons characterized by extensive axonal arborizations. All axonal boutons of these cells contain synaptic vesicles, yet approximately 80% of these boutons do not form synapses. When these cells fire, they release a large, dense axonal cloud of GABA that diffuses through the extracellular space and activates $GABA_B$ receptors as well as $GABA_{A\delta}$, extrasynaptic receptors (which are $GABA_{A\delta}$ receptors containing the δ subunit) found on the neurogliaform cells themselves or on other cells. Neurogliaform cells thus mainly communicate not through synapses but through **volume transmission**. In contrast to wiring transmission referred to earlier, volume transmission involves nonspecific, three-dimensional signal propagation in the extracellular fluid over a considerable distance from the site of origin. Volume transmission could be mediated by neurotransmitters, such as the case of GABA just mentioned, and released into the extracellular space or into the cerebrospinal fluid. Volume transmission could also be mediated by other than chemical agents, such as extracellular field potentials, which will be discussed later.

It should be mentioned that in **ectopic release,** a neurotransmitter is released from sites other than active zones of synapses and activates extrasynaptic receptors. Glutamate is released in this manner from axons and activates extrasynaptic receptors on glial cells. Activation of neuronal receptors by ectopic release seems uncommon.

7.2.4 Electrically Mediated Mechanisms

7.2.4.1 Gap Junctions

Gap junctions (Section 6.6) in the mammalian brain have been observed most frequently between interneurons of the same type, occurring as

dendrodendritic or **dendrosomatic** coupling, the former being between dendrites of different cells, and the latter between dendrites of one cell and the body of another cell. The coupling coefficient between two cells is defined and given in Problem 6.5. The coupling between interneurons is one-to-many, as each interneuron may be electrically coupled to tens of other interneurons, which means that each type of interneuron forms a highly interconnected network over a large cortical area. The one-to-many coupling increases the input conductance of an interneuron by at least 30%, which decreases the passive time constant and increases the current needed to bring the membrane to threshold, as well as the current required to sustain a given frequency of firing.

Of more significance is that strong electrical coupling, which entails rapid propagation and sharing of electrical signals between neurons, increases synchronization of subthreshold oscillations as well as synchronization of firing in neuronal populations, as discussed in Section 8.1.5.

An interesting observation on the effect of electrical coupling is that a subthreshold depolarization in a cell will depolarize an electrically coupled cell, at the same time reducing the depolarization in the first cell, because of movement of charge, as noted earlier (Section 6.6). However, if an AP has a pronounced undershoot, or hyperpolarizing afterpotential (Section 3.1), the time integral of the current due to the spike proper can be less than that of the hyperpolarization of the undershoot. There will be a net transfer of charge from the coupled cell to the first cell undergoing the AP, which leads to a hyperpolarization of the coupled cell. This is illustrated in Figure 7.4, which shows in the top trace an AP in cell 1 followed by a pronounced undershoot.

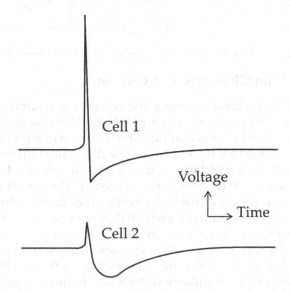

FIGURE 7.4
Voltage variations in electrically coupled cells.

The depolarization of the AP in cell 1 produces an initial rapid depolarization, or **spikelet**, in the coupled cell (cell 2) followed by a hyperpolarization. The net effect is a hyperpolarization that transfers positive charge, carried mainly by K^+ (Section 3.2, Chapter 3) from cell 2 to cell 1. The charge transferred depends on how far E_K is from the resting voltage and on the activation and inactivation of the K^+ channels, which in turn depends on the properties of these channels and on the strength and duration of the stimulus.

7.2.4.2 Field Potentials

Another way by which firing of neurons is directly affected by electrical activity is through so-called field potentials due to extracellular currents caused by gating of ion channels. When neurons are densely packed together without sufficient isolation by surrounding glial cells, these field potentials affect membrane voltages of neighboring neurons, moving them toward, or away from, threshold. This effect, termed **ephatic coupling,** is an example of volume transmission and can be effective in synchronizing the firing of APs among densely packed neurons.

Electrical coupling through field potentials can have an inhibitory effect. This occurs, for example, in the case of inhibitory basket cells that form synapses near the initial segment of cerebellar Purkinje cells (Section 12.2.4.3). The action potentials of a basket cell produce extracellular currents that accumulate positive charge outside the initial segment of a Purkinje cell, thereby hyperpolarizing this segment and resulting in a short-latency inhibition that precedes the inhibition through the chemical synapses.

7.3 Neuronal Ion Channels and Currents

We consider in this section various ionic currents in neurons and some of their most important characteristics and effects. These currents are carried by Na^+, Ca^{2+}, K^+, and Cl^-, the magnitude and direction of any current component being determined by the electrochemical potential difference driving the given ion and the conductance of the ion channel involved. The Na^+ and K^+ currents in axons are relatively stereotyped, as the main function of an axon is rapid propagation of the AP. On the other hand, neurons perform myriad signal-processing functions which they exercise by generating a wide variety of output firing patterns and frequencies in response to diverse input signals. The great variety of neuronal characteristics and responses results from the interplay of many ionic current components flowing through channels whose behavior can be influenced by a multitude of factors. The combinations of possibilities can be dauntingly complex, as can be glimpsed from the discussion in this and the following sections.

A note about nomenclature: ion channel families are designated in the modern literature by adding to the chemical symbol of the ion a subscript that denotes an activating agent of the channel or some distinguishing property. A "v" subscript, for example, denotes voltage-dependency. The subscripted symbol is followed by two numbers separated by a decimal point, such as $Ca_v2.2$. The first number following the subscript refers to a **subfamily** of the ion channel family and the number following the decimal point refers to a particular **subtype** according to the order of discovery. The numbers signify differences in amino acid sequences, particularly the membrane-spanning sequences and the pore loops that join them, as determined by the genes concerned. In contrast to ion channels, ionic currents are often still designated by current symbols or letter designations based on the old channel classifications, as illustrated later.

Generally speaking, channel properties are primarily determined by the main protein subunits of the receptor forming the pore. These subunits, referred to as α subunits, are usually arranged as a tetramer around the pore. In some cases, the pore could be formed by a single polypeptide chain that courses back and forth across the membrane, with each segment spanning the membrane acting as the equivalent of a protein subunit. The same ion channel subfamily can exhibit considerable functional diversity between different nerve tissues, or between different parts of the same neuron, in terms of sensitivity to various activating or blocking agents, voltage dependence, and channel kinetics. This functional diversity can be due to at least the following factors:

1. The α subunits of the channel need not be identical, that is, homomers, but could be heteromers, occurring as combinations of different subunits usually, but not necessarily, from the same subfamily.
2. The presence of auxiliary protein subunits, associated with the α subunits, and referred to as β, γ, and δ subunits, and which modulate the gating of the pore. These auxiliary subunits can span across the membrane, away from the pore, or can be on the inner or on the outer surface of the cell membrane, close to the pore, and in different combinations.
3. The same gene can code for variants of the same protein, referred to as isoforms, through modification of the transcribed RNA by splicing, to produce different mRNAs that are translated into different protein isoforms. The functional diversity serves to fine-tune the channel by conferring specific properties that are required in particular cases.

The following should be noted in connection with the discussion that follows on ion channels:

1. It becomes necessary in many cases to consider the time for deactivation of a given ion channel, that is, removal of activation, in distinction to inactivation, which is a different process, as discussed for

the squid giant axon (Section 3.2.2.2); similarly for deinactivation, or removal of inactivation, which is a different process from activation.

2. It will be recalled that in fitting the responses of the squid giant axon, Hodgkin and Huxley found it convenient to treat activation and inactivation as separate processes. However, there is evidence that, at least in some cases, the voltage-dependence and time-dependence of activation and inactivation are inherently coupled together (Section 3.2.4).

3. An important consideration in the functioning of the channels for a particular ion is their interaction with channels for other ions that are **colocalized**, that is, located in close proximity in the membrane.

7.3.1 Sodium Channels

The voltage-dependent Na^+ channels belong to a single subfamily Na_v1 having various subtypes indicated by numbers after the decimal point. At least nine voltage-gated channels have been identified, designated as $Na_v1.1$ to $Na_v1.9$, with different subtypes, expressed in neurons as well as in cardiac, skeletal, and smooth muscle. At least two non-voltage-gated Na^+ channels have been identified.

Na^+ channels in different neurons, or in different locations in the same neuron, differ in their voltage-dependence and in the time constants of their activation and their inactivation, if this exists, and in their sensitivity to tetrodotoxin (TTX). In certain neurons, including cerebellar Purkinje cells, some of the Na^+ channels that open during the upstroke of the spike inactivate in the usual manner due to a loop of the main channel subunit blocking the channel at its intracellular opening. Other Na^+ channels, however, are blocked by an intracellular, positively charged entity, which could be part of a β subunit of the channel. Both mechanisms stop the influx of Na^+ and contribute to repolarization of the membrane. However, as the membrane repolarizes, the blockage is removed because of the reduction in the outwardly directed electric field acting on the positively charged, intracellular blocking entity. This unblocking action produces a **resurgent Na^+ current** that is observed under certain experimental conditions. The unblocked channels become available to take part in a second AP, thereby contributing to the rapid firing characteristic of Purkinje cells.

Two types of Na^+ currents have been identified in neurons:

$I_{Na,t}$: Voltage-dependent, fast transient current (t for transient), where transient implies the presence of activation and inactivation, as in the squid giant axon. The inactivation may be complete in some cases and partial in others, in which case a substantial Na^+ remains at the resting membrane voltage. The $I_{Na,t}$ current is responsible for the generation of the AP.

$I_{Na,p}$: Voltage-dependent, fast persistent current (p for persistent), where persistent refers to no inactivation or very slow inactivation. Persistent Na^+ currents contribute to a steady level of depolarization, regulate repetitive firing, amplify dendritic depolarization, and produce afterdepolarization (ADP) and plateau potentials, as will be explained later (Section 7.4.5).

7.3.2 Calcium Channels

Ca^{2+} are considered to be the most important of all signaling molecules, as they underlie a wide variety of cellular phenomena such as muscular contraction, neurosecretion, synaptic plasticity, membrane hyperpolarization and subthreshold depolarization, Ca^{2+} action potentials (Section 7.4.4), and modulation of the Na^+-generated action potentials. In addition, Ca^{2+} can act as second messengers (Section 6.3.1) and are involved in gene expression as well. Ca^{2+} enter the cell not only through voltage-dependent Ca^{2+} channels but also through ligand-gated channels, as in the case of NMDA receptors, although a depolarization is required in this case to remove the block by Mg^{2+} (Section 6.2.3).

At least ten different voltage-gated Ca^{2+} channels have been identified in the vertebrate nervous system. They can be divided into two broad classes depending on the voltage threshold of activation:

1. *High-voltage-activated* (HVA), or high-threshold, Ca^{2+} channels. These are divided into two subfamilies, each having various subtypes:

   i) *Ca_v1 subfamily*, characterized by channels having:

      – a relatively large, single-channel conductance of about 25 pS

      – activation voltage >–30 mV, with fast deactivation

      – very slow inactivation in the range –60 mV to –10 mV with a time constant exceeding 500 ms

      First described as L-type channels (L for long-lasting), the current in these channels is referred to as an **L-type Ca^{2+} current**. These currents play an important role in muscle contraction (Section 9.2). $Ca_v1.3$ subfamily is involved in the Ca^{2+} **persistent inward current** (PIC) that is functionally important in motoneurons (Section 11.2.1.2), where they are reported to have a low-voltage threshold. L-type channels are blocked by dihydropyridine nimodipine.

   ii) *Ca_v2 subfamily*. The first of these channels was described as P/Q-type (P for cerebellar Purkinje cells, Q for Ca^{2+} channels in cerebellar granule cells) before it was realized that these channels are essentially the same as P-type channels. These channels are classified as $Ca_v2.1$. Other channels in this subfamily are described as N-type (N for neuronal and classified as $Ca_v2.2$), and R-type

(R for residual or resistant to certain toxins and classified as $Ca_v2.3$). R-type channels are sometimes described as intermediate-voltage-activated. The different subtypes have similar properties, except for their pharmacology. They have:

- – a single-channel conductance of about 10–13 pS
- – activation voltage > –20 mV, with slow deactivation
- – partial, slow inactivation in the range –100 mV to –60 mV with a time constant in the range 50 ms to 80 ms.

The Ca_v2 subfamily channels are influenced by second-messengers and by G proteins (Section 6.3).

As discussed in Chapters 4 and 5, Ca^{2+} play a critical role in the release of neurotransmitters at presynaptic terminals. The channels involved are mainly $Ca_v2.1$, and to some extent $Ca_v2.2$ in the peripheral nervous system (PNS), and both $Ca_v2.2$ and $Ca_v2.3$ in the central nervous system (CNS). These channels are also regulated in different neurons by various substances such as ACh, norepinephrine, GABA, and various opioids, which include the addictive drugs codeine, morphine, and its derivative, heroin. When these drugs bind to what are referred to as opioid receptors in the CNS and the gastrointestinal tract, they reduce Ca^{2+} release at presynaptic terminals and hence depress synaptic transmission through these synapses. The painkilling effect of these drugs is due to depressed synaptic transmission in pathways conveying pain signals to brain centers.

2. *Low-voltage-activated* (LVA), or low-threshold, Ca^{2+} channels, of the Ca_v3 subfamily, having various subtypes. Functionally, these channels have:

   • a single-channel conductance of 5–10 pS
   • activation voltage > –70 mV, with fast deactivation
   • complete, rapid inactivation in the range –100 mV to –60 mV and having a time constant in the range 20–50 ms.

First described as T-type channels (T for tiny), the current in these channels is referred to as a **T-type Ca^{2+} current**. Because they are activated by small changes in membrane voltage, the T-type channels help control excitability at the resting voltage and are an important source of the depolarizing current that drives the rhythmic pacemaker activity of some neurons and cardiac pacemaker cells (Section 10.5.3.2).

Moreover, it is generally true that both LVA and HVA Ca^{2+} channels are found in the dendrites, soma, and initial segments of neurons, their relative distribution between these regions depending on the type of neuron. Being slower to activate than the Na$^+$ transient current, the Ca^{2+} currents make little contribution to the rising phase of the AP but become significant during the falling phase. Depending on the channels involved, Ca^{2+} entry can directly contribute to depolarization, as expected for an inward cationic current, but can also have an opposite effect by activating large-conductance K$^+$ channels. Ca^{2+} currents at the axon initial segment affect excitability and the generation of bursts of spikes, as well as the speed of repolarization, and hence the width of the generated AP. An interesting aspect of LVA Ca^{2+} channels is that they are normally inactivated at resting voltage levels. They are deinactivated by hyperpolarization and activated by a subsequent depolarization. As a result of this activation, Ca^{2+} enter through these channels and further amplify the depolarization. The inhibition–excitation sequence in neurons thus plays an important role under these conditions.

Ca^{2+} channels in dendrites also lead to dendritic Ca^{2+} APs in some neurons, as in cerebellar Purkinje cells, as well as pyramidal cells of the CA1 region of the hippocampus and of layer 5 of the of the neocortex. It is believed that dendritic calcium spikes are due to activation of T- or R-type calcium channels.

7.3.3 Potassium Channels

Potassium channels are the most diverse of all ion channels. They are conventionally divided into four families, based on the genes involved:

1. K_{Ca} *family* of calcium-gated channels that are opened by an increase in [Ca]$_i$.

2. K_v *family* of voltage-gated channels. These are outward rectifier channels (Section 2.6), as in the squid giant axon. This means that the outward current is larger than the inward current, in the direction of increasing depolarization, as a result of the channels opening with depolarization. The general shape of the variation of current with voltage is that of Figure 2.11, but it must be remembered that this figure is based on a non-voltage-gated channel conductance and rectification due to a difference between [K$^+$]$_o$ and [K$^+$]$_i$. The functions of voltage-gated K$^+$ channels include regulation of: the rate of repolarization, the level of afterhyperpolarization, and the firing frequency of neurons. The outward rectifier channels are also referred to as **delayed rectifier channels** because of the relative delay in their activation.

3. K_{ir} *family* of inward rectifying channels. In these channels, the inward current increases in magnitude with increasing hyperpolarization, the general shape of the variation of current with voltage being that of Figure 2.12, but shifted to the left so that zero membrane current is at $E_K < 0$ instead of $E_{Na} > 0$. Again it should be borne in mind that Figure 2.12 is based on a non-voltage-gated channel conductance

and rectification due to a difference between $[Na^+]_o$ and $[Na^+]_i$. Note that in both outward and inward rectification, the positive (outward) current increases with depolarization, and the magnitude of the negative (inward) current increases with hyperpolarization, the membrane conductance being positive in both cases. Inward rectification in potassium channels, also known as **anomalous rectification**, is due to pore blockage by intracellular positively charged entities such as Mg^{2+} and polyamines. Hyperpolarization progressively relieves this blockage because of the increased inwardly directed electric field acting on these positively charged intracellular entities. The behavior of the K_{ir} family is considered in more detail later in this subsection. In humans, at least 15 K_{ir} channels have been identified, divided into 7 subfamilies, and designated as $K_{ir}1$ to $K_{ir}7$. The $K_{ir}3$ subfamily is also known as GIRK (G protein-coupled inwardly-rectifying K^+ channel). In general, K_{ir} channels contribute to the resting membrane voltage of neurons.

4. K_{2P} *family* of channels having two pore-forming P-loops, where a pore-forming P-loop is a pore loop having phosphate binding sites on the loop connecting membrane-spanning segments of the channel protein (Figure 5.9a). The pore is formed by two α subunits, each subunit having four transmembrane segments with two pore-forming P-loops, one between the first and second transmembrane segments and the other between the third and fourth transmembrane segments. The K_{2P} channels are described as "leak channels" in the sense that their conductance is essentially voltage-independent, like that of the leak channel in the HH model (J_L in equation 3.1). However, these channels are potassium selective. They affect the resting membrane voltage and neuronal excitability. The mammalian K_{2P} family has fifteen channels designated as K_{2P1} to K_{2P18}, according to the nomenclature of the IUPHAR (International Union of Basic and Clinical Pharmacology). The numbering matches that of the corresponding KCKN genes, there being no KCKN8, KCKN11, and KCKN14 genes. The channels are also commonly designated according to their biophysical or pharmacological properties by the acronyms: TWIK, TREK, TASK, TALK, TRAAK, THIK, and TRESK, which refer to the gating of some of these subfamilies by physical stimuli such as mechanical stretch, temperature, and pH. The standing outward current in cerebellar granule neurons due to K_{2P} channels is sometimes referred to as I_{Kso}. The resting membrane voltage and conductance of motoneurons (Section 10.2.1.2) are largely determined by channels TASK1 (K_{2P3}) and TASK3 (K_{2P9}), where TASK stands for "two-pore domain, acid-sensitive K^+ channel".

Rather than delve more into the rather elaborate details of subfamilies and subtypes of K^+ channels, it is more appropriate for our purposes to consider, in the following paragraphs, some K$^+$ channels in terms of their functionality and currents.

BK channels (I_C): These have large conductances in the range of 100–250 pS (B for big), are voltage dependent, the conductance increasing e-fold per 9–15 mV, and are *also* activated by [Ca]$_i$ in the range of 1–10 μM. They are maximally open at depolarizations > +10 mV and [Ca]$_i$ > 10 μM. In terms of the gene encoding their α subunits, they are classified as belonging to the K$_{Ca}$1.1 subtype. The current of BK channels is referred to as I_C **current**.

BK channels are widely distributed in the CNS, in the soma, dendrites, and presynaptic terminals of many types of neurons. They are often co-localized with voltage-gated Ca^{2+} channels, so as to control [Ca^{2+}]$_i$ through negative feedback. Thus if [Ca^{2+}]$_i$ increases, activation of BK channels hyperpolarizes the membrane by increasing K$^+$ efflux. This, in turn, reduces Ca^{2+} influx through the voltage-gated Ca^{2+} channels and opposes the rise in [Ca^{2+}]$_i$. The control of [Ca^{2+}]$_i$ regulates muscle contraction and the release of neurotransmitters at presynaptic terminals. Recall that neurotransmitter release is also influenced by the effect of K$^+$ conductance on the shape of the AP. BK channels help repolarize the membrane on the downstroke of the AP and contribute to the afterhyperpolarization (AHP). BK channels in dendrites affect the dendritic Ca^{2+} AP (Section 7.4.4).

BK channels are regulated by the partial pressures of gases in the blood, such as oxygen and nitrous oxide, and by protein kinases, including PKA and PKC (Section 6.3), which couple BK channels to several second-messenger signaling systems. BK channels have been implicated in the intoxicating effects of alcohol and in some brain and smooth muscle disorders. The functional diversity of BK channels is underscored by the fact that, whereas BK channels in smooth muscle and many neurons do not inactivate, BK channels in cells of some sympathetic ganglia of the rat show rapid inactivation.

SK channels (I_{SK}): These have small conductances in the range of 4–20 pS (S for small), are voltage-independent, and are activated by [Ca]$_i$ in the range of 50–900 nM. In terms of encoding genes, they are classified as belonging to the K$_{Ca}$2.1, K$_{Ca}$2.2, and K$_{Ca}$2.3 subtypes. The current of SK channels is referred to as I_{SK} **current**.

SK channels are widely expressed throughout the CNS. SK channels located in or near the soma affect neuronal firing by contributing to the afterhyperpolarization, as discussed later. SK channels in the dendrites can be activated by Ca^{2+} from sources other than voltage-gated Ca^{2+} channels, such as intracellular stores, or through activation of NMDA receptors. The result is a hyperpolarization, due to K$^+$ efflux, and shunting of excitatory inputs because of the increased conductance of SK channels. These effects reduce dendritic excitability and influence synaptic transmission. Moreover, SK channels are co-localized with NMDA channels in the dendritic spines of some neurons, such as hippocampal pyramidal neurons and neurons in parts of the basal ganglia. In these neurons, calcium influx activates SK channels, so that the resultant hyperpolarization enhances the Mg^{2+} block of the NMDA receptor, thereby modulating the induction of synaptic plasticity. SK channels are specifically blocked by apamin, a toxin in bee venom.

IK channels: These have intermediate conductances in the range of 20–80 pS (I for intermediate), are voltage-independent, and are also activated by $[Ca]_i$ in the range of 50–900 nM. In terms of encoding genes, they are classified as belonging to the $K_{Ca}3.1$ subtype. Functionally, they are similar to SK channels, but they do not seem to play a significant role in neurons and muscle. They are found in peripheral tissues, including secretory epithelia and blood cells. They play an important role in supporting T lymphocyte (T cell) proliferation by maintaining large electrical gradients for the sustained transport of ions such as Ca^{2+} influx. Because of this, IK channels may play an important role in autoimmune disorders, such as rheumatoid arthritis, inflammatory bowel disease, and multiple sclerosis.

K_v3 channels: these are high-voltage-activated (HVA), requiring voltages achieved near the peak of the AP. They have steeply voltage-dependent and very fast activation and deactivation. Both of these properties are conducive to rapid repolarization and, hence, a narrow AP. In addition, these channels have a short refractory period as required for fast spiking. Cerebellar Purkinje cells, for example, have APs of half-height-width – that is, width at half the spike height – of about 200 μs at 37°C and can fire at rates of 200–400/s or more. Kv3.3 is strongly expressed in Purkinje cell somata, and Kv3.4 is also expressed both in the dendrites and somata of these cells.

K_v4 channels (I_A): I_A is a fast, transient potassium current that rapidly activates with a time constant of less than 1 ms and inactivates with a time constant of a few milliseconds. The channels involved belong to the K_v4 subfamily, which are low-voltage-activated potassium channels localized in cell bodies and dendrites of some types of neurons, such as neocortical and hippocampal pyramidal cells. I_A is normally inactivated at the resting voltage and requires prior hyperpolarization before it can activate. I_A can contribute to the repolarization of the AP and to the fast afterhyperpolarization (Section 7.3.5). This could facilitate the recovery of Na^+ channels from inactivation, which influences repetitive firing. I_A also plays a role in synaptic modulation (Section 7.4.2), and in the regulation of the backpropagating AP (Section 7.4.3).

K_v7 channels (M current, $I_{K(M)}$ or I_M): The Kv7 channels, also known as M-type K^+ channels, and their various subtypes are slow, low-voltage-activated potassium channels that do not inactivate. They are partially open at the resting voltage, and their conductances are reduced by the binding of ACh to muscarinic, metabotropic receptors (M for muscarinic), as well as the binding of various peptides and hormones to other receptors. The Kv7 channels are known to reduce excitability in autonomic ganglionic neurons and the current through them helps determine the resting membrane voltage and the input resistance of the cell. In the absence of slow synaptic action, a ganglionic neuron normally fires only one or two action potentials in response to prolonged excitatory stimulation that is just above threshold.

This is due, at least in part, to the increase in I_M by the prolonged depolarization, which helps repolarize the membrane below threshold. In the presence of slow synaptic action, the I_M current is reduced because the blocking of fast transmission at the synapses of these neurons is enhanced and repetitive firing can occur. Three factors contribute to this repetitive firing behavior: (i) the depolarizing effect of the reduction in I_M brings the membrane closer to threshold, as discussed in connection with Figure 6.4; (ii) the increase in the cell resistance due to the blocking of the K^+ channels by slow synaptic action means that a given epsp produces a larger depolarization; and (iii) the reduction in I_M during repolarization causes the neuron to remain depolarized above threshold during the prolonged stimulus, thereby firing repetitively in a burst of impulses.

K_v1, K_v4, and K_v7 channels are low-voltage-activated (LVA), opening on small depolarizations from around resting potentials. By opposing the depolarization, they can regulate the number of APs fired by the neuron. K_v2 channels are high-voltage-activated (HVA), like K_v3 channels, but have slower kinetics. In all cases, the large conductance of K^+ channels brings the resting voltage close to E_K.

HCN channels (I_h): HCN stands for **hyperpolarization-activated and cyclic nucleotide-gated.** As their name implies, these channels are activated by hyperpolarization, and their activation is directly facilitated by cAMP levels without involvement of kinases (Section 6.3.1). The properties of these channels vary considerably between various tissues but share many similarities. Although they belong to K_v channels, they are much less selective to K^+, having a permeability to Na^+ that is about 0.25 that of K^+, and are slightly permeable to Ca^{2+} as well. The current of these channels is referred to as I_h, or **hyperpolarization-activated current**. Because the HCN channels are not highly selective for K^+, I_h is described as a cationic current and has an equilibrium voltage of about -20 mV to -40 mV, which is indicative of its Na^+–K^+ mix. It activates and deactivates slowly, in the range of 0.2 s–2 s although, in hippocampal CA1 neurons, the time constant of activation is in the range of 30–60 ms. It is also known as the I_f current, the **funny current**, and the **pacemaker current.**

HCN channels are partially activated at rest, conducting an inward predominantly Na^+ current that slightly depolarizes the resting membrane voltage. The resting membrane voltage is stabilized by the HCN channels, since a small hyperpolarization activates these channels whose inward current then depolarizes the cell. Similarly, a small depolarization deactivates the HCN channels, which hyperpolarizes the membrane back to the resting voltage.

I_h may be activated by ipsps depending on the depolarization level at which the ipsps occur. If activated, the ensuing depolarization affects the amplitude and time course of the ipsps and, hence, the neuronal response. In the case of external tufted cells of the olfactory bulb, ipsps result in classic inhibition when they occur at near normal resting membrane voltages, when I_h is not active. If the ipsps occur at more hyperpolarized membrane voltages,

I_h is activated resulting in a rebound excitation. Effectively in this case, I_h transforms inhibitory inputs to a postsynaptic excitation.

Because they increase the resting membrane conductance, HCN channels decrease the cell's input resistance, membrane time constant, and space constant. I_h in dendrites plays an important role in the modulation of synaptic voltages (Section 6.3.2). I_h can also contribute to the afterdepolarization (ADP) and to the rebound depolarization mentioned later in connection with the ADP.

An important physiological role of I_h is in pacemaker activity in the heart (Section 10.5.3.2) as well as in neurons. Some neurons in the thalamus fire bursts of APs at interburst frequencies of 0.5–4 Hz as a result of an interplay of I_h with a low-threshold, T-type Ca^{2+} current. Thus, the activation of this inward current causes a regenerative action that results in a Ca^{2+} spike with the neuron firing a burst of normal Na^+ APs near the peak of the depolarization due to the Ca^{2+} spike. Subsequently, the T-type Ca^{2+} current inactivates, thereby terminating the Ca^{2+} spike. As the membrane voltage falls below the equilibrium voltage of I_h, this current is activated resulting in a slow depolarization, referred to as the pacemaker voltage. This again activates the T-type Ca^{2+} current and repeats the cycle.

Rather than being directly involved in the generation of intrinsic rhythmic firing, I_h in other neurons governs the rate and regularity of this rhythmic firing, as in the basal ganglia.

Problem 7.2

If a subthreshold current pulse is applied to a patch of membrane in the absence of an I_h current, v_m changes in accordance with the passive properties of the membrane as indicated for the "No I_h" curve in Figure 7.5. In the presence of I_h, a sag and a hyperpolarization are observed upon termination of the pulse. Explain this behavior. Sketch v_m in the presence of I_h when the current pulse is hyperpolarizing.

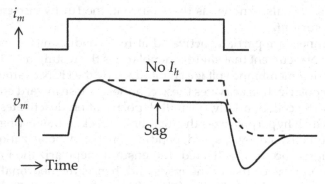

FIGURE 7.5
Figure for Problem 7.1.

We conclude this subsection by considering in more detail the K_{ir} family of inward rectifying potassium channels. The following should be noted concerning the electrical behavior and properties of these channels:

1. Increasing $[K]_o$ decreases the magnitude of E_K (Equation 1.25) and increases the magnitude of the inward current for a given $v_m < E_K$.

2. Since blockage by intracellular, positively charged entities, which causes inward rectification, is subject to statistical variations, it is not complete, which means that K_{ir} channels conduct a small outward current for $v_m > E_K$. This is, in fact, the physiological state of these channels, since v_m of animal cells does not normally become more negative than E_K (Equation 2.26).

3. If the membrane is suddenly hyperpolarized $(v_m < E_K)$, the instantaneous inward current is relatively large and increases, within a few hundred milliseconds, to a steady value of larger magnitude. The instantaneous change in inward current is due to the fast action of Mg^{2+}, whereas the time-dependent change is due to the slower unblocking by polyamines. The blocking action of polyamines is also slower than that of Mg^{2+}.

K_{ir} channels are found in a wide variety of cells, including neurons and cardiac muscle cells (Section 10.5.3). They can be divided into four functional groups:

1. *Classical K_{ir} channels* of the $K_{ir}2$ subfamily, which are strongly rectifying. They serve important functions in cardiac muscle, as discussed in Section 10.5.3, and are also involved in smooth muscle activity and in setting the resting membrane voltage in skeletal muscle. K_{ir} channels are widely distributed in the dendrites and somata of neurons in the brain where they play a critical role in the maintenance of the resting membrane voltage and the regulation of neuronal excitability. K_{ir} channels in parts of Schwann cells near the nodes of Ranvier are believed to remove excess $[K^+]_o$ resulting from nerve activity.

2. *K_G channels*, or G protein-gated K_{ir} channels, of the $K_{ir}3$ subfamily, which are strongly rectifying. These channels are found in presynaptic as well as postsynaptic regions. In presynaptic sites, they modulate synaptic transmission. G subunits in postsynaptic sites open these channels, inducing a slow ipsp.

3. *K_{ATP} channels*. These are ATP-sensitive K^+ channels, of the $K_{ir}6$ subfamily, and are found in neurons as well as in cardiac, smooth, and skeletal muscle. They couple cell metabolism, which produces ATP, to electrical activity. Thus, in neurons in some parts of the hypothalamus, an increase in glucose concentration stimulates metabolism and elevates ATP concentration, which closes K_{ATP} channels resulting in

depolarization and increased firing in these neurons. K_{ATP} channels are believed to play a protective role in neurons under conditions of reduced blood flow (ischemia) or reduced oxygen supply (hypoxia), in which case most neurons undergo massive depolarization culminating in cell death. In parts of the brain that are particularly vulnerable to the effects of ischemia and hypoxia, such as area CA1 of the hippocampus, the reduction in ATP concentrations resulting from ischemia or hypoxia opens K_{ATP} channels, thereby hyperpolarizing the membrane and providing some protection to the neuron against the destructive depolarization.

4. *K+ transport channels*, of the $K_{ir}1$, $K_{ir}4$, $K_{ir}5$, and $K_{ir}7$ subfamilies. The $K_{ir}1$ and $K_{ir}7$ subfamilies are weakly rectifying, whereas the rectification by the $K_{ir}4$ subfamily is intermediate. $K_{ir}1$ channels are found in the cerebral cortex and the hippocampus. Kir4.1 and Kir5.1 channels occur in glial astrocytes (Section 1.2), their main function being spatial buffering of $[K^+]_o$ by transporting K^+ from regions of high $[K^+]_o$ resulting from synaptic activity, to regions of low $[K^+]_o$. A high $[K^+]_o$ is undesirable as it would depolarize neurons and interfere with synaptic functioning. K^+ flow into glial cells in regions of high $[K^+]_o$, move between glial cells through the gap junctions under the influence of their concentration gradient and flow out of glial cells in regions of low $[K^+]_o$.

7.3.4 Chloride Channels

Cl^- channels serve a number of important physiological functions including regulation of pH and cell volume, secretion of fluid from glands and epithelia, transport of some organic solutes, cell migration, cell proliferation and differentiation, and the mediation of inhibition – both postsynaptic and presynaptic. Cl^- channels and cotransporters are implicated in a number of diseases including congenital myotonia, cystic fibrosis, and temporal lobe epilepsy. Cl^- channels can be ligand-gated, voltage-gated, or gated by $[Ca^{2+}]_i$. Ligand-gated Cl^- channels are activated by GABA and glycine, as discussed in Section 6.2.2. $[Ca^{2+}]_i$-activated Cl^- are widely distributed in cardiac muscle and play an important role in regulating cardiac excitability.

Cl^- channels are believed to have evolved from Cl^- transporters and to share some of their properties. Many Cl^- channels are intracellular, that is, they are found inside the cell, in the membranes of various cell organelles such as the nucleus, mitochondria, lysosomes, and endosomes. Cl^- channels are permeable to other anions such as NO^-_3, HCO^-_3, I^-, and SCN^- (thiocyanate).

The ClC family of chloride channels and transporters is divided into three subfamilies according to similarities in the base sequences of the genes encoding their channel proteins. Two members of one subfamily, referred to as **ClC-1** and **CLC-2**, are voltage-gated channels found in the cell membrane. ClC-1 is the major skeletal muscle chloride channel and plays a dominant role in regulating muscle excitability. Muscle cells have a high conductance to Cl⁻ that contributes up to 85% of the membrane conductance at rest and helps to stabilize the membrane voltage.

ClC-2 is expressed in cardiac muscle and in neurons. At least in hippocampal CA1 cells, ClC-2 is more localized in apical dendrites. ClC-2 is activated by hyperpolarization, cell swelling, and a rise in $[Cl^-]_i$ or in extracellular acidity. ClC-2 opens on hyperpolarization but closes only very slowly with depolarization so that it practically does not close during the brief depolarization of an AP. ClC-2 helps to quickly extrude Cl⁻ from neurons in case of high accumulation of Cl⁻ and contributes substantially to the membrane conductance. An increase in Cl⁻ conductance stabilizes the membrane voltage, whereas a reduction in Cl⁻ conductance increases the input resistance and neuronal excitability. It should be noted that hyperpolarization increases the inward current associated with an efflux of Cl⁻ because of their negative charge.

The presumed structure of ClC channels, based on the resolved structure of similarly constituted chloride transporters, is principally a homodimer composed of two identical subunits, as illustrated in Figure 7.6. Each subunit consists of two similar halves, as shown, with each subunit forming its own pore. The channel is thus "double-barreled", having two identical pores, with each pore capable of opening or closing independently of the other. This gives rise to two conductance levels, a smaller conductance when one pore is open, and double this conductance when both pores are open. A slower gating mechanism, however, closes both pores simultaneously. Each subunit has two intracellular subdomains, shown in plain shading in Figure 7.6. Mutations in the genes encoding these subdomains cause various diseases.

A chloride channel of special interest is the **cystic fibrosis transmembrane conductance regulator (CFTR)** channel that is expressed in the membranes of epithelial cells that line the outer surfaces of organs. Mutations in the gene

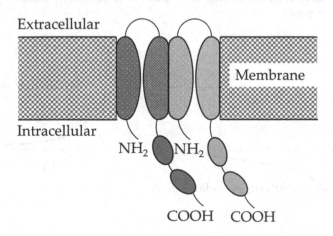

FIGURE 7.6
Chloride channel.

encoding the CFTR channel protein affects the regulation of epithelial fluid transport in the organs involved, resulting in **cystic fibrosis** characterized by abnormally thick mucus secretions mainly in the lungs but also in the pancreas, liver, kidneys, and intestine.

7.3.5 Effects on Afterhyperpolarization and Afterdepolarization

The APs of neurons usually, but not always, exhibit an afterhyperpolarization (AHP) and an afterdepolariztion (ADP) (Figure 7.7). It is important to note that the AP initiated at the initial segment, and which is intended mainly for propagation along the axon, may not show these afterpotentials. On the other hand, the soma-dendritic AP triggered by the backpropagation of the initial segment AP can show these afterpotentials, which can play an important role in the integrative function of the neuron, as discussed below.

The AHP is mainly due to the opening of K^+ channels. It can be divided into three components (Figure 7.7a):

i) a fast AHP (fAHP) that immediately follows the spike and lasts for 1–10 ms. It is mainly due to BK channels and helps repolarize the membrane on the downstroke of the AP. The I_A current can also contribute to the fAHP.

ii) a medium AHP (mAHP) that lasts for several hundred ms, mediated mainly by SK channels with contributions from BK and K_v7 channels in some cases.

iii) a slow AHP (sAHP) that can last for several seconds or more and is usually prominent after the initiation of a train of APs. It is believed to be due, at least in some cases, to SK channels activated by HVA Ca^{2+} currents.

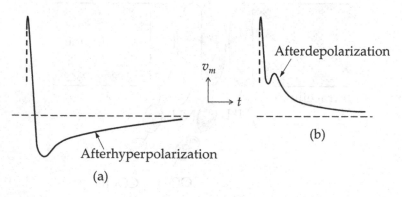

FIGURE 7.7
(a) Afterhyperpolarization; (b) afterdepolarization.

The AHP can have some important modulating effects on neuronal firing. The larger the hyperpolarization, the longer it will take the membrane to return to the resting level and to subsequently depolarize to threshold. The AHP is thus involved in setting the minimum rate of steady firing. If $[Ca^{2+}]_i$ builds up after the initiation of a train of APs, the frequency of the APs in the train will decrease with time due to the hyperpolarization resulting from Ca^{2+}-activated SK channels. This decrease in frequency is known as **spike frequency adaptation (SFA)**. The steady frequency of repetitive firing will then depend on the strength of the AHP. A larger build-up of $[Ca^{2+}]_i$ can lead to a hyperpolarization that is strong enough to terminate the train of APs. As the hyperpolarization subsides, the neuron can fire again leading to another burst of APs and so on. Controlling the AHP can thus cause the neuron to switch between repetitive firing of single APs and repetitive bursts.

Neuronal APs often exhibit an afterdepolarization that ranges from a slow phase of spike repolarization to a prominent rise in membrane voltage following the spike (Figure 7.7b). A suprathreshold ADP can have sufficient strength and duration to cause a burst of firing by the neuron. ADP can be due to one or more of the following factors:

1. *Sodium currents*: At least in the case of pyramidal cells in the CA1 region of the hippocampus, ADP is caused, under certain conditions, by an LVA-persistent sodium current $I_{Na,p}$. The resurgent sodium current, previously described (Section 7.3.1), can contribute to a depolarization in the repolarizing phase of the AP.

2. *Calcium currents*: T-type Ca^{2+} currents contribute to the ADP, particularly in the case of **rebound depolarization (RD)** following inhibition. This occurs, for example, in cerebellar nuclear cells that are inhibited by Purkinje cells. As mentioned previously for LVA Ca^{2+} channels that mediate the T-type current (Section 7.3.2), these channels inactivate at membrane voltages in the vicinity of the resting level. The inhibition of cerebellar nuclear cells by ipsps deinactivates these channels so that, upon release of the hyperpolarization, the channels activate rapidly and inactivate slowly, producing a prominent rebound depolarization. R-type Ca^{2+} current (Section 7.3.2) can also contribute to the ADP, as in the case of pyramidal cells in the CA1 region of the hippocampus.

3. *Hyperpolarization-activated current* (I_h): As a depolarizing current, I_h current can contribute to the ADP as, for example, in the pyramidal cells of layer 5 in the prefrontal cortex. Evidently, I_h may also boost the aforementioned rebound depolarization, as in the case of cerebellar nuclear cells.

4. When a neuron fires, an AP in the soma can depolarize the dendrites, which have a much larger area and hence total capacitance. Even in the case of a purely passive dendritic membrane, charge may

flow back to the soma after its repolarization producing an ADP. This effect is enhanced in the case of an active dendritic membrane containing voltage-dependent Na^+ and Ca^{2+} channels, which can amplify the depolarizations. If the depolarization is strong enough, dendritic spikes may be generated (Section 7.4.4).

SPOTLIGHT ON TECHNIQUES 7A:
ACTION POTENTIAL CLAMP

The Action Potential Clamp (AP-Clamp) is a variant of the voltage-clamp technique (Section 3.2.1) that allows recording of ionic currents during the AP. The following steps are involved: (i) The AP of a cell is recorded following a brief current stimulus. (ii) This recorded AP is applied to the same cell under voltage clamp as the command signal instead of a voltage step, as in Figure 3.2, and the current I_{ref} supplied by the amplifier is recorded. After the short initial transient, this current is zero since the output of the amplifier is the same as the voltage of the AP across the cell membrane. (iii) The process is repeated but with the ionic current of interest eliminated from the normal ionic currents through the membrane by applying a channel blocker for this current. The amplifier will now supply a compensating current I_{comp} that is equal and opposite to the eliminated ionic current, as in the normal voltage clamp. (iv) The ionic current of interest is then $(I_{ref} - I_{comp})$.

Compared to a conventional voltage clamp, the advantage of the AP clamp is that the desired ionic current is recorded under more natural conditions, during an AP, and with all other ionic currents not interfered with. In conventional voltage clamp, all ionic currents other than the desired current are eliminated by changing the ionic composition inside or outside the cell, by using specific, custom-made voltage protocols, or by some other means.

7.4 Dendritic Responses

Dendrites are much more than just a means of extending the membrane area for synapses beyond that of the cell body. Their geometry and membrane properties are diverse and specific to the type of neuron concerned, enabling dendrites to play a crucial role in processing neuronal input signals in myriad ways that befit particular functions. Even as physical structures, dendrites are dynamic: they change during development, in health, and in disease; new dendritic branches can be formed and existing branches eliminated; and dendritic spines can change in size and number depending on activity, as discussed in Section 6.5.3, thereby allowing for long-term changes in synaptic function. As mentioned previously, dendrites have the metabolic machinery for local protein synthesis, as required for fast changes in structure and function, and play a critical role in synaptic plasticity. Moreover, dendritic functions are subject to control and modulation by various physiological and biochemical factors.

When the generation of neuronal APs was first investigated by intracellular recording, it was believed that the somatic and dendritic membranes were inexcitable, so that ipsps and epsps arising in these regions propagated passively, by electrotonic spread, to the initial segment of the axon where an AP is generated once the membrane depolarization exceeded threshold. But electrotonic spread attenuates synaptic signals and slows their time course, an effect commonly referred to as **dendritic filtering**. Theoretically, an epsp generated at 1 mm from the soma may be attenuated to less than one-hundredth of its initial amplitude by the time it reaches the soma, and its time course may be greatly prolonged. So, it was not clear early on how synaptic voltages arising in fine, distal dendrites could have a significant effect on the generation of the AP at the initial segment. It was shown theoretically that distal synapses in passive dendrites can be more effective than what can be expected from electrotonic spread if certain geometrical relations between the sizes of dendritic branches are satisfied (Section 8.2.2.1). Evidently, the question of dendritic filtering does not arise if spikes are generated in the dendrites, so that synaptic inputs do not have to be propagated passively over the relatively long distance to the initial segment. There is now ample evidence for dendritic spikes, particularly in principal neurons having extensive dendritic trees, like cerebellar Purkinje cells. Moreover, there is evidence for mechanisms that enhance the effectiveness of distal synapses compared to synapses that are more proximal to the soma. The enhancement could be due to properties of the synapses themselves, as in the case of CA1 pyramidal neurons where the amplitude of epsps is larger in distal dendrite compared to proximal ones because of an increase in synaptic current resulting from an increased number of receptors and/or the number of quanta of neurotransmitter released. Other mechanisms for the enhancement of the effectiveness of distal synapses are attributed to active properties of the dendritic membrane that arise from the presence of ion channels. These channels are widely distributed in the soma and dendrites, give rise to the neuronal currents mentioned in Section 7.3, and play a critical role in the processing of synaptic inputs. However, the distribution of ion channels between soma and dendrites, or along the dendritic tree, does not follow the same pattern in all neurons. Moreover, the properties of the same type of voltage-gated channel can differ between soma and dendrites. Not surprisingly, the active responses of the soma and dendrites seem to be tailored to the functions of the particular type of neuron.

The active responses of dendrites are involved in three major manifestations of dendritic excitability, namely: (i) modulation of synaptic voltages, (ii) backpropagation, and (iii) dendritic spikes, as will be explained shortly.

Evidently the interplay of these active responses in any particular neuron can be quite complex and can take many forms that change dynamically as inputs change. But before discussing these active responses, it is important to present some basic principles related to electrical signal processing in dendrites, or what is conventionally referred to as "synaptic integration".

7.4.1 Synaptic Integration

The objective of synaptic integration is to combine the electrical signals arising from synaptic activity so as to contribute to neuronal firing in a manner that befits the function of the neuron. Clearly this integration is influenced by the passive properties of the dendrites and by the presence of ligand-gated and voltage-gated ion channels in the dendrites. Two important factors are: (i) the rate of rise of membrane voltage in response to a given synaptic input, as this directly impacts the response time of the neuron to the synaptic input, and (ii) the rate of fall of membrane voltage, as this affects temporal summation and hence the contribution of multiple synaptic inputs to neuronal firing. In particular, there is a time window during which subsequent epsps can effectively summate. The effectiveness of summation is also influenced by its linearity, as mentioned earlier (Section 7.2.1). We will consider, in the remainder of this section, the influence of the passive properties alone on these factors. In the next section we will take into consideration the presence of ion channels.

The passive properties of the dendritic membrane are dictated by the morphology of dendrites and by their cable properties arising from the distributed nature of the cytoplasmic resistance r_a, and membrane conductance g_m and capacitance c_m (Section 4.1.1). Excitatory postsynaptic currents (epscs) of AMPA receptors, believed to be the prevalent excitatory receptors in the CNS, have a very fast rise time that is typically less than 1 ms. It is argued in Section 4.1.2.1 that the voltage at the input of a cable in response to a suddenly applied current is much faster than that expected from the membrane time constant $\tau_m = c_m/g_m$ because of the distributed nature of the electrical cable parameters. The rate of rise of the voltage is therefore determined at the synapse mainly by c_m, and in neighboring regions by c_m and r_a, with g_m playing a minor role. On the other hand, the spread to more distant regions by electrotonic spread is governed by all the cable parameters, including g_m. The same conclusions apply to fast inhibitory inputs. For slower excitatory or inhibitory inputs, the effect of g_m becomes more significant because with slower changes of membrane voltage more current flows in g_m relative to c_m.

When synaptic inputs are considered more realistically, not as brief pulses but as signals having a relatively fast rise time and a much slower decay time as in, for example, the alpha function (Figure 6.3), it is found that the window for temporal summation at the soma is of the order of the membrane time constant, that is, 20–100 ms, as to be expected for an equipotential membrane patch. At distal dendrites, assuming that they can be considered as a semi-infinite cable, cable properties reduce the window for temporal summation to half the membrane time constant or less because of the charge that flows along the cable away from the point of stimulation. In more proximal dendrites, the delay is generally intermediate between these values. But if the dendrites are close to a relatively large soma that can act as an effective sink for electric charge, the decay could be very fast which would severely reduce the window for temporal summation. This would strongly favor synaptic inputs that are highly synchronized rather than spread out over time.

Massive background synaptic activity in some of the 10,000 or more synapses that are typically found on a neuron can effectively increase g_m, and hence reduce τ_m, by a factor of 50 or more. This can have a significant effect on what could be predicted on the basis of the resting value of g_m in terms of the rates of rise and fall of synaptic voltages, temporal summation, and the spread of excitation along the dendrites.

7.4.2 Modulation of Synaptic Voltages

When a synapse is activated, voltage-gated Na^+, K^+, and Ca^{2+} channels that are present nearby are activated by the voltage changes arising from synaptic activity. However, because of the rich variety of responses of ion channels, as discussed previously, the activation of voltage-gated channels could result in inward or outward current flow and an increase or decrease in g_m. As a result, the synaptic signal could be either amplified or reduced in both amplitude and duration. Modulation of synaptic input in this manner can therefore be quite varied depending on the properties of the ion channels, their distribution, and location, as well as the amplitude and time course of the synaptic input. It is to be expected that inward Na^+ or Ca^{2+} currents activated by synaptic depolarization will amplify the magnitudes of epsps. The decay of the epsps will also be prolonged, which facilitates temporal summation. The ion channels also affect dendritic interactions in the immediate vicinity of the synaptic inputs. Thus, voltage-gated channels can alter the local input resistance and time constant, thereby affecting both spatial and temporal summation of epsps and ipsps. The effects of ion channels can therefore be quite varied to suit the functions of particular neurons.

The fast-activating and -inactivating, transient I_A K^+ current (Section 7.3) is commonly present in dendrites. An epsp activates I_A, with the resulting outward current reducing the epsp and the amplifying effects of any nearby Na^+ or Ca^{2+} channels on the epsp. However, I_A can be rapidly inactivated by the depolarization within a few milliseconds, so that epsps that occur a few milliseconds after the first epsp will not be affected by I_A and are amplified by the activated Na^+ or Ca^{2+} channels. Summation of epsps will therefore be enhanced by these channels and the summated epsps will propagate more effectively. On the other hand, synaptic input that is highly synchronized will be subjected to the effect of an activated I_A and will not summate as well.

The hyperpolarization-activated cationic current I_h also affects synaptic input. Depolarization deactivates this cationic current, which reduces the total inward current and hence epsp amplitude and duration, as well as the time window during which subsequent epsps can summate. Modelling studies indicate that the overall effect is a near-equalization of the widths of the epsps as they arrive at the soma, regardless of how far from the soma they originate. This mitigates the effect of dendritic filtering, which increases the widths of the epsps as they spread electrotonically. The normalization of the widths of epsps depends on the dendritic morphology and on the density as well as the distribution of I_h channels along the dendrites. This distribution of I_h channels varies between different neurons. In CA1 hippocampal and layer 5 neocortical neurons, the density of both I_h and I_A channels increases by at least five-fold from the soma to the distal apical dendrites. In cerebellar Purkinje cells, the density of I_h channels is uniform along the dendrites, whereas in the principal neurons of the stratum radiatum region of the hippocampus, the density of I_h channels decreases going from the soma to the dendrites. Evidently, the modulation of synaptic activity is but one aspect of the more global phenomenon of dendritic signaling, which includes other manifestations, such as backpropagation and dendritic spikes and which will be considered in the following subsections.

7.4.3 Backpropagation

When an AP is generated in the region of the axon hillock/initial segment, the AP propagates in both directions: forward, along the axon, and backward toward the soma and dendrites. The efficacy of backpropagation, that is, how far this back-propagating AP progresses along the dendritic tree, varies considerably between different neuron types and between different dendrites of the same neuron. At one extreme are neurons, such as cerebellar Purkinje cells, which have a low dendritic Na+ channel density and in which backpropagation is very limited so that depolarization spreads almost passively along the dendritic tree. At the other extreme are neurons in which the back-propagating AP progresses almost unattenuated along practically the whole dendritic tree, as in the case of dopaminergic neurons of the substantia nigra, apical dendrites of mitral cells of the olfactory bulb, and hippocampal interneurons. Most other neurons, including pyramidal neurons and motoneurons, fall between these extremes.

The efficacy of backpropagation is affected, in general, by the following factors:

1. The density of voltage-gated Na+ channels and, to a lesser extent, Ca²+ channels that produce the regenerative action necessary for sustaining the AP.
2. The density of voltage-gated K+ channels that produce outward currents hindering AP propagation.

3. The shape of the AP, which is affected by the voltage-gated ion channels present and their properties. The broader the AP the less it will be attenuated as it propagates into the dendritic tree, and conversely.

4. Synaptic activity, particularly ipsps. Hyperpolarization and increased membrane conductance hinder AP propagation. Neuromodulators may support or hinder AP propagation, depending on the type of ion channel affected.

5. The number and frequency of APs, both of which tend to enhance the extent of backpropagation. However, when a train of APs propagates into the dendritic tree away from the soma, the amplitudes as well as the number of the APs generally progressively decrease along the train depending on synaptic activity and the properties of the ion channels involved.

6. Dendritic morphology, that is, the branching pattern of the dendritic tree. Simulation results indicate that the efficacy of backpropagation is strongly influenced by the rate of increase of dendritic membrane area with distance from the soma. This depends, in turn, on the number of branch points, that is, the degree of branching, and on the impedance mismatch at branch points, as reflected in the relation between the diameters of parent and daughter dendrites at the branch points. AP propagation is hindered by extensive branching and by a small ratio of the combined input impedance of the daughter branches to the input impedance of the parent branch.

The main functions subserved by the retrograde signal of the backpropagating AP are:

1. The depolarization of the backpropagating AP removes the Mg^{2+} block of NMDA-receptor channels and induces synaptic plasticity when the presynaptic and resulting postsynaptic AP activity occur within a time window not exceeding about 100 ms (Section 6.5.2). This **spike time dependent plasticity (STDP)** can be a potentiation or a depression of the efficacy of the synapses involved, as explained in Section 6.5.

2. Release of neurotransmitter from dendrites, resulting in: (i) activation of dendrodendritic synapses as found, for example, in the olfactory bulb between the principal neurons, mitral cells, and the dominant interneuron class, granule cells, (ii) local synaptic feedback as occurs, for example, when the backpropagating AP in the dendrites of bitufted cortical interneurons releases GABA onto axonal boutons of pyramidal neurons making excitatory glutamatergic synapses on the dendrites of the bitufted interneurons, which reduces the release of glutamate from the boutons of pyramidal neurons, and (iii) release of a neuromodulator (Section 6.3.2) into the

cerebrospinal fluid as occurs, for example, in the dopamine neurons of the substantia nigra, which are the cells of origin of some pathways of the dopamine neuromodulatory system of the brain (Section 6.3.2). These neurons have a broad somadendritic AP which, as in the case of mitral cells of the olfactory bulb, backpropagates very effectively into the dendritic tree.

3. Effect on synaptic signals and synaptic integration in the dendritic tree. Thus, backpropagating APs can lower the threshold for the induction of dendritic regenerative potentials. Activation of dendritic voltage-activated Ca^{2+} channels in dendrites and spines, as well as the activation of Ca^{2+}-dependent K^+ channels, can affect synaptic integration and neuronal firing patterns, as discussed in Sections 7.3 and 8.1. Moreover, the backpropagating AP briefly interrupts synaptic integration and resets the dendritic membrane, which in turn affects the timing of subsequent action potential initiation.

7.4.4 Dendritic Spikes

Three types of spikes can originate in dendrites, namely, Na^+, Ca^{2+}, and NMDA spikes. Na^+ and Ca^{2+} spikes arise in the usual manner through the regenerative action of an inward current through voltage-gated Na^+ and Ca^{2+} channels, respectively. NMDA spikes, on the other hand, are evoked by binding of glutamate to NMDA receptors. Because of random variations, some of these receptors are not blocked by Mg^{2+} at any given time. In the presence of a sufficient concentration of glutamate, Ca^{2+} influx through unblocked NDMA channels initiates a regenerative effect by evoking sufficient depolarization that unblocks more NMDA channels resulting in more influx of Ca^{2+}, more depolarization, more unblocking, and so on.

It should be borne in mind that the properties of the axonal AP, discussed in Section 3.2, result from a strong regenerative effect of a high density of voltage-gated Na^+ channels. In dendrites, the density of the channels involved in regenerative action is much less, and the regenerative action is correspondingly much weaker. Although these APs have a threshold, the all-or-none property is not as robust. Consequently, the amplitudes of these APs can grade with the strength of the stimulus and can decrement with propagation. The refractoriness can be weak, and APs can summate to some extent.

The three types of dendritic spikes are found in some neurons, such as neocortical, layer 5, pyramidal neurons, but not all in the same location. Ca^+ spikes are found in the apical trunk, of diameter of about 3 μm or more, between the soma and the distal apical tufts of these neurons. NMDA spikes, but no Ca^{2+} spikes, occur in the thin dendrites, of diameter about 1 μm or less, of the apical tufts, the thin oblique dendrites, and the distal basal dendrites. Each glutamatergic synapse is found on a dendritic spine, and the synchronous activation of 10–50 neighboring synapses generates an NMDA spike. NMDA spikes are sometimes preceded by a fast Na^+ spikelet that is evoked by glutamate stimulation of the AMPA receptors (Section 6.2.3) and subsequent regeneration through voltage-gated Na$^+$ channels. Figure 7.8 shows a subthreshold epsp evoked by weak glutamatergic inputs. With a stronger input the top trace is the voltage in the basal dendrite featuring rapid onset, a Na$^+$ spikelet, and a "plateau" followed by a rapid decline. The middle trace shows an NMDA spike recorded in the presence of TTX, which blocks voltage-gated Na$^+$ channels, and Cd^{2+}, which blocks voltage-gated Ca^{2+} channels. The top waveform is the summation of the regenerative responses due to Na$^+$, Ca^{2+}, and NMDA channels.

NMDA spikes in distal apical tufts of cortical, layer 5, pyramidal neurons have a threshold of about 10 mV, an amplitude of about 40–50 mV, and a duration of 50–100 ms, the amplitudes and durations of NMDA spikes being less than those of Ca^{2+} spikes, the latter having an amplitude of 50–70 mV, and a duration of about 120–200 ms The NMDA spikes propagate as a depolarization, attenuating by a factor of about 2.25 over a distance of about 200 µm. Although they do not propagate to neighboring tuft branches, NMDA spikes from different branches can summate with one another at branchpoints and with epsps. The resulting depolarization is sufficient to generate Ca$^+$ spikes at the initiation zones for these spikes in the apical dendrites. The Ca$^+$ spikes propagate, in turn, toward the axosomatic spike initiation zone, where they combine with the signals from the basal dendrites to generate the low-threshold, main Na$^+$ output AP of the neuron. The I_h current has a powerful effect in reducing the size and limiting the propagation of NMDA spikes. This serves to restrict the NMDA spikes spatially.

NMDA spikes are a major component of the "plateau potentials" recorded in dendrites of cortical pyramidal neurons following suprathreshold glutamatergic stimulation. Other components of this potential are the Na$^+$ spikelet and active Na$^+$ and Ca^{2+} responses. The plateau potentials depolarize the dendritic segment or the soma, in the case of proximal basal dendrites, by some tens of millivolts for the duration of the plateau, which can last for up to several hundred milliseconds. Strong glutamatergic stimulation of pyramidal neurons occurs during the "Up" states of these neurons, which are observed during sleep and sometimes during the awake state, and are characterized by intervals of strong activity that alternate with "Down" states, or periods of almost complete silence.

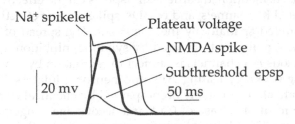

FIGURE 7.8
NMDA spike and plateau potential.

Apart from their role in integrating dendritic signals in separate branches, independently and in parallel, NMDA spikes are associated with massive Ca^{2+} influx and can induce synaptic plasticity, in conjunction with co-activated synapses, without the need for backpropagating APs.

Compared to Na^+ spikes, Ca^{2+} spikes generally have a lower threshold, a slower time course, and can be effectively curtailed by Ca^{2+}-activated K^+ channels, thereby restricting them spatially.

In many cases, each type of dendritic spike, Na^+, Ca^{2+}, or NMDA, remains restricted to parts of the dendritic tree, and does not propagate as such all the way from distal dendrites to the soma. However, the NMDA spike generated in the basal dendrites of pyramidal neurons does propagate to the soma. In the neurons of the globus pallidus, a major component of the basal ganglia, Na^+ spikes propagate to the soma from distal dendrites that are more than 250 µm from the soma. These spikes are the only effective mode of transmitting to the soma information from excitatory synapses in the distal half of the dendritic tree.

As in the case of backpropagation discussed previously, the efficacy of forward propagation is affected by the type and distribution of voltage-gated channels and the dendritic morphology. It should be noted that based on cable properties alone, forward propagation toward the soma is attenuated more than backpropagation due to the different terminal impedances in both cases.

It is seen from the preceding discussion that propagation of different forms of dendritic excitability is, in general, limited to parts of the dendritic tree, which leads to **electrical compartmentalization** of this tree, as determined by the nature and spatiotemporal pattern of the dendritic signals, the types of ion channels present, and dendritic morphology. Dendritic compartmentalization supports powerful computational capabilities by neurons because signals from different compartments can combine in different ways at multiple integration sites to generate a new dendritic signal. We have seen an example of this in the combining of NMDA spikes in the distal apical dendrites of layer 5 pyramidal neurons, with or without epsps, to generate Ca^{2+} spikes at branchpoints downstream toward the soma.

The computational power is enhanced by dynamic regulation of compartmentalization. As mentioned earlier, Ca^{2+} spikes can be effectively curtailed by Ca^{2+}-activated K^+ channels, and NMDA spikes can be effectively reduced in size and limited spatially by the I_h current. The spread of all dendritic depolarization signals can be limited by synaptic inhibition. Moreover, the activity of various ion channels is subject to regulation by a number of factors, including backpropagating APs and neuromodulators. The electrical compartmentalization is usually accompanied by chemical compartmentalization in the form of release of Ca^{2+} and second messengers such as IP3, which in turn can link electrical activity to a variety of cellular transduction systems.

SPOTLIGHT ON TECHNIQUES 7B:
FLUORESCENT $[Ca^{2+}]_i$ IMAGING

The tracking of intracellular concentration of free Ca^{2+} ions using fluorescent calcium indicators has provided an extremely powerful tool for investigating activity of single neurons, or even organelles such as mitochondria, as well as neuronal populations, because of the ubiquitous involvement of Ca^{2+} in so many aspects of cellular activity. The tracking of $[Ca^{2+}]_i$ allows a temporal precision of the order of milliseconds or less and spatial specificity of the order of micrometers.

The principle of Ca^{2+} imaging is to use a **fluorescent calcium indicator**, or dye, that combines a Ca^{2+} **chelator** – a chemical that binds to Ca^{2+} in solution – such as BAPTA and a **fluorophore**, a chemical group that fluoresces, that is, absorbs light of a short wavelength and emits light at a longer wavelength. Various calcium indicators are available to best suit particular applications. Two popular calcium indicators are Fura-2 and Indo-1, both of which are of the ratiometric type that allows determination of $[Ca^{2+}]_i$ based on the ratio of fluorescence intensities at two wavelengths. These are normally chosen so that the difference of fluorescence between bound and free indicator is maximum. Taking the ratio of two measurements largely eliminates the effects of extraneous factors that affect absolute values from a single measurement, such as variations in sample thickness, dye concentration, optical path length, and illumination intensity. Both Fura-2 and Indo-1 are excited in the ultraviolet range, 330 nm–363 nm, Fura-2 is a dual excitation indicator having excitation maxima at 335 nm and 363 nm, whereas Indo-1 is a dual emission indicator having emission maxima at 401 nm and 475 nm. The variation of $[Ca^{2+}]_i$ at various locations can be tracked by measuring the ratio of intensities on digital images recorded by video cameras at two wavelengths.

$[Ca^{2+}]_i$ increases due to activation of voltage-gated Ca^{2+} channels and NMDA channels. It is reduced by binding Ca^{2+} to various intracellular components and through the extrusion of Ca^{2+} by pumps and transporters. The dynamics of $[Ca^{2+}]_i$ are captured by Ca^{2+} imaging using various types of indicators. At the level of single neurons, Ca^{2+} imaging allows the detection of activity that cannot be detected by electrophysiological recording, such as subthreshold epsps in single dendritic spines. Firing of a neuron is indicated by accumulation of Ca^{2+} over large parts of the neuron because of the backpropagation of the Na^+ spikes. It is possible in this manner to detect repetitive firing of a neuron even at fairly high frequencies. Calcium spikes are easily detected as they produce generalized calcium influxes that are much larger than those produced by sodium spikes.

Even more impressive is the Ca^{2+} imaging of populations of neurons *in vivo*. Recording of this activity by electrode arrays is invasive, can sample simultaneously, at most, a few hundred rather superficially located neurons, and provides no information about the underlying anatomical structure. In contrast, Ca^{2+} imaging of neuronal populations is largely noninvasive and can potentially overcome some of the limitations of electrode arrays. A number of enabling technologies have contributed to the growing power of Ca^{2+} imaging. Bulk loading of neuronal tissue with calcium indicators, rather than intracellular loading of individual neurons, is possible using an acetoxymethyl (AM) ester of the calcium indicator which makes the indicator membrane permeant. Once inside the neuron, the ester is de-esterified to yield the free indicator.

There are several variations and modifications of the basic technique. **Scanning confocal microscopy** allows optical sectioning of relatively thick specimens by suppressing contributions from out-of-focus planes that would otherwise blur the image of the desired plane. In **two-photon fluorescence microscopy**, the fluorophore is excited not by photons of a relatively short wavelength but by having two long-wavelength photons, generated by a femtosecond laser, simultaneously absorbed so that their combined energy excites the fluorophore. This allows excitation with infrared light, which results in deeper penetration of neuronal tissue with less damage than may be caused by light of shorter wavelength. Moreover, two-photon microscopy can provide optical sectioning like that of confocal microscopy. A variant of two-photon microscopy is **extended depth of field (EOF) holographic microscopy**, which allows targeting many neurons simultaneously – not just in two dimensions, but at various depths. Continuing improvements and extensions enhance the power of Ca^{2+} imaging, such as computationally supported optics that allow the extraction of the desired image in the presence of scattering and out-of-focus blurring, time multiplexing using several scanning lasers, space multiplexing using multi-anode photomultiplier tubes, and wavelength multiplexing using multiple fluorophores excited by multiple laser beams, each with a different wavelength.

SPOTLIGHT ON TECHNIQUES 7C: OPTOGENETICS

Optical control of neuronal activity through optogenetics offers many unique advantages: noninvasiveness, or minimal invasiveness, millisecond-scale temporal resolution, micron-scale spatial specificity, targeting a particular type of neurons in the brain – and all this in live, freely moving animals.

Neuronal activity can be optically controlled in several ways. Pyramidal cells in brain slices can be stimulated to fire APs by direct application of laser light that is focused on the membrane of the soma or axon hillock. Another technique is that of **photo-uncaging**, in which a signaling molecule, such as a neurotransmitter, is "caged", that is, rendered inactive by combining it with a photocleavable molecule that can be cleaved by illumination with light pulses of appropriate wavelength, thereby releasing the signaling molecule and allowing it to exercise its effect. A variety of signaling molecules have been caged, including glutamate, GABA, and Ca^{2+}. An alternative to photo-uncaging is **photoisomerization**, in which the ligand is attached to a photoisomerizable group, so that the compound can be reversibly switched between active and inactive states, corresponding to two isomers by two different wavelengths of light.

Currently, the most powerful and sophisticated technology for optical control of neuronal activity is **optogenetics**, in which neurons are genetically modified to express light-sensitive proteins, referred to as **opsins**, which can be activated by light of appropriate wavelength. The opsins could be ion channels, pumps, or G-coupled receptors. The opsins are light-sensitive because they are coupled to **retinal**, a form of Vitamin A that is a **chromophore**, that is, a molecule or a chemical group that absorbs light at a specific frequency, thereby giving the compound a characteristic color. When exposed to light of appropriate frequency, retinal changes from one isomer form to another (the *trans* and *cis* forms), which results in a conformational change in the opsin, leading to channel opening, pump activation, or modulation of intracellular signaling cascades.

Naturally occurring opsins were first used in optogenetics, but have since been supplemented by an expanding toolbox of engineered versions that possess some specific properties, such as faster kinetics or a shift in the excitation wavelength. Naturally occurring opsins fall into two major classes: (i) Type I opsins are found in microbial organisms and are membrane-bound proteins that function as ion channels or pumps, and (ii) Type II opsins that are found in animal cells and are G protein-coupled receptors that modify the activity of intracellular signaling pathways. An example of a Type I opsin is Channelrhodopsin-2 (ChR2) found in the green algae *Chlamydomonas reinhardtii*. Absorption of blue light of 480 nm wavelength induces a change from all-*trans*-retinal to 13-*cis*-retinal, opening a channel for cations, which depolarize the membrane. If the depolarization exceeds threshold, the neuron fires APs. Anion-conducting channelrhodopsins (ACRs) also occur naturally. Another Type I opsin is halorhodopsin found in single-celled archaea organisms. When activated by yellow light of about 570 nm wavelength, it functions as an ion pump that pumps Cl⁻ inward which hyperpolarizes the cell and inhibits firing. An example of a Type II opsin is optoXRs obtained by modifying vertebrate rhodopsins. Upon illumination with green light of 500 nm wavelength, they allow for receptor-mediated intracellular signaling.

Opsins are commonly introduced in a desired cell type by using a **viral vector**, that is, a virus that is engineered to carry the desired opsin gene coupled to a **promoter**, which is a DNA sequence where gene expression is initiated. After the virus is injected into the brain region of interest, the opsin gene becomes part of the genetic machinery of the target neuron type and the opsin is expressed in this neuron type. Different neuron types have different promoters, so that using the promoter specific to a particular neuron type will express the opsins only in this neuron type.

Lasers of appropriate wavelength are widely used to activate opsins because lasers have narrow bandwidths and can be efficiently coupled to optical fibers which can deliver light to specific, deep regions of the brain. Having a small diameter of 0.2 mm or less, optical fibers minimize tissue damage. Light-emitting diodes (LEDs) of appropriate color are also used for light activation of opsins because they are inexpensive, of small size, and have low power requirements, although they do not couple to optical fibers as efficiently as lasers. The neuronal responses resulting from optical activation can be read out through electrical recording, calcium imaging, or voltage-sensitive dyes that change color or level of fluorescence according to the voltage across the membrane. An **optrode** is a device that integrates light delivery with electrical recoding, as when an optical fiber is fused with a metal electrode.

Optogenetics provides a powerful tool for investigating neuronal circuits with a high degree of specificity for the types of neurons that are activated within a well-defined region of the brain and at a specified time on a millisecond scale. Optogenetics can identify neuronal projections and has provided a means for establishing cause-and-effect relations between neuronal activity and different forms of behavior, including neurological and psychiatric disorders. Optogenetics may eventually allow the control of behavior using light.

But optogenetics does have its limitations. For example, opsin expression and light delivery are generally not uniform throughout the targeted neuronal population. Optogenetic stimulation may differ significantly from what happens under normal physiological conditions, as by forcing synchronism, inducing rebound excitation upon release of inhibition, or causing antidromic activity when axonal membranes are stimulated.

7.4.5 Bistability in Dendrites

We will begin our discussion of bistability in dendrites by considering a patch of a hypothetical excitable membrane in which sodium and potassium conductances, G_K and G_{Na}, vary instantaneously with membrane voltage, but do not vary with time, and the sodium channels do not inactivate. We will also assume for the sake of simplicity that the sodium-potassium pump is one-to-one. The resulting i-v relation is illustrated in Figure 7.9 in a somewhat exaggerated form to highlight the main points to be emphasized. The vertical axis represents the membrane current i_m, outward positive and carried mainly by K$^+$, inward negative and carried mainly by Na$^+$. The horizontal axis represents v_{mr}, the deviations from the resting membrane voltage V_{m0}, which would correspond to the origin. This i-v relation can be traced point-by-point under voltage-clamp conditions, where steps of depolarization of increasing magnitude are applied to the voltage-clamped patch. At the origin, $i_m=0$ because of the assumption of a one-to-one sodium-potassium pump. For small depolarization steps ($v_{mr}>0$), G_K predominates and i_m is positive. For larger depolarization, G_{Na} increases at a faster rate so that i_m reaches a maximum before decreasing back toward zero at point P, when the net current through the membrane is zero. Beyond point P, i_m is inward, which would trigger a regenerative effect if the patch is not voltage-clamped. Point P therefore represents the voltage threshold. As the depolarization is increased, the inward current decreases as the sodium-equilibrium voltage is approached. i_m reaches a minimum and returns to zero at point Q when the net membrane current is zero. Beyond Q, i_m is again outward.

It is seen that there are now three points of intersection. The resting state and the state corresponding to point Q represent stable equilibrium. A small depolarizing perturbation results in an outward current that repolarizes the membrane. Conversely, a small hyperpolarizing perturbation results in an inward current that again repolarizes the membrane. At point P, however, a small depolarizing perturbation results in an inward current that further

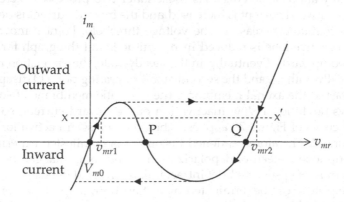

FIGURE 7.9
Bistability in dendrites.

depolarizes the membrane, taking the operating point to Q. Similarly, a small hyperpolarizing perturbation results in an outward current that further hyperpolarizes the membrane, taking the operating point to that of the resting state. Point P, therefore, represents an unstable operating point. The existence of two distinct states of stable equilibrium is termed **bistability**.

The bistability exemplified by the graph of Figure 7.9 is associated with **hysteresis**, which means that the state of a given system depends on how that state is reached. In other words, the path followed by the system when a variable is increased can be different from the path followed when the variable is decreased. Thus, if the patch of membrane is not voltage-clamped and i_m is increased, under current control, from zero in Figure 7.9, v_{mr} increases in accordance with the left-most part of the graph between $i_m = 0$ and the maximum value of i_m. If i_m is reduced back to zero, the same part of the graph is traversed back to the origin. If the current maximum is exceeded, however, the operating point jumps to a higher value of v_{mr} on the right-most part of the graph, as indicated by the upper horizontal dashed line in Figure 7.9. The current maximum therefore represents a current threshold. If i_m is reduced, v_{mr} changes along the right-most part of the graph on Figure 7.9. To traverse the full hysteresis loop, i_m is reversed so that v_{mr} progressively decreases until the minimum of i-v relation is reached. If i_m is reduced further, the operating point jumps to a lower value of v_m on the left-most part of the graph, as indicated by the lower horizontal dashed line in Figure 7.9. As a consequence of hysteresis, a current value, such as that represented by the horizontal line xx' in Figure 7.9 can be associated with two different voltages, v_{mr1} and v_{mr2}.

Before proceeding further, one may wonder how the i-v relation of Figure 7.9 relates to a real axon when the time dependencies of G_{Na} and G_K are taken into account. Since sodium activation is much faster than potassium activation and sodium inactivation, Figure 7.9 would apply momentarily soon after a step of depolarization is applied to a voltage-clamped axon, thereby allowing sodium activation to set in, but before potassium activation and sodium inactivation can have any appreciable effect. As these latter two processes exercise their effects, the outward current is increased and the inward current is decreased, with a concomitant increase in the voltage threshold. Point P moves to the right, the negative lobe is reduced in magnitude, and the graph for positive v_m is shifted upwards. Eventually, in the steady-state, the i-v relation increases monotonically with v_m, and the second stable operating point Q disappears.

In contrast to the axonal membrane, the dendritic membrane of some neurons shows no, little, or slow inactivation of the inward current, so that the N-shaped curve of Figure 7.9 applies, when the potassium activation is sufficiently reduced. The aforementioned bistability is manifested experimentally by an all-or-none plateau of depolarization in the soma that can be induced by a short train of epsps or a brief intracellularly injected depolarizing current pulse. The plateau can be terminated by a short train of ipsps or a brief intracellularly injected hyperpolarizing current pulse. In motoneurons, it has been observed that if the plateau depolarization exceeds the threshold for firing, the motoneuron fires repetitively. If the motoneuron is already firing repetitively, the frequency of firing is increased. The hysteresis described in connection with Figure 7.9 has also been observed in response to a triangular current stimulus. On the rising phase of the triangle, the firing frequency increases with the current at a certain slope, then with an increasing slope, presumably due to the jump from the maximum to the rightmost part of the graph of Figure 7.9. On the falling phase of the triangular current stimulus, the frequency is higher than on the rising phase, for the same current value, because the depolarization is larger, as explained in connection with the line xx' in Figure 7.9, with $v_{m2} > v_{m1}$. This has been referred to as **counterclockwise hysteresis**.

The plateau depolarization arises from a current referred to as a **persistent inward current** (**PIC**) caused by the stable equilibrium point Q of a dendritic membrane having the N-shaped i-v relation illustrated in Figure 7.9. The current is carried by Ca^{2+} that flow into the dendrites through L-type high-voltage-gated Ca^{2+} channels (Section 7.3.2), and by a Na^+ $I_{Na,p}$ current (Section 7.3.1), as discussed in more detail in Section 11.2.1.2 for motoneurons. It is significant that the presence of neuromodulators, such as serotonin or norepinephrine, is required for the PICs to occur. Moreover, anesthetics abolish the PICs because of their inhibitory effect on the cells of origin of these neuromodulators.

Plateau potentials have been observed in neurons other than motoneurons, such as cerebellar Purkinje cells, neocortical pyramidal neurons, and striatal neurons. But not all plateau potentials are due to dendritic instability, as, for example, the plateau potential in Figure 7.8.

Summary of Main Concepts

- Neurons have a great variety of sizes and shapes that serves their particular functions.
- Temporal summation and spatial summation of excitatory and inhibitory synaptic inputs are basic synaptic mechanisms in the generation of neuronal action potentials.
- The inhibition by inhibitory interneurons of their target neurons could be feedforward or feedback.
- Neuronal firing is influenced by nonsynaptic mechanisms, such as extrasynaptic receptors and electrical mechanisms acting through gap junctions or through extracellular field currents.
- The great variety of neuronal characteristics and responses results from the interplay of many ionic current components flowing through channels whose behavior can be influenced by a multitude of factors.

- The same ion channel subfamily can exhibit considerable functional diversity between different nerve tissues, or between different parts of the same neuron, in terms of sensitivity to various activating or blocking agents, voltage dependence, and channel kinetics.

- The afterhyperpolarization is due to the effects of various types of K^+ currents and can have some important modulating effects on neuronal firing.

- The afterdepolarization could be due to inward currents carried by Na^+ and Ca^{2+}, including a contribution from the I_h current.

- The passive properties of the dendritic membrane are dictated by the morphology of dendrites and by their cable properties arising from the distributed nature of the cytoplasmic resistance r_a, membrane conductance g_m, and capacitance c_m. These cable properties result in rates of rise and fall of membrane voltage that are larger than predicted on the basis of the membrane time constant alone.

- The rate of fall of membrane voltage determines the effectiveness of temporal summation.

- The presence of dendritic voltage-gated Na^+, K^+, and Ca^{2+} channels modulates synaptic activity by causing inward or outward current flow and an increase or decrease in g_m, thereby amplifying or reducing synaptic signals in both amplitude and duration.

- The efficacy of backpropagation is affected by the presence of dendritic ion channels.

- Three types of spikes can originate in dendrites, namely, Na^+, Ca^{2+}, and NMDA spikes.

- Bistability, with attendant hysteresis, can occur in dendrites when the inward current shows no, little, or slow inactivation, resulting in an N-shaped i-v characteristic.

8

Neuronal Firing Patterns and Models

Objective and Overview

This chapter is concerned with some theoretical aspects of neuronal behavior. The chapter begins with the classification of neurons based on their excitability and their responses to subthreshold perturbations of their membrane voltages. The classification based on excitability can be interpreted in terms of a dynamical system analysis of neuronal firing, whereas the classification based on responses to subthreshold perturbations of membrane voltage is more relevant to neuro-computational properties. The classification of neurons is followed by a discussion of the six distinctive classes of neuronal firing patterns that have been observed *in vitro* in response to excitation by long-duration current pulses of different magnitudes.

Neuronal models, both dynamical and biophysical, are introduced in the remaining part of the chapter, the goal being to explain the nature of these models and their main characteristic features rather than delve into the mathematical details. The description of dynamical models starts with the basic integrate-and-fire model and progresses through the resonate-and-fire model, the fast–slow reduced Hodgkin–Huxley model, the Fitzhugh–Nagumo model, the quadratic model, and the Morris–Lecar model. As for biophysical models, the basic equivalent cylinder model of a dendritic tree is first presented as an example of a morphological-to-electrotonic transformation, followed by a description of the popular and versatile compartmental models. The chapter ends by considering modelling of neuronal networks, including the firing rate model.

Learning Objectives

To understand:

- The classification of neurons into three types based on their excitability

- The classification of neurons as integrators or resonators and neuro-computational implications of this classification
- The distinctive classes of neuronal firing patterns that have been observed *in vitro* in response to excitation by long-duration current pulses of different magnitudes
- The nature and main characteristic features of various dynamical and biophysical neuronal models

8.1 Neuronal Firing Patterns and Their Modulation

8.1.1 Neuronal Computation

Before considering neuronal firing patterns, and in order to put the discussion in the rest of the chapter in a more meaningful framework, it is necessary to clarify the meaning of "neuronal computation", an expression often encountered in the neuroscience literature. In view of the massive parallelism of neural organization in higher animals, and in view of the variability in the responses of individual neurons even when these are "synchronized", it is generally more meaningful from a functional point of view to consider responses of a population of neurons rather than the response of a single neuron unless this response is considered sufficiently representative of the response of the given population as a whole. **Neuronal computation**, in its most general connotation, can be considered as a *signal processing operation that transforms a spatiotemporal input pattern of neuronal activity to a spatiotemporal output pattern by a population of neurons.* The spatial aspect refers to the spatial distribution of the response arising from a fairly large number of neurons that belong to a given population, whereas the temporal aspect refers to how the activity is patterned in time. Although it may be difficult at present to appreciate the significance of a particular spatiotemporal pattern of activity in the brain, it is not difficult to see that a particular spatiotemporal pattern of activity in a population of motoneurons would result in a corresponding pattern of contractions in a set of muscle fibers.

In a more restricted sense, the term "computation" is also used to describe the functional relation between input and output of a single neuron or even a single synapse or a group of synapses. This more restricted use of the term "computation" is particularly relevant in the case of invertebrates where the number of neurons of a particular type does not justify considering neuronal populations.

8.1.2 Neuronal Excitability

In terms of excitability, neurons can be divided into three classes based on the range of frequencies of APs generated in response to injected current pulses of long duration and of various magnitudes:

1. *Type I neurons* (also referred to as Class 1 neurons): The frequency of APs increases monotonically with the magnitude of the injected current step, above a certain threshold, over a range that extends from a few Hz or less to more than 100 Hz (Figure 8.1a). The sustained, repetitive firing of APs over the duration of the pulse is termed **tonic firing**.

2. *Type II neurons* (also referred to as Class 2 neurons): APs can be generated only over a limited range of frequencies, such as 150–200 Hz, the range being relatively insensitive to the magnitude of the current step above a certain threshold (Figure 8.1b). If the magnitude of the current pulse is increased above this threshold value, the number of APs generated may increase from a few APs to a sustained train of APs.

3. *Type III neurons* (also referred to as Class 3 neurons): Only a single AP is usually generated when the current pulse is applied. Increasing the magnitude of the current pulse reduces the latency of this AP. A limited number of APs may be generated with very large magnitudes of the current pulse.

The aforementioned classification originated with Hodgkin in 1948 based on observations of axonal firing. Since then, the behavior of Type I and Type II neurons has been extensively studied theoretically and from a dynamical system viewpoint, yielding some valuable theoretical understanding of the control of axonal firing. From a dynamical system viewpoint, a Type I neuron is characterized as a saddle node on an invariant circle, whereas a Type II neuron is characterized by Andronov–Hopf type bifurcations. In terms of their neuro-computational properties, Type I neurons are integrators (see Section 8.1.3) and can smoothly encode the strength of a synaptic input in terms of the frequency of firing of the neuron. Type II neurons, on the other hand, are resonators and can only detect when the strength of an input exceeds a certain value. However, unlike Type I neurons, Type II neurons

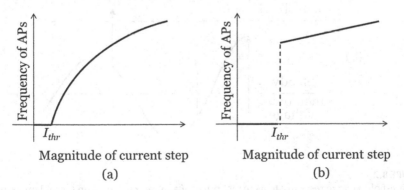

FIGURE 8.1
Frequency of firing characteristic of Type I neurons (a), and Type II neurons (b).

can exhibit subthreshold oscillations of membrane voltage, which allows a rich repertoire of responses, as discussed next.

8.1.3 Resonators and Integrators

Neurons can be classified as resonators or integrators. A neuron is a **resonator** if its membrane voltage exhibits subthreshold oscillations following a perturbation, such as a subthreshold depolarizing pulse or an AP. It is labeled as a resonator because the stimulus for such a neuron is most effective if its frequency is the same as that of the subthreshold oscillations. A neuron whose membrane voltage following a perturbation does not oscillate as it approaches the resting value is an **integrator**, since successive stimuli are more effective in causing firing because of a cumulative, or integrating, effect.

These basic features of integrators and resonators are diagrammatically illustrated in Figure 8.2. In (a), a brief, subthreshold pulse is applied at t_0. The membrane voltage of an integrator is shown returning smoothly, without oscillations, to the steady membrane voltage level, which may or may not be the resting level. A second pulse at t_1 will take the membrane voltage beyond threshold and will cause firing, whereas a pulse at t_2 will not. The effect is similar to that of temporal and spatial summation discussed in Section 7.2.1.

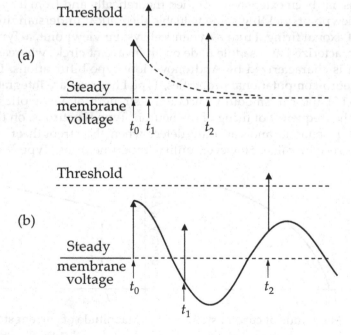

FIGURE 8.2
Subthreshold voltage response to a current pulse of an integrator neuron (a), and an oscillator neuron (b).

In (b) a brief, subthreshold pulse is also applied at t_0. However, the membrane voltage of a resonator is shown returning to the steady level with damped oscillations. A second pulse at t_1 will not cause firing, whereas a pulse at t_2 will.

The following may be noted concerning resonators and integrators:

1. The stimulus is most effective in firing a resonator if its frequency is the same as, or close to, that of the subthreshold oscillations and the timing of the stimulus is such that the pulses occur at or near the maxima of the oscillations. Increasing or decreasing the stimulus frequency outside the optimum range makes the stimulus less effective. In contrast, as the frequency of the stimulus to an integrator is increased, the stimulus becomes more effective in firing, and firing occurs sooner.

2. A high-frequency burst of pulses applied to a resonator may not cause firing at all if the pulses occur near the minimum of the subthreshold oscillations. On the other hand, if the burst of pulses is applied at the crests of the oscillations, and the pulses surpass threshold, the neuron may, in turn, fire a burst of APs.

3. If the pulse at t_1 occurs at an oscillation minimum, the oscillations that it produces will tend to cancel out the oscillations due to the pulse at t_0 in which case a pulse at t_2 may not cause firing. Effectively, the pulse at t_1 will thus inhibit firing due to the pulse at t_2, although the pulse at t_1 is depolarizing.

4. Switching between resonators can be accomplished by frequency selectivity. Thus, if neuron A excites two resonators B and C having different frequencies of subthreshold oscillations, then neuron B, but not C, can be fired when neuron A fires at a frequency equal to that of the subthreshold oscillations of neuron B. Similarly, neuron C can be selectively fired at a different frequency by A.

5. In integrators, firing is determined by excitatory and inhibitory synaptic inputs at any instant. In resonators, synaptic input can modulate oscillatory firing by the neuron.

Evidently, a larger repertoire of responses is possible with resonators compared to integrators.

The frequency of subthreshold oscillations in resonators ranges between a few Hz and about 200 Hz, the amplitude being generally a few millivolts at the resting voltage. However, both the amplitude and the frequency of the oscillations increase with background depolarization, the amplitude increasing to more than 10 mV peak-to-peak in some cases. Moreover, background synaptic activity continuously perturbs the membrane voltage and can make the subthreshold oscillations effectively continuous. In some neurons the subthreshold oscillations may be due to a periodic synaptic input, whereas in other neurons, as in some classes of cortical interneurons, the subthreshold oscillations are intrinsically generated.

Subthreshold oscillations could be generated intrinsically by some form of coupling between inward and outward currents through the membrane and are influenced, in general, by the membrane time constant. In some neurons, such as those of the inferior olivary nucleus that excite cerebellar Purkinje cells, the inward current is due to voltage-gated Ca^{2+} channels, whereas the outward current is due to Ca^{2+}-dependent K^+ channels. A membrane depolarization activates voltage-gated Ca^{2+} channels, which increases $[Ca^{2+}]_i$. This in turn activates Ca^{2+}-dependent K^+ channels, leading to a hyperpolarization that reduces Ca^{2+} influx through the voltage-gated Ca^{2+} and lowers $[Ca^{2+}]_i$. The reduction in $[Ca^{2+}]_i$ decreases the outflow of K^+ through the Ca^{2+}-dependent K^+ channels, thereby causing a depolarization, and so on. Note that Ca^{2+} influx through the voltage-gated Ca^{2+} has a depolarizing effect and can be partly regenerative, which can provide sufficient amplification to sustain the oscillations.

In some neurons, the depolarization involved in subthreshold oscillations is due to a Na^+ rather than a Ca^{2+} current. Pyramidal cells of region CA1 of the hippocampus have subthreshold oscillations at two frequencies in the theta range, 2–7 Hz: (i) one that occurs at depolarized membrane voltages, between the resting voltage and the spike threshold (–70 to –50 mV) and is mainly due to $I_{Na,p}$ and I_M, and (ii) the other occurring at hyperpolarized membrane voltages (–75 to –95 mV) and is mainly due to $I_{Na,p}$ and I_h. The currents I_M, $I_{Na,p}$, and I_h all do not inactivate. At membrane voltages of –70 to –50 mV, I_h is essentially deactivated, whereas I_M is dominant and in opposition to $I_{Na,p}$. A depolarization activates $I_{Na,p}$, which tends to augment the depolarization, and also activates I_M, causing a hyperpolarization. The hyperpolarization deactivates $I_{Na,p}$, which tends to augment the hyperpolarization and also deactivates I_M, leading to a depolarization and a repetition of the cycle. At membrane voltages of –75 to –95 mV, I_M is deactivated and I_h is dominant, but a similar situation occurs because I_h is an inward current. A depolarization activates $I_{Na,p}$ but deactivates I_h, leading to a hyperpolarization which deactivates $I_{Na,p}$ but activates I_h, leading to a depolarization and a repetition of the cycle. Both I_M and I_h have rather slow kinetics, whereas $I_{Na,p}$ is fast. The time variation of currents depends on the channel kinetics, whereas the ensuing changes in membrane voltage are affected by the membrane time constant. Evidently, subthreshold voltage oscillations can conceivably be caused by the interplay of other combinations of ion channels.

A neuron can behave as an integrator when a reasonably fast, high-threshold K^+ current dominates. Recall from the discussion of Section 3.2, Chapter 3, on the squid giant axon that an afterhyperpolarization is caused by a relatively slow K^+ conductance that lags behind the membrane voltage, so that when the resting membrane voltage is reached, G_K is still high, resulting in K^+ efflux and consequent hyperpolarization. Hence, a reasonably fast K^+ current will track the membrane voltage during the repolarization phase of an

AP, allowing the membrane voltage to return smoothly to the resting value under the influence of Na^+ inactivation without any subsequent hyperpolarization. Moreover, such a fast K^+ current should have a high threshold, so it is not activated by small depolarizations; otherwise, activation of the fast K^+ current will lead to a hyperpolarization and consequent oscillations in membrane voltage.

Channel kinetics can be changed by various means, causing the same neuron to switch between behaving as an integrator or a resonator. For example, a type of neuron, the mitral cell in the rat olfactory bulb, has two resting voltage states, one at −50 mV (the upstate), the other at −60 mV (the downstate), and can be rapidly switched between these states by a strong enough synaptic input. Membrane oscillations occur in the upstate, but not the downstate, so that the neuron can behave as an integrator in the downstate and as a resonator in the upstate. CA1 hippocampal pyramidal cells can switch from integrators *in vivo* to resonators *in vitro*, presumably due to the activity of the I_M potassium current.

Problem 8.1

The membrane voltage of a neuron is measured in response to a subthreshold current of constant amplitude and slowly increasing frequency. (a) How would the voltage magnitude vary with frequency for: (i) an integrator, (ii) a resonator? (b) What type of filter would be represented by each type of neuron?

ANS.: (a) The response will vary with frequency according to the filter type in (b); (b) (i) low-pass filter, (ii) band-pass filter.

8.1.4 Neuronal Firing Patterns

Neuronal firing patterns are fundamental for information coding in the brain. Some distinctive neuronal firing patterns have been observed in the mammalian brain and classified into a small number of classes that are referred to in the literature by special designations. It must be kept in mind, however, that these classifications are largely based on responses of neurons *in vitro*, that is, in an artificial environment outside the living body. Slices of neural tissue, generally ranging in thickness between 100 and 500 µm, are bathed in artificial cerebrospinal fluid, and intracellular or extracellular recordings are obtained from the neurons under investigation in response to applied current pulses of long duration and of various magnitudes. Using slices of neural tissue is very convenient because: (i) the slices can be rapidly prepared and are mechanically stable, in the absence of heart beat and respiratory pulsations, (ii) the extracellular environment can be readily altered, as may be desired, without the impediment of a blood-brain barrier, and (iii) neurons can be directly observed under a microscope, allowing accurate placement of recording and stimulating electrodes in the desired locations. In some cases, either neuron cultures are used instead of neural slices, or isolated, dissociated neurons having a cell body, a stump of axon and some of the proximal dendrites.

In vitro responses under controlled conditions are useful for gaining some basic understanding of the relation between electrical responses of neurons and their morphology, connections to other neurons, and synaptic properties. However, they differ from responses of neurons *in vivo*, that is in the living body, in the following respects: (i) they lack inputs from neurons other than those in the relatively thin slices used, (ii) the artificial CSF used lacks blood-borne substances and other substances normally present in the extracellular medium, particularly neuromodulators, and (iii) real-life excitations to neurons are not current pulses of long duration, but synaptic inputs of considerable variability, or "noise", so that a much broader range of responses can be expected in practice than is described by the "standard" classifications. Nevertheless, these classifications are sometimes used in discussing the possible functional significance of some neuronal responses.

The first three classes of firing patterns considered in what follows, and illustrated in Figure 8.3, are those of excitatory cortical neurons, whereas the firing patterns illustrated in Figure 8.4 are those of inhibitory interneurons. The fact that interneurons have some distinct firing patterns implies that these interneurons have some underlying, particular ionic channels of different properties and kinetics.

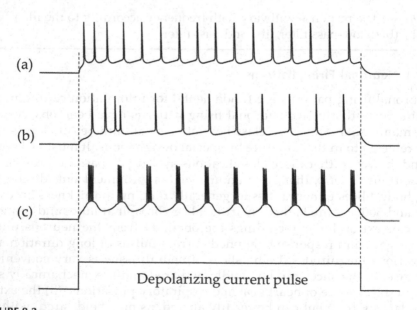

FIGURE 8.3
Firing patterns of excitatory cortical neurons in response to a current pulse; (a) regular spiking; (b) intrinsically bursting; and (c) fast rhythmic bursting.

8.1.4.1 Regular Spiking Neurons

Regular spiking (RS) neurons respond to depolarizing current pulses of long duration by tonic firing, that is, repetitive firing for the duration of the pulse with some spike frequency adaptation (Figure 8.3a). The mechanism of this adaptation was explained earlier in connection with the AHP (Section 7.3.5). The firing frequency generally increases with the magnitude of the applied pulse over a wide range, as in the case of Type I excitability discussed previously. RS neurons are the major class of cortical excitatory neurons and include pyramidal cells in layers 2, 3, 5, and 6 and spiny stellate cells in layer 4.

8.1.4.2 Intrinsically Bursting Neurons

Intrinsically bursting (IB) neurons respond to strong depolarizing pulses by generating an initial burst of high-frequency spikes followed by tonic firing (Figure 8.3b). The interspike intervals within the initial burst may be increasing or decreasing, depending on the underlying ionic mechanisms. The interspike interval following the initial burst is usually longer than the intervals that follow. As the strength of the depolarizing pulse is reduced, the initial burst may disappear altogether in many cases – the response becoming a regular train of pulses of relatively low frequency. Compared to RS neurons, IB neurons generally have a lower firing frequency, a higher threshold of the depolarizing pulse, and a shorter latency of the first spike. IB neurons include pyramidal cells found in all cortical layers, particularly layer 5.

8.1.4.3 Fast Rhythmic Bursting Neurons

Fast rhythmic bursting (FRB) neurons respond to depolarizing pulses with bursts of high-frequency spikes having relatively short interburst intervals (Figure 8.3c). Also referred to as **chattering neurons (CH)**, they include spiny stellate cells and pyramidal cells in layers 2–4, particularly layer 3. The frequency of the spikes within a burst can be several hundred per second.

Many roles have been ascribed to neuronal bursting, a common theme being that, compared to a single spike, a burst produces a desired response more effectively and reliably. Thus, bursts overcome failure of synaptic transmission, and their epsps summate and may exceed threshold. They are more effective in producing short-term facilitation and long-term potentiation. Bursts may resonate with subthreshold oscillations. They have a greater informational content that may be encoded in the burst duration, frequency, and the interspike intervals. In some sensory systems of lower animals, bursts from the same neuronal network encode sensory inputs that are different from those encoded by single spikes.

8.1.4.4 Fast Spiking Interneurons

Fast spiking (FS) interneurons respond to depolarizing pulses with high-frequency tonic firing for the duration of the applied current pulse, with no

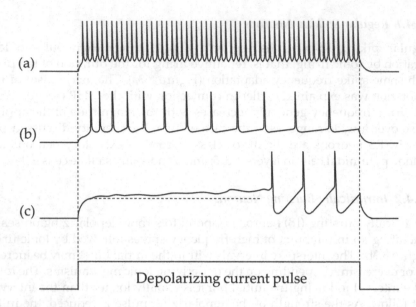

FIGURE 8.4
Firing patterns of inhibitory interneurons in response to a current pulse; (a) fast spiking; (b) low threshold spiking; and (c) late spiking.

or very slow frequency adaptation (Figure 8.4a). The frequency can reach 1 kHz, the spike width being less than 0.5 ms at half amplitude. Pulse trains of long duration, that is more than 20 s or so, may be followed by a slow AHP that decreases excitability for tens of seconds. FS interneurons exhibit Type II excitability in that when the magnitude of the depolarizing pulse is reduced below a critical value, the spikes become irregular and the interneuron may switch randomly between spiking and non-spiking states.

FS interneurons are a subgroup of cortical GABAergic inhibitory interneurons that includes basket and chandelier cells (Section 7.1). They have a powerful inhibitory effect on pyramidal cells; basket cells innervate the soma and proximal dendrites of these cells, whereas the chandelier or axo-axonic neurons make synapses on the initial segment of the pyramidal cell axon. FS interneurons are found in all cortical layers except layer 1, their inhibition being prominent in the intralaminar, or horizontal, direction. FS interneurons have **autapses**, which are synapses formed by a neuron on the neuron itself. As mentioned earlier (Section 7.3.3), K_v3 channels are well-suited for the generation of narrow spikes at high frequencies.

8.1.4.5 Low Threshold Spiking Interneurons

Compared to FS interneurons, low threshold spiking (LTS) neurons fire at a lower tonic frequency with pronounced spike frequency adaptation (Figure 8.4b). They are designated "low-threshold" because, unlike FS interneurons, they exhibit rebound firing of one or more spikes at the end of a *hyperpolarizing* current pulse. Compared to FS neurons, the firing frequency decreases with reduced stimulus strength down to lower frequencies.

Morphologically, LTS neurons are GABAergic interneurons that innervate pyramidal cell dendrites more distally from the pyramidal cell somata. These include neurogliaform cells and cells whose inhibition is more prominent in the interlaminar, or vertical, direction, such as double-bouquet cells, Martinotti cells, and bitufted cells (Section 7.1).

8.1.4.6 Late Spiking Interneurons

When the amplitude of the depolarizing current pulse is just above threshold, the membrane voltage of a late spiking (LS) interneuron increases in a ramp-like manner, culminating in low-frequency firing with a latency that can be as long as 1 s or up to several seconds in some cases (Figure 8.4c). The latency decreases as the amplitude of the depolarizing pulse is increased, the firing becoming tonic, with little or no spike frequency adaptation. The firing frequency, however, is much less than that of FS interneurons. When the duration of the depolarizing current pulse is long, that is 0.5 s or more, the individual spikes are followed by a fast AHP, as in Figure 8.4c. But if the pulse duration is reduced so as to generate a single AP, the spike is often followed by a prominent ADP.

Morphologically, LS interneurons include neurogliaform cells (Section 7.1) and are found in all cortical layers, especially layer 1, which lacks pyramidal neurons.

Although the aforementioned are the distinct firing patterns of the majority of interneurons, some minor subsets of interneurons exhibit other firing patterns, such as the **stuttering interneurons** (ST) of the mouse lateral amygdala. The firing pattern of these interneurons are characterized by bursts of spikes intermingled with variable quiescent periods in which no spikes occur.

8.1.5 Rhythmic and Synchronized Firing

It is believed that rhythmic firing of neurons and the attendant synchronization of activity in an ensemble, or population, of neurons is a fundamental mechanism that allows the coordinated activity that is necessary for normal brain functioning; this includes motor control, awareness, memory, and perception, particularly the binding of sensory features such as the shape and color of an object, so that these two aspects appear unified, although shape and color are processed by different neuronal populations. Neuronal synchrony is seemingly impaired in brain disorders such as schizophrenia and autism, and is excessive in epileptic seizures and Parkinson's disease. Of particular interest in this book is the role of rhythmic and synchronized firing in motor control, which is discussed in detail in Section 13.4. Some general aspects of rhythmic and synchronized firing are discussed in what follows.

A striking example of synchronized activity by a large neuronal ensemble is exemplified by the **electroencephalogram (EEG)**, which is a recording by electrodes placed on the scalp of field potentials arising from underlying brain activity, mainly that of cortical pyramidal cells (Spotlight on Techniques 12A). The rhythms identified in the EEG have been divided into frequency bands, conventionally referred to as: **delta** (up to 4 Hz), **theta** (4–8 Hz), **alpha** (8–13 Hz), **beta** (13–30 Hz), **low gamma** (30–70 Hz), and **high gamma** (70–150 Hz) rhythms.

Rhythmic activity may occur at three levels: (i) single neurons, (ii) a group of neurons in a given location, or (iii) neurons in different areas of the brain.

Several mechanisms involving the interplay between ionic channels can cause single neurons to fire spontaneously and rhythmically in the absence of synaptic inputs. As noted previously (Subsection 7.3.3), I_h, in conjunction with a low-threshold, T-type Ca^{2+} current, plays a major role in this pace-making activity of some neurons, such as thalamic neurons, neurons of the inferior olive, and interneurons in region CA1 of the hippocampus.

Spontaneous firing in cerebellar nuclear cells appears to depend on the interplay of the usual Na^+ and K^+ currents involved in the generation of the AP with a tonic, voltage-independent cationic current that depolarizes the membrane to threshold. On the other hand, in dopaminergic (DA) neurons of the substantia nigra, rhythmic firing is initiated by a slow depolarization to threshold, which triggers a broad AP. The Ca^{2+} influx through voltage-gated Ca^{2+} channels activates calcium-gated K^+ (SK) channels, resulting in a large afterhyperpolarization (AHP). This deinactivates the voltage-gated Ca^{2+} channels, so that the rebound from the AHP initiates another slow depolarization to threshold, and the cycle is repeated.

Inhibitory interneurons play a prominent role in producing synchronized rhythmic firing in an ensemble of neurons by generating a narrow window for effective excitation and by rhythmically modulating the firing rate of excitatory neurons. GABAergic interneurons in the cortex are particularly important in this respect. These interneurons constitute about 25% of neurons in the neocortex and form distinct networks in which interneurons belonging to a particular subtype are more strongly coupled to one another through chemical and electrical synapses than to interneurons of other networks. When GABAergic interneurons are coupled through both types of synapses, an AP in a given interneuron will inhibit its target interneuron via chemical synapses but will excite it via electrical synapses. Moreover, repetitive stimulation can cause facilitation or depression in chemical synapses (Section 6.5). Given the differences in gains and temporal relations in synaptic pathways, it is to be expected that a network of interconnected GABAergic interneurons can generate oscillations, as in any feedback connection of elements having gains and phase shifts or time delays. In fact, networks of fast-spiking (FS) GABAergic interneurons are believed to be responsible for high-frequency oscillations in the brain in the gamma range (30–80 or more Hz), whereby the duration of ipsps of the chemical synapses determines the oscillation frequency, and the electrical synapses ensure a high degree of synchrony in the firing of interneurons in the network by rapidly propagating electrical activity between the interneurons. There is also evidence that some background excitation is needed to sustain oscillations in networks of GABAergic interneurons. Some of this excitation may be provided by terminations of pyramidal cell axons on GABAergic interneurons.

Another type of synchrony may be forced on target cells of inhibitory interneurons. In Figure 8.5a, a single GABAergic interneuron is shown innervating several pyramidal cells. If the inhibition is strong enough and if generated at the somata, it would inhibit the three pyramidal cells, but, at the end of the ipsps, the pyramidal cells will fire in near synchrony (Figure 8.5b). The firing of the target cells following an ipsp can be due to release of inhibition or to rebound depolarization, as mentioned earlier in connection with T-type Ca^{2+} currents and the I_h current (Section 7.3). The synchronizing action of a single inhibitory interneuron, illustrated in Figure 8.5, is potentiated by the aforementioned synchronized firing of networks of GABAergic interneurons because several interneurons of the same network may converge on a given pyramidal cell, and the synchrony of a network of interneurons will, in turn, synchronize the firing of a larger population of pyramidal cells. The pyramidal cells are thus **entrained** to fire synchronously at a rate determined by that of the GABAergic network. In general, entrainment is a phenomenon by which some external stimulation synchronizes the firing of a group of neurons at a frequency that is different from any native rhythm that the group of neurons may have.

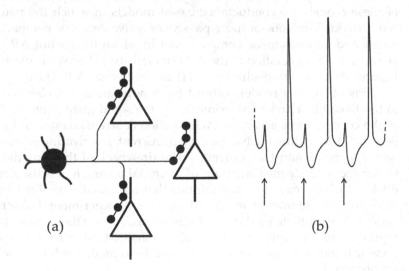

(a)   (b)

FIGURE 8.5
(a) Inhibitory interneuron innervating pyramidal cells; (b) synchronous firing of pyramidal cells at the end of ipsps whose start is indicated by arrows.

An important example of synchronized oscillatory activity between different areas of the brain is the **thalamocortical network**, involving feedback loops between the thalamus and various areas of the cerebral cortex either directly or through other brain structures, such as the basal ganglia or the cerebellum (Chapter 12). Different frequencies of oscillation are prominent, depending on mental activity and the state of the brain. For example, gamma-band frequencies are associated with conscious perception, selective concentration on tasks, learning, and short-term memory. The alpha rhythm is prominent, particularly in the occipital-parietal regions of the cortex in the awake, relaxed state, with the eyes closed and in tasks that heavily involve working memory. Alpha activity is diminished in the attentive, visually-dependent state. Gamma-band oscillations are prominent in rapid-eye-movement (REM) sleep, whereas lower frequencies, in the delta and theta ranges, are prominent in non-rapid-eye-movement (NREM) sleep.

8.2 Neuronal Models

Neuronal models can be classified into two broad categories:

1. *Biophysical models*, based on presumed behavior of the neuronal membrane expressed in terms of relations involving parameters that, at least in principle, can be measured experimentally. Examples of these models are conductance-based models in which the neuron is divided into one or more passive or active dendritic compartments and an axosomatic compartment in which the output AP is generated. The generation of the AP is usually based on some modification of the Hodgkin–Huxley (HH) model (Section 3.2). The modifications of the HH model account for conductances not observed in the Hodgkin–Huxley experiments on the squid giant axon, such as the conductances associated with various currents mentioned in Section 7.3, including the Na^+ persistent current, inactivating K^+ current, I_h current, and Ca^{2+} currents. The drawback of these models is that the accurate measurement of neuronal parameters is usually difficult, which leads to uncertainties that are usually resolved by adjusting the parameters to reproduce a set of experimental observations. The ultimate vindication of these models is in their ability to reproduce experimental results in situations that differ from those on which they are based and in their ability to predict behavior not yet observed.

2. *Dynamical models* that reproduce the computational properties of neurons without recourse to detailed or rigorous biophysical

justifications. Such models are useful for understanding the overall functions of neurons in terms of the nature of the computations performed and the essential principles governing these computations – both for single neurons and neuronal networks.

Some basic dynamical models are presented in the next section and biophysical models in the section that follows the next section.

8.2.1 Dynamical Neuronal Models

8.2.1.1 Integrate-and-Fire Model

The simplest neuronal model is the **integrate-and-fire (IaF)** model or, more specifically, the leaky integrate-and-fire model in which the membrane voltage, following a perturbation, decays towards its resting value with a characteristic time constant. In the perfect integrate-and-fire model, this decay is neglected, as in a perfect, or ideal, integrator.

In the IaF model, the state of the neuron is characterized by the voltage v_m of its membrane. From Figure 8.6, the governing equation is:

$$I_S + I_{\text{leak}} + I_C - I_{\text{in}} = 0 \tag{8.1}$$

where each term represents a total neuronal current rather than per-unit-area quantities. Similarly, the capacitance and conductances in Figure 8.6 refer to the whole neuron. $I_C = C dv_m/dt$ is the passive capacitive current. $I_{\text{leak}} = G_{\text{leak}}(v_m - V_{m0})$ is a passive leak current through a constant membrane conductance G_{leak}, and the equilibrium voltage for the leak current is assumed to be the same as the resting membrane voltage V_{m0}. I_S is the synaptic current,

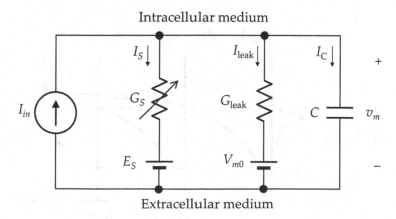

FIGURE 8.6
Equivalent circuit for the integrate-and-fire model.

and I_{in} is the current applied intracellularly to the neuron. Equation 8.1 is usually expressed as:

$$C\frac{dv_m}{dt} = -G_{leak}\left(v_m - V_{m0}\right) - I_S + I_{in} \tag{8.2}$$

The essence of the IaF model is that when v_m reaches a threshold value V_{thr}, an AP is *assumed to be generated* and is usually represented by an impulse, or a Dirac delta function, $\delta(t - t_{out})$, where t_{out} is the time at which V_{thr} is reached, and the output impulse is generated. Following the impulse, v_m is reset to a voltage V_{reset} that is less than V_{thr} but need not be the same as V_{m0}. v_m remains at V_{reset} for an absolute refractory period, τ_{ref}, before it begins to change again in accordance with Equation 8.2.

The basic behavior of the IaF model can be illustrated by considering a constant current $I_{in} = I_A$ applied, with $I_S = 0$. In the presence of I_A, the çapacitor charges towards a steady value V_{SS} that can be obtained from Equation 8.2 by setting $dv_m / dt = 0$ and $I_S = 0$, which gives:

$$V_{SS} = V_{m0} + \frac{I_A}{G_{leak}} \tag{8.3}$$

The value of I_A that will make $V_{SS} = V_{thr}$ is, from Equation 8.3, $I_{Athr} = G_{leak}(V_{thr} - V_{m0})$. If $I_A > I_{Athr}$, then v_m will reach threshold before reaching V_{SS} (Figure 8.7). When threshold is reached, an AP is generated at $t = t_{out1}$, for example, and v_m is reset to V_{reset}. After a delay τ_{ref}, the capacitor charges again, so that v_m increases exponentially towards V_{SS}. When V_{th} is again reached, an AP is generated, and the cycle is repeated. The neuron fires repetitively, as illustrated in Figure 8.7, at a frequency $1/(\tau_{exp} + \tau_{ref})$, where τ_{exp} is given by (Problem 8.1):

$$\tau_{exp} = \tau_m \ln\frac{V_{SS} - V_{reset}}{V_{SS} - V_{thr}} \tag{8.4}$$

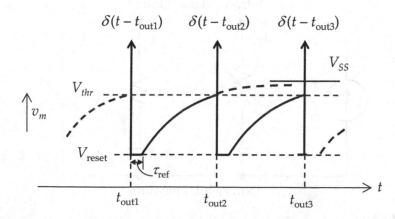

FIGURE 8.7
Generation of action potentials in the integrate-and-fire model.

with $\tau_m = C_m/G_{leak}$. Note that if $V_{SS} = V_{thr}$, $\tau_{exp} \longrightarrow \infty$ because, theoretically, v_m approaches V_{thr} but never reaches it.

Synaptic input can be formally added in the same manner as the leak conductance (Figure 8.6). Equation 8.2 becomes, with I_{in} neglected:

$$C\frac{dv_m}{dt} = -G_{leak}\left(v_m - V_{m0}\right) - \sum_n G_{Sn}\left(v_m - E_{Sn}\right) \tag{8.5}$$

where the summation extends over all excitatory and inhibitory synapses. The synaptic conductances can be modeled as impulses of appropriate strengths that occur in a specified manner in time, such as a homogeneous or inhomogeneous Poisson process.

The IaF model is simple, computationally efficient, and reproduces some basic neuro-computational properties such as all-or-none response, threshold, effects of excitation and inhibition, and Type I excitability (Subsection 8.1.2). The IaF model is not a **spiking model** in which v_m actually reproduces the time course of an AP, since it is assumed that an AP in the form of an impulse is generated once v_m equals V_{thr}. Moreover, the fact that v_m is reset after each impulse implies that the neuron has no memory of the effect of previous firing. The simple IaF model does not reproduce the firing patterns depicted in Figures 8.3 and 8.4.

Properties of the AP, such as adaptation of spike frequency and relative refractoriness can be added to the basic IaF model. Spike rate adaptation involves a memory of the effect of previous firing and can be included by making G_{leak} time-dependent, or by replacing V_{m0} in Equation 8.2 by a time-dependent voltage. Alternatively, a time-dependent spike rate adaptation conductance g_{srad} can be added to Equation 8.2:

$$C\frac{dv_m}{dt} = -G_{leak}\left(v_m - V_{m0}\right) - g_{srad}\left(v_m - E_{srad}\right) - I_S + I_{in} \tag{8.6}$$

$$\tau_{srad}\frac{dg_{srad}}{dt} = -g_{srad} \tag{8.7}$$

where E_{srad} is more negative than V_{m0}, so that the current $-g_{srad}(v_M - E_{srad})$, when activated, is an outward current that hyperpolarizes the neuron away from V_{thr} and reduces the spiking frequency. g_{srad} is zero in the resting state, and when increased, it returns to zero exponentially with a time constant τ_{srad}, in accordance with Equation 8.7. When the neuron fires, g_{srad} is incremented by Δg_{srad}, so that during repetitive firing g_{srad} increases in a sequence of steps that depend on τ_{srad} and the interspike interval. The hyperpolarizing current increases, causing spike rate adaptation.

A relative refractory period can be included by making the threshold time dependent. Alternatively, a conductance term like g_{srad} can be added to the equation. When the neuron fires, the conductance is incremented by a large value, which effectively clamps v_m at V_{reset} and simulates absolute refractoriness. The conductance then decays to zero at a certain rate, which simulates relative refractoriness.

Problem 8.2

Derive Equation 8.4.

Problem 8.3

Consider an ideal, that is, a non-leaky integrate-and-fire neuron which simply acts as a capacitor C, thereby integrating the applied current I_{in}, generates a spike when the threshold voltage V_{thr} is reached, and instantly resets the voltage to zero. (a) Show that for a constant I_{in}, the firing frequency is $\dfrac{I_{in}}{CV_{thr}}$; (b) show that if a refractory period τ_{ref} is added after the spike, the firing frequency becomes $\dfrac{I_{in}}{CV_{thr} + \tau_{ref}I_{in}}$, which saturates at $1/\tau_{ref}$ for large I_{in}.

8.2.1.2 Resonate-and-Fire Model

In the resonate-and-fire (RaF) model, a recovery variable u is added whose effect is the same as the K$^+$ current in the HH model, which aids in the recovery from the spike. The recovery variable also has the same effect as g_{srad} in Equations 8.6 and 8.7. The equations that describe the system become:

$$C\frac{dv_m}{dt} = -G_{leak}(v_m - V_{m0}) - u - I_S + I_{in} \tag{8.8}$$

$$\frac{du}{dt} = \frac{(v_m - E_u)}{k} - u \tag{8.9}$$

where k and E_u are parameters of the model, k being of the nature of a time constant and E_u being an equilibrium voltage (Equation 1.26). When v_m reaches threshold, a spike is said to be fired, as in the IaF model, without an AP being generated by the equations. Following the spike, both v_m and u are reset to two parameter values, V_{reset} and u_{reset}, respectively.

The RaF model has the same limitations as the IaF model. Its main distinguishing feature, compared to the IaF model is that, following a subthreshold perturbation, v_m does not decay monotonically towards V_{m0} but approaches this value as a damped oscillation characteristic of a resonator neuron (Figure 8.2b). The RaF neuron is incapable of generating sustained subthreshold oscillations of v_m. Computationally, the RaF model is not as efficient as the IaF model.

Whereas the simple IaF model is described by a single differential equation involving v_m, the RaF model is described by two differential equations that express the first derivatives with respect to time of an **excitation variable** v_m and a **recovery variable** u as functions of v_m and u. This is characteristic of a two-dimensional **phase-plane representation of dynamical systems**, where v_m and u are the state variables, which evolve as a function of time in a manner dependent on the system under consideration. In the phase plane, v_m is plotted along one axis and u along the other axis. At any given time t, the values of v_m and u define a state point in the phase plane. As time evolves, the state point moves along a trajectory in the phase plane that defines the response of the system. An extensive theory has been developed on the trajectories of dynamical systems in multi-dimensional phase space. The neurodynamical models commonly used are mostly two-dimensional, since trajectories in a plane are easier to visualize and interpret than trajectories in multi-dimensional space. Nevertheless, these neurodynamical models have provided valuable insight into the theoretical aspects of neuronal firing patterns. Some basic models are briefly presented in the following subsections. A more detailed treatment of neurodynamical models is beyond the scope of this book but can be found in the bibliography at the end of the book.

8.2.1.3 Fast-Slow Reduced HH Model

The Hodgkin–Huxley model (Section 3.2) involves four variables: v_m, m, n, and h. These can be reduced to two by making some simplifying assumptions that will not have a major impact on the behavior of the system. The first assumption relies on the fact that τ_m (Figure 3.8) is an order of magnitude smaller than τ_n and τ_h (Figures 3.6 and 3.9). It can therefore be assumed that m instantaneously reaches its steady-state value m_∞, as a function of v_m. This is referred to as a **quasi-steady state approximation**. The second approximation relies on the fact that the graphs of n and $(1 - h)$ are quite similar (Figures 3.6 and 3.9). Hence, they can be considered as a single recovery variable u where $n = u$ and $h = 1 - u$. Equation 8.2 can be expressed as:

$$C\frac{dv_m}{dt} = -G_{Na}^0 m_\infty^3\left(v_m\right)\left(1-u\right)\left(v_m - E_{Na}\right) - G_K^0 u^4\left(v_m - E_K\right) - G_{leak}\left(v_m - V_{m0}\right) + I_{in} \quad (8.10)$$

Equation 3.4 becomes:

$$\tau_u\left(v_m\right)\frac{du}{dt} = u_\infty\left(v_m\right) - u \quad (8.11)$$

Equations 8.10 and 8.11 define what is sometimes referred to as the **fast-slow reduced HH model**. The model reproduces threshold behavior, generation of an action potential in response to a suprathreshold stimulus, and repetitive firing when a sustained current is applied. With $I_{in} = 0$, a small perturbation from a stable state point in the phase plane, representing the resting state, results in a trajectory that returns to the stable state point of the resting state without deviating too far from this point. However, if the perturbation is large enough, the trajectory follows a prescribed path in the phase plane that involves a large excursion away from the stable state point before returning to this point. The trajectory now corresponds to the generation of an action potential in response to a suprathreshold stimulus. If a steady I_{in} is applied, the trajectory becomes a closed loop that is repeatedly traversed in the phase plane. Such a loop is referred to as a **limit cycle** and corresponds to repetitive firing of action potentials.

8.2.1.4 Fitzhugh–Nagumo Model

In 1961 R. Fitzhugh introduced a simplified, two-dimensional neurodynamical model that reproduced many of the properties of the action potential as described by the HH equations. The model was derived not by reducing the number of variables, as in the reduced HH model described in the preceding section, but by modifying the well-known van der Pol, two-dimensional state-space equations for a relaxation oscillator. The aim was to derive a model that embodied the essential mathematical properties of excitation independently of the underlying biophysical phenomena involving ion channels. A year later, J. Nagumo and his associates described an equivalent electronic circuit that obeys the same set of equations. The model became known as the Fitzhugh–Nagumo model.

The Fitzhugh–Nagumo model can be described by the following equations:

$$\frac{dv_m}{dt} = v_m - \frac{v_m^3}{3} - u + I_{in} \tag{8.12}$$

$$\frac{du}{dt} = \phi\left(v_m + a - bu\right) \tag{8.13}$$

where ϕ, a, and b are parameters. In the van der Pol equations, $a = b = 0$. Equations 8.12 and 8.13 are simpler in form than Equations 8.10 and 8.11, but both Equations 8.10 and 8.12 are cubic in v_m, where a cubic function is generally N-shaped, having a region of negative slope.

The circuit that obeys equations of similar form incorporates a tunnel diode, as shown in Figure 8.8. From Kirchoff's current law,

$$I_{in} = I_C + f(v_m) + u \tag{8.14}$$

where $f(v_m)$ is the current through the tunnel diode as a function of the voltage v_m across the diode, and u is the current in the inductive branch. From Kirchoff's voltage law applied to this branch:

$$v_m = L\frac{du}{dt} + Ru - E \tag{8.15}$$

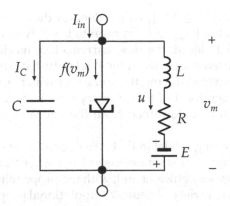

FIGURE 8.8
Equivalent circuit for the Fitzhugh–Nagumo model.

Substituting $J_C = C dv_m/dt$ and rearranging, Equations 8.14 and 8.15 become:

$$C\frac{dv_m}{dt} = I_{in} - f(v_m) - u \qquad (8.16)$$

$$L\frac{du}{dt} = v_m + E - Ru \qquad (8.17)$$

which are the same form as Equations 8.12 and 8.13, bearing in mind that $f(v_m)$ is of the form of an N-shaped cubic equation passing through the origin and having a negative-resistance region.

The Fitzhugh–Nagumo model is able to reproduce the threshold behavior of the HH model, including accommodation (Section 3.3.2), repetitive firing for a range of values of I_{in} and excitation block for large values of I_{in}, as well as anode break excitation (Problem 3.2).

8.2.1.5 Quadratic Model

In the quadratic-type model introduced by E. M. Izhikevitch (2007) v_m appears as v_m^2 in the equation for dv_m/dt, in contrast to the reduced HH and Fitzhugh–Nagumo models, which are cubic in v_m. The equations of the model can be expressed in the following form, using the same symbols previously used for the IaF and RaF models:

$$C\frac{dv_m}{dt} = k(v_m - V_{m0})(v_m - V_{thr}) - u + I_{in} \qquad (8.18)$$

$$\frac{du}{dt} = a[b(v_m - V_{m0}) - u] \qquad (8.19)$$

with the condition: if $v_m \geq V_{peak}$, which denotes the peak of the spike, v_m is reset to a chosen value V_{reset} and u is reset to $u + d$, where u is a slow recovery variable that includes all the slow currents that modulate spike generation, and d is the difference between the total outward currents and the total inward currents activated during the spike and which affect the after-spike behavior. a is the reciprocal of the recovery time constant, and k and b are parameters that can be determined from the neuron's rheobase and input resistance.

The model is a spiking model in that APs of peak value V_{peak} are generated by the equations, and v_m instantaneously drops after the peak to V_{reset}. The model is computationally efficient and, with the proper choice of parameters, can reproduce a wide variety of neuro-computational properties of biological neurons including behavior as integrators and resonators, repetitive firing, spike frequency adaptation, accommodation, after-depolarization, phasic and tonic bursting, bistability, rebound spiking, and rebound bursting.

8.2.1.6 Morris–Lecar Model

In 1981 C. Morris and H. Lecar proposed a mathematical model to emulate the electrical activity of the giant muscle fiber of the barnacle. The model is based on two currents: (i) an excitatory Ca^{2+} or Na^+ current, or both, depending on the system being modelled, and having an equilibrium voltage that is more depolarized than the resting voltage; the excitatory current is assumed to activate instantly, without inactivation; and (ii) a recovery K^+ current having an equilibrium voltage that is more hyperpolarized than the resting voltage. This current is slower than the excitatory current but does not inactivate either. The equations that describe the model can therefore be expressed as:

$$C_m \frac{dv_m}{dt} = -G_{Na}^0 m_\infty(v_m)\left(v_m - E_{Na}\right) - G_K^0 n\left(v_m - E_{Na}\right) - G_{leak}\left(v_m - E_L\right) + I_{in} \quad (8.20)$$

$$\tau_n\left(v_m\right)\frac{du}{dt} = n_\infty(v_m) - n \quad (8.21)$$

with $m_\infty(v_m) = \frac{1}{2}\left[1 + \tanh\left(\dfrac{v_m - V_1}{V_2}\right)\right]$, $n_\infty(v_m) = \frac{1}{2}\left[1 + \tanh\left(\dfrac{v_m - V_3}{V_4}\right)\right]$, and $\tau_n(v_m)$

$= \tau_0 \,\text{sech}\left(\dfrac{v_m - V_3}{2V_4}\right)$, where V_1, V_2, V_3, V_4, and τ_0 are parameters.

All of the parameters of the Morris–Lecar model are experimentally measurable, which makes it useful for simulating a variety of excitable systems. The model reproduces the observed electrical activity of the giant muscle fiber of the barnacle. When applied to neurons, with the proper choice of parameters, the model can exhibit Type I and Type II excitability and can reproduce many of the features of neuronal firing including subthreshold oscillations, firing of single APs with a quasi-threshold, repetitive firing, and bistability.

8.2.2 Biophysical Neuronal Models

8.2.2.1 *Morphoelectrotonic Transformations*

Morphological-to-electrotonic transformation, or morphoelectrotonic transformation (MET), in its most general sense as used here, denotes any technique for transforming a given dendritic geometry to an equivalent electrical representation assuming that dendrites conduct signals only passively, as in an *RC* cable. The simplest of these transformations, which will be considered in what follows, replaces a dendritic tree by a single **equivalent cylinder**, subject to certain restrictions, and was introduced by W. Rall. More elaborate transformations will be briefly mentioned at the end of this section.

The derivation of the equivalent cylinder begins with Equation 4.14 for the steady state of an *RC* cable subjected to a step of depolarization, based on the assumptions stated in Section 3.4. The general solution of this equation is:

$$v_{mr} = A_1 e^{-X} + A_2 e^{X} = B_1 \cosh(X) + B_2 \sinh(X) \tag{8.22}$$

where the A's and B's are arbitrary constants, $X = x/\lambda$, and $\lambda = 1/\sqrt{g_m r_i}$ is the space constant, assuming, for simplicity, that $r_e = 0$. The expression in terms of hyperbolic functions follows from the fact that $e^{X} = \cosh(X) + \sinh(X)$ and $e^{-X} = \cosh(X) - \sinh(X)$.

Consider a dendritic segment of length l extending from $X = 0$ to $X = L$, where $L = l/\lambda$ is the electrotonic length. It is required to derive the input conductance g_{in} of this segment when a voltage v_{m0} is applied at $X = 0$, with the segment terminated by a conductance g_L at $X = L$.

From Equation 8.22 and Equation 4.7, with $x = X/\lambda$ and $r_e = 0$, so that v_a becomes identified with v_{mr},

$$i_a = -\frac{1}{\lambda r_a} \frac{\partial v_{mr}}{\partial X} = -\frac{1}{\lambda r_a} [B_1 \sinh(X) + B_2 \sinh(X)] \tag{8.23}$$

It follows from Equations 8.22 and 8.23 that at $X = 0$, $i_a = i_{ain}$, $v_{mr} = v_{m0} = B_1$, and

$$i_{ain} = -\frac{B_2}{\lambda r_a} \tag{8.24}$$

This gives:

$$g_{in} = \frac{i_{ain}}{v_{m0}} = -\frac{1}{\lambda r_a} \frac{B_2}{B_1} \tag{8.25}$$

At $X = L$,

$$i_L = -\frac{1}{\lambda r_a}\left[B_1 \sinh(L) + B_2 \cosh(L)\right], \text{ and } v_L = B_1 \cosh(L) + B_2 \sinh(L) \quad (8.26)$$

It follows that:

$$g_L = \frac{i_L}{v_L} = -\frac{1}{\lambda r_a}\left[\frac{B_1 \sinh(L) + B_2 \cosh(L)}{B_1 \cosh(L) + B_2 \sinh(L)}\right] = -\frac{1}{\lambda r_a}\left[\frac{\tanh(L) + (B_2 / B_1)}{1 + (B_2 / B_1)\tanh(L)}\right] \quad (8.27)$$

Solving for B_2/B_1 from Equation 8.27 and substituting in Equation 8.25,

$$g_{in} = \frac{1}{\lambda r_a}\left[\frac{\lambda r_a g_L + \tanh(L)}{1 + \lambda r_a g_L \tanh(L)}\right] \quad (8.28)$$

If the dendritic segment under consideration is a terminal segment, it will be terminated at $X = L$ by a membrane seal that closes off the intracellular medium. Electrically, the membrane seal presents a very high impedance at the end of the RC cable, so that the cable is effectively terminated with an open circuit (see Problem 4.12). $g_L \longrightarrow 0$ in this case, so that Equation 8.28 becomes:

$$g_{in\infty} = \frac{1}{\lambda r_a}\tanh(L) \quad (8.29)$$

where $\lambda r_a = \sqrt{r_a / g_m} = \left[\left(\rho_a / \pi(d/2)^2\right)\left(\rho_m \delta / \pi d\right)\right]^{1/2}$, ρ_a and ρ_m are the resistivities of the intracellular medium and the membrane, respectively, δ is the membrane thickness, and d is the dendrite diameter. $(1/\lambda r_a)$ can be expressed as $Kd^{3/2}$, where $K = (\pi / 2)(\rho R_m)^{-1/2}$ is a constant and R_m, the membrane resistance per unit area, is $\rho_m \delta$. The input conductances become:

$$g_{in\infty} = Kd^{3/2} \tanh(L), \text{ and } g_{in} = Kd^{3/2}\left[\frac{g_L + Kd^{3/2} \tanh(L)}{Kd^{3/2} + g_L \tanh(L)}\right] \quad (8.30)$$

Note that if $L \longrightarrow \infty$, $\tanh(L) \longrightarrow 1$, and $g_{in} = Kd^{3/2}$ irrespective of g_L.

Consider the dendritic tree diagrammatically illustrated in Figure 8.9, where the diameters of the various dendritic segments are indicated. Two key assumptions are now made:

1. At every branch point, the diameter of the parent dendrite d_P raised to the power 3/2 is equal to the sum of the diameters of all the daughter dendrites d_{Di} also raised to the power 3/2:

$$d_P^{3/2} = \sum_{i=1}^{n} d_{Di}^{3/2} \quad (8.31)$$

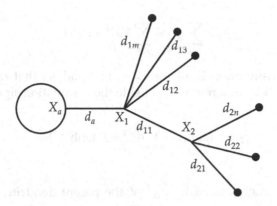

FIGURE 8.9
Dendritic branching pattern for equivalent cylinder.

This is referred to as the **3/2 power rule** and is a conductance matching condition for segments of infinite length.

2. All terminal branches of the dendritic tree connected to a given branch point have the same electrotonic length. Thus, in Figure 8.9, assuming that d_{21}, d_{22},..., d_{2n} are terminal branches, as indicated by the dots, and connected to the common branch point X_2, they would all have the same electrotonic length. Similarly, for terminal branches d_{12}, d_{13},..., d_{1m}.

It will be shown that subject to these assumptions, the electrotonic length is the same for all paths from the soma, or any branch point of the dendritic tree, to the tip of any terminal dendritic branch. Thus, the electrotonic length from X_1 to the tip of terminal branch d_{1m} is the same as that to the tip of terminal branch d_{12} and is the same as that through d_{11} and to the tip of any of the terminal branches d_{21} to d_{2n}. Similarly, the electrotonic length from the soma (X_a) to the tip of any terminal dendritic branch is the same as that through the root branch d_a and any of the branches d_{21} to d_{2n}, or through d_a, d_{11}, and any of the terminal branches d_{21} to d_{2n}. This defines an equivalent cylinder for the whole dendritic tree. The diameter of the equivalent cylinder is that of the dendritic branch d_a, and its electrotonic length is that from the soma to the tip of any of the terminal dendritic branches.

To derive the electrotonic length of the equivalent cylinder, consider, to begin with, the branch point X_2. The conductance of one of the terminal branches, say d_{21}, is from Equation 8.30 for g_{in}:

$$g_{21} = Kd_{21}^{3/2} \tanh(L_{21}) \qquad (8.32)$$

where L_{21} is the electrotonic length of branch d_{21}. The total conductance of all the daughter branches at X_2 is:

$$g_2 = \sum_{i=1}^{n} g_{2i} = \sum_{i=1}^{n} K d_{2i}^{3/2} \tanh(L_{2i}) \tag{8.33}$$

According to assumption 2, all the L_{2i}'s are equal, so that $\tanh(L_{2i})$ can be replaced by $\tanh(L_2)$ and removed outside the summation sign. Thus:

$$g_2 = K \tanh(L_2) \sum_{i=1}^{n} d_{2i}^{3/2} = K \tanh(L_2) d_{11}^{3/2} \tag{8.34}$$

where $\sum_{i=1}^{n} d_{2i}^{3/2}$ was replaced by $d_{11}^{3/2}$ of the parent dendrite, in accordance with assumption 1.

g_2 is now the load conductance g_L of the d_{11} segment. The input conductance g_{11} of this segment is, from Equation 8.30 for g_{in}:

$$g_{11} = K d_{11}^{3/2} \left[\frac{K d_{11}^{3/2} \tanh(L_2) + K d_{11}^{3/2} \tanh(L_{11})}{K d_{11}^{3/2} + K d_{11}^{3/2} \tanh(L_2) \tanh(L_{11})} \right] = K d_{11}^{3/2} \tanh(L_{11} + L_2) \tag{8.35}$$

The expression for g_{11} from Equation 8.35 is of the same form as that of the single terminal branch g_{21} of Equation 8.32, but with d_{21} replaced by d_{11} and with $L_{21} = L_2$ replaced by the total electrotonic length $(L_{11} + L_2)$ from node X_1 to any of the tips of terminal branches $d_{21}, d_{22} \ldots d_{2n}$.

Assuming the same electrotonic length $(L_{11} + L_2)$ from node X_1 to any of the ends of the other terminal branches through $d_{12} \ldots d_{1m}$, in accordance with assumption 2, and that at X_1, $d_a^{3/2} = \sum_{j=1}^{m} d_{1j}^{3/2}$, in accordance with assumption 1, the preceding procedure can be repeated to give the input conductance at X_a through the root branch d_a, which gives:

$$g_a = K d_a^{3/2} \tanh(L_a + L_{11} + L_2) \tag{8.36}$$

where $(L_a + L_{11} + L_2)$ is the electrotonic length from X_a through the root branch d_a and to the tip of any of the terminal branches of the dendritic tree through any of the intermediate paths. Once the electrotonic length of the equivalent cylinder is found, the problem is reduced to that of a single cable or dendrite of finite length and a sealed end. The steady-state voltage at any point X, for example, can be readily determined (see Problem 4.15). A simplified case is considered in Problem 8.9.

If the neuron has more than k dendritic trees emanating from the soma through k root branches, then each dendritic tree can be represented by an equivalent cylinder as long as the aforementioned assumptions are satisfied. The k equivalent cylinders can be reduced to a single cylinder of diameter

$\left(\sum\limits_{j=1}^{k} d_j^{3/2}\right)^{2/3}$. The equivalent cylinder representation preserves membrane area. That is, the area of the equivalent cylinder is the same as that of the branching dendritic tree that it represents (Problem 8.10).

Evidently, the derivation of the equivalent cylinder involves many assumptions that cannot all be satisfied in real dendritic trees, namely, equality of electrotonic lengths, the 3/2 power rule, uniformity of diameter of dendritic segments, and constancy and equality of all the electrical parameters of the cytoplasm and membrane throughout the dendritic tree. Although some forms of synaptic excitation can be added to the equivalent cylinder representation, this representation is of limited usefulness when it comes to: (i) mapping between arbitrary configurations of synaptic inputs to the dendritic tree and corresponding inputs to the equivalent cylinder, and (ii) evaluation of interactions between arbitrary configurations of inputs to the dendritic tree. Nevertheless, the equivalent cylinder was a very useful concept that provided considerable insight into the electrical behavior of dendrites and which paved the way to more elaborate and less restrictive representations. One such generalized representation is the **equivalent cable** for dendritic trees of arbitrary morphology, including the relaxation of some of the aforementioned restrictions. Other approaches are based on two-port circuit analysis and on transforming dendrite morphology to another geometry based on the electrotonic length from the soma, or on the logarithm of voltage attenuation between the soma and dendrites, or on the delay of signal propagation between the soma and dendrites. Although these representations will not be pursued here, interested readers are referred to books and articles listed in the bibliography at the end of the book. We will consider in the following subsection the more versatile analysis of dendritic behavior based on numerical computations and simulations.

Problem 8.4

Deduce from Equations 4.29 and 8.29 that the input resistance R_{inL} of a cable of electrotonic length L that is terminated with an open circuit can be expressed as: $R_{inL\infty} = R_{in\infty} \coth(L)$.

Problem 8.5

Consider a cable of electrotonic length L that is terminated by a resistance R_T. Use the result of Problem 4.15 and the boundary condition:

$\dfrac{dv_{mr}(X)}{dX}\bigg|_{X=0} = -\lambda r_a I_0 = -R_{in\infty} I_0$, where $R_{in} = \lambda r_a$ (Equation 4.31) and I_0 is the

input current at $X = 0$ to show that the input resistance of the cable is:

$$R_{inL} = R_{in\infty} \frac{R_T + R_{in\infty} \tanh(L)}{R_{in\infty} + R_T \tanh(L)}.$$

Problem 8.6

Verify, using the expression for R_{inL} of Problem 8.5, that the input resistance of an infinitely long cable is R_{in}, that of an open-circuited cable of electrotonic length L is $R_{in}\coth(L)$, and that of a short-circuited cable of electrotonic length L is $R_{in}\tanh(L)$.

Problem 8.7

Using the results of the preceding problems, derive the expressions for the input resistance R_{inp} of a parent dendrite of length L_p that branches into two terminal dendrites of lengths L_1 and L_2 (Figure 8.10).

ANS.: $R_{inL1} = R_{in1}\coth(L_1)$, $R_{inL2} = R_{in2}\coth(L_2)$, $\dfrac{1}{R_{Tp}} = \dfrac{1}{R_{inL1}} + \dfrac{1}{R_{inL2}}$, $R_{inp} = R_{in\infty p}\dfrac{R_{Tp} + R_{in\infty p}\tanh(L_p)}{R_{in\infty p} + R_{Tp}\tanh(L_p)}$.

Problem 8.8

Assume, for simplicity, that in Figure 8.10, the two terminal dendrites have the same diameter d_1 and electrotonic length L_1 and that the 3/2 power rule applies, that is, $d_p^{3/2} = 2d_1^{3/2}$. Using the relation that the input conductance of an infinite cable is $Kd^{3/2}$ (Equation 8.30), show that at the branchpoint, the infinite-cable conductance of the parent dendrite is matched to the combined infinite-cable conductances of the two terminal dendrites, assuming the same membrane and cytoplasmic resistances.

Problem 8.9

Show that in the case considered in Problem 8.8, if a current step is applied at the input of the parent dendrite, the steady-state voltage at any point X along the equivalent cylinder is given by: $V_{m0}\dfrac{\cosh(L_p + L_1 - X)}{\cosh(L_p + L_1)}$, as for an open-circuited cable of electrotonic length $(L_p + L_1)$.

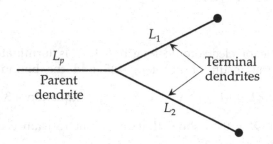

FIGURE 8.10
Figure for Problem 8.7.

Problem 8.10

Assume that in Figure 8.10 the conditions for the equivalent cylinder are satisfied, that is, the electronic lengths are the same ($L_1 = L_2 = L_t$) and the 3/2 power rule applies ($d_p^{3/2} = d_1^{3/2} + d_2^{3/2}$). The additional electrotonic length that is added to that of the parent dendrite to obtain the equivalent cylinder is L_t, the additional physical length is area $\lambda_p L_t$, and the additional area is proportional to $d_p \lambda_p L_t$. Show that this is the same as the sum of the physical areas of the two terminal dendrites, assuming the same properties of the membrane and axoplasm. This shows that the physical area of the equivalent cylinder is the same as that of the branching dendritic tree that it represents.

Problem 8.11

Show that if the diameter of a terminal dendrite is $1/\alpha$ ($\alpha > 1$) that of the parent dendrite, the additional physical length that is added to that of the parent dendrite to obtain the equivalent cylinder is $\sqrt{\alpha}$ times the physical length of the terminal dendrite.

8.2.2.2 Compartmental Models

In compartmental modeling, various parts of the neuron are divided into individual compartments, each represented by an appropriate combination of electric circuit elements. For good spatial resolution and accuracy, several thousand compartments may be used to model a complex dendritic tree. The differential equations describing the electrical behavior of the compartmental model are solved numerically using an appropriate digital computer program. The solutions are of course not subject to the restrictions of the equivalent cylinder representation.

Figure 8.11a illustrates the division of a dendritic tree into a number of small cylindrical segments, each of appropriate diameter and length. It may be assumed, for example, that a dendritic segment of length l over which the diameter is $d \pm 0.1$ µm, say, is assumed to be a cylinder of length l and diameter d. In Figure 8.11b, a cylinder j is shown modeled by a T-equivalent circuit in which each series branch is $R_{ij}/2$, where $R_{ij} = \rho_{ij} l_j / \pi (d_j/2)^2$ is the internal resistance of the cylinder, l_j its length, d_j its diameter, and ρ_{ij} the cytoplasmic resistivity. In modeling the shunt branch, the cylinder is assumed to be isopotential of voltage v_{mj} and capacitance $C_{Mj} = C_{mj} A_j$, where C_{mj} is the membrane capacitance per unit area, and $A_j = \pi d_j l_j$ is the surface area of the cylinder. The ohmic, or leak, channels are modeled by a constant resistance $R_{Mj} = R_{mj}/A_j$ in series with a battery E_{Mj}, where R_{mj} is the membrane resistance per unit area and E_{Mj} is the equilibrium voltage of the passive channels. Synaptic inputs via ligand-gated channels are modeled by one or more shunt branches having a time-dependent conductance g_{sj} in series with a battery E_{sj} representing the equilibrium voltage for the given channel. Voltage-gated, or active,

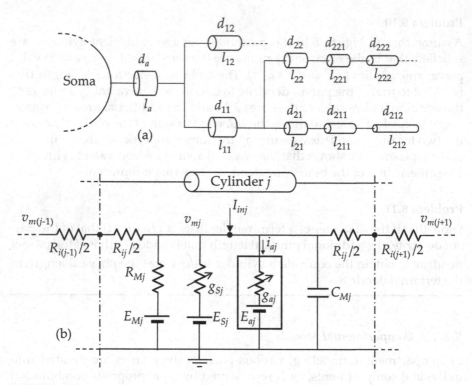

FIGURE 8.11
Compartmental modelling; (a) division of a dendritic tree into a number of small cylindrical segments; (b) equivalent circuit of a segment.

channels are represented by one or more shunt branches like that enclosed in the rectangle in Figure 8.10b. An injected current I_{inj} could also be added, as in Figure 8.6.

The voltage equation for the cylinder is derived as in Section 8.2.1.1 from the conservation of current equation at the node of a given cylinder. The voltage equation for cylinder j (Figure 8.11b) is of the general form:

$$C_{Mj}\frac{dv_m}{dt} = -\frac{(v_{mj}-E_{Mj})}{R_{Mj}} - \sum_n (v_{mj}-E_{Sjn})g_{Sjn} - \sum_k i_{ajk} -$$

$$\frac{v_{mj}-v_{m(j+1)}}{(R_{ij}/2)+(R_{i(j+1)}/2)} - \frac{v_{mj}-v_{m(j-1)}}{(R_{ij}/2)+(R_{i(j-1)}/2)} + I_{inj} \tag{8.37}$$

where the summations are for n synaptic channels and k voltage-gated channels, which could be for different ions. The terms in $(v_{mj}-v_{m(j+1)})$ and $(v_{mj}-v_{m(j-1)})$ represent the longitudinal current flowing from cylinder j to the two adjoining cylinders. If cylinder j is the terminal cylinder of a dendritic branch, the term in $(v_{mj}-v_{m(j+1)})$ is missing. On the other hand, if cylinder j is

followed by cylinders representing dendritic branches, there will be similar additional terms due to the longitudinal currents in these branches. The
soma is usually modelled as an isopotential sphere or spheroid. The axon
hillock/initial segment, where the neuronal action potential is generated can
be modelled as a compartment.

Mathematically, Equation 8.37 is a finite-difference approximation of the
cable equation in which the continuous spatial variation of v_m has been dis-
cretized by assuming a series of isopotential cylinders, with v_m considered
constant over the length of each cylinder. The mathematical description is
thereby reduced to a system of N ordinary differential equations, where N
is the number of compartments and could be in the thousands. The system
of N simultaneous differential equations are solved numerically for v_{mj}, $j = 1$,
$2, ..., N$ for successive time steps Δt that are sufficiently small. A number of
computer simulation programs have been developed to facilitate compartmental modeling, the most commonly used being NEURON and GENESIS.

As for the details of the branches, the conductance g_{Sj} in the synaptic
branch can be expressed as a function of time, as in Equations 6.2 and 6.3.

In the voltage-gated branch, the conductance g_{aj} is both voltage- and time-
dependent and may be modeled as such. More commonly, the branch current i_{aj} carried by Na⁺ or K⁺ is expressed in a generalized form, similar to that
of the Na⁺ current in the HH model, as:

$$i_{aj} = G_{Si}m^x h^y \left(v_j - E_{aj}\right) \tag{8.38}$$

where G_{Si} is a constant equal to the maximum conductance, and x and y are
powers of the activation and inactivation variables m and h, each of which
is a function of an α, a β, and a τ, as is the HH model. Currents that do not
inactivate have $y = 0$. A potassium current such as I_{SK} (Section 7.3.3) that is
voltage-independent but calcium-dependent will have its activation variable
a function of $[Ca^{2+}]_i$, whereas a potassium current such as I_C (Section 7.3.3)
that is both voltage- and calcium-dependent will have its activation variable
a function of both v_m and $[Ca^{2+}]_i$. For Ca^{2+} currents, the GHK current equation (Equation 2.24) is used, as explained, in connection with Equation 3.19.
NMDA channels can be modeled according to the expression for conductance given by Equation 6.6.

The variation of channel densities with distance from the soma can be
taken into account in an approximate manner in compartmental modelling
as well as the existence of dendritic spines. When current flows from a dendrite into a spine, the spine can be considered to be essentially equipotential
and is usually accounted for by multiplying C_M by a factor F, and dividing R_M
by F. But if current is generated at the spine head by a synaptic input, there
will be a relatively large voltage drop from the spine head to the spine base,
in which case the spine is modelled by a few compartments.

$[Ca^{2+}]_i$, the intracellular concentration of free Ca^{2+}, plays an important role
in controlling neuronal excitability because of the many roles that Ca^{2+} play

in modulating neuronal behavior, including activation of some K^+ channels, Ca^{2+} spikes, and synaptic plasticity. Realistic neuronal models should therefore include Ca^{2+} dynamics in some appropriate manner. $[Ca^{2+}]_i$ is affected by several factors, namely: (i) Ca^{2+} currents through voltage-gated Ca^{2+} channels and through ligand-gated channels, such as NMDA channels in the presence of depolarization, (ii) diffusion of Ca^{2+} through the intracellular regions of interest, (iii) buffering action of intracellular binding proteins, (iv) the uptake and release of Ca^{2+} by some cell organelles, such as mitochondria and the endoplasmic reticulum, and (v) activity of Ca^{2+} pumps and transporters. Inclusion of such factors in compartmental modelling significantly complicates the models but has been implemented to varying degrees.

The choice of parameter values in compartmental modeling is evidently of critical importance. Most parameter values are derived from experimental measurements using appropriate voltage and current excitation and chemical agents for selective blocking of ionic channels. Nevertheless, there is always a degree of uncertainty concerning these values, which necessitates some fine-tuning to bring the results of the simulation closer to experimental observation. Passive membrane parameters are usually determined from the responses to subthreshold voltage or current perturbations applied at the soma. It is usually assumed that the myelinated axon presents a negligible passive load to the soma. The parameters of voltage-gated channels can be determined from voltage-clamp or patch-clamp experiments in the presence of various channel blockers, usually on cells *in vitro*. Some parameters are derived from published experimental results under controlled conditions.

8.3 Models of Neuronal Networks

8.3.1 Compartmental Models

In principle, compartmental models of individual neurons can be coupled together to simulate a network of interconnected neurons. Both NEURON and GENESIS modelling programs, previously mentioned, allow such coupling. Whereas this is feasible for a network consisting of a small number of neurons, it is impractical for a large neuronal network comprising several populations of neurons of different types, with each population of neurons of the same type numbering in the tens of thousands. The computational complexity becomes enormous and must be reduced to a manageable level.

The computational complexity can be reduced by restricting the number of neurons in each population to a couple of hundreds at most, with the output of each neuron representing the average activity of a larger number of neurons in a given region. This involves scaling of the number of synapses on a particular neuron and of the strengths of these synapses. The number of compartments representing each neuron type is drastically reduced to a small number. The details of synaptic operation, such as the kinetics of neu-rotransmitter release and binding, are omitted. Instead, the shapes of psps observed experimentally are approximated by modelling changes in synap-tic conductances using combinations of appropriate exponential functions. Moreover, the membrane properties of the neurons are simplified by neglect-ing some ion channels altogether. In particular, the details of spike genera-tion, as in the Hodgkin–Huxley equations, are neglected. Discrete spikes are usually generated by applying a simple threshold criterion to membrane voltage. However, the AP waveform is often superimposed on the membrane voltage to generate the appropriate membrane voltages and currents associ-ated with real APs.

8.3.2 Firing Rate Models

An alternative to compartmental modelling of a large neuronal population is some form of a firing rate model in which the neuronal input and output spike trains are represented in terms of a firing rate that is a continuous function of time.

The starting point in a basic firing rate model is to consider the membrane voltage $v_{mi}(t)$ at the spike generating site of a given ith neuron in the popu-lation and perform the following: (i) consider the various inputs to the ith neuron as spike trains and account for their influence on $v_{mi}(t)$ in terms of impulse responses, (ii) express $f_i(l)$, the firing rate of neuron i, in terms of $v_{mi}(t)$, (iii) generalize $v_{mi}(t)$ to an ensemble average over the given population, and (iv) derive an equation for $v_{mi}(t)$ as an excitation variable. These steps will be elaborated in what follows.

The inputs to the ith neuron can be divided into two groups: (i) intrinsic excitation or inhibition due to other neurons in the given population, and (ii) extrinsic excitation or inhibition from sources that are external to the given population. To evaluate the contribution of the input from the jth neuron of the population to $v_{mi}(t)$, this input is considered as a train of spikes $q_j(t)$, with a spike at time τ_k being represented as an impulse or delta function, $\delta(t - \tau_k)$, so that:

$$q_j(t) = \sum_k \delta(t - \tau_k) \tag{8.39}$$

Assuming linearity, the effect of this input on $v_{mi}(t)$ is the convolution of $q_j(t)$ and the impulse response $h_{ij}(t)$,

$$v_{mij}(t) = h_{ij}(t) * q_j(t) \tag{8.40}$$

Mathematically, the convolution of a function $f(t)$ with an impulse that is delayed by τ_k is simply $f(t - \tau_k)$, that is, $f(t)$ delayed by τ_k. Evidently, according to Equations 8.39 and 8.40, the contribution to $v_{mij}(t)$ of the input $q_j(t)$ is simply the sum of impulse responses $h_{ij}(t)$, with each response being delayed by the delay of the impulse that causes it.

$h_{ij}(t)$ is supposed to account for all the influences of the spike input from the jth neuron on the output firing of the ith neuron, through $v_{mij}(t)$, including the efficacy of synaptic transmission, the temporal variation of the post synaptic potential (psp), and its spatial transformation as it propagates through the dendritic branches to the site of generation of the action potentials of the ith neuron. $h_{ij}(t)$ is not to be interpreted as embodying the details of all these phenomena but as a rather abstract and highly simplified functional description of the overall effect of the input from the jth neuron on the output of the ith neuron. $h_{ij}(t)$ can be expressed as:

$$h_{ij}(t) = w_{ij}(t)h_{io}(t) \tag{8.41}$$

where w_{ij} is a scalar, synaptic weight that accounts for the relative synaptic strength of the jth input and is positive for excitatory synapses and negative for inhibitory synapses. $h_{io}(t)$ is a function that can take one of several forms, depending on the level of detail that is required in the analysis. $h_{io}(t)$ can be: (i) simply a time delay, $h_{io}(t) = \delta(t - \tau_{io})$, with $w_{ij} = 1$, or (ii) an exponential, $h_{ij}(t) = (1/\tau_{io})e^{-t/\tau_{io}}$, like that of a parallel RC circuit, that is, a leaky integrator, which neglects the rising phase of the psp, or (iii) an alpha function (Equation 6.3) that accounts in an approximate manner for the rising phase of the psp, or iv) more elaborate difference-of-exponentials functions.

Assuming that the inputs to the ith neuron from all the other neurons of the given population of N neurons do not interact with each other, and that they summate essentially linearly at the level of $v_{mi}(t)$, the $v_{mij}(t)$ due to all the other neurons of the population can be summated to give:

$$v_{mi}(t) = \sum_{j=1, j \neq i}^{N} h_{ij}(t) * q_j(t) \tag{8.42}$$

The next step is to transform $v_{mi}(t)$ to an "instantaneous" firing rate $f_i(t)$ of the ith neuron that is a continuous function of time. The form of $f_i(t)$ must reflect two essential neuronal properties, namely: (i) refractoriness, which limits the firing frequency to some maximum rate F_{max} and (ii) some kind of threshold, since the firing rate cannot be negative and begins to increase only after some significant depolarization. The function that is commonly used is a sigmoidal function:

$$f_i(t) = \frac{F_{max}}{1 + e^{-b(v_{mi}(t) - V_{m1/2})}} \tag{8.43}$$

where $V_{m1/2}$ and b are constants. $V_{m1/2}$ is the half-voltage of the membrane, at which the firing rate is $F_{max}/2$, and b determines the maximum rate of change of $f_i(t)$ with respect to $v_{mi}(t)$.

In generalizing from the modeling of a single ith neuron to the population of N neurons, it is considered that the output train of spikes $q_{io}(t)$ of the ith neuron is, like the input spikes, a nonstationary point process whose instantaneous firing rate $f_i(t)$ is given by the expected value $E(q_{io}(t))$. This allows elimination of explicit references to the stochastic firings of all cells by redefining $v_{mi}(t)$ as an ensemble voltage for spike generation. Thus,

$$v_{mi}(t) = \sum_{j=1, j \neq i}^{N} h_{ij}(t) * E(q_j(t)) = \sum_{j=1, j \neq i}^{N} h_{ij}(t) * f_j(t) \tag{8.44}$$

When the particular form chosen for $h_{ij}(t)$ is selected, Equation 8.44 leads to an equation of $v_{mi}(t)$ as an excitation variable, similar to Equation 8.2 with both sides divided by a conductance:

$$\tau_i \frac{dv_{mi}(t)}{dt} = -v_{mi}(t) + \sum_{m=1}^{N} w_{ij} f_j(t) + \sum_{m=1}^{N'} z_{im} f_{em}(t) \tag{8.45}$$

τ_i is a time constant that is, in general, different from the membrane time constant, and the last term on the RHS has been added to account for the extrinsic inputs in a manner similar to that of Equation 8.44, where $f_{em}(t)$ is the firing rate of the mth external input, z_{im} is the weight assigned to the mth input, and the summation extends over all the N' external inputs to the given population.

The repertoire of responses of the firing rate model can be considerably expanded by adding a second equation for a recovery variable involving the firing rate.

Summary of Main Concepts

- Neuronal computation is essentially a signal processing operation that transforms a spatiotemporal input pattern of neuronal activity to a spatiotemporal output pattern by a population of neurons.

- Based on the range of frequencies of APs generated in response to injected current pulses of long duration and of various magnitudes, neurons can be divided into three classes: (i) Type I neurons whose frequency of APs increases monotonically with the magnitude of the injected current step, above a certain threshold, (ii) Type II neurons in which APs can be generated only over a limited range of frequencies that is relatively insensitive to the magnitude of the current step above a certain threshold, and (iii) Type III neurons in which only a single AP is usually generated, whose latency is reduced as the magnitude of the current pulse is increased.

- Neurons can also be classified as resonators or integrators, whereby a neuron is a resonator if its membrane potential exhibits subthreshold oscillations following a perturbation and is an integrator if its membrane voltage following a perturbation does not oscillate as it approaches the resting value.

- The subthreshold oscillations of resonators may be due to a periodic synaptic input, or may be intrinsically generated by some form of coupling between inward and outward currents through the membrane. A larger repertoire of responses is possible with resonators compared to integrators.

- Several distinctive neuronal firing patterns are observed *in vitro* in slices of the mammalian brain in response to applied current pulses of long duration and of various magnitudes. According to these firing patterns, excitatory cortical neurons can be classified as regular spiking (RS), intrinsically bursting (IB), and fast rhythmic bursting (FRB). On the other hand, inhibitory interneurons can be classified as fast spiking (FS), low threshold spiking (LTS), and late spiking (LS).

- Rhythmic, synchronized activity of a neuronal population is a fundamental mechanism that allows the coordinated activity that is necessary for normal brain functioning, including motor control, awareness, memory, and perception, particularly the binding of sensory features.

- Rhythmic activity may occur at the levels of: (i) single neurons, (ii) a group of neurons in a given location, or (iii) neurons in different areas of the brain.

- Neuronal models can be classified into two categories: (i) biophysical models based on the postulated behavior of the neuronal membrane, and (ii) dynamical models that reproduce the computational properties of neurons without rigorous or detailed biophysical justifications.

- Examples of dynamical models are: integrate-and-fire model, resonate-and-fire model, reduced HH model, Fitzhugh–Nagomo model, quadratic model, and Morris–Lecar model.

- Examples of biophysical models are the equivalent cylinder and equivalent cable models of a dendritic tree and compartmental models. The latter have become very useful and versatile in the form of simulation programs, such as NEURON and GENESIS.

- Neuronal networks can be modeled in terms of simplified compartmental models, or firing rate models.

9

Skeletal Muscle

Objective and Overview

We considered in preceding chapters the generation and propagation of the AP and its processing by neurons. The present chapter is concerned with muscular contraction, how it is elicited by the APs generated in muscle, and many of its features.

There are three types of muscle: (i) **skeletal muscle**, so called because practically all of these muscles attach to the bony skeleton, (ii) **smooth muscle** that lines internal cavities, such as the gastrointestinal tract, the uterus, some blood vessels, and air passages in the lungs, and (iii) **cardiac muscle**, the basic constituent of the heart. The mechanism of contraction is essentially the same in the three types of muscle but differs in some important respects that are adapted to the particular functions. In keeping with the scope of the book, the present chapter focuses on contraction of skeletal muscle; the other types of muscle are briefly discussed, mainly for comparison purposes, at the end of the following chapter.

The chapter begins by presenting the gross structure and the detailed microstructure of skeletal muscle, followed by a discussion of excitation-contraction coupling – that is, the sequence of events that occur between the generation of a muscle AP and muscle contraction – and of associated ATP synthesis, heat production, and muscular fatigue. Organization of muscle fibers is considered next including motor units, muscle fiber types and their interactions with motoneurons, and the different muscle architectures. The chapter ends with a discussion on muscle receptors, mainly Golgi tendon organs and muscle spindles, with a detailed explanation of the responses and innervation of the latter.

Learning Objectives

To understand:

- The macrostructure and microstructure of skeletal muscle
- Excitation-contraction coupling and phenomena associated with muscle contraction
- Organization of muscle into motor units
- The different types of muscle fibers
- The different types of muscle architectures
- Responses of Golgi tendon organs
- Structure, innervation, and responses of muscle spindles

9.1 Structure of Skeletal Muscle

9.1.1 Gross Structure

Skeletal muscle, which together with cardiac muscle is also known as **striated muscle** because of its distinctive appearance under the microscope as explained later, is the largest tissue in the body, accounting in humans for more than 40% of total body weight of a lean adult. Most of the approximately 640 skeletal muscles of the human body directly attach to the bony skeleton – but not all, as in the case of some tongue and abdominal muscles which attach to other muscles. Muscle size varies widely, from the tiny stapedius muscle of the middle ear to the gluteus maximus of the buttocks, whose volume is about 1000 ml in a human adult. The primary function of skeletal muscle is in the execution of movement and the maintenance of posture. In addition, skeletal muscle can serve important communicative functions, as in making facial expressions and in speech production, as well as some important secondary functions. For example, layers of skeletal muscle in the abdominal wall and the floor of the pelvic cavity protect and support the weight of visceral organs. Muscular contraction in the form of shivering is very effective in rapidly increasing heat production in the body to combat outside cold. Although skeletal muscle is essentially under voluntary control, it can be involved in many involuntary activities, such as eye blinking, breathing, shivering, and reflex action.

A skeletal muscle is basically a grouping of several thousand to several million individual **muscle fibers** of 10–100 μm diameter and of length in the range of one millimeter or so in the smallest skeletal muscle, the stapedius muscle of the inner ear, but can be as long as 60 cm in the human sartorius muscle that extends obliquely over the thigh. Although they are often described as single muscle cells, individual muscle fibers are in fact formed during embryonic development from the fusion of progenitor cells known as **myoblasts**. They are therefore multinucleated, with hundreds to thousands of nuclei in longer fibers. Muscle fibers are typically cylindrical in shape, circular or oval in cross section, with conical ends. It is estimated that the human body contains more than 10^8 muscle fibers. Muscle fibers are also known as **myocytes** or **myofibers**.

The cell membrane of the muscle fiber is the **sarcolemma** and has the ionic properties characteristic of excitable cells, manifested as a resting membrane voltage of about −90 mV and the ability to generate and propagate a muscle action potential. The sarcolemma is coated on the outside by a **basement membrane** formed largely of glycoproteins, and each muscle fiber is surrounded by a delicate layer of connective tissue, the **endomysium** (Figure 9.1). Groups of about 10 to more than 100 muscle fibers are bundled together into **fascicles**, the number of muscle fibers in a fascicle being larger in muscles that produce greater force, with less fineness of control. Fascicles are surrounded, in turn, by another layer of connective tissue, the **perimysium**. The whole muscle is ensheathed by a dense layer of irregular connective tissue, the **epimysium**.

The endomysium, perimysium, and epimysium are made up mostly of the protein collagen, with some elastic connective tissue fibers consisting of the protein elastin, both proteins being made by cells referred to as **fibroblasts**. The endomysium conveys blood capillaries and nerve terminals to the muscle fibers and encloses fibroblasts and satellite cells involved in the repair of muscle tissue and in the formation of new muscle fibers to replace

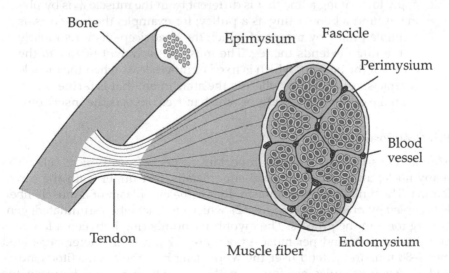

FIGURE 9.1
Skeletal muscle structure.

those that are irreversibly damaged. The epimysium separates the muscle from adjacent tissues and organs and reduces friction with other muscles and with bone.

The endomysium, perimysium, and epimysium are interconnected together and blend with **tendons** and **aponeuroses.** Tendon is a tough, fibrous, whitish, cord-like tissue that connects muscle to bone and is composed of parallel arrays of closely packed fibers that are mostly collagen. The collagen fibers aggregate to form fascicles that are surrounded by connective tissue. Bundles of fascicles, as well as the tendon as a whole, are surrounded in turn by connective tissue. Some tendons are surrounded, in addition, by a sheath that encloses synovial fluid, which acts as a lubricant that reduces friction associated with tendon movements. Tendons include blood vessels and fibroblasts for maintaining tendon tissue, as well as a type of receptor, the **Golgi tendon organ** (Section 9.4.1) that responds to tension in the tendon.

Aponeuroses are similar in composition to tendon but are flat and broad. They are associated with sheet-like muscles having a wide area of attachment rather than the restricted area of attachment of tendons (Figure 9.1). Examples are abdominal muscles, as well as intercostal muscles of the ribs, and muscles of the hand and foot. Tendons and aponeuroses convey muscular activity, whether it is mainly force or movement, to the body parts acted upon by the muscles. They contribute to the viscoelastic properties of muscle, as discussed in Section 10.3.2, and allow skeletal muscle to:

1. be conveniently located some distance away from its point of action; for example, some of the muscles that move the fingers are located in the forearm, not the hand, and act on the fingers through long tendons;

2. apply force along a line that is different from the muscle axis by plying around a bone acting as a pulley; for example, the knee acts as a simple pulley by means of which the quadriceps femoris muscle in the thigh extends the leg. The muscle attachment nearer to the center of the body, or which is fixed or moves least when the muscle contracts, is the **origin**, whereas the attachment that is farther away from the center of the body, or which moves more, is the **insertion**.

9.1.2 Microstructure

Muscle fibers differ in several important respects from typical cells. The many nuclei are typically located superficially, just underneath the sarcolemma. The bulk of the **sarcoplasm**, that is the cytoplasm of a muscle fiber, is occupied by **myofibrils** (Figure 9.2), which contract when stimulated, generating force in the process. The myofibrils, numbering between a few tens and several thousand per muscle fiber, are 1–2 μm in diameter, separated by 40–80 nm, and extend over the whole length of the fiber. Mitochondria and glycogen granules are found in the sarcoplasmic space between the myofibrils.

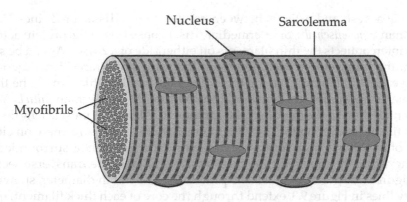

FIGURE 9.2
Basic structure of a muscle fiber.

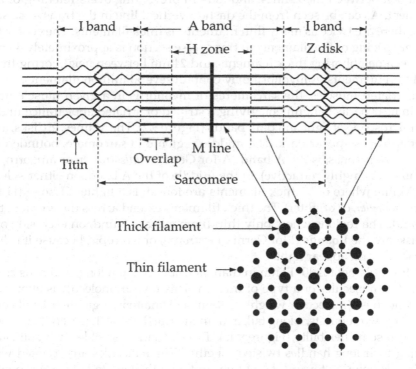

FIGURE 9.3
Diagrammatic ultrastructure of a muscle fiber in longitudinal and transverse sections.

A myofibril consists mainly of two types of filaments, the **thick filaments**, composed principally of the protein myosin, and the **thin filaments**, composed of the protein actin with smaller amounts of three other proteins: nebulin, troponin, and tropomyosin. The thick and thin filaments are arranged, as illustrated diagrammatically in the longitudinal view of Figure 9.3, in repeating units known as **sarcomeres**. Each sarcomere is 2–4 μm long in a muscle at rest, and extends between consecutive **Z disks,** or Z lines (Z for German *zwichenscheibe,* or intermediate disc), consisting of the protein actinin that interconnects the thin filaments on either side of a Z disk. As can be seen from the transverse section, the thick filaments are arranged hexagonally, with each thin filament having three neighboring thick filaments. The thick filaments are connected together at the **M line** (M for German *mittelscheibe,* or central disc) in the middle of the sarcomere. At normal muscle lengths the thin filaments overlap the thick filaments in parts of the sarcomere on either side of the M line. In the overlap zone, every thick filament is surrounded by six hexagonally arranged thin filaments, as shown in the transverse section in Figure 9.3. Strands of an elastic protein, titin, 1 nm in diameter, shown as wavy lines in Figure 9.3, extend through the core of each thick filament, from an M line to the Z disks on both sides. Titin stabilizes the thick filament, centers it between the Z-lines, and aids in preventing overstretching of the filament. As can be seen from the dashed vertical line in the transverse section, there are twice as many thin filaments as thick filaments. The center-to-center spacing of the filaments in transverse section is approximately 40 nm between neighboring thick filaments and 24 nm between neighboring thick and thin filaments. A myofibril may contain several million filaments.

In striated muscles, the sarcomeres of myofibrils in a given muscle fiber are in near perfect alignment, giving a striped or striated appearance under the microscope of light and dark bands (Figure 9.2). The M lines, Z disks, and overlap zones appear dark. The middle region of a sarcomere, bounded by two overlap zones, is the **A band** (A for German *anisotropen,* or anisotropic, because it is highly refractive). In the middle of the A band, on either side of the M line where only thick filaments are found, is a lighter **H zone** (H for German *heller,* or brighter). The thick filaments extend across the width of the A band. The regions where only thin filaments are found, on each side of a Z disk, are the **I bands** (I for German *isotropen,* or isotropic, because it is less refractive) and appear light.

A thick filament is 10–12 nm in diameter, about 1.6 μm long, and consists of some 300 molecules of a type of myosin. This myosin molecule is about 150 nm long, has a molecular weight of about 480 kdaltons, a globular head composed of two subunits, a long tail, and an intermediate neck region that allows the head some flexibility (Figure 9.4a). The molecule resembles two golf clubs having their long handles twisted together. The molecules are oriented with their tails pointing toward the M line, and with the long tail of a myosin molecule bound to the tails of other myosin molecules to form a thick filament. The globular heads of the myosin molecules protrude from the body of the thick filament, successive heads being displaced about 14 nm longitudinally and rotated 60° around the filament, in accordance with the hexagonal arrangement in Figure 9.3. The orientation of the myosin molecules, and their shape, leaves about 100 nm of the thick filament, on either side of the M line, devoid of any heads, which makes this region somewhat lighter than the rest of the A band.

A thin filament is about 5–6 nm in diameter, 1 μm long, and consists mainly of a strand of F-actin (F for fibrous), as illustrated in Figure 9.4b. A strand

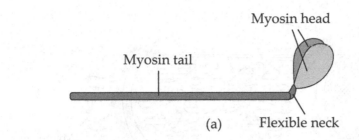

Myosin head

Myosin tail

(a)   Flexible neck

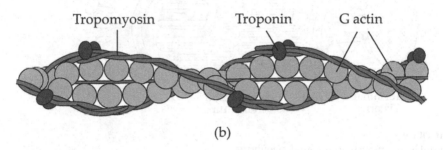

Tropomyosin   Troponin   G actin

(b)

FIGURE 9.4
(a) Thick filament; (b) thin filament.

of F-actin is a polymer composed of two twisted rows of 300–400 individual molecules of G-actin (G for globular), each molecule having a diameter of about 5 nm and a molecular weight of about 42 kdaltons. The F-actin strand is held together by a thread of nebulin that extends along the F-actin between the two rows of G-actin molecules. Each G-actin molecule has an **active site** that can bind to the head of a myosin molecule but is prevented from doing so under resting conditions, when there is no contraction, by tropomyosin molecules that cover the active sites. A tropomyosin molecule is a double strand that joins head-to-tail with other tropomyosin molecules to form a twisted strand over the length of the F-actin. Each tropomyosin molecule covers seven active sites and is bound to a troponin molecule. A troponin molecule is composed of three largely globular subunits: (i) troponin T (tropomyosin-binding troponin) that forms a troponin-tropomyosin complex, (ii) troponin C (Ca^{2+} binding troponin) that plays a major role in contraction, as explained later, and (iii) troponin I (inhibitory troponin) that is attached to G-actin in the absence of Ca^{2+} and holds the tropomyosin in a position that blocks myosin from reaching the active sites on G-actin. When Ca^{2+} bind to troponin C, troponin I detaches from the actin, thereby allowing the tropomyosin to move over the surface of the thin filament.

Myofibrils are surrounded by two distinctive structures (Figure 9.5). One is a system of transverse tubules, or **T tubules**, that encircle each myofibril at regular intervals. In mammals, the T tubules are found near the junction of the A bands and the I bands. The T tubules have a lumen, 2–5 nm across, that is open to the external medium at the surface of the fiber. They are therefore

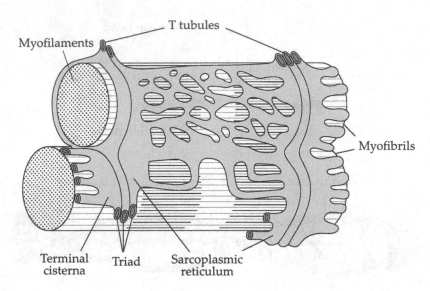

FIGURE 9.5
Tubular structures that surround myofibrils.

filled with extracellular fluid, and their membrane lining is a continuation of the sarcolemma, thereby allowing the muscle AP to propagate deep into the muscle fiber. Because of their extensive nature, the T tubules increase the effective capacitance per unit area of the muscle fiber to more than 5 $\mu F/cm^2$, compared to about 0.8 $\mu F/cm^2$ for the membranes of other cells.

The other system is the **sarcoplasmic reticulum** (SR), which is analogous to the endoplasmic reticulum of other cells (Section 1.1). The SR consists of a network of tubes or channels that run mostly longitudinally around each myofibril. The SR channels enlarge in the vicinity of the T tubules to form chambers, or **terminal cisternae**, whose membrane is in close apposition to that of the T tubule. The combination of a T tubule and the two terminal cisternae on either side is a **triad**.

Problem 9.1

Select the appropriate word or phrase from among those in parentheses in each of the following sentences. As a muscle contracts,

(a) the Z disks move (closer together, further apart).

(b) the width of the A bands (increases, decreases, stays the same).

(c) the width of the H zones (increases, decreases, stays the same).

(d) the width of the I bands (increases, decreases, stays the same).

ANS.: (a) closer together; (b) stays the same; (c) decreases; (d) decreases.

9.2 Contraction of Skeletal Muscle

9.2.1 Excitation-Contraction Coupling

Excitation-contraction coupling is the sequence of events that occur between the generation of a muscle AP and muscle contraction. As explained below, force is produced when the myosin heads strongly bind to the active sites of the G-actin molecules. However, these molecules also have sites to which the myosin heads can bind weakly by electrostatic forces but without force generation. Thus, there are three states of the thin filament with respect to the heads of the myosin molecule: (i) when there is no binding between the G-actin molecules and the myosin heads, (ii) when there is a weak electrostatic binding between the G-actin molecules and the myosin heads, and (iii) when the binding between the G-actin molecules and the myosin heads is strong and force can be generated. The tropomyosin molecule is not tightly held in position but dynamically rolls around its axis and slides back and forth over the actin molecules. In the absence of Ca^{2+}, state (i) predominates, so that active sites on the G actin molecules are effectively covered by tropomyosin. As the concentration of Ca^{2+} increases, there is a progression to the strong-binding, force-generating state (iii), as explained next.

Following the arrival of a nerve AP, a muscle AP is initiated in regions of the muscle membrane adjoining the endplate, as explained in Section 5.2. Excitation-contraction coupling then proceeds as follows:

1. The AP propagates from the endplate region in both directions along the length of the muscle fiber, invading the T tubules along the way and resulting in the depolarization of the T tubule membrane in the triads of the muscle fiber. The velocity of the AP is in the range 2–10 m/s.

2. The depolarization activates $Ca_v1.1$ channels (Section 7.3.2) in the T, tubule membrane, causing some influx of Ca^{2+} from the external fluid into the cytoplasm of the muscle fiber. However, this influx is not very significant in the case of skeletal muscle. Of much greater significance is the mechanical coupling between the receptors of the Ca^{2+} channels of the T, tubule membrane and the receptors of Ca^{2+} channels of the SR membrane. This mechanical coupling is subserved by the close apposition of these two types of membrane in the triad, as mentioned previously. In this manner, activation of the L-type channels opens the Ca^{2+} channels of the SR membrane, leading to a massive influx of Ca^{2+} from the terminal cisternae into the cytoplasm of the muscle fiber. The concentration of Ca^{2+} in the cytosol of the muscle fiber at rest is roughly 50 nM, whereas that in the terminal cisternae is at least 10,000 higher, both as free Ca^{2+} and Ca^{2+} reversibly bound to the protein calsequestrin. The influx of Ca^{2+}

from the SR into the cytoplasm rapidly increases Ca^{2+} concentration locally by at least a few-hundred-fold.

3. Ca^{2+} bind to troponin C, detaching the troponin I from actin. The tropomyosin moves over the G-actin molecules, thereby exposing the strong-binding sites and allowing a process of **cross-bridge cycling** involving the following steps 4–6.

4. The myosin head in the thick filament has ATPase activity, that is, it has enzymatic action that hydrolyzes ATP into ADP and phosphate (P_i) in the presence of Mg^{2+}. This occurs whenever the myosin head is detached from actin, as in state (i) mentioned above. The energy of the hydrolysis "charges" the myosin head, which becomes like a coiled spring primed for action and oriented at about 90° with respect to the body of the thick filament. The myosin weakly binds to the actin, as in state (ii) mentioned above.

5. With the strong-binding actin site exposed in the presence of Ca^{2+} (step 3), the energized myosin head binds to an actin molecule, forming a **cross bridge**, as illustrated diagrammatically in Figure 9.6a. The binding frees P_i first, then ADP, releasing the energy stored in the myosin head. This causes the head, while still bound to the actin

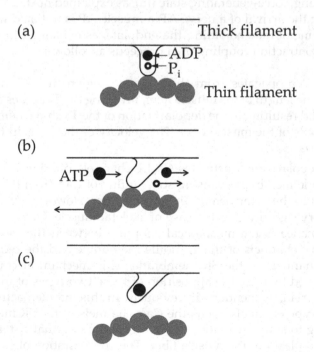

FIGURE 9.6
Cross-bridge recycling; (a) formation of a cross-bridge; (b) swiveling of the myosin head; and (c) myosin head detached.

molecule, to swivel through an angle of 45–50°, thereby moving the actin filament in the direction that increases the overlap between the thick and thin filaments and shortens the sarcomere (Figure 9.6b). If the muscle is prevented from shortening (isometric conditions), force is developed, of about 2 pN/cross bridge, with little or no change in sarcomere length. On the other hand, if the muscle is free to shorten, the myosin head swivels toward the M line, moving the thin filament by about 10 nM. This action is the **power stroke**. To avoid confusion, it must be borne in mind that "muscle contraction" is used in the literature to signify generation of force by the action of the thick filaments on the thin filaments, and does not necessarily imply shortening of the muscle. Thus, a muscle may not change its length while developing force, or it may be compelled to lengthen by an externally applied force, while developing a weaker force that opposes the applied force.

6. ATP, in the presence of Mg^{2+}, binds to the myosin head, causing it to detach from the actin molecule (Figure 9.6c). The binding of ATP to myosin is very rapid and irreversible at the normal ATP concentration of 3–5 mM. The subsequent detachment of the myosin-ATP from the actin is also very rapid. The state now reverts back to step 4 and the myosin head can now bind to another actin molecule further along the thin filament in the direction of the Z disk. This process of cross bridge cycling repeats in the presence of ATP and a Ca^{2+} concentration of a least 1 μM. The state in which Ca^{2+} concentration is high enough to maintain cross bridge cycling, in the presence of ATP, is the **active state**.

7. As soon as Ca^{2+} are released from the terminal cisternae, they begin to be extruded from the cytoplasm, mainly to the tubules of the SR. This is effected by an efficient Ca^{2+}-ATPase pump that transports 2 Ca^{2+} for each ATP molecule hydrolyzed. The concentration of Ca^{2+} in the cytoplasm rapidly drops, so that, typically, the active state does not last for more than a few tens of milliseconds. The Ca^{2+} in the tubules of the SR are eventually transferred to the terminal cisternae for subsequent influx into the cytoplasm.

8. The muscle relaxes when Ca^{2+} in the cytoplasm falls below 1 μM and ATP and Mg^{2+} are available to dissociate the myosin heads from the actin molecules.

The following should be noted:

1. At any instant, the myosin heads in a given fiber are in different phases of the excitation-contraction coupling process. Consequently, the movement of the thin filaments past the thick filaments is smooth rather than jerky. Contraction of a muscle fiber due to the movement

of the thick and thin filaments past one another is the **sliding filament model** of contraction.

2. The myofibrils are attached to the sarcolemma at Z disks and at the ends of the muscle fiber. Hence, when the myofibrils shorten so does the muscle fiber. The contraction-relaxation sequence of a muscle fiber in response to a single AP is a **twitch**. The tension developed is transmitted to the muscle ends via the connective tissue and tendons. However, the tension appearing at the ends of a muscle is only a fraction of the maximum tension that the muscle fibers are capable of developing because of the viscoelastic properties of the muscle and the short duration of the active state, as will be explained in Section 10.2.

3. Both ATP and Mg^{2+} are required for muscle relaxation. A deficiency of Mg^{2+} leaves the myosin heads bound to actin molecules, resulting in muscle cramps and pain. Following death, ATP production ceases. Ca^{2+} flow down their electrochemical potential gradient from the terminal cisternae and the extracellular fluid into the sarcoplasm, where they accumulate in the absence of extrusion by the Ca^{2+}-ATPase pump. The myosin heads bind to actin; but without ATP, the bond is not broken, resulting in muscle rigidity referred to as **rigor mortis**.

4. The swiveling of the cross bridge is a fast process that may take less than 1 ms.

5. There are at least two important reasons for the elaborate T tubule/triad system. First, it speeds up contraction by reducing the diffusion distance for Ca^{2+}. In the absence of the Ca^{2+} stores in the terminal cisternae, it would take Ca^{2+} a few tens of milliseconds to diffuse from the extracellular fluid to the troponin binding sites of the sarcomeres. The triad system brings the Ca^{2+} stores to within a fraction of a micrometer from the binding sites, thereby reducing the diffusion time to less than a few milliseconds or so. Second, it aids in the synchronization of contraction of all the sarcomeres in a muscle fiber by bringing the AP to the triads, thereby synchronizing the release of Ca^{2+} from the terminal cisternae. If the contraction of the sarcomeres in a muscle fiber is not adequately synchronized, then contracting segments of the myofibrils would stretch non-contracting segments, which reduces the force transmitted to the ends of the fiber. However, because of the relatively slow conduction velocity of the muscle AP, the requirement for synchronization places a limit on the length of the muscle fiber, even in the presence of the T tubules, as will be explained later under parallel vs. pennate muscle in Section 10.4.

6. If a depolarization is applied locally to the muscle membrane, it is found that contraction begins at a depolarization of about −50 mV,

from a resting membrane voltage of about −90 mV. The smallest depolarization that initiates contraction is the **mechanical threshold**, and corresponds to a Ca^{2+} concentration that is just high enough to start cross bridge recycling between the thick and thin filaments. The contraction increases with depolarization but saturates at about −20 mV, when the Ca^{2+} concentration is high enough so that all the cross bridges available for recycling are activated. Increasing the depolarization further, and hence the Ca^{2+} concentration, cannot activate more cross bridges. In a normal muscle AP, the membrane voltage reverses and reaches a peak of about +30 mV. It follows that with the normal muscle AP, the Ca^{2+} concentration in the cytoplasm rises to a value that is well above that required to activate all the cross bridges that are available for a given overlap of thick and thin filaments, and hence for a given sarcomere length. Thus, a muscle fiber is normally either relaxed, in the absence of stimulation, or is fully activated when stimulated by the muscle AP. This is referred to as the **all-or-none principle** for muscle.

9.2.2 ATP Synthesis

As in most cells, ATP is the energy source that is immediately available to fuel cell activity – in this case, muscle contraction. However, the number of ATP molecules required for even moderate muscle activity can be enormous. The contraction of a single thick filament may involve the hydrolysis of a few thousand ATP molecules per second, and a muscle may contain tens of billions of thick filaments. The ATP immediately available can only sustain about 10 twitches or so. More ATP is readily available in the form of creatine phosphate (CP), also known as phosphorylcreatine. Creatine is a small molecule that is produced from three amino acids (L-arginine, glycine, and L-methionine), mainly in the liver and kidneys, and transported by the blood for use by various cells including muscle. Creatine undergoes in muscle the following reversible reaction that is catalyzed by the enzyme creatine phosphokinase (CPK):

$$ATP + creatine \xrightleftharpoons{\quad CPK \quad} ADP + creatine\ phosphate$$

When the concentration of ATP falls and that of ADP rises, the reaction is driven to the left. The ATP that becomes available in this manner is about 6 times what was available as free ATP. Even then, this energy store cannot sustain muscle activity for more than about 10 s or so. For longer activity, other sources of ATP are needed.

In practically all body cells, the main source of ATP is the **citric acid cycle**, also known as the **tricarboxylic acid cycle** or the **Krebs cycle**, which can metabolize all forms of nutrients, that is, carbohydrates, fats, and proteins.

The input to the cycle is from **glycolysis,** and the output feeds **oxidative phosphorylation,** which provides most of the ATP, using oxygen, ADP, and phosphate (Figure 9.7). Both the citric acid cycle and oxidative phosphorylation occur in the mitochondria. Glycolysis is the metabolic pathway that breaks down one glucose molecule into two pyruvate molecules, the ionized form of pyruvic acid, and occurs in the cytoplasm outside the mitochondria. Under **aerobic** conditions, that is in the presence of oxygen, pyruvate feeds into the citric acid cycle, but under **anaerobic** conditions, that is in the absence of oxygen, pyruvate is converted to lactate.

The glucose involved in glycolysis usually comes from the bloodstream, but in muscle cells, in particular, it can come from glycogen. This is an important form of energy storage in the body, second only to fat cells, or adipose tissue. Glycogen is a large, branched polymer of glucose and is synthesized using ATP, mainly in liver and muscle cells. Under aerobic conditions, one glucose molecule produces 6 ATP molecules from glycolysis and 30 ATP molecules in two turns of the citric acid cycle, one turn for each pyruvate molecule, which makes a total of 36 ATP molecules. Under anaerobic conditions, one glucose molecule produces 2 ATP and two lactate molecules. The rate of production of ATP is regulated by the concentrations of ATP and ADP. As the concentration of ATP falls, and that of ADP rises, the citric acid cycle moves at a faster rate. As more pyruvate is used, glycolysis also proceeds at a faster rate.

A muscle at rest produces a surplus of ATP, which goes into building CP and glycogen stores. The use of CP at the start of muscle activity provides the few seconds necessary for the mobilization of the slower processes of oxidative phosphorylation and glycolysis to increase their rates of ATP production to match the rates of ATP hydrolysis in the muscle. In low-level motor activity, such as that involved in the maintenance of posture or slow movement

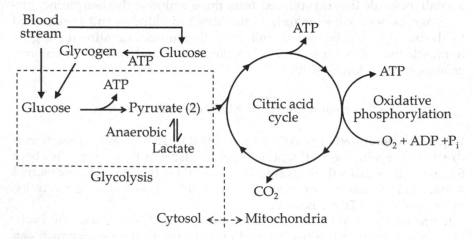

FIGURE 9.7
The citric acid cycle.

or walking, the energy needs are provided by oxidative phosphorylation, as described previously. At moderate levels of muscular activity, most of the ATP used for muscle also derives from oxidative phosphorylation, but fueled through the citric acid cycle by pyruvate derived from glycogen. Beyond about 5–10 minutes, the glycogen reserves are nearly depleted, and blood-borne glucose and fatty acids become the dominant source of energy, with the balance shifting toward fatty acids after about 30 min or so of activity. As the demand for oxygen increases, breathing becomes more rapid and deeper. At high levels of muscular activity, the oxygen supply is insufficient and more ATP is produced by anaerobic glycolysis. Eventually, the muscle fatigues, as described later.

At the end of muscle activity, the demand for oxygen persists so as to produce the ATP required to replenish CP and glycogen stores and in order to oxidize the lactic acid produced. This required oxygen is the **oxygen debt** and is the reason why breathing continues for some time at a faster than normal rate after the activity has stopped.

9.2.3 Heat Production

From conservation of energy, the total rate of energy production from metabolism in muscle is equal to the sum of the rate at which heat is liberated and the rate at which mechanical work is done by the muscle. The latter is the instantaneous power developed by the muscle and is equal to the product of the force generated and the velocity of shortening at any instant. **Work efficiency**, defined as the ratio of mechanical work output to the increase in total metabolic cost for a given activity, as measured by oxygen consumption, depends on the type of activity, the muscle composition, and the individual concerned. In the case of cycling, for example, the efficiency is 25–30%, depending on the individual.

The heat produced in muscle plays an essential role in maintaining body temperature and accounts for about 70% of total body heat production. Several components of muscle heat can be identified. **Resting heat** is the heat produced by the basic metabolic processes of muscle at rest, as in sitting still or lying down, and accounts for about 25% of the total oxygen consumption. **Activation heat** is the heat produced initially when the muscle is excited. It is due to Ca^{2+} release and uptake and can amount to 25–30% of the total energy consumed during muscle activity. **Maintenance heat** is the heat associated with cross-bridge recycling during sustained activity and accounts for the largest fraction of total heat liberated during muscle activity. It depends on the total number of cross bridges involved, which depends in turn on the overlap between the thick and thin filaments. If the muscle is allowed to shorten, there is, in addition, a **shortening heat** associated with the work performed and which depends on the shortening distance. After the activity ends, there is a **recovery heat**, associated with replenishing of the energy stores of CP and glycogen.

Shivering in response to a drop in ambient temperature is involuntary, with rhythmic muscular contractions at a rate of 8 Hz–10 Hz in humans. Because no external work is performed, practically all the energy generated by the muscles appears as heat. This mechanism is effective in increasing body heat production by up to threefold within minutes.

It is of interest to note that in the resting state, muscle temperature may be several degrees less than the body core temperature of about 37°C. For example, the temperature of thigh muscles could be around 34°C at an ambient temperature of 27°C. The temperature of hand muscles could drop below 20°C at an ambient temperature of 15°C or less. On the other hand, the temperature of an active muscle may be well over 37°C. Raising the temperature of a muscle increases the developed force and, more strikingly, the speed, as reflected by the force-velocity relation and, hence, the power. The effect is more marked in slow units compared to faster units. Additionally, a warm-up before exercising improves muscle elasticity and reduces the risk of over-stretching a muscle and causing injury.

9.2.4 Muscle Fatigue

Muscle fatigue refers to the decline in muscle performance with continued activity, as evidenced, for example, by reduced force, decreased shortening velocity, and a slower rate of relaxation. Recovery from fatigue requires a period of rest that depends on the type of muscle and on the duration and intensity of the activity. ATP concentration only slightly decreases in muscle fatigue. Otherwise, the cross bridges will remain in a locked state, which can be very damaging to muscle fibers.

Many factors can contribute to muscle fatigue. In low-intensity, long-duration activity, a major factor is depletion of fuel substrates in the form of a decrease in muscle glycogen and level of blood glucose. Fatigue from high-intensity, short-duration activity is associated with an increase in $[K^+]_o$, $[Na^+]_i$, ADP, P_i, $[H^+]$, lactate, and other metabolic products and a decrease in $[Ca^{2+}]_i$. Lactate and H^+ produced in the muscles diffuse into the blood stream, thereby reducing blood pH. The metabolic products and reduced cytoplasmic pH affect proteins and enzymes, including actin, myosin, and various ionic pumps. The overall consequences are:

1. Reduced amplitude and conduction velocity, as well as broadening, of the muscle AP due to the depolarization resulting from the increase in $[K^+]_o$ and changes in membrane conductance.

2. Disruption of normal excitation-contraction coupling. This occurs because of persistent depolarization of the T tubule membrane and consequent accommodation of AP generation (Section 3.3.2) as a result of accumulation of K^+ in the small extracellular volume inside the T tubules. This may eventually lead to failure of conduction of the AP along the T tubules. Recovery with rest is rapid, as

accumulated K^+ diffuse out of the T tubules and as the Na^+–K^+ pump becomes much more active because of the increase in $[Na^+]_i$.

3. Reduction in the force developed, due to the drop in $[Ca^{2+}]_i$ and a decrease in the Ca^{2+} sensitivity of the myofilaments.

4. Reduction in shortening velocity because of a reduced rate of detachment of cross bridges, due to the buildup of ADP and P_i.

5. Slower relaxation time, mainly because of a reduced rate of re-uptake of Ca^{2+} into the sarcoplasmic reticulum and changes in the time it takes the cross bridges to detach.

6. Behavioral and physiological factors that affect brain centers due, for example, to a reduced pH or a sensation pain, resulting in a feeling of weariness and reduced desire to continue the activity.

Increased muscular activity requires a substantial increase in blood flow to the muscles involved, which is met by increased circulation and by vasodilation arising from central mechanisms as well as local factors such as a decrease in pH and an increase in the concentration of metabolic products. However, as muscle fibers contract and expand sideways, they exert pressure on the blood vessels in the muscle, which can seriously interfere with blood flow. This enhances fatigue in maintained, strong contractions.

The NMJ junction and the motor nerve can normally handle a high rate of AP under voluntary activity. However, at very high rates, exceeding 80 Hz for example, as may occur with electrical stimulation, transmission through the NMJ may be impaired. Moreover, a motor axon may have many tens of terminal branches as it approaches its target muscle fibers. A conduction block may occur at a large number of branch points at high rates of stimulation because of the increased conductive and capacitive membrane load.

The neuromuscular system can adapt to fatigue in some ways:

1. Mean firing rates of motoneurons under conditions of brief maximal voluntary contraction (MVC) are normally in the range of 25–40 Hz. The firing rate drops substantially during prolonged activity at high force levels, as for example, by 50% in a few tens of seconds. Interestingly, this drop is not detrimental to force generation for two reasons. First, the lower rate of firing prevents transmission block that may occur at the NMJ or at the axonal terminal branches, as noted previously. Second, as noted earlier, relaxation time is prolonged during extended activity at high force levels. This makes summation of contractions more effective (Figures 10.3 and 10.4) so that the same maximal force can be produced at a lower firing rate. This is advantageous in another respect because the force vs. frequency curve saturates at high rates of firing. Reducing the firing rate brings the operating point closer to the knee of the curve, where

the force varies with frequency, which allows gradation of muscle force by varying the frequency (Section 10.2.4). The drop in the mean firing rate of motoneurons during prolonged activity results from a reduction in the excitatory input to these neurons, partly due to the inherent decline in the firing rate of muscle spindle afferents during isometric contractions and partly due to the inhibitory effect of small-fiber muscle afferents (Groups III and IV), many of which are nociceptive and would respond to the products of muscle metabolism.

2. Where a long-lasting, sub-maximal torque at a joint can be produced by several synergist muscles (Section 9.3.4), the contributions of the individual muscles are rotated amongst them so that fatigued muscles can rest without a drop in the overall torque.

3. A similar effect is observed at the level of motor unit recruitment (Section 9.3.1), whereby during a prolonged submaximal contraction, motor units are deactivated and replaced by other motor units.

Before ending this section on fatigue, it should be pointed out that potentiation of the force of twitch contraction occurs following a sub-maximal voluntary contraction lasting for a few seconds. The potentiation can persist for minutes and is more marked in fast-twitch motor units or muscles than in their slow-twitch counterparts (Section 9.3.2). This potentiation is believed to arise from the intramuscular kinetics of Ca^{2+} and from intrinsic characteristics of the contractile elements. Because potentiation and fatigue-induced depression of the force of contraction are due to different mechanisms, they can occur simultaneously in a prolonged contraction, the potentiation being predominant early in the contraction, the fatigue later in the contraction.

9.3 Organization of Muscle Fibers

9.3.1 Motor Unit

A **motor unit** consists of a motoneuron and the muscle fibers it innervates in a given skeletal muscle. The motoneuron is referred to as an **alpha-motoneuron** (α-motoneuron), or a lower motoneuron, to distinguish it from other types of motoneurons discussed later (Section 9.4.2.1 and Section 11.2.1.1). Figure 9.8 illustrates two motor units in a skeletal muscle. The muscle fibers of a motor unit are referred to as the **muscle unit**. A muscle fiber is normally innervated, through a neuromuscular junction (Section 5.1), by one – and only one – α-motoneuron, and a given α-motoneuron only innervates the muscle fibers of a single motor unit. All the motoneurons that innervate a single muscle are collectively referred to as the motoneuron pool for that muscle.

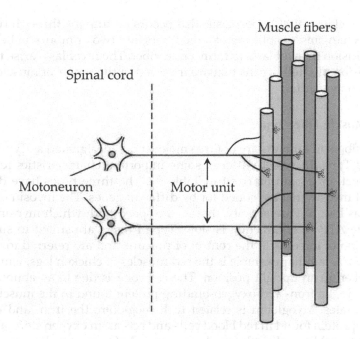

FIGURE 9.8
Motor units.

The motor unit is the basic unit of contraction of skeletal muscle. An AP along the axon of an α-motoneuron generates single APs in all the muscle fibers of the motor unit, producing a single twitch contraction in each of these fibers. The muscle fibers of each motor unit are dispersed among fibers of other motor units generally in the same region of the muscle or through-out the muscle, depending on muscle size and number of fibers. This ensures that the direction of the force exerted does not depend on the number of motor units activated. It is also advantageous to have active and inactive muscle fibers intermingled at low levels of muscle contraction so as not to compromise the blood supply to the muscle which, as noted earlier, can occur at high levels of contraction.

The number of muscle fibers in a motor unit, referred to as the **innervation ratio**, ranges from relatively few – five to ten, for example, in the extraocular muscles – to more than a thousand in the gastrocnemius muscle of the calf. In general, the innervation ratio depends on the finesse of control of muscle action: the smaller the innervation ratio, the more precise is muscle action, and conversely.

The increase in fiber size is called **hypertrophy**. Muscle disuse and loss of innervation, as in paralysis, causes a severe loss of muscle tissue, referred to as **muscle atrophy**. The increase in the number of muscle fibers is **hyperpla-sia**. This has been demonstrated in laboratory animals, such as birds, rats, and cats, as a result of some form of exercise or prolonged electrical stimulation.

There are claims that hyperplasia also occurs in humans through two possible mechanisms: (i) splitting of large fibers into two or more smaller fibers, and (ii) fusion of myoblasts to form a new fiber. The myoblasts arise through cell division of satellite cells that are involved in the repair of muscle tissue, as mentioned earlier.

9.3.2 Muscle Fiber Types

Muscle fibers in humans are of three major types, designated as Type I, Type IIA, and Type IIB, which differ in some important characteristics related to their function, as summarized in Table 9.1. The three types have different forms of myosin that are coded for by different genes. The myosin of Type I fibers has low ATPase activity, that is, the speed with which myosin is able to utilize ATP, so contraction is slow. Type I fibers are suited to sustained production of force, as in the control of posture, and are referred to as slow-twitch (S). A familiar example is the red muscles of chicken legs, which keep the chicken in an upright position. The red color is due to an abundance of **myoglobin**, an iron- and oxygen-binding protein found in the muscle tissue of vertebrates. Myoglobin is related to hemoglobin, the iron- and oxygen-binding protein found in red blood cells and acts as an oxygen store, allowing animals to hold their breath for a longer period of time. It is thus particularly abundant in muscles of diving mammals, such as whales and seals.

TABLE 9.1

Muscle Fiber Types

Property	Type I	Type IIA	Type IIB
Fiber diameter, innervation ratio, and motoneuron size	Small	Intermediate	Large
Myoglobin content, capillary supply, and mitochondrial density	High	Intermediate	Low
Color	Red	Pink	White
Glycolytic capacity	Low	High	High
Oxidative capacity	High	Intermediate	Low
Main fuel for ATP generation	Fats, glucose	Creatine phosphate, glycogen	Creatine phosphate, glycogen
Contraction speed	Slow	Moderately fast (about 5 times Type I)	Fast (about 10 times Type I)
Power produced	Low	Moderate	High
Fatigue resistance	High	Intermediate	Low
Maximum duration of use	Hours	<30 min	<1 min
Other designations	Slow oxidative, (SO), slow twitch (S)	Fast oxidative/glycolytic (FOG), fast-twitch fatigue-resistant (FR)	Fast glycolytic (FG), fast-twitch fatigable (FF)

Type IIB fibers provide rapid acceleration and short-lasting maximal contraction, relying mainly on anaerobic glycolysis to generate the ATP needed for contraction. They are referred to as fast-twitch fatigable (FF). A familiar example is the white muscles of chicken breasts and wings. These muscles are involved in powering the wings, which are used by chickens only sporadically for fast "hopping", as when escaping danger. On the other hand, flight muscles of birds that fly long distances must sustain contraction for a long time without fatigue. Type IIA fibers have intermediate properties and are referred to as fast-twitch fatigue-resistant (FR). It is estimated that the average person has approximately 60% type II muscle fibers and 40% type I.

Most skeletal muscles of the body have a different mix of the three types of muscle fibers, depending on their type of activity. Thus, postural muscles of the leg, back, and neck have a higher proportion of type I fibers. The soleus, a powerful muscle in the back of the calf that is involved in standing and walking, contains 60%–100% slow fibers in humans. On the other hand, muscles of the shoulder and arms have a higher proportion of type IIB fibers because they are usually used only intermittently, and for short periods, to produce a relatively large force for activities such as lifting.

The muscle fibers of a given motor unit are all of the same type. Type I fibers are of the smallest size (Table 9.1), are innervated by the smallest α-motoneurons and their motor units are the slowest and have the smallest number of fibers. Type IIB fibers are of the largest size, are innervated by the largest α-motoneurons, and their motor units are the fastest and have the largest number of fibers. Type IIA fibers and their motor units have correspondingly intermediate values.

It is possible, through appropriate exercise, to induce muscle fibers to change type and size, a feature generally referred to as **muscle plasticity**. Animal experiments have demonstrated that fast muscle fibers can change to slow muscle fibers as a result of a change in the innervation activity from a fast pattern consisting of occasional high-frequency bursts of 30–60 impulses/s to steady, slow firing at 10–20 impulses/s. Conversely, slow muscle fibers can change to fast muscle fibers by an opposite switch in the pattern of innervation activity. The changes in muscle type are believed to be mediated through mechanisms controlled by synaptic activity. In addition, motoneurons and muscle fibers are believed to influence each other through ongoing bidirectional movement of various types of molecules along motor axons (Section 9.3.3).

In humans, endurance type exercises, such as swimming and running, cause a gradual transformation of type IIB fibers to type IIA fibers, with attendant increase in diameter, mitochondria, and blood capillaries. On the other hand, strength type exercises that require the exertion of a large force for a short period, such as weight lifting, increase the size and strength of type IIB fibers through synthesis of new thin and thick filaments. In addition to the enlargement of muscle fibers (hypertrophy), training may also increase the number of muscle fibers (hyperplasia).

It is difficult to achieve maximal muscle force through voluntary contraction. The maximum achievable muscle force can be increased by training, partly because of increased cross-sectional area of the muscle but also because of increased central drive to motoneurons, which increases their discharge rate.

Beyond the age of about 35–40 years muscle tissue begins to be lost with age; muscle fibers become thinner and fewer in number, at least in part because of the death of motoneurons.

9.3.3 Motoneuron-Muscle Fiber Interactions

Motoneurons and muscle fibers exert far-reaching trophic influences over one another. If a peripheral nerve is cut not too close to the cell body, the distal part of a motor axon will degenerate because of its separation from the cell body, but the axon will regrow. The motoneuron whose innervation of muscle fibers has been interrupted will undergo some alterations, such as changes in the rough endoplasmic reticulum; changes in dendritic length, arborization patterns, and synaptic organization; decrease in size and speed of Ia epsps; reduction in voltage threshold for AP generation; and increase in input resistance, membrane resistivity, and time constant. These changes are mostly reversed after reinnervation and are believed to be due to the loss of some trophic factors normally brought, through retrograde transport, from muscle fibers to the motoneuron. If the axon is cut close to the cell body, the motoneuron will most likely die.

The denervated muscle fiber undergoes some drastic alterations, apparently due to the appearance of new types of protein as a result of altered DNA transcription. As the distal part of the axon degenerates, the presynaptic part of the NMJ breaks up. On the postsynaptic side of the NMJ, the folds disappear and the endplate concentration of AChE drops. The muscle membrane becomes less depolarized, mainly because of reduced activity of the Na^+–K^+ pump. The membrane permeabilities to K^+ and Cl^- decrease and the membrane resistivity increases. The membrane capacitance decreases, mainly due to reduced muscle fiber diameter and changes in the surface area of the sarcoplasmic-T tubule system relative to fiber volume. Newly synthesized, fast turnover type of AChRs (Section 5.4.1) are inserted all over the fiber, starting from the endplate region and proceeding to the ends of the fiber. The membrane AP becomes wider and of reduced rate of rise, positive afterpotential, and conduction velocity, mainly because of the insertion of a different type of voltage-gated Na^+ channels that are resistant to tetrodotoxin. About a week after denervation, **fibrillation potentials** can be recorded in an electromyogram (EMG, Spotlight on Techniques 9A) as spontaneous discharges in single muscle fibers of about 10–100 µV amplitude and frequency in the range 0.5–10 Hz. They do not produce any visible contractions under the skin and are believed to occur because of alterations in electrophysiological properties of the muscle membrane, such as increased excitability and changes in permeabilities to various ions. There is a rapid loss of contractile and other elements of muscle fibers leading to a drop in the maximum tetanic force. The atrophied muscle fibers may eventually die if not reinnervated. However, all the changes in denervated muscle fibers are reversible after reinnervation.

The stump of the cut nerve regrows at a rate of 2–3 mm/day, guided along its path by "landmarks" such as nerve sheaths containing surviving Schwann cells. The motor axons, in contrast to sensory axons, will grow toward muscles, but not necessarily the same muscle fibers they innervated before the nerve was cut. Upon reaching muscle fibers, the growing axon terminals recognize the sites of the old endplate, because of the existence of some marker molecules, and will reestablish the NMJ at the old site. Typically, a given endplate may be contacted by several axons during reinnervation, but only one of these connections will eventually remain. After reinnervation, motoneurons and muscle fibers become matched in terms of slow or fast type, mainly by muscle fibers changing type, if necessary, to match that of the newly innervating neuron.

If some of the neurons innervating a muscle are destroyed, as happens for example in poliomyelitis, and some muscle fibers lose their innervation as a consequence, intact axons will sprout collaterals that will innervate denervated muscle fibers at the locations of the old endplates. If necessary, the newly innervated fibers will also change type, fast or slow, to match that of the newly innervating motoneuron. As a result of this collateral innervation, motor unit size may be multiplied by a factor of 5 or more, resulting in "giant" motor units.

9.3.4 Muscle Action

Before proceeding further with our discussion on skeletal muscle, it is useful to define some terms that are commonly used to describe muscle action. Muscle action can result in a variety of movements, common types of which are the following:

1. **Flexion** is a bending movement at a joint that decreases the angle between the body parts attached to the joint. The muscles involved are **flexors**. For example, the biceps brachii muscle is an elbow flexor that allows flexion at the elbow, as when lifting a weight held by the hand, which decreases the angle between the upper arm and the forearm. Flexion of the hip is moving the thigh forward; similarly, flexion of the shoulder is moving the arm forward. Flexion of the neck is bending the head forward; note that these movements occur in a sagittal plane.

2. **Extension** is the opposite of flexion, that is, a straightening movement at a joint that increases the angle between the body parts attached to the joint. The muscles involved are **extensors**. For example, the

triceps brachii muscle is an elbow extensor that allows extension at the elbow, which increases the angle between the upper arm and the forearm.

3. **Abduction** is a movement that draws a limb away from the midline of the body, or away from the centerline of the hand or foot in the case of the fingers or toes, respectively. The muscles involved are **abductors**. For example, the deltoid muscle at the shoulder is an arm abductor that raises the arm laterally, on the sides of the body. Abduction of the fingers is spreading them out by increasing the angle between adjacent fingers.

4. **Adduction** is the opposite of abduction, that is, a movement that pulls a limb toward the midline of the body or closes together the fingers or the toes in the case of the hand or foot, respectively. The muscles involved are **adductors**. For example, the latissimus dorsi muscle at the side of the chest is an arm adductor that lowers the arm laterally, on the sides of the body.

5. **Dorsiflexion** is the upward movement of the foot at the ankle joint, which decreases the angle between the top of the foot, or dorsum, and the leg. The opposite movement is **plantarflexion**, which decreases the angle between the sole of the foot (adjective, plantar) and the line of the leg.

6. **Pronation** is the counterclockwise rotation of the right forearm, or the clockwise rotation of the left forearm, when looking from the elbow toward the wrist. It is also the rotation of the ankle that makes the sole face laterally from a downward position. The opposite movement is **supination**.

7. **Internal rotation,** or **medial rotation**, is rotation toward the center of the body. For example, if the elbow is set at 90°, internal rotation of the shoulder brings the hand closer to the chest. **External rotation** or **lateral rotation** is rotation away from the center of the body.

Muscles can play the following roles:

1. **Agonist**, or **prime mover**, is a muscle that is the main muscle, or member of a group of muscles, responsible for a particular movement. For example, the quadriceps femoris muscle of the thigh is a prime mover in extension at the knee joint.

2. **Antagonist** is a muscle that opposes the movement of an agonist. Since muscles can only pull when they contract, and not push, agonists and antagonists have to act at any joint to move it in opposite directions. Thus, in the flexion at a joint, the agonists are the flexors and the antagonists are the extensors. On the other hand, in the extension of a joint, the agonists are the extensors and the antagonists

are the flexors. Moreover, in the movement of a limb, whether in flexion or extension, sets of muscles at different joints have to act together so as to move the whole limb in the same direction. The sets of muscles that act together in a given movement are called **synergists**. For example, the biceps femoris, the semitendinosus, and the semimembranosus muscles are synergists in knee flexion. In any movement, the nervous system has to not only excite the synergist muscles, but it also must relax the antagonist muscles at the same time.

3. **Fixator** is a muscle that stabilizes one part of the body during movement of another part so as to prevent any unnecessary movement. Fixators are usually required because when a muscle contracts it contracts at both ends. Hence, a desired movement at only one end of a muscle requires stabilization at the other end. For example, during elbow flexion, fixator muscles prevent unwanted movement of the shoulder and wrist.

It should be borne in mind that muscles that extend between two joints will cause flexion at one joint and extension at the other joint when they contract. For example, the quadriceps femoris flexes the thigh at the hip and extends the leg at the knee.

9.3.5 Muscle Architecture

Skeletal muscle architecture refers to the arrangement of fibers in a muscle and has an important bearing on muscle function. Whereas the muscle fibers in fascicles (Figure 9.1) are parallel to one another, the fascicles themselves can have different orientations relative to one another and to tendons, which lie on the line of action of the muscles. There are basically four different muscle architectures:

1. **Parallel muscles**, in which the fascicles run parallel to the line of action of the muscle and generally extend from one end of the muscle to the other (see Section 10.4). A good example is the sartorius muscle, a ribbon-shaped muscle, of about 40 cm length in humans, that runs obliquely along the thigh and is involved in flexion of the knee as well as flexion, abduction, and lateral rotation of the hip. Parallel muscles could also be spindle shaped, or fusiform, as in the biceps brachii muscle of the upper arm.

2. **Convergent muscles**, in which the fascicles extend over a fairly wide area at one end of the muscle and converge to a common attachment site at the other end, as in the pectoralis muscles of the upper chest. The distinguishing feature of a convergent muscle is that it can pull in different directions, depending on the parts of the muscle that are activated.

3. **Pennate muscles,** in which the fascicles are oblique to the force-generating axis of the muscle as a whole. In **unipennate** muscles, the muscle fibers are oriented at a single angle relative to the force-generating axis, this angle being generally between 0°–30°. An example is the extensor digitorum longus muscle of the leg, which extends the small toes and dorsiflexes the foot. Typically, the fascicles extend from an aponeurosis on one side to a tendon on the other side, as illustrated diagrammatically in Figure 9.9a. In **bipennate** muscle, the fascicles converge toward a central tendon from both sides (Figure 9.9b), as in the rectus femoris, a large muscle in the quadriceps group of muscles that extend the knee. In **multipennate** muscle, the fascicles are oriented at several angles relative to the axis of force generation (Figure 9.9c), as in the deltoid muscle that controls shoulder movement. The important characteristics of pennation are discussed in Section 10.4.

4. **Sphincter muscles,** which are circular muscles that surround an opening or recess and perform some controlling function upon contracting. An example is the orbicularis oculi muscle around the eye, which closes the eyelids and whose contraction can be involuntary, as in sleeping and blinking.

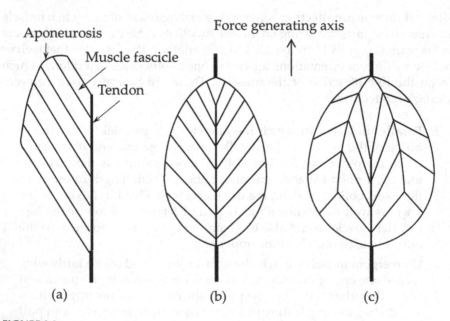

FIGURE 9.9
(a) Pennate muscle. (b) bipennate muscle; and (c) multipennate muscle.

9.4 Muscle Receptors

There are three types of receptors in skeletal muscle, namely: **Golgi tendon organs (GTOs)**, **primary muscle spindle receptors**, and **secondary muscle spindle receptors**. The structure and properties of these receptors are considered in the present section, and their role in reflexes is discussed in Section 11.3.

Muscle also contains other types of receptors that are not believed to play a primary role in the control of movement. These include paciniform corpuscles in the form of rapidly adapting lamellar nerve endings located near the junction between muscle and tendon. They respond to muscle stretch, light pressure, and are sensitive to high-frequency vibrations of about 100 Hz. There are also free sensory nerve endings distributed throughout the muscle and which respond to strong mechanical stimuli, such as pinching, as well as to some chemicals.

All peripheral sensory fibers below neck level are part of sensory neurons whose cell bodies are located in dorsal root ganglia close to the spinal cord, as mentioned earlier (Section 7.1). Figure 9.10 diagrammatically illustrates such a neuron. The neuron has no dendrites, only a single process that emanates from the cell body and divides into two main branches, one that enters the spinal cord via the dorsal roots and one that extends to the periphery and terminates in sensory receptors, as those of skeletal muscle, or in specialized sensory nerve endings. The sensory, or afferent, AP propagates from the sensory termination, past the branch point, and into the spinal cord. The AP in the part between the branch point and the spinal cord propagates away from the cell body of the sensory neuron, as in an axon. However, the AP in the part between the sensory termination and the branch point propagates toward the cell body of the sensory neuron. Hence, this part is properly referred to as a "nerve fiber" and not an axon.

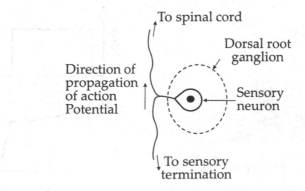

FIGURE 9.10
Sensory neuron.

9.4.1 Golgi Tendon Organ

A skeletal muscle typically has 40–60 golgi tendon organs (GTOs), most of which are located at the junction between the skeletal muscle and its tendon, although about 10% of GTOs may be found in the body of the tendon. A GTO consists of a capsule about 1 mm long and 0.1 mm in diameter, made up of collagen. Inside this capsule, branches of a sensory fiber intertwine with the collagen strands of the tendon (Figure 9.11a). Each GTO is typically in series with 3–25 muscle fibers belonging to different motor units. A motor unit contributes one or two muscle fibers that insert into tendon strands in a GTO, and only rarely are more than 25 muscle fibers associated with a GTO. When the muscle contracts, the nerve terminals of the sensory fibers are stretched, opening stretch-sensitive cation channels. The resulting depolarization generates APs, generally in more than one site, in the terminals of the sensory nerve fiber. The APs from the various terminal branches propagate along the parent sensory fiber to the spinal cord. The firing rate increases nonlinearly with the force developed by the muscle, typically reaching about 100 Hz for a force of 20 N. A step change in muscle force produces a step change in firing rate, with a relatively small, sagging dynamic component that depends on the rate of change of tension (Figure 9.11b). Not all motor units in a muscle activate GTOs lying at the muscle-tendon junction. However, GTOs further along the tendon would be in series with most of the muscle and would respond more to the mean force developed by the muscle rather than the force in a restricted number of motor units.

GTOs have a much higher threshold to a force due to muscle stretch compared to active contraction, due to the following factors:

1. When a muscle is stretched, the force is divided among all the large number of muscle fibers, so the fraction of the force that activates the muscle fibers associated with a GTO is small; but when a motor

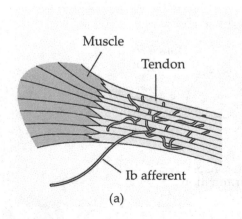

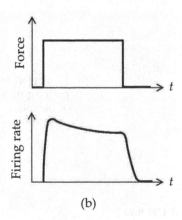

(a)   (b)

FIGURE 9.11
(a) Golgi tendon organ; (b) response to muscle force.

unit contracts, it directly activates the GTOs connected to its muscle fibers, although the force appearing at the muscle ends is small.

2. In pennate muscles, only the component of force that is in parallel with the muscle fibers is effective in activating the GTOs.

3. When a muscle is stretched, only a small fraction of the stretch appears across the relatively stiff tendon, and most of the stretch appears in the rest of the muscle, which is much more compliant. For a stretch of a given length, muscles having longer bodies will generate less passive tension.

Sensory fibers that innervate GTOs are referred to as Ib fibers according to the classification of sensory nerve fibers. They are amongst the largest myelinated nerve fibers in the body, having a diameter of 11–19 μm and an AP conduction velocity of 65–110 m/s.

9.4.2 Muscle Spindle

9.4.2.1 Structure and General Properties

The muscle spindle is a fusiform capsule that is 4–10 mm long and 80–200 μm in diameter. It is a complex sensory organ that is richly innervated with both sensory and motor nerve fibers, accounting for more than two thirds of the myelinated fibers that innervate muscle. Human muscles contain 35–500 muscle spindles per muscle, with few exceptions, such as the muscles of the middle ear, which do not have any muscle spindles. The number of muscle spindles relative to the muscle mass is high in the small muscles of the neck, the muscles of the hand, and in the extraocular muscles of the eyes, reflecting, respectively, the need for fine movements in positioning of the head, manipulating objects, and in moving the eyes. Muscle spindles are generally scattered widely in the muscle body and are attached at both ends to the muscle connective tissue so as to be in parallel with the muscle fibers.

The anatomy and physiology of the muscle spindle has been most extensively studied in the cat. It is believed, nevertheless, that all mammalian muscle spindles share the same essential features although they differ in some of the details. Moreover, muscle spindles differ somewhat between various muscles of the same species, particularly in humans. This section, therefore, highlights the salient features of the muscle spindle without dwelling too much on detailed variations.

Muscle spindles contain three types of muscle fibers, described as: **nuclear bag$_1$ fibers**, **nuclear bag$_2$ fibers**, and **nuclear chain fibers**, collectively referred to as **intrafusal fibers**. This is to distinguish them from the muscle fibers proper of the muscle itself, which are referred to in the same context as **extrafusal fibers**. Nuclear bag$_1$ fibers are also termed **dynamic bag$_1$ fibers (DB$_1$)** and nuclear bag$_2$ fibers are termed **static bag$_2$ fibers (SB$_2$)** for reasons that will become apparent shortly. The intrafusal fibers are illustrated in

Figure 9.12, where the nuclear chain fibers are denoted by Ch. Some of the nuclear chain fibers are longer than the other chain fibers.

As their name implies, nuclear bag fibers are so called because they have 100 or more nuclei concentrated in a bulge in the equatorial, or middle, region of the fiber. Nuclear chain fibers, on the other hand, have a row of nuclei in this region. Nuclear chain fibers are generally thinner than nuclear bag fibers, about half as long, and 2–3 times as numerous. In a human muscle spindle, there usually are 3 or 4 nuclear bag fibers of both types and up to 10 nuclear chain fibers.

Intrafusal fibers are innervated by two groups of sensory fibers denoted as Ia and II. The nerve endings of group Ia fibers spiral around the equatorial regions of all intrafusal fibers (Figure 9.12), forming **annulospiral endings** that constitute the primary muscle spindle receptors. The Ia fibers are therefore referred to as **primary afferents**. The nerve endings of group II fibers in human muscle spindles are spray-like and located to one side of the equatorial region, toward the pole, or end, of the fiber. These terminations are generally found on all intrafusal fibers and constitute the secondary muscle spindle receptors. Hence, group II fibers are referred to as **secondary afferents**. Every muscle spindle has a single primary afferent and, typically, a single secondary afferent. However, complex muscle spindles may have 4 or 5 secondary afferents, whereas simple spindles may not have any secondary afferents. There does not appear to be essential, fine-structural differences between primary and secondary endings. Both are in close apposition to the membrane of the intrafusal fiber, the separation being in the range of 15–35 nm, and contain mitochondria, vesicles, granules, and tubules. Some endings, particularly those of primary afferents, are deeply invaginated into the fiber surface.

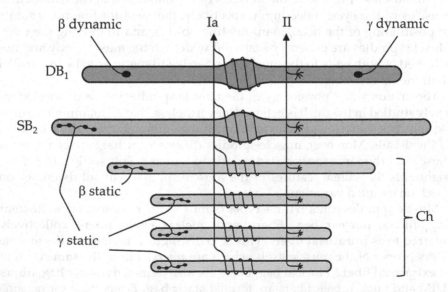

FIGURE 9.12
Muscle fibers of muscle spindle and their innervation.

Group Ia fibers are the largest myelinated fibers in the body, having a diameter of 13–20 µm and an AP conduction velocity of 75–120 m/s. In contrast, group II fibers have a diameter of 6–12 µm and an AP conduction velocity of 30–65 m/s.

The intrafusal fibers also have **fusimotor innervation**, that is, they are innervated by axons of motoneurons. These are of two types: γ, and β, each of which is also of two types, dynamic and static because of their effects on the muscle spindle response, as will be described later. Compared to α-motoneurons, whose axons are of 9–17 µm diameter and conduct APs at 55–105 m/s, γ-motoneurons are the smallest motoneurons, having axons of 2–8 µm and a conduction velocity of 10–45 m/s. β-motoneurons are of intermediate size, having axons of 6–12 µm and a conduction velocity of 35–70 m/s. Whereas γ axons terminate exclusively on intrafusal fibers, β axons terminate on both intrafusal and extrafusal fibers. Although there is considerable overlap, dynamic type axons tend to fall in the lower part of the ranges of diameters and conduction velocities, whereas static type axons tend to fall in the upper part.

All DB_1 fibers are innervated by dynamic γ axons, and some are innervated by dynamic β axons as well (Figure 9.12). Static β axons innervate long chain fibers. SB_2 and Ch fibers are innervated by static γ axons, either selectively, whereby a γ axon terminates on only one fiber, or collectively, with a γ axon terminating on more than one fiber. Occasionally, a static γ axon may terminate on a DB_1 fiber and dynamic γ axon may terminate on a Ch fiber. All fusimotor terminations are located toward the pole regions of intrafusal fibers. The dynamic fibers end in plate-like terminations, whereas the static fibers end in what are referred to as trail endings because lengths of unmyelinated axon often link several of the plate terminations together. Spindles also have some sympathetic and parasympathetic innervation.

The plate endings of all fusimotor fibers are essentially neuromuscular junctions that differ somewhat in ultrastructure between the various types of fibers, particularly in the degree of invagination into the surface of the intrafusal fiber and in the extent of postsynaptic folding. Dynamic axons always elicit local, nonpropagating voltages in DB_1 fibers. Static axons almost always elicit local, nonpropagating voltages in SB_2 fibers and occasionally APs. In Ch fibers, static axons mostly elicit APs. Nonpropagating potentials produce localized contractions. In the case of APs, both the APs and the contractions they produce are confined to the pole that is stimulated.

There are some striking differences between intrafusal fibers. The myosin differs in form between extrafusal and intrafusal fibers and between the intrafusal fibers themselves. Intrafusal fibers also differ in their glycogen content and ATPase activity. Not surprisingly, nuclear bag regions are practically devoid of myofibrils because the large number of closely packed nuclei does not leave room for myofibrils. In terms of speed of contraction, Ch fibers resemble fast extrafusal fibers; they exhibit twitch responses that fuse into a maximal tetanic contraction at stimulation frequencies of 150–200 Hz, applied for about 0.4 s. On the other hand, DB_1 fibers resemble very slow or tonic muscle that does not produce twitches but responds with a slowly developing contraction that increases with frequency, reaching a maximum at frequencies of 75–100 Hz, applied for about 1 s. There is also evidence that DB_1 fibers are activated by stretch in the absence of any fusimotor stimulation. SB_2 fibers are faster than DB_1 fibers but not as fast as Ch fibers. They reach maximum tetanic contraction at a frequency of about 100 Hz, applied for about 0.6 s.

9.4.2.2 Sensory Responses

Like GTOs, the adequate stimulus for both primary and secondary nerve endings is stretch, which activates stretch-sensitive cation channels and results in the generation of APs in the sensory endings. However, whereas the response of the secondary afferents depends essentially on the amount of stretch, the response of the primary afferents depends on both the amount of stretch and its rate of change, as illustrated diagrammatically in Figure 9.13 for a trapezoidal stretch lasting for several seconds.

During the positive ramp stretch, the instantaneous rate of firing in primary afferents rises rapidly at first and then more slowly to a peak value at the end of the ramp. For a ramp height of 2 mm, the peak rate is about 70/s.

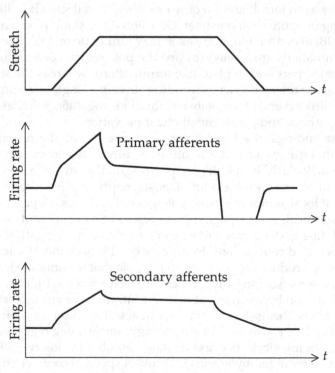

FIGURE 9.13
Responses of primary and secondary afferents to stretch.

The instantaneous rate of firing falls rapidly at the beginning of the constant phase of the stretch, followed by a slow decay. A **dynamic index** can be derived as the difference between the peak instantaneous rate at the end of the ramp and its value 0.5 s later. When the stretch is released by the negative ramp, firing ceases altogether, then slowly returns, before the end of the negative ramp, to the initial firing rate before the stretch is applied.

The response of the secondary afferents is different. Compared to the response of the primary afferents, the initial, fast rise is much less noticeable, the firing rate rising almost linearly to a much smaller peak value, so that the dynamic index is much smaller. Firing does not cease during the negative ramp. The dynamic index of the primary afferents increases considerably with the velocity of stretch, whereas that of the secondary afferents shows much smaller variation. Moreover, the resting rate of firing of secondary afferents is more regular than that of the primary afferents, and the latter are much more sensitive to very small stretches.

A purely static response to stretch can be very simply explained assuming: (i) a fiber of uniform structure, and (ii) a stretch-sensitive sensory response due to stretch over part of the fiber. This is illustrated in Figure 9.14, where an idealized Ch fiber is modeled by viscoelastic elements consisting of paralleled

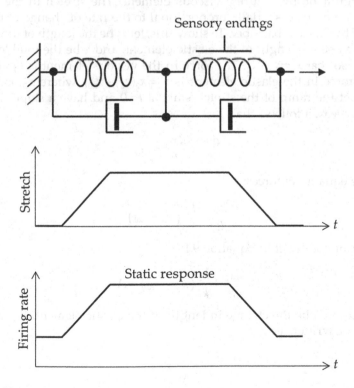

FIGURE 9.14
Modelling of the static response of muscle spindle.

springs and dashpots, with sensory endings over part of the fiber. The viscous damping, represented by the dashpot, is due to the decrease of the force developed by the sliding filaments with the velocity of shortening, as explained in Section 10.3.2. The dashpots in Figure 9.14 produce an opposing force that is proportional to velocity of stretch and will have no effect under static conditions. The spring allows for the elastic properties of the fiber, mainly due to the connective tissue and, to a lesser extent, the cell membrane of the fiber (Section 10.3.1). When the fiber is stretched, the sensory endings are stretched in direct proportion to the total stretch, assuming perfect uniformity of structure. If the firing rate of the sensory endings is directly proportional to stretch, the firing rate under static conditions will have the same waveform as the applied stretch, as shown. If the transduction is nonlinear, then the rising and falling parts of the waveform will also be nonlinear, with a constant middle part of the same onset and duration as in the applied stretch.

Figure 9.15a illustrates a purely dynamic response that arises from an idealized nuclear bag fiber which is devoid of any contractile elements in the equatorial region. Hence, this region is represented by a purely elastic element having a sensory ending, whereas the contractile elements in the rest of the fiber are represented, for simplicity, and without invalidating the essence of the argument, by a purely viscous element. The stretch in the equatorial region will be essentially proportional to the rate of change of the total length of the nuclear bag fiber. To show this, let x_1 be the length of the viscous element, x_2 be the length of the elastic element, and x be the total length of the fiber, so that $x = x_1 + x_2$. The force in the viscous element is $F_1 = k_1 dx_1/dt$, and the force in the elastic element is $F_2 = k_2(x_2 - x_{20})$, where x_{20} is the rest length. Let the ramp of the stretch start at $t = 0$ and have a slope $A = dx/dt$. Since $x = x_1 + x_2$, it follows that:

$$\frac{dx_1}{dt} + \frac{dx_2}{dt} = A \tag{9.1}$$

From the equality of forces:

$$\frac{dx_1}{dt} = \frac{k_2}{k_1}(x_2 - x_{20}) \tag{9.2}$$

Substituting for dx_1/dt in Equation 9.1,

$$\frac{dx_2}{dt} + \frac{k_2}{k_1}(x_2 - x_{20}) = A \tag{9.3}$$

Let $x_2' = x_2 - x_{20}$ be the change in length of the elastic element. Equation 9.3 can then be written as:

$$\frac{dx_2'}{dt} + \frac{k_2}{k_1}x_2' = A \tag{9.4}$$

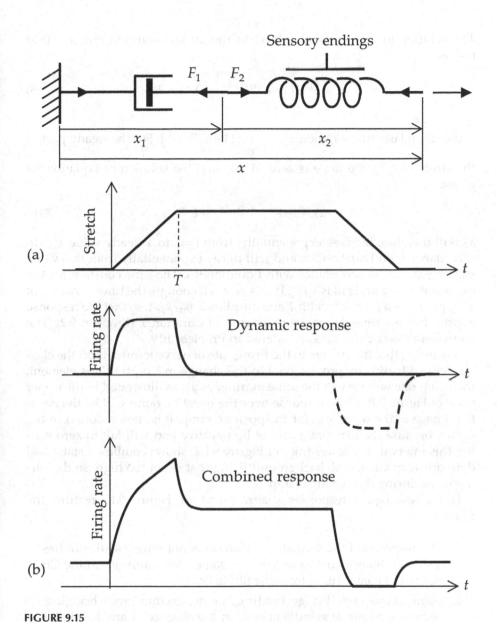

FIGURE 9.15
(a) Modelling of the dynamic response of a nuclear bag fiber; (b) dynamic response (upper trace) and combined static and dynamic responses (lower trace).

The solution to this equation subject to the initial condition that $x_2' = 0$ at $t = 0$ is:

$$x_2' = \frac{k_1}{k_2} A\left(1 - e^{-(k_2/k_1)t}\right) \quad 0 \leq t \leq T \tag{9.5}$$

If the ramp lasts till $t = T$, then $x_{2T}' = \frac{k_1}{k_2} A\left(1 - e^{-(k_2/k_1)T}\right)$. For the steady part of the stretch $(t \geq T)$, the slope is zero $(A = 0)$, and the solution of Equation 9.4 gives:

$$x_2' = x_{2T}' e^{-(k_2/k_1)(t-T)} \quad t \geq T \tag{9.6}$$

x_2' will therefore increase exponentially, from $t = 0$, to a steady value x_{2T}', in accordance with Equation 9.5, and will decay exponentially from this value starting at $t = T$, in accordance with Equation 9.6. The time constant for the exponential rise and fall is k_1/k_2. If this is small enough, the time variation of x_2' approaches a pulse of width T and amplitude $(k_1/k_2)A$, so that the response approaches the time derivative of a ramp of duration T (Problem 9.2). The equivalent electric circuit is considered in Problem 10.9.

Assuming that the change in the firing rate of nerve terminals on the elastic element is directly proportional to the change in length of this element, the firing rate will vary in the same manner as x_2', as illustrated in the upper trace of Figure 9.15b. The response over the negative ramp will be the negative image of the waveform for the positive ramp. It is shown dotted in the figure, because the firing rate cannot be negative and will fall to zero during this interval. The lower trace of Figure 9.15b shows combined static and dynamic responses, with background firing that is not too high, so that firing stops during the negative ramp.

The physiological responses diagrammed in Figure 9.13 feature the following:

1. The response of the secondary afferents is not purely static but has a small dynamic component due to some nonuniformity in the Ch fibers and contribution from the SB$_2$ fiber.

2. Some stable cross-bridge binding, or noncycling cross bridging, occurs in a muscle spindle at rest, in the absence of any fusimotor drive, which considerably increases the stiffness of the regions containing myofibrils. When a small stretch is applied that does not significantly disrupt these bridges, most of the stretch appears across the more compliant equatorial regions of nuclear bag fibers, which are practically devoid of myofibrils. The larger stretch will cause a high initial rate of firing from the sensory endings at the beginning of the positive ramp. As mentioned earlier, this effect is very

noticeable in the case of the primary afferents and much less notable in the case of the secondary afferents, mainly because the equatorial regions of the Ch fibers have nearly the same number of myofibrils as the polar regions. As a result of this stable cross-bridge binding at rest, the muscle spindle is highly sensitive to small stretches. A larger stretch disrupts the cross bridges and reduces the stiffness of the regions containing myofibrils. The effect of stable cross bridging at rest is sometimes described as **stiction**.

3. During the constant-length phase following the positive ramp, both the primary and secondary afferent responses show some adaptation, or a decrease in the firing rate, a phenomenon described as **creep**. In the case of DB_1 fibers, creep is believed to be due to stretch-activation (Section 10.3.1) causing local shortening and stiffening, during the positive ramp, of the myofibril-containing regions on either side of the equatorial region. After the ramp ends, the fibers in these regions slowly relax and lengthen, which shortens the equatorial region and reduces the firing rate. A similar effect occurs in the chain fibers due to the aforementioned nonuniformity in these fibers manifested by less myofibrils in equatorial regions compared to polar regions.

It should be mentioned before ending this section on muscle sensory responses that skeletal muscle is also innervated by Type III thinly myelinated sensory fibers and Type IV unmyelinated fibers. These fibers are activated by "nociceptive type" stimuli and facilitate "central fatigue", which is manifested by inhibitory influences on central motor drive during exercise. Thus, they play an important role in the susceptibility to fatigue and the capacity for endurance exercise.

Problem 9.2

Argue that if the time constant is large, that is, $k_1/k_2 \gg 1$, then the firing rate is nearly directly proportional to the stretch. On the other hand, if the time constant is small, that is, $k_1/k_2 \ll 1$, the firing rate is nearly directly proportional to the rate of change of stretch.

Problem 9.3

The voltage response of a stretch receptor to a step of stretch L_0 applied at $t=0$ is: $v(t)=L_0(1 - e^{-t/\tau})$. If an additional step of stretch L_0 is applied at $t=T$, determine the response for $t \geq T+\tau$. Check the result by setting $T=0$.

ANS.: $L_0\left(2 - \dfrac{1+e^{-T/\tau}}{e}\right)$

9.4.2.3 Fusimotor Effects

The intrafusal fibers are connected in parallel with the extrafusal fibers, so when the latter are stretched, the former will also be stretched and will fire accordingly. But when the extrafusal fibers contract, the intrafusal fibers will be relieved of stretch and the muscle spindles may go silent, so they will not signal any information about muscle length or speed of contraction. Activation of the fusimotor fibers will cause contraction of the intrafusal muscle fibers, so they will send the appropriate signals under these conditions. The role that this fusimotor drive plays in the control of movement is discussed in Section 13.5.1.

In general, γ or β fusimotor drive causes both contraction and stiffening of the polar regions of intrafusal muscle fibers (Figure 9.12). At constant length of the intrafusal fibers, the contraction elongates the equatorial region. If the fiber is stretched, the effect of stiffening of the polar regions is to make more of the stretch appear across the equatorial region. In both cases, the sensory endings will be stimulated and will increase their firing rate, in accordance with the viscoelastic effects and the various factors discussed previously. Consequently, stimulation of dynamic or static fusimotor axons produces some distinctive effects that can be summarized as follows:

1. Dynamic γ and β axons almost exclusively innervate DB_1 fibers. Hence, activation of these fibers, or **fusidynamic** activation, markedly enhances the dynamic response of the primary afferents, with a relatively small increase in the static response, that is, the firing rate at constant length of intrafusal fibers.

2. Activation of static axons, or **fusistatic** activation, significantly increases the static response of both primary and secondary afferents. However, fusistatic activation has little effect on the dynamic response of primary afferents, or may even depress it.

The following should be noted:

1. At very small stretches of less than 1 μm, or so, the stiffness of all intrafusal fibers is relatively high in the regions containing contractile elements because of stable cross-bridging at rest referred to earlier. The stiffness is decreased at larger stretches because of the breaking of these cross bridges except in DB1 fibers, where the stiffness can increase because of stretch activation (Section 10.3.1) in these fibers.

2. Primary endings are located in equatorial regions, which contain less myofibrils than the rest of the fiber and are almost devoid of myofibrils in the case of nuclear bag fibers. Secondary endings are

located in regions adjacent to the equator, which contain myofibrils. When the myofibril-containing regions are stiff, as in the case of stable cross bridges at small stretches, the secondary endings are less sensitive than the primary endings. But when the myofibril-containing regions are less stiff, as in the case of large stretches, the secondary endings can be more sensitive.

3. The various types of intrafusal fibers are in parallel, which means that activation of some fibers will affect the responses of other fibers. For example, fusistatic activation of SB_2 or Ch fibers tends to shorten, or unload, DB_1 fibers, thereby reducing their dynamic sensitivity. Similarly, the firing rate of secondary endings can be reduced when contraction of SB_2 fibers unloads Ch fibers.

4. A certain degree of overlap in motor innervation exists. For example, a dynamic axon may innervate DB_1 fibers in one muscle spindle but may terminate on SB_2 or Ch fibers in another spindle, so that its overall action on a number of primary afferents is mixed static-dynamic. However, such admixture seemingly occurs in not more than 15–20% of the cases examined.

5. Fusistatic stimulation at relatively high frequencies such as 100 Hz, can result in **driving of the primary afferent response**, whereby the response becomes locked to the stimulus, impulse for impulse. Driving is believed to be due to mechanical oscillation of Ch fibers.

6. Combined fusistatic and fusidynamic stimulation produces an afferent response that is greater than the larger of the individual responses but smaller than their sum, an effect described as **partial occlusion**. It is believed that the primary afferent has two impulse-generating sites, one due to the receptor voltage from nerve endings on the DB_1 fiber and the other due to the receptor voltages from nerve endings on both SB_2 and Ch fibers (Figure 9.12). The primary afferent firing results from whichever of these two sites dominates and suppresses the firing by the other generator. However, the two sites influence one another through electrotonic spread as well as mechanical coupling between intrafusal fibers.

Summary of Main Concepts

- According to the sliding filament model, contraction is due to the movement of the thick and thin filaments past one another because of cross-bridge cycling in the presence of a sufficiently high concentration of Ca^{2+}.

- The state in which Ca^{2+} concentration is high enough to maintain cross bridge cycling, in the presence of ATP, is the active state.
- The contraction-relaxation sequence of a muscle fiber in response to a single AP is a twitch.
- Contraction is sustained first by the readily available ATP, then by the ATP that becomes available from creatine phosphate, and lastly by the ATP made available through the citric acid cycle.
- Heat is produced in a muscle at rest, during the release and uptake of Ca^{2+}, during cross bridge recycling, during shortening, and during replenishing of the energy stores of creatine phosphate and glycogen.
- Muscle fatigue occurs because of the depletion of glycogen and glucose, a decrease of $[Ca^{2+}]_i$, and an increase in $[K^+]_o$, $[Na^+]_i$, ADP, P_i, $[H^+]$, lactate, and other metabolic products.
- A motor unit consists of an α-motoneuron and all the muscle fibers it innervates. It is the basic unit of contraction of skeletal muscle. The muscle fibers of a motor unit are homogenous, dispersed throughout a given muscle, and innervated by a single α-motoneuron.
- The number of muscle fibers in a motor unit is known as the innervation ratio and is reduced in muscles involved in fine, precise action.
- Muscle fibers are divided into three main types that differ in their speed of contraction, power produced, and resistance to fatigue.
- Motoneurons and muscle fibers exert far-reaching trophic influences over one another.
- There are basically four muscle architectures – parallel, convergent, pennate, and sphincter – that differ in the arrangement of the fibers in the muscle.
- Golgi tendon organs respond to the force of muscle contraction and are much less sensitive to force due to muscle stretch.
- Muscle spindles have three main types of intrafusal fibers – nuclear bag$_1$, nuclear bag$_2$, and nuclear chain fibers, with some of the latter type being longer than typical. Muscle spindles have two types of sensory innervation: primary via group Ia afferents and secondary via group II afferents. Secondary afferents respond mainly to the amount of stretch, whereas primary afferents respond mainly to the rate of change of stretch. Muscle spindles also have motor innervation from γ- and β-motoneurons. Each of these innervations is of two types: dynamic and static. Activation of the dynamic motor input markedly enhances the dynamic response of the primary afferents, with a relatively small increase in the static response of the intrafusal fibers, whereas activation of the static motor input significantly increases the static response of both primary and secondary afferents, with effect on the dynamic response of primary afferents.

10

Functional Properties of Muscle

Objective and Overview

This chapter continues the discussion of the preceding chapter on skeletal muscle by considering various aspects of skeletal muscle functioning. The chapter begins with examining the four main types of contraction encountered, followed by the time courses of isometric and isotonic twitch contractions, the summation of contractions in a tetanus, and the means by which muscle force is graded to meet load requirements. The features of the two basic relations of muscle mechanics, namely, length-tension and force-velocity relations, are examined and the basic underlying mechanisms explained in terms of the sliding-filament model of muscle contraction. A basic kinetic model of contraction is presented as well a basic mechanical model consisting of elastic and viscoelastic elements that is used to explain the twitch/tetanus ratio and some aspects of the force-velocity relation. The salient features of pennate muscles compared to parallel muscles are examined.

The chapter ends with a brief discussion of cardiac and smooth muscle, mainly to highlight their similarities and differences compared to skeletal muscle.

Learning Objectives

To understand:

- The types of contractions that a muscle undergoes
- The time course of isometric and isotonic twitch contractions
- The gradation of the force developed by a muscle
- The main features and underlying mechanisms of the length-tension and force-velocity relations

- The basic kinetic and mechanical models of muscle contraction
- The salient features of pennate muscle
- How cardiac and smooth muscle differ from skeletal muscle

10.1 Types of Contraction

Muscle contraction can take one of the following forms, depending on how the length of the muscle changes while force is developed by the muscle:

1. **Isometric** contraction is the development of force by the muscle without a change in length, as when one tries to lift too heavy a load, or when holding or gripping an object without moving it.

2. **Isotonic** contraction is shortening of the muscle while a constant force is developed by the muscle. Strictly isotonic contractions can only be produced in the laboratory. In the body, the eyeball is rotated by extraocular muscles. Since the rotation occurs at practically constant resistance, particularly for small movements, the contraction of the extraocular muscles is isotonic under these conditions.

3. **Concentric** contraction is shortening of the muscle while developing a varying force of contraction. This is the usual case. For example, the force diagram for lifting a weight W by a force F developed by the biceps brachii muscle is illustrated in Figure 10.1, where O represents the elbow joint, segment Oc the upper arm, segment Ob the forearm from the elbow to where the weight W is applied to the hand, and "a" is the point of insertion of the biceps muscle on the radius bone of the forearm. Equating moments about the fulcrum O: $(F\cos\alpha)d_1 = (W\cos\theta)d_2$, where F is the force developed by the muscle, d_1 is the length Oa, and d_2 is the length Ob. As the weight is lifted, the muscle shortens, and α, θ, and F change.

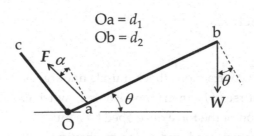

FIGURE 10.1
Diagram of muscular force applied to a load.

4. **Eccentric** contraction is lengthening of the muscle while developing a force of contraction. In Figure 10.1, for example, if W is increased beyond what the arm muscles can support, the weight moves downward, and the biceps brachii will lengthen while still developing force. Eccentric contraction plays important roles in the control of movement, as in braking and control of joint stiffness (Section 13.1.2). The force developed during eccentric contraction is larger than that for an isometric contraction at the same muscle length. Thus, one can set down a heavier weight that one can lift. Eccentric contractions are important for muscle strengthening during athletic training and rehabilitation.

10.2 Twitch Contractions

For the purpose of analyzing muscle mechanics, it is convenient to assume isometric conditions. Isometric twitches are therefore considered in what follows and contrasted with isotonic twitches.

10.2.1 Isometric Twitch

If a single AP is generated in a muscle while the muscle length is kept constant, an isometric twitch occurs. The variation of muscle force with time has the general shape illustrated in Figure 10.2a, where the muscle AP is assumed to occur at $t=0$. The start of the force is delayed with respect to the this AP by a **latent period,** lasting for few milliseconds or so, during which the AP propagates along the T tubules to the triads, Ca^{2+} are released from the terminal cisternae, and bridge recycling is initiated (Section 10.2.1). The

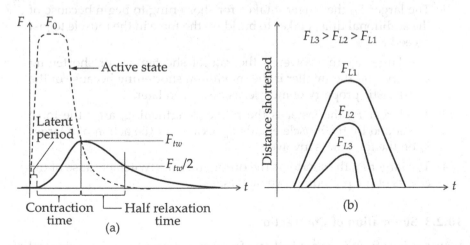

FIGURE 10.2
(a) Isometric twitch; (b) isotonic twitch at various loads.

curve labeled "active state" represents the force that the muscle is capable of developing based on the concentration of Ca^{2+} in the sarcoplasm. This force rises rapidly to a plateau F_0 before decaying to zero due to the action of the Ca^{2+}-ATPase pump. The force at the muscle terminations builds up to a maximum F_{tw} during the active state, then decays more slowly to zero. Because of the uncertainty in determining the total relaxation time due to the slow decay of the relaxation phase, the half-relaxation time is usually measured as the time that the force decays from F_{tw} to $0.5F_{tw}$.

F_{tw} is only a fraction of F_0 because of the viscoelastic property of muscle, as will be explained later. The contraction time is nearly equal to the duration of the plateau of the active state. The contraction and relaxation times of single muscle fibers or single motor units depend on the rate of ATPase activity, and hence on the type of fiber. ATPase activity influences contraction through its effect on cross-bridge recycling, and it influences relaxation time through its effect on the Ca^{2+} extrusion pump. The contraction time of fast fibers is about 10 ms, whereas that of slow fibers can exceed 100 ms. The half-relaxation time is roughly equal to the contraction time. The contraction and relaxation times of a whole muscle would then depend on the proportion of fast and slow fibers in the muscle.

10.2.2 Isotonic Twitch

In contrast, Figure 10.2b illustrates an isotonic twitch contraction, or the shortening of a muscle at a fixed load force F_L following a single muscle AP. Compared to an isometric twitch, the latent period is longer because of the additional time required to build up the force in the muscle to just exceed F_L before the muscle shortens. Increasing F_L has the following effects:

1. The larger F_L, the longer it takes for shortening to begin because of the additional time it takes to build up the force in the muscle to just exceed F_L.

2. The larger F_L, the slower is the rate of shortening, or shortening velocity, and the smaller is the maximum shortening because of the viscoelastic property of muscle, as discussed later.

3. The larger F_L, the faster is the rate of lengthening, after the force developed by the muscle equals F_L because of the action of a larger F_L on the total mass involved.

4. The larger F_L, the shorter is the duration of the twitch because of the effects of the preceding two factors.

10.2.3 Summation of Contractions

If successive stimuli are applied at a frequency larger than the reciprocal of the contraction time, the individual isometric twitches summate nonlinearly and the force builds up to the level F_0 of the active state (Figure 10.3). In

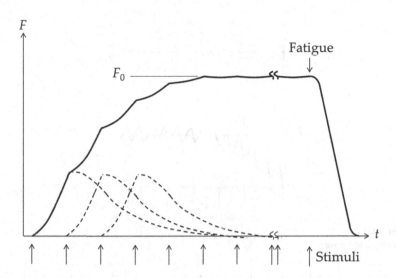

FIGURE 10.3
Summation of isometric twitches to give a complete tetanus.

effect, the active state is prolonged and the force F_0 is maintained until the onset of fatigue. This state of maximal contraction with repetitive stimulation is described as **tetanus**, a **complete tetanus**, or a **fused tetanus**. Thus, if the contraction time is 20 ms, the frequency for a fused tetanus is at least 50 Hz. Normally, the contraction time is considerably longer than the refractory period of the muscle AP, so that fused tetanus is reached at a frequency well below the limit set by refractoriness of the muscle membrane.

The **twitch/tetanus ratio**, F_{tw}/F_0, depends on the velocity of contraction of the muscle fibers and on the viscoelastic properties of the muscle. The F_{tw}/F_0 of motor units is generally in the range, 0.1–0.4, the mean ratio for fast fibers being larger than that for slow fibers.

It may be noted that when motoneurons are firing asynchronously, as in voluntary contraction, the same force can be generated more smoothly and at substantially lower rates of firing than with synchronous firing of motoneurons because of more effective summation of the asynchronous contractions.

10.2.4 Gradation of Muscular Contraction

The smallest force that can be developed by a muscle is that of a single twitch in the smallest motor unit, that is, the motor unit having the smallest number of muscle fibers. On the other hand, the maximum force is the sum of the fused tetanic contractions of all the motor units in the muscle. In between these limits, muscle force can be graded by two mechanisms:

1. **Frequency summation**, by varying the frequency of APs of motoneurons. If the frequency of APs in a given motor unit is less than that of fused tetanus (Figure 10.3) so that successive stimuli occur

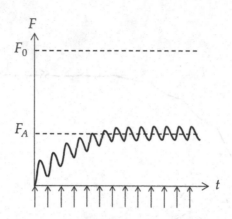

FIGURE 10.4
An incomplete tetanus.

during the relaxation phase of the twitch, an incomplete tetanus is produced having an average force F_A that is less than F_0 (Figure 10.4). F_A increases as the average frequency of the APs increases from low values that produce individual twitches to that of fused tetanus. At any intermediate frequency, the force developed by the motor unit fluctuates about the average value because of the force variation during successive twitches. In practice, the fluctuations are smoothed out mainly by the asynchronous firing of APs in a number of activated motor units. At low subtetanic frequencies the average force decreases with time, an effect referred to as **sag**. The sag increases with decreasing frequency and is believed to be due to an increase in the relative effect of the uptake of sarcoplasmic Ca^{2+}.

It should be noted that APs in motoneurons are fired, in general, at irregular intervals and not in the regular manner indicated in Figure 10.4 for illustration only.

2. **Recruitment of motor units,** or activation of additional motor units to meet the demand for more force by a given muscle. Motor units are recruited through excitation of the corresponding motoneurons, as discussed in Section 11.2.1.3.

Both of these mechanisms occur concurrently. At lower levels of force, motor unit recruitment is dominant, with some increase in the firing rates of already recruited motor units. After nearly all the motor units have been recruited, force is increased by increasing the rate of firing.

When only a weak contraction is required, the smallest motor units are activated first. As the contraction gets stronger, the motor units of intermediate size are activated next, then those of the largest size. This order of recruitment, referred to as the **size principle**, or **Henneman principle**, has some attractive features discussed later (Section 11.2.1.3). As mentioned in connection with Table 9.1, the smallest motor units have the smallest size motoneurons, the smallest innervation ratio, and their muscle fibers are the slow type I fibers. Conversely, the largest motor units have the largest size motoneurons, the largest innervation ratio, and their muscle fibers are the fast type IIB fibers. The order of recruitment from the smallest to the largest motor units makes sense from two points of view:

1. The small contractions are likely to be required for fine movements; they last longer and would therefore require muscle fibers that are more resistant to fatigue.

2. The smaller motor units produce initial, relatively small increments of force. As the force increases, the larger motor units result in larger absolute increments of force, but with an almost constant fractional increase.

In accordance with the size principle, reduction in force levels proceeds in the reverse sequence, that is, motor units that are recruited last are the first to be deactivated. However, the size principle does not always apply, as explained in Section 11.2.1.3.

10.3 Mechanics of Contraction

10.3.1 Length-Tension Relation

Consider a muscle that is stimulated under isometric conditions to produce a fused tetanus at a series of different muscle lengths. It is desired to determine first how the active force developed by the contractile elements varies with the length of the muscle, according to the sliding-filament model (Section 9.2.1). This variation can be divided into three main regions denoted in Figure 10.5a as ab, bc, and cd.

Region ab is that of normal overlap between thick and thin filaments. At a, there is no overlap, which means that no cross bridges can form, and hence the force is zero. At shorter lengths, the force increases with the overlap until the ends of the thin filaments reach the edge of the central region of the thick filaments that is devoid of myosin heads (point b). At this point, the force is in the peak region, when all available myosin heads are taking part in cross-bridge recycling. This peak region extends to point c, with further shortening, when the edges of the thin filaments reach the M line. Over this region of peak force, there is nominally no increase in muscle force because no additional cross bridges form as the thin filaments move over the central region of the thick filaments that is devoid of myosin heads.

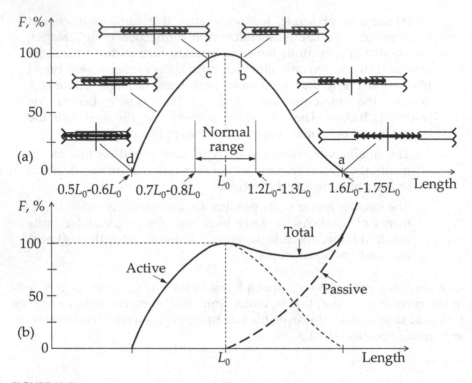

FIGURE 10.5
(a) Active force developed by the contractile elements as a function of muscle length; (b) total muscle force as a function of muscle length.

Further shortening leads to region cd, where thin filaments cross the M line. The force now decreases with shortening due to several factors:

1. The "contrarian" thin filaments that cross the M line not only disrupt the normal three-dimensional pattern of thick and thin filaments, thereby reducing the number of cross bridges previously formed on this side, but also form cross bridges that generate a force that opposes that produced by the cross bridges on the other side of the M line.

2. The thick filaments eventually butt against the Z disks, which prevents any sliding movement. The force is reduced to zero.

The **optimal length** for maximum active force is denoted as L_0 in Figure 10.5a, and is close to the resting length of muscles in the body. During normal muscular action, the smallest muscle length is about $0.7L_0$–$0.8L_0$ and the largest is $1.2L_0$–$1.3L_0$. The lengths at which the force drops to zero are in the ranges $0.5L_0$–$0.6L_0$ and $1.6L_0$–$1.75L_0$ (Figure 10.5a).

It may be noted that the maximum stress, or force per unit area, that a muscle can develop depends, strictly speaking, on the length of the A bands, that is the length of the thick filaments, which correlates linearly with the resting sarcomere length (Section 9.1.2). Hence, the longer this length is, the larger is the number of cross bridges that can form within each sarcomere and the greater is the maximum stress produced. Vertebrate and insect flight muscles have sarcomeres of 2–4 μm length, which yield stresses of 100–300 kN/m^2. On the other hand, muscles that close the claws of some crustaceans have sarcomeres of 14–18 μm length and a stress as high as 2000 kN/m^2, the largest in the animal kingdom. Crustaceans need very strong claws because they feed on hard-shelled prey.

The force that appears at the muscle terminations is the sum of the active force developed by the contractile elements and the passive force due to the elastic elements that are effectively in parallel with the contractile elements (Figure 10.5b). These are mainly due to the membranes – epimysium, perimysium, endomysium, sarcolemma, and SR membranes. The elastic force increases steeply with muscle length in a nearly quadratic manner. The total force is what appears in the tendons, as will be elaborated later in connection with the mechanical model of the muscle. The shape of the total force curve depends on the relative amount of parallel connective tissue in the muscle and on the muscle length at which the elastic force begins to increase. In Figure 10.5b, it is assumed that this force begins to increase at the optimal muscle length L_0. If the elastic force does not begin to increase until the muscle length exceeds L_0, the elastic force curve is shifted to the right, and the maximum and minimum of the total force become more pronounced. Conversely, if the elastic force begins to increase at muscle lengths considerably less than L_0, the elastic force curve is shifted to the left, and the maximum and minimum of the total force will disappear altogether. Isolated single muscle fibers have very little parallel elastic elements because they lack the epimysium and perimysium. Moreover, the amount of parallel connective tissue differs between different muscles; it is, for example, relatively high in the gastrocnemius muscle and is less in the sartorius muscle. The variation of the total muscle force with muscle length, depicted in Figure 10.5b is referred to as the **length-tension relation** of the muscle.

The preceding discussion focuses on the basic mechanisms underlying the length-tension relation based on the sliding filament model. Other factors may contribute to the variation of force developed with muscle length. It is believed that the increase in sarcomere length with stretch, within the physiological range, increases the sensitivity of force to Ca^{2+}, that is, the amount of force produced at a given Ca^{2+} concentration, due to several factors:

1. The reduction in the lateral spacing between thick and thin filaments with increase in sarcomere length at constant sarcomere volume, which would increase the probability of cross bridging.

2. Enhanced Ca^{2+} binding to troponin C at longer sarcomere lengths, although this effect is believed to be small in skeletal muscle but is significant in cardiac muscle (Section 10.5). In the case of skeletal muscle contraction, the amount of Ca^{2+} released into the cytoplasm

is more than that needed for saturation of troponin C, so that variations in Ca^{2+} binding affinity to troponin C has a relatively small effect in skeletal muscle.

3. Cooperative activation of thin filaments by the formation of strong-binding cross bridges. It is believed that the strong-binding state (Section 9.2.1) enhances the activation of thin filaments by holding the tropomyosin molecules more firmly, thereby exposing strong-binding sites on actin for a longer duration and increasing the number of force-generating cross bridges.

An important factor contributing to the increase of the strength of contraction with stretch is the phenomenon of **stretch activation,** where the contraction of the muscle is triggered by stretching. The underlying mechanism of stretch activation is the formation of myosin-troponin bridges, in the presence of a relatively low concentration of Ca^{2+}, which mechanically pull the tropomyosin aside, thereby exposing the actin binding sites (Figure 9.4). Moreover, the twisting of the thick filaments with stretch, concomitantly with the elongation of the thin filaments, brings more myosin heads in contact with actin binding sites, further enhancing force production. Interestingly, a muscle fiber adapts to prolonged stretch by adding new sarcomere at its ends. Conversely, it adapts to prolonged shortening by deleting sarcomeres.

Although stretch activation does not significantly contribute to the force of concentric contraction in skeletal muscle, it is more significant in cardiac muscle (Section 10.5) and plays an essential role in the indirect flight muscles of insects, where contraction of a given set of flight muscles stretch-activates, after a short delay, the antagonist set. When activated, this set, in turn, stretch-activates the original set, and so on, resulting in continuous rhythmic contraction. Such a mechanism is advantageous in several respects. First, it provides an intrinsic and automatic means of contracting and relaxing the flight muscles with every wing beat, which allows beat frequencies as high as 1 kHz, higher than can be produced by neural motor control. Second, it is energy efficient because nervous stimulation is only required to maintain a low, steady level of concentration of Ca^{2+} and because the resulting beat frequency allows proper matching between the energy expended and the wing-thorax aerodynamic load. It is noteworthy that at a higher concentration of Ca^{2+}, the indirect flight muscles behave like other skeletal muscle, developing force continuously until the concentration of Ca^{2+} is reduced.

Problem 10.1

If the width of the region of peak force, or "plateau" bc in Figure 10.5, extends for 25 μm, then according to the sliding-filament model, what should be the length of the middle part of the thick filaments that is devoid of cross bridges?

ANS.: 25 μm.

10.3.2 Force-Velocity Relation

The total force developed by a muscle decreases as the velocity of shortening increases. This is borne by everyday experience: a light load can be lifted faster than a heavy load. There is thus an inverse relation between the load, F, applied to the muscle and the velocity of shortening, v. This active force developed by the contractile elements as a function of muscle length has the general shape illustrated in Figure 10.6a. Experimentally, the muscle is stimulated to develop a fused tetanic force F_0 under isometric conditions for about a fraction of a second, after which a load F is suddenly applied to the muscle. If $F = F_0$, then evidently no shortening occurs. If $F < F_0$, a **quick release** is said to occur. There is an initial rapid shortening, followed by a velocity transient that lasts for 10–20 ms, after which the muscle shortens at a constant velocity v. The plot of v against F gives the F-v relation (Figure 10.6a) for a given muscle length, usually L_0 (Figure 10.5a). For other initial lengths, there will

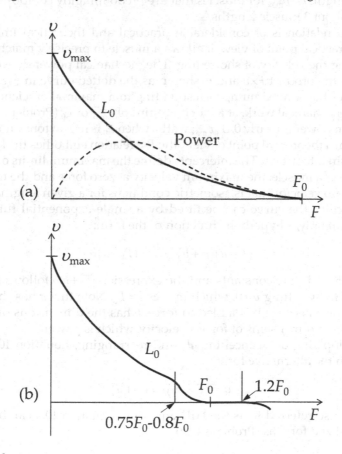

FIGURE 10.6
(a) General shape of the force-velocity relation of a muscle; (b) a more detailed force-velocity relationship.

be similar curves having different values of maximal isometric contraction F_0 for the different initial lengths, in accordance with Figure 10.5b, so that the length-tension relation follows from the F-v plots for different lengths. All of these plots for different lengths will converge to the same value of v_{max}, the maximum shortening velocity at no load. In other words, v_{max} is independent of the degree of overlap between the thick and thin filaments because if the force developed by a single cross bridge at v_{max} is zero, it will be zero for any number of cross bridges in the overlap zone. Having a common v_{max} for F-v relations for different lengths implies a difference in the slopes of the plots, at least at low and moderate velocities. This difference in slope is ascribed to the effects on cross-bridge kinetics of the changes in spacing of the filaments which result from changes in sarcomere length. Compared to slow muscles, the F-v curve of fast muscles has less pronounced curvature, with a larger v_{max}. For the human biceps brachii, v_{max} is about 8 m/s, which is about 20 muscle lengths/s. v_{max} for muscles that are predominantly composed of slow fibers is about 3 muscle lengths/s.

The F-v relation is of considerable practical and theoretical importance. From a practical point of view, it allows a muscle to precisely match the load by varying the velocity of shortening. The mechanical power delivered by a muscle is the product $F{\times}v$ and is shown as the dotted curve in Figure 10.6a. The power has a maximum, corresponding to a maximal efficiency of performing mechanical work, at about one-third of v_{max} or F_0 (Problem 10.5). The maximum power is nearly $0.1F_0v_{max}$ watts, when F_0 is in newtons and v_{max} is in m/s. From a theoretical point of view, the F-v relation embodies the kinetics of actomyosin interaction. The intercepts define the maximum limits on the performance of a muscle: the maximum velocity at zero force and the maximum force at zero velocity, that is, isometric conditions for a given length.

Empirically, the curve can be fitted by a single-exponential function or, more commonly, a hyperbolic function of the form:

$$(F+a)(v+b) = c = (F_0+a)b \tag{10.1}$$

where a, b, and c are constants and the expression $(F_0+a)b$ follows from that on the LHS by setting $v=0$, which makes $F=F_0$. Note that a has the dimensions of force because it is added to force, b has the dimensions of velocity, and c has the dimensions of force$\times$velocity, which is power.

By multiplying out, cancelling ab, and rearranging, Equation 10.1 can be written in the alternative form:

$$v = (F_0 - F)b / (F+a) \tag{10.2}$$

which is also referred to as the **Hill equation**. Equation 10.2 can be written in normalized form as (Problem 10.2).

$$v' = (1 - F') / (1 + F'/k) \tag{10.3}$$

where $v' = v/v_{max}$, $F' = F/F_0$, and $k = a/F_0 = b/v_{max}$. The dimensionless constant $k = a/F_0$ is a measure of the curvature of the hyperbola; the smaller this ratio the more pronounced is the curvature.

The hyperbola gives a good fit of the F-v relation for a whole muscle over the observed range if the effect of inertia in the experimental setup is negligible or if this effect is accounted for, particularly for small loads and hence high velocities. However, careful recordings with single fibers have revealed, at least in the frog, that:

1. The F-v relation varies along the fiber and between different fibers of the same type, that is fast or slow, with F_0 varying by about 4% and v_{max} by about +10 and –45% of the mean values for the fiber. This variation reflects differences in the kinetics of actomyosin interaction along the fiber.

2. The F-v relation is biphasic, having a breakpoint at $F = 0.75F_0$–$0.8F_0$, beyond which the F-v relation follows a different curve that can also be fitted with a single exponential or a hyperbola (Figure 10.6b). The overall F-v relation becomes that of Equation 10.2 multiplied by an additional expression:

$$v = \frac{(F_0 - F)b}{(F + a)}\left(1 - \frac{1}{1 + e^{-k_1(F - k_2 F_0)}}\right) \tag{10.4}$$

where k_1 and k_2 are constants. The significance of the biphasic shape of the F-v curve is that a nearly flat region of zero velocity is introduced over the range $0.8F_0$–$1.2F_0$, which means that over this range contraction is practically isometric. This would protect the muscle from overstretching if suddenly subjected to a load greater than F_0, as may happen in the leg muscles, for example, during running or jumping. Other protection mechanisms, such as the recruitment of additional motor units to support the extra load would be too slow. The flat region would also minimize the redistribution of sarcomere lengths along the muscle fibers at high loads because of the aforementioned variation of the F-v relation in different parts of the muscle fibers.

The reduction in force on a load with velocity, as exemplified by the F-v relation, is basically due to **viscous damping**. The defining characteristic of viscous damping is the development of a force that opposes motion and which increases with velocity in a specified manner. In the case of linear viscous damping, the opposing force is directly proportional to velocity. The effect of viscous damping can be illustrated quite simply by a force generator, F_0, that is opposed by a force, $k_D v$, due to viscous damping, represented by a dashpot

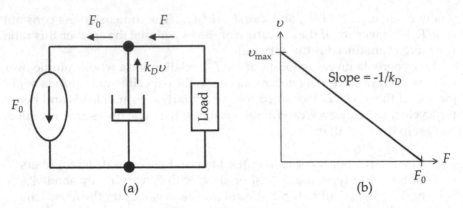

FIGURE 10.7
(a) A basic model of the force-velocity relation; (b) linear force-velocity relation of basic model.

as in Figure 10.7a, where v is the velocity. The net force developed, $(F_0 - k_D v)$, is applied to a load which experiences a force $F = (F_0 - k_D v)$. It follows that:

$$v = (F_0 - F) / k_D \qquad (10.5)$$

If k_D is a constant, the F-v relation will be a straight line of slope $-1/k_D$, with $v_{max} = F_0 / k_D$ (Figure 10.7b). The case of contractile elements in muscle is of course much more complicated. Not only is the system nonlinear, but the actomyosin interaction is believed to be stochastic, that is, each step in cross-bridge cycling, whether physical or chemical in nature, affects the probability of occurrence of the following steps. The force F_0 generated at any instant is proportional to the number of cross bridges undergoing the power stroke at that instant. The viscous effect is due to the fact that at low shortening velocities, each myosin head will have time to attach to each potential site of cross-bridging on the thin filament. At high shortening velocities, the probability of attachment is reduced, as some myosin heads will miss some of the opportunities for forming cross-bridges; force is reduced as the shortening velocity increases. It is also believed that some elasticity exists within the neck joining the myosin head to the body of the thick filament. This elasticity, together with the compliance of the filaments, affects force production in a manner that depends on the velocity.

Problem 10.2

(a) The standard parabola, $xy = c$, where c is a constant, has the x and y axes as asymptotes. What are the asymptotes of the F-v relation (Equation 10.1). (b) Derive Equation 10.3 starting from Equation 10.1.

ANS.: $F = -a$ is the vertical asymptote and $v = -b$ is the horizontal asymptote.

Problem 10.3

Considering v' to be given by Equation 10.3, show that maximum power occurs at $\Gamma'_{\max p} = k\left(\sqrt{1+1/k} - 1\right)$.

Problem 10.4

(a) Show that Equation 10.3 can be expressed as: $F' = (1 - v')/(1 + v'/k)$, where F' and v' are interchanged. (b) Show that maximum power occurs at $v'_{\max p} = k\left(\sqrt{1+1/k} - 1\right)$, where the function on the RHS is the same as that of $F'_{\max p}$ in the preceding problem.

Problem 10.5

Deduce from the preceding two problems that if maximum power occurs at $v'_{\max p} = 1/3$, then this maximum power also occurs at $F'_{\max p} = 1/3$.

10.3.3 Kinetics of Contraction

The basic kinetics of actomyosin interaction that would account for many of the experimentally observed behavior of muscle can be represented by a three-state model, as illustrated in Figure 10.8, in which A denotes actin and M denotes myosin. The state on the left, denoted by M.ADP.P is that of the relaxed state, with the myosin head energized, but the Ca^{2+} concentration is low, so that no binding with actin filaments can occur. The state AM.ADP.P is the active state in which the cross bridge has been formed (Figure 9.6a), where the letters AM refer to an actomyosin complex. In state AM the myosin head has given up its ADP and P_i but is still bound to actin (Figure 9.6b). ATP dissociates the myosin head from the actin and the system subsequently reverts to the energized, unbound state.

Let the fractions of the myosin heads in each of the three states be denoted by α, β, and γ, as indicated in Figure 10.8. It follows that:

$$\alpha + \beta + \gamma = 1 \tag{10.6}$$

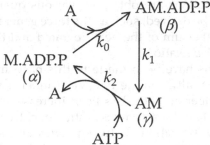

FIGURE 10.8
Three-state kinetic diagram of actomyosin interaction.

$$\frac{d\beta}{dt} = k_0\alpha - k_1\beta = k_0\left(1-\beta-\gamma\right)-k_1\beta \qquad (10.7)$$

$$\frac{d\gamma}{dt} = k_1\beta - k_2\gamma \qquad (10.8)$$

If the muscle is developing a steady force under isometric conditions, $d\beta/dt=0$ and $d\gamma/dt=0$. Equations 10.7 and 10.8 can be solved to give:

$$\beta = \frac{k_0k_2}{k_0k_1 + k_1k_2 + k_2k_0} \qquad (10.9)$$

$$\gamma = \frac{k_0k_1}{k_0k_1 + k_1k_2 + k_2k_0} \qquad (10.10)$$

Knowing the rate constants, the fraction of the myosin heads in each of the three states can be found.

Under conditions of a constant velocity of contraction v, following a quick release to a fixed load, $v=dx/dt$, so that $d\beta/dt = (d\beta/dx)(dx/dt) = v(d\beta/dx)$. Equation 10.7 becomes:

$$v\frac{d\beta}{dx} = k_0\alpha - k_1\beta = k_0\left(1-\beta-\gamma\right)-k_1\beta \qquad (10.11)$$

Similarly, Equation 10.8 becomes:

$$v\frac{d\gamma}{dx} = k_1\beta - k_2\gamma \qquad (10.12)$$

where x denotes distance of the actin binding site from the equilibrium position of the myosin head. Cross bridges are therefore formed at a maximum rate near $x=0$. The rate constants k_0, k_1, and k_2 now depend on x under the assumed isotonic conditions. Some forms of these dependencies have to be assumed in order to solve Equations 10.10 and 10.11, analytically or numerically. Once these solutions are obtained, various quantities pertaining to muscle behavior can be derived, such as the force generated; the total energy expended, which is the sum of the heat liberated and the mechanical work performed; and the F-v relation.

Many modifications have been made to this basic model in attempts to reproduce, more faithfully, muscle behavior under a variety of operating conditions. The number of states has been increased to seven or more by postulating various states for actomyosin attachment, considering reversible, rather than one-way, transitions between states and stochastic modeling. Stretch activation involving myosin-troponin bridges can also be represented by a kinetic model.

10.3.4 Mechanical Model

In the simplest, approximate mechanical model for muscle (Figure 10.9), the cross bridges are considered to be force generators that generate a constant force F_0 during the active state plateau. S_{S1} in series with F_0 represents mainly the elasticity of the titin molecules and the elasticity associated with the myosin head. In parallel with this combination is a dashpot D that accounts for the viscous property of contraction, as in Figure 10.7. The spring S_P represents the elasticity of the connective tissue of the epimysium, perimysium, and endomysium and, to a lesser extent, the sarcolemma and the SR membranes. The perimysium is a stronger connective tissue than the endomysium, and the epimysium is stronger still. Another spring S_{S2} represents the elastic element of the tendons and aponeuroses. Although tendons and aponeuroses possess some viscosity, this is usually neglected in the simple mechanical model, as is the mass of the muscle. To make sense of the direction of forces in Figure 10.9, imagine that the muscle fiber and tendon are being stretched by a force F while the contractile elements are contracting and developing a force F_0. F is then opposed by F_0 and by the force F_P in the parallel elastic elements which are also being stretched. Since the fiber is elongating, the viscous damping is also opposing the applied force. The force balance equation is $F = F_0 + F_P + F_D$.

Under isometric conditions, the two ends of the muscle are held fixed, and when the muscle is not moving, $F_D - 0$. The force in S_{S1} is F_0, and $F = F_0 + F_P$, as assumed in Figure 10.5b, where the active force F_0 is added to the passive force F_P due to S_P. That is, the force appearing at the muscle ends is the sum of the force developed by the contractile elements and the force in the elastic elements in parallel with the contractile elements.

The mechanical model can be used to explain why the peak twitch force under isometric conditions is smaller than that of fused tetanus. It is convenient for this purpose to neglect S_P, as in the case of an isolated muscle fiber which has very little parallel elastic elements. The small S_{S1} does not affect

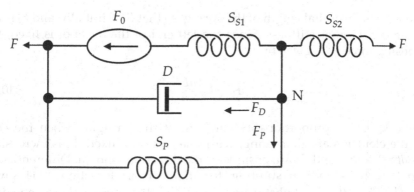

FIGURE 10.9
Basic mechanical model for muscle.

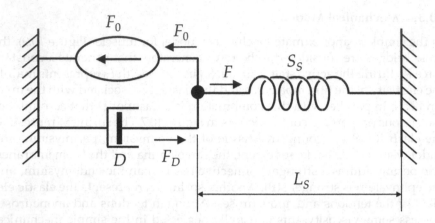

FIGURE 10.10
Simplified mechanical model for muscle.

the force F_0 in the force equation. It only affects the length of the filaments for different values of F_0. It is sometimes lumped with S_{S2} to give S_S in the simplified model of Figure 10.10. Assume that the muscle fiber is contracting under isometric conditions, so that the node in the middle is moving to the left under the influence of F_0, the force developed by the contractile elements. F_0 is opposed by the viscous damping force F_D, and the net force $(F_0 - F_D)$ stretches the series elastic element, producing an equal and opposite force F in this element. The force balance equation is therefore $F_0 - F_D = F$, or, $F_0 = F + F_D$. Note that although the total length of the fiber is constant under isometric conditions, the length of the contractile elements decreases while the length of the series elastic element increases.

The force F in the spring is given by:

$$F = k_S (L_S - L_{S0}) \tag{10.13}$$

where L_{S0} is the initial length of the spring, so that $F = 0$ initially, and k_S is the spring constant, or stiffness. Since the left end of the dashpot is fixed, the opposing force F_D is given by:

$$F_D = k_D \frac{dL_S}{dt} \tag{10.14}$$

where k_D is the proportionality constant of the dashpot. When the contractile elements are shortening, with one end of S_S fixed, L_S is increasing, $dL_S/dt > 0$, and F and F_D are positive in the directions shown. Differentiating Equation 10.13 and substituting for dL_S/dt from Equation 10.14 gives: $F_D = (k_D / k_S)(dF / dt)$. Substituting for F_D in the force balance equation $F_0 = F + F_D$,

$$F + \frac{k_D}{k_S}\frac{dF}{dt} = F_0, \quad 0 \le t \le T_A \tag{10.15}$$

where the active state is assumed to extend from $t=0$ to $t=T_A$, during which F_0 is constant. The solution to this equation, with the initial condition $F=0$ at $t=0$, is:

$$F = F_0\left(1 - e^{-(k_S/k_D)t}\right), \quad 0 \le t \le T_A \tag{10.16}$$

At $t=T_A$, $F=F_{tw}$, the peak force of the isometric twitch, so that:

$$F_{tw} = F_0\left(1 - e^{-(k_S/k_D)T_A}\right) \tag{10.17}$$

For $t > T_A$, $F_0=0$, and the solution to Equation 10.15 is:

$$F = F_{tw}e^{-(k_S/k_D)(t-T_A)}, \quad t \ge T_A \tag{10.18}$$

From Equation 10.17, the twitch/tetanus ratio is:

$$\frac{F_{tw}}{F_0} = 1 - e^{-(k_S/k_D)T_A} \tag{10.19}$$

and is small if $(k_S/k_D)T_A$ is small. It approaches unity if $(k_S/k_D)T_A \gg 1$. The interpretation is that at the onset of the active state, in response to a single AP, F increases exponentially in accordance with Equation 10.16 because of the interplay of the parallel damping and series elastic components, which slows down the increase in F depending on the time constant k_D/k_S. Normally, the active state is over well before F_0 is approached, which makes F_{tw} only a fraction of F_0, as illustrated in Figure 10.11 for a linear system. The effect is similar to the charging of a capacitor through a resistor from a dc voltage supply V_0, but with the supply reduced to zero well before the capacitor voltage approaches V_0. In a fused tetanus, the active state is effectively prolonged so that the force developed reaches F_0.

The mechanical model of Figure 10.9 can be used to explain the basic features of quick release used to derive the F-v relation, as described earlier. Under isometric conditions, the force balance at node N in Figure 10.9 is illustrated in Figure 10.12a, where the force at the muscle terminations is F_{iso} and $F_D=0$. When the load on the muscle is suddenly reduced to $F_L=F_{L1}$ (Figure 10.12b), the following happens instantly:

1. The series elastic element S_{S2} shortens by an amount proportional to $(F_{iso} - F_L)$ so that force in the series elastic element is equal to F_L, which is less than F_{iso}. The muscle as a whole shortens accordingly. The smaller F_L, the larger is the amount of shortening (Figure 10.12c).

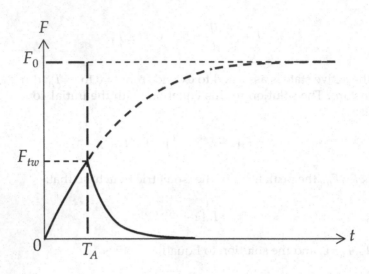

FIGURE 10.11
Isometric twitch contraction from the simplified model.

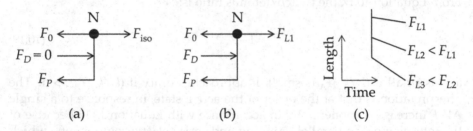

FIGURE 10.12
Force diagram: (a) under isometric condition, (b) when load is suddenly reduced, and (c) when shortening at various loads.

2. Theoretically, F_P does not change at this instant because a finite change in F_P implies an equal change in length of the dashpot, which, during an infinitesimal interval, would require an infinite change in velocity and hence in F_D. Force balance now requires that:

$$F_L = F_0 + F_P - F_D \qquad (10.20)$$

as indicated in Figure 10.12b. The reduction of force at the terminals from F_{iso} to F_L requires a balancing force F_D in the direction that opposes shortening. This F_D appears before any movement of the dashpot occurs.

3. The dashpot starts to move at a velocity F_D/k_D. The movement of the dashpot shortens the muscle further and F_P decreases. If F_P is small compared with $(F_0 - F_L)$, as is the case with single fibers, it follows

from Equation 10.20 that F_D remains constant with time, since F_L and F_0 are constant, so that shortening occurs at constant velocity. The lighter the load, the larger is $(F_0 - F_L)$ and the larger is the shortening velocity (Figure 10.12c).

Problem 10.6

Show that v_C in Figure 10.13 obeys the differential equation:

$$v_C + RC\frac{dv_C}{dt} = V_0, \quad 0 \leq t \leq T_A$$

where $v_{SRC} = V_0$, $0 \leq t \leq T_A$, and is zero elsewhere. This differential equation is exactly analogous to Equation 10.15, which makes the circuit of Figure 10.13 the electrical equivalent circuit of the mechanical model of Figure 10.10. The mechanical $\equiv$ electrical analogy is based on the following analogies: force $\equiv$ voltage, velocity $\equiv$ current, distance (length) $\equiv$ electric charge, dashpot $\equiv$ resistor $(F = kdv/dt \equiv v = Ri)$, spring $\equiv$ capacitor $(F = kx \equiv v = (1/C)q$, where k is analogous to $1/C$, that is, the reciprocal of capacitance, which is the **elastance**. Capacitance is analogous to the reciprocal of stiffness, which is the **compliance**. It follows that $k_D \equiv R$, $1/k_S \equiv C$, and the time constants $(k_D/k_S) \equiv (RC)$. In this analogy, $F = mdv/dt \equiv v = Ldi/dt$, so that inductance is analogous to mass, or inertia. An alternative electrical equivalent circuit is given in Problem 10.7.

Problem 10.7

Show that i_L in Figure 10.14 obeys the differential equation:

$$i_L + GL\frac{di_L}{dt} = I_0, \quad 0 \leq t \leq T_A$$

where $i_{SRC} = I_0$, $0 \leq t \leq T_A$, and is zero elsewhere. This differential equation is exactly analogous to Equation 10.15, which makes the circuit of Figure 10.14

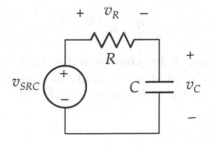

FIGURE 10.13
Figure for Problem 10.2.

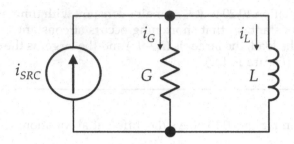

FIGURE 10.14
Figure for Problem 10.3.

the electrical equivalent circuit of the mechanical model of Figure 10.10. The mechanical ≡ electrical analogy is based on the following analogies: force ≡ current, velocity ≡ voltage, distance (length) ≡ flux linkage (time integral of voltage), dashpot ≡ resistor (represented by its conductance, $F = k \, dv/dt \equiv i = Gv$), spring ≡ inductor ($F = kx \equiv i = (1/L)\lambda$). It follows that $k_D \equiv G$, $1/k_S \equiv L$, and the time constants $(k_D/k_S) \equiv GL$. In this analogy, $F = m \, dv/dt \equiv i = C \, dv/dt$, so that capacitance is analogous to mass, or inertia. Note that the circuits of Figures 10.13 and 10.14 are duals.

Problem 10.8
Assume that in the mechanical model of Figure 10.10, a second stimulus is applied at $t = \nu \geq T_A$ in Figure 10.11. Express the ratio of the amplitude of the second twitch to that of the first twitch. Check the result by substituting, $\nu = T_A$, which effectively doubles T_A.

ANS.: $1 + e^{-(k_S/k_D)\nu}$

Problem 10.9
Show that the circuits of Figure 10.15a and 10.15b are the electrical equivalent circuits of the mechanical model of Figure 9.15a because they reproduce Equation 9.4 in the form:

$$\frac{dq_C'}{dt} + \frac{1}{RC} q_C' = A$$

(Figure 10.15a), where q_C' is the change in the charge on the capacitor, R is analogous to k_1, C is analogous to $1/k_2$, and A is analogous to the source current i. The voltage v, analogous to the force, is the same across the paralleled elements.

$$\frac{d\lambda_L'}{dt} + \frac{R}{L} \lambda_L' = A$$

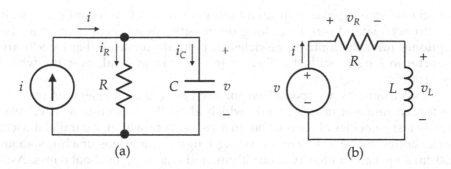

FIGURE 10.15
Figure for Problem 10.4.

(Figure 10.15b), where λ'_L is the change in the flux linkage of the inductor, $1/R$, or G is analogous to k_1, L is analogous to $1/k_2$, and A is analogous to the source voltage. The current, i, analogous to force, is the same through all the series-connected elements. Note that the circuits of Figures 10.15a and 10.15b are duals.

Problem 10.10

Interpret the effects of a large or small time constant of Problem 9.2 in terms of the equivalent electric circuit of Figure 10.15a.

10.4 Pennate vs. Parallel Muscles

As mentioned previously, the fascicles in a parallel muscle generally extend from one end of the muscle to the other. In long, parallel muscles, however, the muscle fibers in the fascicles do not extend over the whole length of the muscle because this would not allow proper synchronization of contraction of the sarcomeres in the fiber. Conduction velocities of the muscle AP are typically 2–10 m/s. If initiated in the middle of a muscle fiber 20 cm long, an AP may take few tens of milliseconds to reach the end of the fiber. This may be in excess of twitch contraction time, so that while the sarcomeres in the middle of the fiber contract, the sarcomeres at the ends of the fiber will still be relaxed and will be passively stretched. The overall shortening of the fiber will be reduced, as will be the force developed at the muscle terminations. To avoid this problem, individual muscle fibers are 2–4 cm long, with each fiber innervated by a terminal branch of the motor axon. In a long fascicle, individual short fibers are arranged serially in groups extending from one end of the fascicle to the other, the serial groups being staggered with respect to one another. Force is passed through both end-to-end fiber coupling as well as through lateral coupling along the lengths of adjacent fibers and the endomysium. By contrast, fascicles in pennate muscles (Figure 9.9) are shorter in length, with the fibers extending, in general, over the whole length of the fascicles.

Parallel muscles are specialized for excursion, that is, operation over an extended range of muscle length, which also allows them to achieve relatively fast velocities of contraction. In a strong contraction, a parallel muscle can shorten by 30%–40% of its resting length. The biceps brachii is about 30 cm long, can shorten by about 10 cm, and has a v_{max} of about 8 m/s. As a muscle shortens, its volume remains constant, which means that the cross-sectional area increases by redistribution of cytoplasmic elements. The increase in cross-sectional area is evident by the bulging of the biceps brachii with elbow flexion.

The salient feature of pennation is an increase in the force developed by the muscle for a given volume. This can be demonstrated with reference to Figure 10.16a, which shows a block of pennate muscle of width w, depth d, length L, and pennation angle ϕ. The muscle fibers extend between two vertically oriented aponeuroses to the right and left of the fibers in Figure 10.16a, as exemplified by the two dashed arrows. The cross-sectional area of the fibers is $dL\sin\phi$, as illustrated by the rectangle in Figure 10.16b. The force developed by the fibers is some function of this area $f(dL\sin\phi)$, and the component of force in the direction of the line of action of the muscle is $F_{pen} = f(dL\sin\phi)\cos\phi$. If the muscle were a parallel muscle of the same volume, with the fibers oriented vertically along the direction of the line of action of the muscle, the cross-sectional area would be dw, and the force developed in

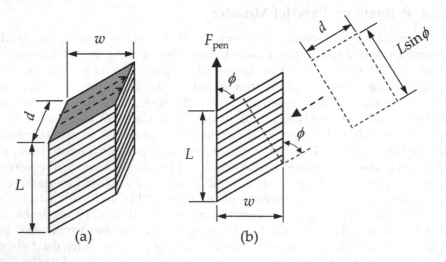

FIGURE 10.16
(a) Block of pennate muscle; (b) section showing cross-sectional area.

the direction of the line of action of the muscle would be $F_{par} = f(dw)$. The ratio of the two forces is:

$$\frac{F_{pen}}{F_{par}} = \frac{f(dL \sin\varphi)}{f(dw)} \cos\varphi = \frac{(dL \sin\varphi)^n}{(dw)^n} \cos\varphi = \left(\frac{L}{w}\right)^n \sin^n\varphi \cos\varphi \quad (10.21)$$

where the function is assumed to be some power law. It would seem logical to assume, as has been the case in most of the literature, that the force developed by a muscle fiber is directly proportional to the cross-sectional area of the fiber. However, more recent work on chemically skinned human muscle fibers has indicated that the force is more nearly proportional to fiber diameter rather than cross-sectional area, the force per unit diameter being approximately 6.5 N/m for both type I and type IIA fibers. Thus, $n = 1$ in Equation 10.21 if the force is directly proportional to the cross-sectional area, and $n = 1/2$ if the force is proportional to a linear dimension of the cross-sectional area. In either case, the term $\sin^n\phi\cos\phi$ is always less than 1. But the ratio $(L/w)^n$ can be large, so that the ratio $(F_{pen}/F_{par}) > 1$. The force developed by a pennate muscle is therefore larger than that developed by a parallel muscle of the same volume, but the excursion in the line of action of the muscle will be smaller. In human limb muscles, the mean pennation angles are usually between 5° and 15°. On the other hand, the lateral gastrocnemius muscle of the wild turkey is a unipennate muscle having an average pennation angle of 25°.

An interesting property of pennate muscle is that it allows a force-dependent, variable **architectural gear ratio** (**AGR**), defined as the ratio of the change in length of the whole muscle to the change in length of the muscle fibers. Since the changes in lengths occur during the same time interval, the ratio is the same as whole-muscle velocity to muscle fiber velocity. It is a gear ratio due to pennation architecture rather than due to mechanics of the connection of the muscle to bones (Section 13.1.1). The AGR arises from the way the muscle width and muscle depth change with shortening of the muscle fibers so as to keep the muscle volume constant. This is diagrammatically illustrated in Figure 10.17, where (a) shows a longitudinal section and a transverse section of a pennate muscle at rest. In (b) the fibers shorten by about 20%, the muscle width w increases to w', and the muscle depth d decreases to d'. The distance between the aponeuroses increases, and the muscle fibers rotate as they shorten, increasing the pennation angle to ϕ'. The whole muscle shortens by more than 20%, as indicated by the vertical arrow. However, because of the increase in pennation angle, a smaller fraction of the force developed by the individual fibers appears along the line of action of the muscle (F_{pen} in Figure 10.16b). In (c) the fibers again shorten by about 20%, compared to (a), but the contraction decreases the muscle width to w'', and the muscle depth increases to d''. The distance between the aponeuroses decreases, which counteracts fiber rotation, so that the pennation angle increases only slightly and may in fact decrease. The shortening is

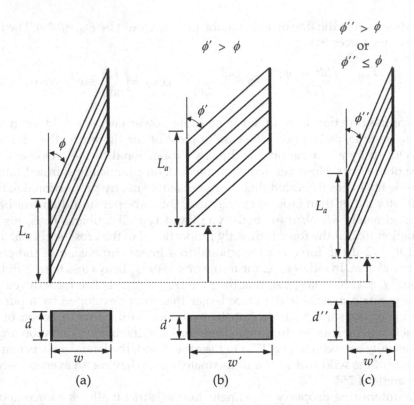

FIGURE 10.17
(a) Pennate muscle at rest; (b) and (c) shortening of the muscle, with different combinations of muscle depth and width, at constant muscle volume.

considerably less than in (b), but a larger fraction of the force developed by the individual fibers appears along the line of action of the muscle. Because the muscle shortens in (b) more than the fibers, AGR is considerably larger than unity. On the other hand, the muscle shortening in (c) is less than that of the fibers, so AGR is less than unity. Thus, case (b) favors muscle velocity rather than force, whereas case (c) favors muscle force rather than velocity. The change between (b) and (c) is believed to occur automatically during contraction because of the passive effects of intramuscular forces on connective tissue elements of the muscle.

The variable AGR enhances the F-v relation of the whole muscle compared to that of the muscle fiber, by boosting the muscle force at low velocities and by allowing a higher v_{max}, and hence higher velocities at low values of force. The effect is analogous to that of an automatic transmission system that shifts from high gear during rapid contractions to low gear during forceful contractions. The tradeoff between force and velocity that is represented by the F-v relation is extended to better match the demands of muscular contraction.

Another interesting property of pennate muscle is encountered in the aforementioned claw muscle of crustaceans. This is a bipennate muscle that cannot change its width and depth as it contracts because it is surrounded by the rigid shell of the claw. Pennation makes this contraction possible. The individual fibers can shorten but with the pennation angle increasing so as to keep the muscle width w constant. Since L_a, the length of the aponeuroses, is constant, a constant w implies a constant muscle depth d, since the muscle volume is constant.

10.5 Cardiac Muscle

The purpose of this section is to highlight the main differences between cardiac and skeletal muscle that underlies the specialization of the several billion cardiac cells to subserve the functioning of the heart. It is assumed that the reader is familiar with the basic physiology of the heart.

10.5.1 Cardiac Cells

Cardiac muscle cells, also called **cardiocytes** or **cardiac myocytes**, are relatively small muscle cells 20–35 μm across and 100–150 μm long. The cells are packed with contractile myofilaments whose alignment gives a striated appearance, as in skeletal muscle (Figure 10.18). Most cardiocytes have a single nucleus, although a few may have two or more nuclei. Many cardiocytes

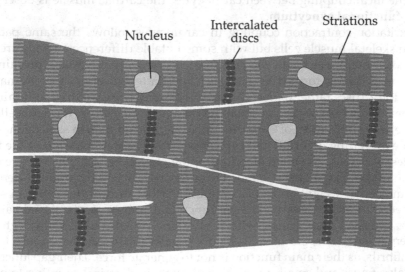

FIGURE 10.18
Cardiac muscle cells.

show variable Y-shaped branching that allows a higher degree of interconnection between cardiocytes in the longitudinal direction, as illustrated by the two myocytes at the top of Figure 10.18.

An important functional requirement of the heart is that the cardiocytes should contract in a highly coordinated, near-synchronous manner. The squeezing action required for pumping blood by the heart is the result of a force of contraction produced by cardiocytes forming the outer walls of the atria and ventricles. This force is to be exerted over the whole of the cardiocyte body and is not to be transmitted just to tendons or aponeuroses at the muscle ends, as in skeletal muscle. Specialized structures, known as **intercalated discs**, are found at the junctions where cardiocytes are joined together end-to-end (Figure 10.18). The discs contain gap junctions (Section 6.6), which allow fast spread of depolarization from a cardiocyte undergoing an AP to other cardiocytes. This ensures a high degree of synchrony in the electrical excitation of cardiocytes in the atria and in the ventricles, as well as transmission of chemical messengers between cardiocytes. At the intercalated discs, the cell membranes of adjacent cardiocytes become extensively interdigitated and bound together by **desmosomes**, which consist of dense protein plaques in each adjacent cell separated by a thin layer of extracellular material. These connections help stabilize the three-dimensional structure and the relative positions of adjacent cells, thereby also protecting the fragile gap junctions from the mechanical forces resulting from contraction. The mechanical coupling between cardiocytes is enhanced in the lateral direction, as opposed to the end-to-end direction, by an extensive web of protein filaments that interconnect the myofibrils to the plasma membrane and an extracellular skeleton matrix. Because of this tight mechanical, electrical, and biochemical coupling between cardiocytes, the cardiac muscle is described as a **functional syncytium**.

Excitation-contraction coupling in cardiocytes follows the same pattern as in skeletal muscle cells but with some notable differences. Triads are not found in cardiocytes and the terminal cisternae are much less prominent. Instead, a diad is formed by T tubules and flattened sacs of the SR, mainly near the Z discs rather than at the zone of overlap between thick and thin filaments. Influx of Ca^{2+} occurs from the diads as well as from the extracellular space, the latter source of Ca^{2+} being more significant than in skeletal muscle.

The preceding description is typical of cardiocytes in the walls of the ventricles, which are the main blood pumping chambers of the heart. The connexin of these cardiocytes is mostly connexin43, where the number denotes the molecular weight of the protein in kD. Other cardiocytes differ in some respects, in accordance with their functional requirements. Atrial cardiocytes are long and slender with less extensive T tubules. Cardiocytes of pacemaking nodes, discussed later, are smaller in size and have much fewer myofibrils, as their main function is not to generate force. Their gap junctions are also fewer and smaller, which reduces their electrical coupling to other cardiocytes. This enhances the ability of the pacemaking cardiocytes of the sinoatrial (SA) node to drive a relatively large number of atrial cardiocytes without being subject to too much hyperpolarization, as occurs in electrical synapses (Section 6.6). In the case of cardiocytes of the atrioventricular (AV) node, the reduced electrical coupling introduces the required delay of 0.1–0.15 s between atrial and ventricular contraction. Cardiocytes of the **bundle of His** and its branches, as well as those of the **Purkinje fiber network**, are required to conduct APs at high speeds. They have an abundance of large gap junctions whose channels are formed from connexin40, a form of connexin that is associated with large-conductance channels.

Cardiocytes rely almost exclusively on aerobic metabolism, which makes them particularly vulnerable to a reduced blood supply. A sudden occlusion of the coronary arteries, as occurs during a heart attack, can quickly lead to the death of the cardiocytes affected.

10.5.2 Starling's Law

An important intrinsic property of cardiac muscle is that the force of ventricular contraction increases with the end-diastolic volume, that is, the volume of blood in the ventricles just before the beginning of contraction (Figure 10.19), an effect referred to as **Starling's law** of the heart, or the **Frank–Starling relation**. The larger the end-diastolic volume, the larger is the initial stretch, or **preload**, of the ventricles, and the larger is the **stroke volume**, or the volume of blood ejected by a ventricle because of the increased force of contraction. This effect regulates the input-output of the heart, beat-by-beat,

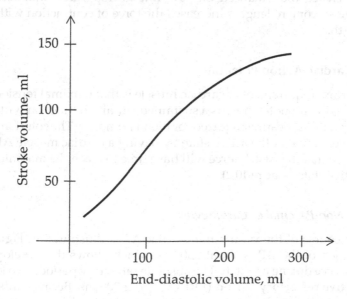

FIGURE 10.19
Variation of stroke volume with end-diastolic volume.

without the involvement of control mechanisms from outside the heart and is of physiological importance. For example, the two halves of the heart beat together, like two pumps in parallel. Yet, blood flows in series in the two circuits: the pulmonary circuit, in which blood is pumped through the lungs by the right half of the heart and the systemic circuit, in which blood is pumped through the rest of the body by the left half of the heart. The flow in the two circuits must be equalized to prevent accumulation of blood in the heart. If, for example, the peripheral resistance increases due to vasoconstriction in the systemic circuit, the reduced outflow from the left ventricle will cause accumulation of blood in this ventricle. However, the resulting increase in end-diastolic volume will automatically increase the force of contraction, so as to pump more blood from the left ventricle and equalize the flow in the two circuits.

The relation between stoke volume and end-diastolic volume (Figure 10.19) is essentially the same as the length-tension relation for skeletal muscle (Figure 10.5) but with the axes relabeled in terms of parameters appropriate for the heart. The resting length in the case of cardiac muscle is on the steeper part of the curve, where the active contraction increases rapidly with sarcomere length, rather than near the maximum of curve as in the case of skeletal muscle. In addition, a significant factor in increasing the force of contraction with sarcomere length in the case of cardiac muscle, which makes the curve of Figure 10.19 steeper, is stretch activation (Section 10.3.1). Another factor is that the amount of Ca^{2+} released into the cytoplasm during excitation-contraction coupling is less than that needed for saturation of troponin C. Hence, the enhanced Ca^{2+} binding to troponin C that occurs with increasing sarcomere lengths increases the force of contraction with muscle fiber length.

10.5.3 Cardiac Action Potential

An important requirement of cardiac muscle is that maximal tension should be produced without tetanus, as a sustained tetanic contraction will not produce the rhythmic beating required for blood pumping. The solution adopted is the prolongation of the active state by having a cardiac muscle AP of long duration, so that the twitch force will have time to reach the maximum value of the active state (Figure 10.2).

10.5.3.1 Non-Pacemaker Cardiocytes

The time course of the ventricular cardiac AP is illustrated in Figure 10.20. The duration of the AP is about 250 ms, which allows the development of maximal force during a twitch. The absolute refractory period, also known as the **effective refractory period** (ERP), is about 200 ms. Because this exceeds the contraction time, it is not possible to tetanize cardiac muscle. The ERP limits the maximum frequency to about 250–300 beats/min.

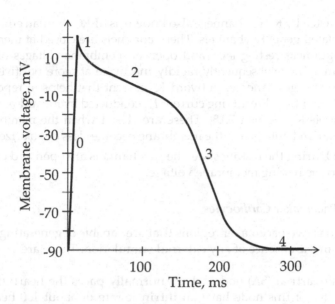

FIGURE 10.20
Cardiac action potential.

The time course of the AP can be divided into five phases:

Phase 0: From a resting membrane voltage of –85 to –95 mV, close to the K⁺ equilibrium voltage, the upstroke of the AP is due to a Na⁺ inward current, I_{Na}, conducted by fast Na⁺ channels as in the case of APs of nerve and skeletal muscle. The threshold is about –70 mV and the Na⁺ equilibrium voltage about +50 mV.

Phase 1: This is a small rapid, notch-like repolarization due to a transient outward K⁺ current, I_{Kto}, conducted by K_{to} channels. This repolarization sets the level for the slow change in the following phase.

Phase 2: By this time, the short-duration currents, I_{Na} and I_{Kto}, are practically over and the resulting depolarization initiates an inward Ca²⁺ current and an outward K⁺ current. The near-equality of these currents results in a slow depolarization phase, or "plateau", that is the salient feature of the cardiac AP. The inward current is carried by Ca²⁺ conducted by channels of the Ca_v1 subfamily (Section 7.3.2), which are large, slow channels having an activation voltage > –30 mV, fast deactivation, and very slow inactivation. The outward current is mainly the slow, delayed rectifier K⁺ current, I_{Ks}. There are also small contributions during this phase from the Na⁺-K⁺ pump and the Ca²⁺-Na⁺ ion exchanger (Section 2.3.2), which exports one Ca²⁺ for three Na⁺ imported.

Phase 3: The membrane is rapidly repolarized at this stage, mainly by an outward K⁺ current referred to as a rapid delayed rectifier K⁺ current, I_{Kr}, conducted by $K_v11.1$ channels, also known as hERG (human ether-a-go-go-related gene) K^+ channels. These channels are closed at membrane voltages near resting level and open at membrane voltages of about -60 mV. They subsequently rapidly inactivate at more positive membrane voltages. Another outward K^+ current that helps in repolarization is an inward-rectifying current I_{K1} conducted by inward-rectifying channels K_{ir} (Section 7.3.3). These are closed when the membrane is depolarized but open as the membrane becomes less depolarized.

Phase 4: During the resting state, the K_{ir} channels are open and contribute to the resting membrane voltage.

10.5.3.2 Pacemaker Cardiocytes

The heart has two pacemaker regions that are capable of generating APs continuously, in the absence of any external stimulation. These are:

1. the sinoartrial (SA) node, which normally paces the heart; the cardiocytes of this node have an intrinsic rate of about 100 beats per minute, but are nevertheless under the influence of the autonomic nervous system, as discussed later, and

2. the atrioventricular (AV) node, which has a slow intrinsic beat of 25–40 beats per minute, so that it is normally driven by the faster APs generated by the SA node. Pacemaking cells of the heart are also referred to as **autorhythmic cells**.

The waveform of the membrane voltage of a pacemaker cardiocyte is illustrated in Figure 10.21a. Its most salient feature is a slow depolarization, the **prepotential**, between successive APs. The early part of the prepotential is due to an I_h current (Section 7.3.3) carried by Na^+, K^+, and Ca^{2+}. In the later phase of the prepotential, channels of the Ca_v3 subfamily (Section 6.3.1) open, which further depolarizes the cell and brings the membrane voltage closer to threshold. Eventually, channels of the Ca_v1 subfamily (Section 7.3.2) are activated, driving the membrane voltage to threshold and initiating an AP. Whereas the upstroke of the AP is due to Ca^{2+} and not Na^+, repolarization is mainly due to K^+ efflux, as usual.

Bearing in mind that a node consists of a large number of pacemaker cells, each having its own intrinsic frequency because of inevitable variations between individual neurons, the firing frequency of the node as a whole is that of the fastest firing cell. This is diagrammatically illustrated in Figure 10.21b showing the waveforms for two cells A and B. Assuming that they both fire together at a certain time, cell A, having a higher firing frequency than cell B, will reach threshold sooner and will fire first, driving cell B through the gap junctions coupling the two cells. Cell B is thus forced to fire at the frequency of cell A and not at its own, slower rhythm. The same is true of any number of coupled pacemaker cells.

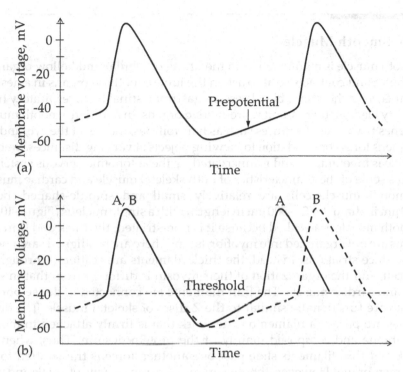

FIGURE 10.21
(a) Action potentials of cardiac pacemaker cell; (b) a faster pacemaker cell driving a slower pacemaker cell.

The rate of firing of pacemaker cells is controlled by the autonomic nervous system. ACh released by terminals of parasympathetic nerves in the heart binds to muscarinic receptors in pacemaker cells, thereby activating a G protein and reducing the intracellular concentration of cyclic cAMP (Section 5.3). The overall result is to:

1. reduce the slope of the prepotential by decreasing the I_h current, and
2. reduce the slope of the upstroke of the AP by slowing the opening of the Ca_v1 channels.

Both effects reduce the heart rate. Conversely, norepinephrine released by terminals of sympathetic nerves in the heart binds to beta-adrenoreceptors, mainly $\alpha 1$ and $\beta 1$, and produces the opposite effect. Under normal resting conditions, the parasympathetic effect dominates, resulting in a heart beat of 60–70 beats per minute. In addition, norepinephrine increases the force of contraction of the heart and speeds the spread of excitation along conducting pathways, which speeds up the contraction and relaxation times of the ventricles.

10.6 Smooth Muscle

Smooth muscle is mainly found in the lining of tubular and hollow organs of the body, and contracts so as to act on the lumens of these organs in a desired manner, as in the case of blood vessels, gastrointestinal tract, respiratory tract, urinary bladder, uterus, and reproductive organs. In addition, smooth muscle attaches to various structures such as hair follicles, the iris of the eye, and the eye's lens for accommodation to viewing objects at varying distances. Smooth muscle is involuntary, and is innervated by the autonomic nervous system. It shares some of the characteristics of both skeletal muscle and cardiac muscle.

Smooth muscle cells are relatively small and spindle-shaped, being 2–10 µm in diameter 20–500 µm in length, with a single nucleus (Figure 10.22). Smooth muscle is so called because it is non-striated, the thick and thin filaments are not organized into myofibrils, and there are no aligned sarcomeres to produce striations. Instead, the thick filaments are scattered throughout the cell, and the organization of their myosin is different from that in skeletal and cardiac muscle. The thin filaments are connected to **dense bodies**, which are functionally similar to the Z discs of skeletal muscle. The dense bodies are part of a filamentous network that is firmly attached to the cell membrane and composed mainly of the protein desmin. Thus, when the thick and thin filaments slide past one another, force is transmitted to the cell membrane. However, the unorganized arrangement of thick and thin filaments allows the development of force over a range of lengths that can be four times that in skeletal muscle. The amount of myosin in smooth muscle, in mg/g of muscle, is roughly a third to a quarter of that in skeletal muscle, while the amount of actin can be up to twice as much.

Smooth muscle is conventionally divided into two subgroups: single unit and multiunit. In single unit smooth muscle, the cells are packed, roughly in parallel, in dense sheets or bundles. The muscle has pacemaker regions, and the generated AP is propagated to neighboring cells through gap junctions that interconnect the cells, so that the whole muscle behaves electrically and mechanically as a syncytium as in the heart. The pacemaking activity is produced by specialized cells that generate **slow waves** of depolarization and repolarization. The slow waves periodically reach threshold, through temporal summation, and generate APs that result in periodic contractions.

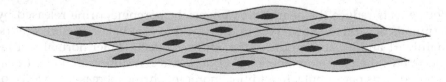

FIGURE 10.22
Smooth muscle cells.

In the gastrointestinal tract, the frequency of the contractions is about 3 per minute in the stomach and colon and 10–12 per minute in the small intestine.

The fibers of multiunit smooth muscle are richly innervated by autonomic nerve fibers and contract independently of one another upon nerve stimulation. The force generated depends on the frequency of nerve stimulation and on the number of fibers activated, as in skeletal muscle. Contraction occurs upon sufficient depolarization of the muscle fiber, often without the occurrence of a muscle AP. Multi-unit skeletal muscle is found in the trachea, in large elastic arteries, and in connections to the iris of the eye and to hair follicles.

It should be noted that smooth muscle shows considerable diversity that ranges between the aforementioned characteristics of single-unit and multiunit muscle. Because of this, it has become customary to refer to smooth muscle according to its location, such as vascular smooth muscle, visceral smooth muscle, or uterine smooth muscle.

The following are notable features of smooth muscle in general:

1. Under normal conditions, smooth muscle has some background level of activity, or **smooth muscle tone**, which can be regulated up or down by neural, hormonal, or chemical factors.

2. Smooth muscle can be made to contract or relax not only by autonomic nerve stimulation but also in response to hormones, chemicals, and local concentrations of oxygen, carbon dioxide, nitric oxide, and various ions. All these stimuli can act in unison, reinforcing or opposing one another.

3. Autonomic nerves exercise their effects not through a specialized endplate but through varicosities at nerve branch endings which contain vesicles filled with neurotransmitter that is released by the depolarization of an AP arriving at the nerve endings neurotransmitter. The varicosities are often restricted to pacemaker regions, where the neurotransmitter can directly affect pacemaker activity. In general, a given varicosity can affect several muscle fibers, and a given muscle fiber may be affected by several varicosities. Moreover, a given neurotransmitter can have opposite effects in different smooth muscles by acting on different types of receptors. For example, norepinephrine released by sympathetic nerves causes contraction of most smooth muscle by acting on alpha-adrenergic receptors but causes relaxation of airway smooth muscle by acting on beta-2-adrenergic receptors.

4. Stretch can also activate smooth muscle, with the force of contraction usually increasing with stretch, as in cardiac muscle. This is useful, for example, in the case of the stomach, whereby a fuller stomach will automatically contract more forcefully.

5. The ATPase activity of smooth muscle, accounting for both ATP splitting by the cross bridges and the action of the pump that extrudes Ca^{2+} to the SR and the extracellular medium, is only a small fraction of that in skeletal muscle. This makes smooth muscle both slow and resistant to fatigue. A single twitch of smooth muscle may last several seconds.

6. Smooth muscle relies mostly on aerobic metabolism, although anaerobic metabolism can occur during peak activity.

7. Despite its slower action, smooth muscle can develop force per cross-sectional area, under isometric conditions, that is comparable to that of skeletal muscle.

8. Smooth muscle cells are capable of cell division, so that smooth muscle tissue can regenerate following injury.

Excitation-contraction coupling is different in smooth muscle from that in skeletal muscle. Smooth muscle lacks T tubules, since the small cell size, the slow contraction, the lack of organization of myofibrils, and the absence of APs in many cases, do not warrant a T tubule system. Smooth muscle lacks troponin but has, instead, two other thin filament proteins, caldesmon and calponin. On stimulation, contraction is initiated by Ca^{2+} influx from the extracellular fluid as well as the SR, the relative contribution of these sources differs between different smooth muscle. Ca^{2+} influx from the SR occurs in some locations through a close association between the SR membrane and the sarcolemma, analogous to that of T tubules in skeletal muscle, as well as through second messengers (Section 6.3) that are released upon stimulation and which cause opening of Ca^{2+} channels in the SR membrane. The total amount of Ca^{2+} released by a single stimulus is usually sufficient to activate only a fraction of the cross bridges available, which allows variation in Ca^{2+} concentration to grade the force produced by the muscle. In the absence of an AP, Ca^{2+} concentration can be graded by membrane depolarization or hyperpolarization.

Ca^{2+} bind to the protein calmodulin (Section 6.3.1), which activates the enzyme **myosin light chain kinase (MLCK)**. This enzyme phosphorylates the myosin light chain in the myosin head, in the presence of ATP. Only when the myosin head is phosphorylated can it combine with actin to form cross bridges and initiate cross-bridge recycling through ATP splitting. To relax the muscle, the myosin is dephosphorylated by the enzyme **myosin light chain phosphatase**, which is continuously active in smooth muscle. However, when the concentration of Ca^{2+} rises, the rate of phosphorylation exceeds that of dephosphorylation and cross-bridge recycling occurs. The converse applies when the concentration of Ca^{2+} falls.

Some smooth muscle have a **latch state** in which the rate of ATP splitting declines, with the cross bridges maintained in a rigor-like manner, so that force is developed under isometric conditions. This is advantageous in blood vessels, as it allows maintaining the vessel diameter against blood pressure, with little energy expenditure.

Summary of Main Concepts

- There are four main types of muscle contraction: isometric, isotonic, concentric, and eccentric.

- Twitches can be characterized under isometric conditions, where the muscle length is fixed, or under isotonic conditions, where muscle load is fixed.

- Because of the viscoelastic properties of muscle, the peak force developed in an isometric twitch is only a small fraction of the maximum force that can be developed by the muscle fiber.

- If action potentials are applied in quick succession to a muscle fiber under isometric conditions, the individual twitches summate and the force developed by the fiber increases. At a sufficiently high frequency of action potentials, the twitches fuse to give an almost steady force that approaches the maximum force that the muscle fiber can develop – a condition known as tetanus.

- The force developed by a whole muscle is graded by two mechanisms: (i) recruitment of new motor units, and (ii) increase in the average frequency of action potentials in motor units already recruited.

- The length-tension relation of a muscle is the sum of an active component due to the force developed by the myofilaments and a passive component due to elastic elements effectively in parallel with the myofilaments. The active component varies with the muscle length according to the degree of overlap between the thick and thin filaments.

- The length-tension relation is modified by factors such as stretch activation and by the increase, due to stretch, of Ca^{2+} sensitivity to force.

- According to the force-velocity relation, the total force developed by a muscle decreases as the velocity of shortening increases. This variation is basically due to viscous damping.

- Many of the observed behaviors of muscle can be accounted for by a simplified three-state kinetic model and by a simplified mechanical model using a dashpot and series and parallel elastic elements. The mechanical model can readily account for the twitch/tetanus ratio and for the basic features of quick release of force developed by a tetanized muscle.

- Pennation offers important advantages, namely, a larger force for a given muscle volume, and the possibility of a variable architectural gear ratio that better caters to the force-velocity requirements of the muscle load.

- Cardiac cells are specialized for developing maximum tension with every heartbeat, with a high degree of synchronism. Maximal tension with every heartbeat is achieved through prolonging the active state by having a plateau of the action potential that extends its duration to about 250 ms. A high degree of synchronism is provided by gap junctions between adjacent cardiac cells.

- The length-tension relation of cardiac muscle is exemplified by Starling's law relating the stroke volume to the end-diastolic volume, and which has important physiological implications. The variation of tension with length is accentuated by operating on the rising part of the length-tension relation and by stretch activation.

- Both the non-pacemaker and pacemaker action potentials of the heart are based on interplay of Na^+, K^+, and Ca^{2+} channels.

- Unlike skeletal and cardiac muscle, smooth muscle is unstriated and is adapted for the development of force over a range of lengths that can be four times that in skeletal muscle. Some smooth muscle has intrinsically generated slow waves of depolarization and repolarization that generate action potentials upon reaching threshold. Other smooth muscles generate force upon sufficient depolarization of the muscle fiber, often without the occurrence of muscle action potentials.

- Smooth muscle contracts or relaxes not only due to autonomic nerve stimulation but also in response to hormones, chemicals, and local concentrations of oxygen, carbon dioxide, nitric oxide, and various ions.

11

Spinal Cord and Reflexes

Objective and Overview

After having discussed neurons and muscle in previous chapters, we consider next the motor system, starting with the lowest hierarchical level of this system, which consists of motoneurons of the spinal cord that directly control skeletal muscle together with their associated interneurons and neuronal circuitry.

The chapter begins with a gross anatomical overview of the vertebral column and its main subdivisions and a discussion of peripheral nerves, both cranial and spinal, and some of their primary features. The general neural organization of the spinal cord is considered starting with the transverse division of the gray matter, containing cell bodies, into laminae, with the motoneurons in the most ventral lamina and other neurons and interneurons in other laminae. The gray matter is surrounded by white matter comprising ascending and descending nerve fiber tracts.

The discussion then focuses on the somatomotor neurons of the spinal cord, highlighting various aspects of the morphology and physiology of α-motoneurons, including persistent inward currents (PICs) and the size principle. Renshaw cells and Ia inhibitory interneurons are considered and various schemes for controlling α-motoneurons by interneurons are presented. The effects of neuromodulatory inputs from higher centers are expounded.

The second part of the chapter is concerned with explaining the general features of spinal reflexes, followed by a detailed consideration of the flexion reflex, the stretch reflex, and the tendon organ reflex, including supraspinal influences. The chapter ends by considering reflexes elicited by stimulation, namely, the H-Reflex and the tonic vibration reflex.

Learning Objectives

To understand:

- The gross anatomical features of the vertebral column and peripheral nerves
- The general neural organization of the spinal cord and its main ascending and descending tracts
- The main features of α-motoneurons, their inhibitory interneurons, and their interconnection schemes
- The nature and role of the persistent input current in α-motoneurons
- The size principle governing recruitment and derecruitment of α-motoneurons
- The effects of neuromodulatory inputs from higher centers
- The nature and main features of spinal reflexes and, in particular, the flexion reflex, the stretch reflex, the tendon organ reflex, the H-Reflex, and the tonic vibration reflex.

11.1 Gross Features

11.1.1 Vertebral Column

Vertebrates are distinguished from other animal species by having a bony vertebral column that encloses the spinal cord, which is bathed in cerebrospinal fluid (CSF) inside the column. The vertebral column provides mechanical protection to the spinal cord in the same manner as the skull provides mechanical protection to the brain. The CSF adds a cushioning effect to both the brain and the spinal cord and serves other functions, such as transporting chemical substances between various parts of the nervous system, including the supply of nutrients and the removal of waste products.

In humans, the vertebral column normally consists of 33 vertebrae divided into five subdivisions (Figure 11.1), based on some distinguishing characteristics of the vertebrae in each subdivision. The five subdivisions are, rostrally to caudally:

1. Cervical, or neck, region comprising 7 vertebrae, denoted as C1 to C7. The C1 vertebra supports the skull, and the C2 vertebra serves as a pivot for C1.
2. Thoracic, or chest, region comprising 12 vertebrae, denoted as T1 to T12. These vertebrae support the 12 pairs of ribs, which are joined ventrally at the **sternum**, or breastbone.

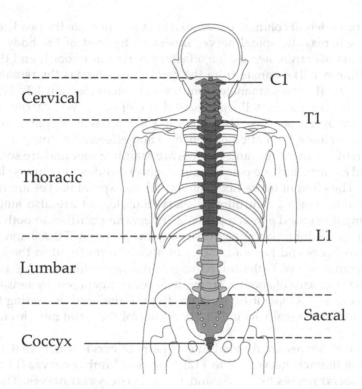

Cervical — C1

— T1

Thoracic

Lumbar — L1

Sacral

Coccyx

FIGURE 11.1
Vertebral column and its subdivisions.

3. Lumbar, or lower back, region comprising 5 vertebrae, denoted as L1 to L5.
4. Sacral, or thigh, region comprising 5 fused vertebrae (the **sacrum**) having a roughly triangular shape. The sacrum articulates laterally with the hip bones. The vertebrae in this region are denoted as S1 to S5.
5. **Coccyx**, or tailbone, region comprising 4 fused, rudimentary vertebrae.

All adjacent vertebrae, except the fused vertebrae of the sacrum and the coccyx, are separated by an **intervertebral disc** that forms a fibro-cartilaginous joint, which provides some slight flexibility to the vertebral column.

The gray matter of the spinal cord normally ends at the L1 vertebra, beyond which there exist only nerve fibers that form the dorsal and ventral roots of the lower spinal nerves.

11.1.2 Peripheral Nerves

In humans, there are 12 pairs of cranial nerves, bilaterally divided on each side of the skull, and 31 pairs of spinal nerves, bilaterally divided on each side of the vertebral column. The cranial nerves innervate the head and neck regions, whereas the spinal nerves innervate the rest of the body. Two of the 12 pairs of cranial nerves, the olfactory nerves (number I) and the optic nerves (number II) emanate from the cerebrum, whereas the remaining 10 pairs of cranial nerves emanate from the brainstem (Section 1.3). The olfactory nerves, carrying smell signals, and the optic nerves, carrying visual signals, are purely sensory. The remaining 10 cranial nerve pairs are mixed nerves, containing both afferent (sensory) and efferent (motor) nerve fibers, although five of these 10 cranial nerves are mainly motor and are sometimes classified as such, and one pair is mainly sensory and is sometimes classified as such. The efferent nerve fibers of cranial and spinal nerves are not only somatomotor, that is, controlling skeletal muscle, but are also autonomic, both sympathetic and parasympathetic, innervating cardiac, smooth muscle, and glands. Each human optic nerve contains between 770,000 and 1.7 million nerve fibers and has a diameter of about 1.6 mm within the eye. The largest cranial nerve is the mixed trigeminal nerve (number V), having a cross-sectional area of about 4.7 mm^2. Its sensory fibers convey sensations of touch, pain, and temperature from the front of the head, including the face and the top of the scalp. Its motor fibers control the facial muscles involved in chewing.

The spinal nerves are divided into: 8 pairs of **cervical nerves** (C1 to C8), 12 pairs of **thoracic nerves** (T1 to T12), 5 pairs of **lumbar nerves** (L1 to L5), 5 pairs of **sacral nerves** (S1 to S5), and 1 pair of **coccygeal nerves.** The spinal nerves emerge from the vertebral column through an opening between adjacent vertebrae, except for C1, which emerges between the first vertebra and the occipital bone of the skull. Thus, cervical nerves C1 to C7 emerge above the vertebra having the same name, whereas cervical nerve C8 emerges below vertebra C7, as there is no vertebra C8, and all the other spinal nerves emerge below the vertebra having the same name. All the spinal nerves are mixed nerves. The **cauda equina,** or horse's tail, refers to the bundle of spinal nerves and spinal nerve roots comprising the nerve pairs L2 to L5, S1 to S5, and the coccygeal nerves.

The human spinal cord is divided into 31 **spinal segments**, in transverse sections, so that a pair of spinal nerves emerges from each segment. Each segment of the spinal cord innervates, that is, receives afferent signals from and sends efferent signals to, a well-defined area of the body. The spinal segments and the body areas they innervate are homologous, that is, they correspond in position along the axis of the body. Thus, the segments that innervate the leg, for example, are below the segments that innervate the thigh.

The largest and longest **peripheral nerve,** that is, a nerve outside the central nervous system, is the sciatic nerve. In humans, it is a flat thick band, about 2 cm wide, formed by the grouping of spinal nerves L4 to S3. It originates in the lower back, runs through the buttock and thigh, and divides, usually at the back of the knee joint, into the tibial nerve and the common fibular (or peroneal) nerve. The sciatic nerve directly controls the muscles of the posterior thigh and the hamstring portion of the adductor magnus muscle. Its branches control the muscles of the leg and foot. These branches also convey signals from the skin of the lateral leg and the foot.

In a manner exactly analogous to skeletal muscle (Figure 9.1), the individual nerve fibers of a peripheral nerve are surrounded by a thin layer of connective tissue, the **endoneurium**. Groups of nerve fibers are bundled together into fascicles that are surrounded in turn by another layer of connective tissue, the **perineurium**. The whole nerve is ensheathed by a layer of connective tissue, the **epineurium**. Blood vessels run between the fascicles.

11.1.3 Neural Organization

The spinal cord has a butterfly-shaped gray region surrounded by white matter. As mentioned in Section 1.3, the gray region contains predominantly the cell bodies of neurons, whereas the white matter consists largely of bundles of myelinated nerve fibers running longitudinally through the spinal cord and forming tracts, as will be described later. Afferent fibers, whose cell bodies are in the dorsal root ganglion, enter the spinal cord via the dorsal roots, whereas efferent fibers leave the spinal cord via the ventral roots (Figure 11.2). The ventral roots of the human spinal cord contain a total of about 200,000 axons, including preganglionic sympathetic fibers. A short distance from the spinal cord, the ventral and dorsal roots from the same level combine to form a pair of spinal nerves, one on each side of the spinal cord.

Figure 11.2 illustrates, in cross section of the spinal cord, the ten **Rexed laminae**, or layers, of the spinal cord, defined mainly by their cell structure. The laminae are shown in the left half of the cross-section of Figure 11.2. Layers I to VI are in the dorsal horn of the gray matter, layers VII and X

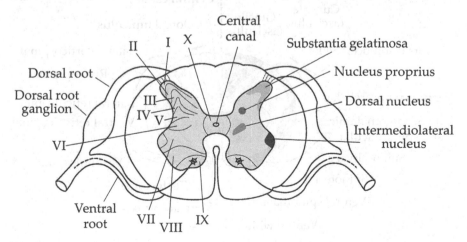

FIGURE 11.2
Cross section view of the spinal cord.

are in the intermediate zone, and layers VIII and IX are in the ventral horn. Motoneurons are located mainly in layer IX, interneurons are in layers V to VIII. Layer X is also known as the **gray commissure**. The gray matter of the spinal cord also contains various nuclei, shown in the right-half of Figure 11.2, which will be discussed later.

The gray matter of the spinal cord is surrounded by massive ascending and descending fiber tracts (Figure 11.3). The right-half of the figure shows the main tracts through the spinal cord that are involved in motor function, which will be referred to later (Section 12.2.5). The left half of the figure shows the main ascending tracts, that is, the tracts that carry signals to the brain. The ascending and descending tracts course through the spinal cord as three major longitudinal funiculi referred to as the **anterior, or ventral, funiculus,** the **lateral funiculus** and the **posterior, or dorsal, funiculus.** Just anterior to the layer X is the **ventral white commissure,** which is a fiber tract that connects the two halves of the spinal cord (Figure 11.3).

The sensory fibers that enter the spinal cord through the dorsal roots are part of what are referred to as the **first-order neurons** in the pathway from the periphery to the cerebral cortex. As mentioned before, the cell bodies of these first-order neurons are in the dorsal root ganglia. The **dorsal column–medial lemniscus pathway (DCML)**, also known as the **posterior column-medial lemniscus pathway (PCML)**, contains first-order afferents that convey sensations of fine touch, vibration, and proprioception. The pathway is divided into two components: (i) the **gracile fasciculus** that carries information from the lower limb, below T6, and terminates in the **gracile nuclei** in the medulla (Figure 12.17), and (ii) the **cuneate fasciculus** that

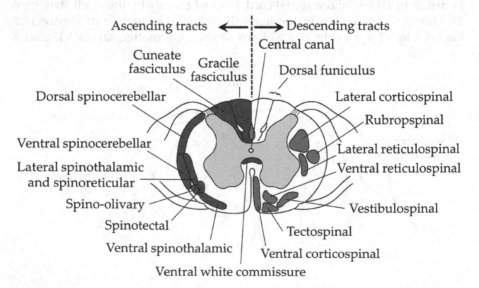

FIGURE 11.3
Ascending and descending fiber tracts of the spinal cord.

carries information from the upper limb, above T6, and terminates in the **cuneate nuclei** in the medulla. The neurons of the gracile and cuneate nuclei are second-order neurons in the pathway from the periphery to the cerebral cortex. Their axons decussate at the level of the medulla, travel up the brainstem as the **medial lemniscus** on the contralateral side, and terminate in the thalamus, which relays the signals to the cerebral cortex.

The gray matter of the spinal cord also contains nuclei of second-order sensory neurons that give rise to ascending tracts. The **spinothalamic tract** has two components: the **ventral**, or **anterior, spinothalamic tract**, which carries signals of crude touch and firm pressure, and the **lateral spinothalamic tract**, which carries signals of pain and temperature. After entering the spinal cord, the first-order afferents ascend 1–2 vertebral levels before synapsing with the second-order neurons, which are the neurons of origin of the spinothalamic tract in the **substantia gelatinosa** or the **nucleus proprius** (Figure 11.2). The axons of the tract decussate to the other side of the spinal cord via the ventral white commissure (Figure 11.3) and ascend the spinal cord and the brainstem to terminate in the thalamic nuclei. The sensations transmitted by the spinothalamic tract call for action, such as scratching in response to an itch, or the withdrawal from a painful stimulus.

Positioned closely to the lateral spinothalamic tract, and similarly having its neurons of origin in the dorsal horn of the spinal cord, is the **spinoreticular tract**. The axons of this tract decussate to the other side of the spinal cord and ascend the spinal cord to terminate on third-order neurons in the medullary-pontine reticular formation. The third-order neurons project to **intralaminar nuclei** of the thalamus, which in turn project diffusely to many parts of the cerebral cortex. In this way, pain reaches consciousness and results in behavioral arousal and a memory of the pain. In fact, this pathway through the reticular formation is considered part of the **ascending reticular arousal system (ARAS)**, which is responsible for regulating states of consciousness, alertness, and sleep-wake transitions.

Proprioceptive information is conveyed to the cerebellum, mostly ipsilaterally, through a number of tracts. Group Ia and Ib muscle afferents from the lower extremities make synapses on neurons in the **dorsal nucleus** (Figure 11.2), also known as **Clarke's column**, which extends from T1 through L2. The fast-conducting axons of the neurons of Clarke's column form the major part of the **dorsal spinocerebellar tract**, which also carries information from touch and pressure receptors and from secondary muscle spindle afferents. The axons of this tract ascend the spinal cord and project to the cerebellum through the inferior cerebellar peduncle.

The **ventral spinocerebellar tract** originates at lumbosacral spinal levels mainly in the medial part of lamina VII in Clarke's column. Axons first cross the midline in the spinal cord then ascend to the pons, where they enter the cerebellum through the superior cerebellar peduncle, then mostly cross again to end in the ipsilateral cerebellum. A small part of the tract, originating from the sacrococcygeal region, enters the cerebellum through the inferior cerebellar peduncle. The tract carries proprioceptive information for coordinated movement and posture of the entire lower limb. The tracts that covey the same type of information as the dorsal spinocerebellar tract and the ventral spinocerebellar tract, but from the upper part of the body, are, respectively, the **cuneocerebellar tract** and the **rostral spinocerebellar tract**. The cuneocerebellar tract originates in the **accessory cuneate nucleus**, which is located laterally to the cuneate nucleus in the medulla. Another tract that conveys information from proprioceptive and cutaneous receptors to the cerebellum is the **spino-olivary tract**. The tract originates in the medial part of laminae III and IV and the **central cervical nucleus**, located in the upper regions of the spinal cord. The tract terminates in the nuclei of the inferior olivary complex, which is the source of climbing fibers to the cerebellum (Section 12.2.4).

Another ascending tract is the **spinotectal tract**, or the **spinomesencephalic tract**, that originates in the anterolateral parts of the spinal cord and terminates in the inferior and superior colliculi (Section 12.2.4.2). It is believed to be involved in inhibiting or controlling pain sensations. The spinothalamic tract, the spinoreticular tract, and the spinotectal tract constitute what is designated as the **anterolateral system** that carries information from the skin to the thalamus.

Lamina VII of the spinal cord contains the **intermediolateral nucleus** (Figure 11.2), which extends from vertebral levels T1 to L2, and harbors the autonomic motor neurons of the entire sympathetic innervation of the body. In addition, the various laminae contain **propriospinal neurons** that transmit signals between segments of the spinal cord. The axons of these neurons form **propriospinal tracts** that could be ascending, descending, crossed, or uncrossed. Propriospinal neurons may receive strong peripheral input but generally little supraspinal input. The **C3-C4 propriospinal system** is a special descending system discussed in Section 12.2.5.5.

11.2 Somatomotor Neurons

We will discuss in this section some characteristics of motoneurons and their associated interneurons.

11.2.1 Motoneurons

11.2.1.1 General

Motoneurons are located in the ventral horn of the spinal cord (Figure 11.2) or in the cranial nerve nuclei of the brainstem. Their axons leave the central nervous system to innervate skeletal muscle. Hence, they are designated as "the final common path". They are also known as **lower motoneurons** in contrast to **upper motoneurons**, which originate in the motor cortex and brainstem and which activate lower motoneurons and associated interneurons.

The axons of lower motoneurons, or simply motoneurons, exit the brainstem and the spinal cord via the ventral roots to become eventually part of cranial and spinal peripheral nerves, respectively. There are three types of motoneurons: α-motoneurons, which innervate extrafusal muscle fibers; γ-motoneurons, which innervate intrafusal muscle fibers; and β-motoneurons, which innervate both types of muscle fibers (Figure 9.12). Of these three types of motoneurons, α-motoneurons are the most abundant, totaling about 120,000 in the human spinal cord. γ-motoneurons are roughly half as numerous as α-motoneurons. α-motoneurons are divisible into three subtypes according to the type of muscle fibers they innervate (Table 9.1), those innervating type IIB fibers being the largest, and those innervating type I fibers being the smallest.

The motoneurons that innervate a given muscle are grouped together into a **motoneuron pool** for that muscle. Motor pools are spatially organized in a manner that roughly reflects the peripheral distribution of the corresponding muscles. Thus, motoneurons are distributed longitudinally along the spinal cord according to the level along the body of the muscles they control; motoneurons innervating proximal muscles, such as trunk muscles, are medial to those innervating distal muscles, such as hand muscles; and motoneurons innervating flexor muscles are dorsal to those innervating extensor muscles. Motoneurons are illustrated in Figures 1.4 and 1.5.

The morphology and electrophysiology of α-motoneurons have been studied most extensively in the cat. These α-motoneurons have ellipsoidal bodies of varying size, the mean diameter varying from about 15 μm to about 80 μm. The largest α-motoneurons have a major axis of about 100 μm, a minor axis of about 70 μm, and a surface area of about 24×10^3 μm². About 10 primary dendrites, on average, emanate from the cell body, the diameter of a primary dendrite being as large as 20 μm. The radial extent of the dendritic tree is in the range of 1–3 mm, the total length of dendrites is about 20 mm, and the total dendritic surface area is about 0.15 mm². Axon diameters of α-motoneurons are in the range 10–16 μm. By contrast, axon diameters of γ-motoneurons are in the range of 3–8 μm.

An α-motoneuron receives input from over 10,000 different nerve fibers through more than 50,000 synapses. Most of the input to α-motoneurons is from interneurons. However, α-motoneurons receive monosynaptic inputs from Ia primary afferent fibers, as illustrated in Figure 6.13, as well as monosynaptic inputs from pyramidal cortical cells (Section 12.2.2.1) in primates but not in lower animals. When stimulated repetitively at interspike intervals of less than about 10 ms, the pyramidal cell-motoneuron synapse shows pronounced facilitation, that is, the amplitude of epsps due to successive spikes increases. This facilitation is believed to be functionally important in that it ensures that the α-motoneuron reaches threshold and fires in response to a train of high-frequency spikes from a pyramidal neuron. By contrast, the

Ia synapses of motoneurons of fast fatigable muscle fibers tend to become facilitated after repetitive stimulation, whereas the Ia synapses of motoneurons of slow muscle fibers tend to become depressed.

Conventionally, the resting membrane voltage of α-motoneurons is considered to be about −70 mV, the threshold around −60 mV, the equilibrium voltage for glutamate epsps about 0 mV, and the equilibrium voltage for GABA or glycine epsps around −80 mV. Unitary epsps and ipsps are 0.1–0.2 mV in amplitude. Monosynaptic epsps evoked by Ia fibers are largely produced via AMPA receptors and may include a slower component evoked via NMDA receptors. Motoneuron firing rates in the range 80–120 Hz have been recorded in human muscles during the onset of ballistic voluntary contractions.

Motoneurons are destroyed by the polio virus, by **motor neuron disease (MND)** (also known as **amyotrophic lateral sclerosis (ALS)**, or **Lou Gehrig's disease)**, and by **spinal muscular atrophy (SMA)**. In all cases, destruction of motoneurons results in partial or complete paralysis of the muscles involved. When the intercostal muscles responsible for breathing are destroyed, death by asphyxiation ensues.

In around 1 percent of cases of **poliomyelitis**, the polio virus spreads along certain nerve fiber pathways, preferentially replicating in and destroying motoneurons within the spinal cord, brain stem, or motor cortex, leading to the development of paralytic poliomyelitis. In the spinal form of paralytic poliomyelitis, α-motoneurons are destroyed, their axons degenerate, and the muscle fibers lose their innervation, resulting in muscle weakness and poor muscle control. If all the motoneurons of a muscle are destroyed, the muscle is paralyzed and will atrophy. Any limb or combination of limbs may be affected, and proximal muscles are usually more affected than distal muscles.

In ALS, motoneurons die, eventually leading to inability to speak, swallow, walk, use the hands, and breathe. The exact cause of ALS is not established. Contributing factors are believed to include exposure to environmental toxins and chemicals, infection by viral agents, auto-immune effects, loss of growth factors required for the survival of motoneurons, and genetic predisposition. SMA is also characterized by destruction of motoneurons and muscle wasting. It is a genetic disorder associated with mutation in the SMN1 gene.

11.2.1.2 Persistent Inward Current

Persistent inward current (PIC) is a depolarizing current that inactivates, only very slowly. As will become apparent from the following discussion, the PIC plays a crucial role in shaping the firing patterns of motoneurons in a manner that depends on both the state of the motoneuron, that is, its level of excitability, and the task being executed. PIC significantly amplifies the effects of synaptic input and can result in self-sustained firing of motoneurons. This is advantageous in some cases, as in the maintenance of posture.

The PIC was discussed in Section 7.4.5 in connection with dendritic bistability, which allows a short pulse of excitatory input to produce self-sustained firing that can be terminated by a short pulse of inhibition The PIC and its effect are illustrated in Figure 11.4 for a cat motoneuron. An afferent input, such as vibration applied to the tendon of the muscle that provides an excitatory sensory input to the given motoneuron, produces a synaptic input that activates the PIC when the soma is sufficiently depolarized under voltage clamp. A typical time course of the PIC recorded in the cell soma is shown. It is seen that the effect of the PIC is twofold: (i) it amplifies synaptic input several-fold, presumably due to an increase in $[Ca^{2+}]_i$, and (ii) it introduces an inward tail current that persists. In addition, the resulting depolarization reduces the threshold for motoneuron firing. The tail of the PIC can cause self-sustained firing of the motoneuron.

The PIC has two components: a Na^+PIC mediated by Na^+ channels in the soma and proximal dendrites, and a $Ca^{2+}PIC$ mediated by $Ca_v1.3$ channels (Section 7.3.2) in the mid-dendritic regions. The Na^+PIC activates and inactivates relatively fast, during which it contributes about half the PIC in spinal motoneurons. The $Ca^{2+}PIC$ activates more slowly, with a time constant of about 50 ms, and shows little or no time-dependent inactivation. The $Ca_v1.3$ channels that mediate the $Ca^{2+}PIC$ are facilitated by repeated depolarization due to increased $[Ca^{2+}]_i$. The Na^+PIC shows little or no hysteresis, that is, it activates and deactivates at nearly the same voltage on an ascending and descending voltage ramp, and has small tail currents. The $Ca^{2+}PIC$ shows strong hysteresis and has long tail currents. The Na^+PIC plays an essential role in the initiation of APs during rhythmic firing of motoneurons and may serve to amplify and prolong transient synaptic inputs that would otherwise be too fast to activate the $Ca^{2+}PIC$.

The PIC is due to the action of serotonin (5-HT) and norepinephrine (NE) neuromodulatory descending pathways (Section 6.3.2) that activate

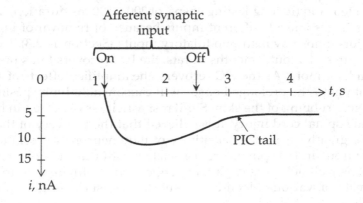

FIGURE 11.4
Time course of PIC.

metabotropic receptors on motoneurons. More will be said about these pathways and their various effects on motoneurons in Section 11.2.3. In fact, the effect of neuromodulatory inputs on motoneurons is so strong that, in their absence, all the excitatory inputs to a motoneuron will barely be sufficient to generate an action potential, let alone cause the intense firing that tetanizes the muscle fibers of the motor unit. Therefore, the loss of these inputs due to spinal injury, for example, produces an immediate, dramatic decrease in motoneuron excitability, manifested by almost complete loss of some reflexes and severe weakening of other reflexes.

The following effects of PIC on motoneuron firing should be noted:

1. PIC in high-threshold (FF) motoneurons decay with time, but are more stable in low-threshold (S) motoneurons, resulting in self-sustained firing. This makes sense, since self-sustained firing is appropriate for highly fatigue-resistant slow-twitch motor units.

2. Although the firing rate of a motoneuron increases markedly with the synaptic input in the presence of PICs, the firing rate "saturates", or increases very little, when the PIC is fully activated. This is because when the PIC in the dendrites is large enough, additional synaptic input will not significantly increase the dendritic current.

3. Because of hysteresis, PIC is deactivated at a more hyperpolarized level than it is activated. This results in a similar hysteresis in the recruitment and derecruitment of motoneurons.

4. Because of the facilitation of $Ca_v1.3$ channels by repeated depolarization, recruitment thresholds of motoneurons can be decreased with repeated stimuli.

PIC shows plasticity. There are two immediate effects of a loss of descending neuromodulatory inputs, as happens with spinal transection or severe spinal cord injury: (i) a sharp reduction in the magnitude of the PIC, as to be expected, and (ii) long-lasting epsps, of 200 – 500 ms duration, in motoneurons in response to afferent inputs because of removal of inhibition in the dorsal horn by neuromodulatory inputs (Section 11.2.3). However, in the course of about 3 months in rats, the PIC recovers to its pre-injury amplitude or more. As the PIC recovers, the amplified effects of afferent stimulation on the prolonged epsps will cause whole-limb spasms upon even gentle rubbing of the skin. Similar spasticity is observed in humans following spinal cord injury. It is believed that the recovery of the PIC is due to a greatly increased sensitivity of motoneurons to residual 5-HT from neurons in the spinal cord associated with the autonomic nervous system, as well adrenergic, glutamatergic, and cholinergic neurons and the effects of various blood-borne substances, such as hormones and peptides.

11.2.1.3 Size Principle

The recruitment of motor units was discussed in Section 10.2.4, and the size principle was introduced, according to which the order of recruitment normally proceeds from the smallest to the largest motor units of a given motoneuron pool. Since the size of a motor unit is directly related to the size of the cell body of the α-motoneuron of the given motor unit, the size principle implies that the smaller α-motoneurons discharge before the larger ones.

From a physiological viewpoint, the order of recruitment is dictated by which neurons reach threshold first. *For the same synaptic input*, that is, the same change in synaptic conductance, it follows from voltage division (Figure 6.2) that the neuron with the largest input impedance, will have the largest epsps and will reach voltage threshold first. This threshold is not significantly different between motoneurons of a given motoneuron pool. The input resistance varies inversely as the surface area of the neuron, assuming the same membrane resistivity, and the input capacitance varies directly as the surface area of the neuron. The smallest neuron will therefore have the largest input resistance and the smallest input capacitance. Its input impedance will be largest, so it will reach voltage threshold first. If current is injected at the soma, the range of currents required to reach threshold in different motoneurons is approximately 3 – 30 nA. However, this wide range of about 10:1 is not only due to variations in input resistance but is also due to the potent effects of neuromodulatory inputs on motoneuron excitability (Section 11.2.3). There is also some evidence that smaller motoneurons have a larger membrane time constant, which is presumably due to a higher membrane resistivity.

The size principle greatly simplifies the task of control of recruitment of motoneurons by higher centers, as it only becomes necessary to send approximately the same synaptic input to the target motoneurons irrespective of their sizes, and the required order of recruitment is automatically followed. However, as noted earlier, it is sometimes necessary not to follow the order of recruitment according to the size principle. This is particularly true of rapid, repetitive movements, where the prolonged relaxation time of slow muscles would impede the movement. There is also evidence that in some muscles, motor units are grouped into functional groups that are preferentially activated during the performance of a certain task such as a stride during locomotion (Section 13.4).

11.2.2 Interneurons

The importance of interneurons is highlighted by the fact that spinal interneurons are much more numerous than principal neurons such as motoneurons. Interneurons receive synaptic input from sensory axons, from descending axons from higher centers, from other interneurons, and from collaterals, or branches, of motoneuron axons. Before they leave the spinal cord, most axons of α-motoneurons give off one or more such collaterals that make synaptic contact with interneurons in the spinal cord.

The axons of interneurons may be restricted to one segment, or vertebral level, of the spinal cord, or they may extend over many segments. We consider, in what follows, two examples of interneurons involved in motor function: Renshaw cells and Ia inhibitory interneurons. In addition, there are Ib inhibitory neurons that are involved in the tendon organ reflex (Section 11.3.4).

11.2.2.1 Renshaw Cells

The Renshaw cell (Figure 1.5) is an inhibitory interneuron that has been studied quite intensively. Originally, Renshaw cells were shown to be excited by branches of axons of motoneurons within the gray matter of the spinal cord (Figure 1.5). These branches are known as **recurrent collaterals**. It has since been established that Renshaw cells also receive sensory inputs, mainly from segmental sensory fibers, as well as inputs from other interneurons and descending systems. A Renshaw cell may be contacted by more than one α-motoneuron axon collateral and may synapse on multiple α-motoneurons and γ-motoneurons, in the same motoneuron pool, or in the motoneuron pools of synergist muscles. The inhibitory feedback connection to the same, or **homonymous**, motoneuron pool is referred to as **homonymous recurrent inhibition,** whereas the inhibition of motor neuron pools of other muscles is referred to as **heteronymous recurrent inhibition**. Figure 11.5 illustrates these two types of inhibition.

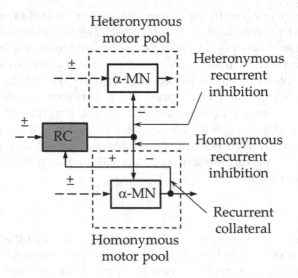

FIGURE 11.5
Homonymous and heteronymous recurrent inhibition of α-motoneurons (α-MN) involving a Renshaw cell (RC).

Excitation by motoneuron axon collaterals elicits prolonged epsps in Renshaw cells, lasting more than 50 ms and causing them to fire a high-frequency burst of spikes. The Renshaw cell synapses on motoneurons are located more remotely on the dendrites, and the Renshaw cell ipsp in a motoneuron is due to the combined action of many Renshaw cells. The ipsps last about 40 ms and are apparently due to co-release of both glycine and GABA, with strychnine as a reversible competitive antagonist to glycine and picrotoxin as a noncompetitive antagonist at $GABA_A$ receptors. The release of glycine is inhibited by *Clostridium tetani*, bacteria that live in the soil, resulting in tetanus disease, characterized by convulsions because of the removal of inhibition.

In addition to their terminations on α- and γ-motoneurons, Renshaw cells synapse on other Renshaw cells, on neurons of the ventral spinocerebellar tract, and on other interneurons.

Renshaw cells have excitatory nicotinic as well as muscarinic (ACh) and AMPAergic (glutamate) synapses. They also have inhibitory glycinergic and GABAergic synapses, the GABA-induced ipsps being of longer duration than the glycine-induced ipsps.

It is believed that the negative feedback of homonymous recurrent inhibition protects a muscle against damage from excessive force of contraction. It has also been suggested that the dynamics of recurrent inhibition results in pauses in the firing of α-motoneurons, rather than continuous firing, which reduces motor unit fatigue. Renshaw cells may also contribute to the desynchronization of motoneuron discharges, particularly at low rates of discharge, which enhances the smoothness of the force developed by muscles. Renshaw cells are also involved in the modulation of the activity of spinal circuits by higher centers, as may be required during voluntary movement or rhythmic activity such as locomotion.

Problem 11.1

Will the response of an α-motoneuron to peripheral afferent excitation increase or decrease immediately following a volley of peripheral electrical excitation of its axon? Explain why.

ANS.: Decrease because of activation of Renshaw cells by antidromic APs along the axon of the α-motoneuron.

11.2.2.2 Ia Interneurons

Because muscles can only pull, and cannot push, motion in opposite directions at a given joint requires the action of two opposing sets of muscle, the agonists and the antagonists for the motion in question (Section 9.3.4). Thus, the agonists and the antagonists are alternately activated in the execution of rhythmic movements. In the simplest form of walking, for example, the hips are alternately flexed and extended. It may seem natural to assume that when movement in a given direction is required, only the agonists should be

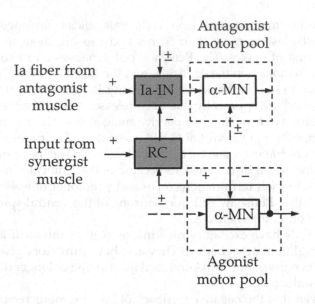

FIGURE 11.6
Reciprocal inhibition of α-motoneurons (α-MN) involving an Ia inhibitory interneuron (IN) and a Renshaw cell (RC).

activated, with the antagonists inactivated, a state referred to as **reciprocal inhibition**. For a given net or resultant force, inhibiting the antagonists minimizes the force that needs to be developed by the agonists and hence reduces the energy expended. In fact, training to increase the force developed at a given joint, such as the knee, involves learning to deactivate the antagonists. This is not always the case, however, as is discussed in Section 13.1.2.

Reciprocal inhibition involves Ia inhibitory interneurons and is illustrated in Figure 11.6. Suppose that an agonist muscle for a given movement is contracting, so that its antagonist muscle, together with the muscle spindles in this muscle, are being stretched, which activates the Ia primary afferents of these muscle spindles. These afferents make excitatory synapses on Ia-interneurons that are in turn inhibitory to the α-motoneurons of the antagonist muscle (Figure 11.6). The contraction of the agonist muscle therefore relaxes the antagonist muscle though inhibition mediated by the Ia inhibitory interneurons of the antagonist muscle. A Renshaw cell is also shown in Figure 11.6 that is excited by the α-motoneurons of the synergist muscle and which in turn inhibits the Ia interneuron that inhibits the α-motoneurons of the antagonist muscle, as may be required in some cases. This inhibition of a neuron that inhibits a target neuron is an example of **disinhibition** of the target neuron.

11.2.2.3 Interneuronal Circuits

We will next consider in this section some hypothetical schemes involving spinal interneurons (Figure 11.7), which serve to illustrate the roles that these

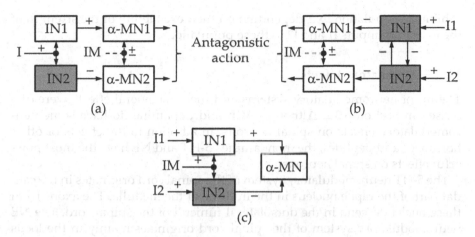

(a)   (b)

(c)

FIGURE 11.7
Hypothetical schemes for the control of α-motoneurons (α-MN) by inhibitory interneurons (IN); (a) turning on and off α-MN1 and α-MN2 by the level of an input I to the INs; (b) reciprocal activation of α-MN1 and α-MN2 by inputs I1 and I2; (c) the same input IM can have opposite effects on α-MN depending on the state of interneurons IN1 and IN2.

can play in controlling motor behavior. It should be emphasized that these schemes are highly simplified, and that a given population of interneurons may be involved in more than one scheme.

Figure 11.7a illustrates two populations of α-motoneurons α-MN1 and α-MN2 that have antagonistic actions. These motoneuron populations receive inputs from two populations of interneurons, respectively: an excitatory population IN1 and an inhibitory population IN2, in addition to other inputs IM denoted by the dashed line. Assuming that IM alone is sufficient to activate α-MN2 but not α-MN1, then a high level of input I will turn on α-MN1 and turn off α-MN2. Conversely, a low level of input I will turn off α-MN1 and turn on α-MN2. The antagonistic actions of two populations of α-motoneurons can thus be controlled by the input I.

In Figure 11.7b, IN1 and IN2 are both inhibitory interneuron populations that mutually inhibit one another. These could be, for example, a population of Renshaw cells and a population of Ia inhibitory interneurons considered in the preceding sections. Activation of I1 reduces the excitation of α-MN1 and increases the excitation of α-MN2 through disinhibition. Conversely, activation of I2 reduces the excitation of α-MN2 and increases the excitation of α-MN1. Controlling the levels of excitation of α-MN1 and α-MN2 allows controlling the force developed by each of the two sets of antagonist muscles and their degree of coactivation (Section 13.1.2).

Figure 11.7c illustrates a basic scheme that can be used to select one of two opposite actions on a motoneuron population α-MN, depending on the states of the two interneuron populations IN1 or IN2. The common input IM to these populations may activate either the excitatory population IN1 or the inhibitory population IN2, depending on their excitability or the strength of the excitatory inputs I1 and I2 to these populations.

11.2.3 Modulatory Effects

The major neuromodulatory systems and their behavioral effects were discussed in Section 6.3.2. Although ACh and dopamine do have some neuromodulatory effects on spinal neurons, in addition to the effects of other hormones and peptides, the monoamines 5-HT and NE have the most powerful effects on spinal neurons.

The 5-HT neuromodulatory system of the spinal cord originates in the caudal part of the raphe nucleus in the midline of the medulla. The axons from these nuclei descend in the dorsolateral funiculi of the spinal cord. The NE neuromodulatory system of the spinal cord originates mainly in the locus coeruleus in the pons and in the lateral tegmentum. Both systems terminate in a diffuse manner at practically all levels of the cord and have many similar effects on spinal neurons.

Monoamines exercise their effects through G protein second-messenger systems (Section 6.3). Generally speaking, these effects vary progressively from predominantly excitatory in the ventral horn to predominantly inhibitory in the dorsal horn, depending on the receptors involved. The excitatory effects involve $G_{\alpha q}$ protein subunits and 5-HT2 and NEα1 receptors, whereas the inhibitory effects involve $G_{i/o}$ protein and 5-HT1 and NEa2 receptors and is mostly extrasynaptic (Section 6.2.3), affecting both interneurons and astrocytes. On the sensory side, the inhibition in the dorsal horn is mostly presynaptic inhibition of high-threshold afferents and is linked to the suppression of pain pathways, as may be required during exposure to a high-stress environment or stimulus. On the motor side, the monoamines strongly suppress the flexion reflex (Section 11.3.2). In the intermediate regions of the spinal cord, there is moderate facilitation of interneurons receiving inputs from Ia, II, and Ib afferents.

In the ventral horn, motoneurons are densely covered with excitatory 5-HT and NE receptors. Usually, these neuromodulatory inputs alone do not generate an output from motoneurons. They strongly facilitate PICs (Section 11.2.1.2) and modify the properties of ionic channels. They increase the intrinsic excitability of motoneurons by virtue of the following effects:

1. Inhibiting K^+ channels, such as $K_{ir}3$ channels, that largely determine the resting membrane voltage and conductance (Section 7.3.3). Inhibition of these channels depolarizes the membrane and increases its time constant, thereby increasing the amplitude and duration of epsps.

2. Reducing the medium afterhyperpolarization (mAHP, Section 7.3.5) by inhibiting $Ca_v2.1$, $Ca_v2.2$ channels (Section 7.3.2), as well as SK

channels (Section 7.3.3). The reduction in mAHP increases the firing frequency of motoneurons.

3. Increasing the I_h current (Section 7.3.3), which depolarizes the membrane and increases excitability.

The excitatory effects of neuromodulation are quite diffuse and affect both agonists and antagonists. Localized inhibition can turn off these excitatory effects, including PICs, when they are not needed. The inhibition could be through Ia reciprocal inhibition synapses (Section 11.2.2.2) that are located close to the soma. Moreover, the interneurons that mediate this type of inhibition receive strong inputs from descending systems via the cortico-, rubro-, and vestibulospinal tracts (Section 12.2.5). Thus, these interneurons could be used by descending motor commands to adjust motoneuron excitability, as may be needed for any given movement. Other sources of inhibition include presynaptic and recurrent inhibition, as well as inhibitory neuromodulatory effects exercised through $GABA_B$ receptors on motoneurons.

Although the excitatory effects of 5-HT and NE on motoneurons are similar, these neuromodulatory systems are active under different conditions (Section 6.3.2). Hence, the effects of these monoamines are said to be task-dependent. 5-HT activity increases with the intensity of motor outflow, as with increasing speed of locomotion. NE activity increases with the state of arousal. Both systems have low activity during sleep, particularly during rapid-eye-movement (REM) sleep, and both systems are strongly active during fight-or-flight situations, where large forces may be required.

11.3 Spinal Reflexes

11.3.1 General

A **reflex** is a largely involuntary and stereotyped action that occurs relatively fast in response to a certain stimulus. An example is the flexion reflex discussed below and which results in rapid withdrawal of a limb, such as the hand, upon touching a hot object. A large number of reflexes can occur in the body under physiological conditions, which could involve glands, internal organs, the autonomic nervous system, the somatic nervous system, the brain, or the spinal cord. An example of a purely endocrine reflex that involves endocrine glands without mediation of the nervous system is the regulation of glucose level in the blood. Thus, an increase in the level of glucose in the blood causes the pancreas to release more insulin, which allows more glucose to enter cells, thereby reducing glucose level in the blood. Some reflexes can be conditioned through learning, so that the reflex action is elicited by a stimulus other than the natural stimulus for the reflex.

Neural reflexes could be somatic, affecting skeletal muscles, or autonomic, involving internal organs. The pathway from the receptor that normally initiates the reflex to the effector that executes the reflex action is the **reflex arc**. Depending on the type of neural reflex, the effector could be muscle – skeletal, smooth, or cardiac – or a gland. An example of an autonomic reflex is the control of blood pressure. When this falls, the pressure receptors in the aorta and carotid arteries cause the sympathetic system to increase the heart rate and the cardiac output so as to restore the blood pressure to its normal level. Neural reflexes are "wired" in the neuronal circuitry, but can be influenced by the action of inputs from other parts of the nervous system on various neurons in the reflex arc. Reflexes are thus modifiable, and almost all somatic reflexes can be overridden by voluntary control. The eyeblink reflex and the flexion reflex discussed below are examples of protective reflexes, whereas other reflexes, such as the aforementioned blood pressure and blood glucose level serve a regulatory function through negative feedback control.

Somatic reflexes are classified as **spinal reflexes** or **cranial reflexes**, according to whether they mainly involve the spinal cord or the brain, respectively. Examples of cranial reflexes are the eyeblink reflex and the vestibulo-ocular reflex (Section 12.2.5.3) that stabilizes the gaze on a particular object as the head is rotated.

This section discusses some spinal reflexes that are important in the control of movement and posture. Spinal reflexes occur at the level of the spinal cord and are fast because the reflex arc does not extend beyond the spinal cord and involves a minimal number of neurons. A spinal reflex arc typically consists of five components:

1. A type of receptor that generates APs in response to the stimulus characteristic of the given reflex. Receptors can be divided into three broad classes: **proprioceptors**, **exteroceptors**, and **interoceptors**. Proprioceptors signal information about the position and motion of skeletal muscle and joints. They comprise the muscle spindle receptors that signal muscle length and velocity (Section 9.4.2), the Golgi tendon organs that signal muscle tension (Section 9.4.1), and the joint receptors found in the connective tissue of joints, particularly within the fibrous tissue surrounding the joint capsules and ligaments. Joint receptors respond to changes in the angle, direction, and rate of movement of the joint. However, they are rapidly adapting, so they provide little information about the resting state of the joint. The vestibular apparatus and proprioceptors in muscles and joints convey information about the position of the body in space, muscle tension, muscle length and its rate of change, and joint movement. Exteroceptors are sensory nerve endings and specialized receptors that respond to external stimuli and are located in the skin, mouth (taste receptors), eyes (photoreceptors), ears (auditory receptors),

and nose (smell receptors). Exteroceptors in the skin, that is, cutaneous receptors, convey information about touch, pressure, changes in temperature, and harmful stimuli. The latter type of receptors is referred to as **nociceptors** and is associated with sensation of pain. Interoceptors convey information about conditions within the body. Examples are the pressure receptors, or baroreceptors, that monitor blood pressure, chemoreceptors that respond to the partial pressure of oxygen in the blood, and chemoreceptors that respond to the concentration of hydrogen ions in the cerebrospinal fluid.

2. Sensory neurons having processes in the form of nerve fibers and axons, which form afferent pathways that conduct sensory information from the skin, muscles and joints to the spinal cord. The cell bodies of these sensory neurons are located in the dorsal root ganglia, as previously mentioned (Section 9.4).

3. Signal-processing neuronal structures between the sensory axons and the α-motoneurons. These comprise synaptic inputs to the α-motoneurons and to any intervening interneurons that synapse on α-motoneurons. If the sensory axons synapse directly on α-motoneurons, the reflex is **monosynaptic**. The pathway from a given sensory axon to an α-motoneuron is **disynaptic** if it involves two synapses, is **trisynasptic** if it involves three synapses, etc. If the pathway of all the sensory axons of a given reflex to the α-motoneurons has a fixed number of more than one synapse, the reflex is **multisynaptic**. The term multisynaptic, or **oligosynaptic**, is also used to denote reflexes involving two or three synapses. If the sensory fibers of a given reflex have different pathways with different numbers of synapses, the reflex is **polysynaptic**. The term polysynaptic is also used to denote reflexes involving more than three synapses.

4. The final neuronal common pathway, namely, the α-motoneurons and their axons, which constitute the somatic efferent fibers innervating skeletal muscle.

5. Skeletal muscle, discussed in Chapters 9 and 10.

11.3.2 Flexion Reflex

The **flexion reflex**, also known as the **withdrawal reflex**, moves a limb away from a harmful, or noxious, stimulus. Suppose, for example, that one accidentally touches a hot plate (Figure 11.8). This activates cutaneous thermoreceptors, resulting in APs traveling to the spinal cord over afferent nerve fibers and innervating interneurons in the spinal cord. Although, for simplicity, a single interneuron is shown in the reflex arc in Figure 11.8, several populations of both excitatory and inhibitory interneurons are in fact

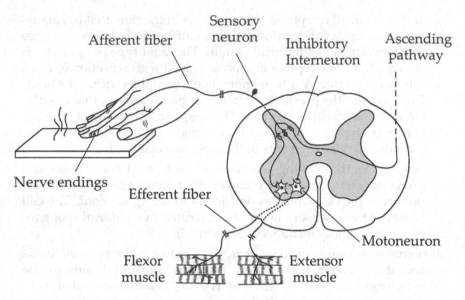

FIGURE 11.8
Flexion reflex pathways.

involved, making the reflex polysynaptic. The afferent input, acting through interneurons, eventually reaches α-motoneurons that mediate the following responses:

1. Excitation of the α-motoneurons that innervate the homonymous flexor muscles of the arm, where the homonymous muscles are those that are directly associated with the receptors that initiate the reflex. This leads to the withdrawal of the arm away from the hot plate. A more intense stimulus may excite the α-motoneurons that cause internal rotation of the shoulder (Section 9.3.4). In withdrawing the arm from the painful stimulus, the muscles of the arm and shoulder act as synergist muscles. It may be noted that the intensity of the stimulus is coded both in the mean frequency of APs as well as the number of afferent fibers activated. Thus, a more intense stimulus will stimulate receptors over a larger area of the skin, which activates a larger number of afferent fibers in parallel. This parallelism in the organization of the nervous system is not portrayed by single-line diagrams such as Figure 11.8.

2. Concomitant with the excitation of the flexor α-motoneurons, the interneurons inhibit the α-motoneurons innervating the extensor muscles (Figure 11.8), as explained in connection with Ia interneurons (Section 11.2.2.2). This leads to relaxation of the extensor muscles that are antagonists to the contracting flexor muscles – an example of reciprocal inhibition referred to earlier.

3. Withdrawal of a limb from a noxious stimulus generally involves some postural adjustments. This is most easily seen when the foot is withdrawn from a noxious stimulus. Elevation of the foot shifts the weight of the body to the contralateral leg on the other side of the body. The extensor muscles of this leg, which are the antigravity muscles, should contract, and the flexor muscles inhibited in order to support the additional load. Therefore, the afferent input should also activate interneurons that excite extensor α-motoneurons, and inhibit α-motoneurons of the contralateral leg. This is an additional reflex, resulting from the flexion reflex, and is referred to as the **crossed extension reflex**, or the **crossed extensor reflex**.

The flexion reflex withdraws the limb from the noxious stimulus within 0.5 s. A variety of sensory receptors could be involved in the stretch reflex such as cutaneous receptors sensitive to pain or to changes in temperature, as when the noxious stimulus is due to touching a hot object. All the afferents involved in the flexion reflex are grouped together as **flexor reflex afferents**. These involve Type III thinly myelinated fibers and Type IV unmyelinated fibers. Because of the involvement of a relatively large number of muscles that are innervated by α-motoneurons at various longitudinal levels of the spinal cord, the flexion reflex is intersegmental, that is, it is not confined to a single segment or level between adjacent vertebrae. The brain is not directly involved in the reflex movement, but is informed through ascending pathways, so one becomes aware of what happened and feels the pain, which helps to "ingrain" the stimulus as harmful.

11.3.3 Stretch Reflex

The block diagram of the **stretch reflex**, also known as the **myotatic reflex**, is illustrated in Figure 11.9. The reflex arc comprises the following:

1. Primary and secondary muscle spindle receptors.

2. The Ia and II afferent fibers innervating, respectively, the primary and secondary muscle spindle receptors.

3. The synaptic terminations of these afferent fibers in the spinal cord. The connections of Ia fibers with α-motoneurons are monosynaptic, that is, the Ia fibers directly synapse with α-motoneurons without any intervening interneurons. The α-motoneurons with which the Ia fibers synapse are mainly those innervating the homonymous muscle, that is, the same muscle from which the Ia fibers emanate and, to a lesser extent, the synergist muscles. The group II fibers from the secondary muscle spindle receptors also connect to α-motoneurons but via one or two interneurons, which makes the connection disynaptic or trisynasptic, respectively.

4. The last two elements of the reflex arc are the α-motoneurons and the muscles.

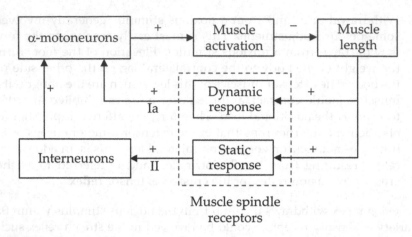

FIGURE 11.9
Block diagram of stretch reflex.

A pathway in a block diagram such as that of Figure 11.9 is marked with a plus or a minus sign according to whether an increase in the magnitude of the variable representing a given block increases or decreases, respectively, the magnitude of the variable represented by the block on which the pathway terminates. Consider what happens in the **tendon tap reflex,** elicited, for example, by a brisk tap on the tendon just below the kneecap. The tap stretches the quadriceps muscles and their muscle spindles, but the rapid, brief stretch evokes a strong, synchronous response mainly from the primary muscle spindle receptors because of their sensitivity to the rate of change of stretch (Figure 9.13). The increased activity in these fibers increases the excitation in α-motoneurons of the homonymous and synergist muscles, as indicated by the plus sign in this pathway. For a relatively weak stimulus, the excitation of some of the α-motoneurons of the homonymous muscle will be sufficient to generate APs in these neurons, whereas in other α-motoneurons of the homonymous muscle, and in the α-motoneurons of the synergist muscles, the excitability is increased but not sufficiently to generate APs. These neurons constitute what is referred to as a **subliminal fringe,** and would discharge APs in case of a stronger stimulus, or in case of excitatory inputs from other parts of the spinal cord or from higher centers. This would potentiate the reflex, producing a stronger than normal response.

The discharge of the α-motoneurons activates the muscle, as indicated by the plus sign in the pathway. The muscle contracts, decreasing the muscle length, so that the pathway from the muscle activation block to the muscle length block in Figure 11.9 has a negative sign. An increase in muscle length activates the muscle spindle receptors, so the pathways from the muscle length block to those of the muscle spindle receptors have a plus sign. When the homonymous muscle, the quadriceps in this case, contracts, the knee is extended, resulting in a knee jerk. The quick shortening of the muscle unloads the muscle spindles, which turns off the spindle afferents and relaxes the muscle. The stretch reflex elicited by a rapid muscle stretch and mediated through the velocity-dependent component of the primary muscle spindle response is referred to as a **phasic stretch reflex** because of its transient nature. The delay involved is in the range of 20–30 ms and is mainly due to the conduction times along afferent Ia fibers and efferent motoneuron axons.

The stretch reflex exemplified in Figure 11.9 is also activated by a slow, sustained stretch that would evoke discharges from both the primary and secondary spindle receptors, thereby exciting α-motoneurons of the muscles involved. The increased force of contraction opposes the stretch and restores the muscle length, as well as the activity of the spindle receptors, back to their pre-stretch levels. This slow stretch reflex, mediated by the length-dependent component of the response of muscle spindles, is referred to as the **tonic stretch reflex**.

It is seen that the stretch reflex can potentially function as an automatic mechanism, or a negative feedback control system (servomechanism), for keeping the muscle length constant, and hence fixing the position of the joint associated with the muscle. A disturbance of muscle length from its equilibrium value, or **set point**, tends to restore the equilibrium muscle length through negative feedback. That the feedback is negative feedback can be ascertained by multiplying the signs of all the pathways in Figure 11.9 around a closed loop, say, from the spindle receptors, to α-motoneurons, to muscle activation, to muscle length, and back to the spindle receptors. The sign of the product is negative. However, to strictly maintain a constant muscle length, despite load changes, requires that the ratio of a change in force to a change in length ($\Delta F/\Delta L$) is very large, since ΔL would be very small if the length is being held constant. This means that the stiffness of the system would be very high, or, in control system terminology, the loop gain would be very large. This is not in fact the case, and will be referred to in Section 13.5.1. The loop gain is affected by the activity of the γ-motoneurons and by the excitability of α-motoneurons, which is governed by neuromodulatory influences (Section 11.2.3).

Although the stretch reflex is encountered in most skeletal muscle, it is particularly strong in antigravity muscles, where it would tend to automatically maintain posture, this being an important function of the stretch reflex. The antigravity muscles of the back and legs are extensor muscles because the joints involved will tend to flex, or buckle, under the influence of gravity. Any such movement will stretch the extensor muscles, thereby eliciting a stretch reflex that would restore posture. The antigravity muscles of the arm, however, are the flexors, not the extensors because gravity would tend to extend the arm joints. The monosynaptic component of the stretch reflex can play an important physiological role during fast movements or for a fast readjustment of posture upon loss of balance. For example, in jumping from a moderate height to land with both feet on the ground, if the stiffness of the knee is insufficient, the knee extensors would be rapidly stretched, and the monosynaptic component of the reflex would produce a fast contraction of the extensors that would prevent the buttocks from hitting the ground. The monosynaptic component of the stretch reflex is fast, having a response time of few tens of milliseconds because: (i) there is only one synaptic delay, (ii) the Ia fibers are the largest and hence the fastest conducting peripheral fibers, and (iii) the monosynaptic connections of the Ia fibers are on the larger α-motoneurons, which in turn have large, fast conducting axons.

An important point to note in Figure 11.9 is the effect of activation of γ-motoneurons. Activation of these neurons increases the activity of the muscle spindle receptors, as explained in Section 9.4.2.3. This, in turn, would excite α-motoneurons and cause the muscle to contract so as to offset the increased activation of the muscle spindles, with the following important implications: (i) a new set point for muscle length is established, so that the stretch reflex will now maintain constant a different value of muscle length; and (ii) if the stretch reflex is to remain operative during muscular contraction, then the γ-motoneurons must be activated as well to prevent muscle shortening from unloading and hence inactivating the muscle spindles. This concomitant activation of both types of motoneurons is referred to as **alpha-gamma coactivation**.

The following should be noted concerning the stretch reflex:

1. Excitation of α-motoneurons of agonists automatically leads to inhibition of α-motoneurons of antagonists, in accordance with reciprocal inhibition discussed earlier.

2. The stretch reflex is confined to one or two segments of the spinal cord.

3. Even a monosynaptic reflex is modifiable, through presynaptic inhibition on Ia fibers (Section 6.4) as well as through changes in inputs to α-motoneurons from higher centers.

4. The tendon tap reflex is of clinical value. A weak reflex is generally indicative of some abnormality in the components of the reflex arc, whereas an exaggerated reflex is usually indicative of excessive excitatory inputs from higher centers to the spinal cord.

11.3.4 Tendon Organ Reflex

Also known as the **inverse myotatic reflex** and **autogenic inhibition**, the reflex arc comprises the Golgi tendon organs (GTOs) as the receptors (Section 9.4.1), Ib afferent fibers terminating on Ib-interneurons, the α-motoneurons of the homonymous muscle, and the muscle itself, as illustrated in Figure 11.10. The Ib afferents excite Ib-interneurons that inhibit α-motoneurons, making the reflex disynaptic. As in the case of the stretch reflex, the effect of the sensory fibers, which is inhibitory in this case, is exerted on synergist muscles

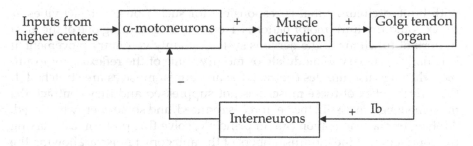

FIGURE 11.10
Block diagram of tendon organ reflex.

as well, and inhibition of the α-motoneurons of the agonist muscles leads to excitation of the α-motoneurons of the antagonist muscles. This occurs through excitation by Ib fibers of excitatory interneurons that excite the α-motoneurons of the antagonist muscles, and is an example of **reciprocal innervation**, a special case of which being reciprocal inhibition.

As can be seen from Figure 11.10, the GTO reflex is also a negative feedback system that tends to maintain a constant tension in a muscle. If muscle tension drops due to fatigue, for example, the discharge of GTOs is reduced, which reduces the inhibition of the α-motoneurons of the homonymous muscle, which causes increased contraction of the muscle and restores the tension to its original level As explained in connection with Figure 11.8, the product of the signs of the pathways in Figure 11.10 is negative, as it should be for a negative feedback system. Again, as in the case of the stretch reflex, the maintenance of a nearly constant tension requires a very high loop gain in the system, which is also not the case.

The GTO reflex can also protect the muscle from excessive tension and is believed to be important in the execution of fine movements at low levels of tension, as a finer control is possible through opposing actions of excitation and inhibition. The GTOs are also involved in the **clasp-knife reflex** that occurs under certain pathological conditions, as in spasticity, that is, abnormally large, stretch-resistant tension of some extensor muscles, such as those of the arm. If the arm is forcibly flexed by a gradually increasing, external force, the resistance to flexion suddenly collapses at some point, similar to the snapping closure of a clasp-type pocket knife. This sudden relaxation of the muscle occurs when the inhibition through GTOs becomes strong enough to overcome the muscle excitation.

11.3.5 Supraspinal Influences

In accordance with the hierarchical organization of the motor system (Section 12.1), the lowest hierarchical level consists of motoneurons of the spinal cord and brainstem that directly control skeletal muscle, together with their associated interneurons and neuronal circuitry. Thus, the spinal cord is endowed with built-in circuitry that is responsible for such "low-level" activities as reflexes and rhythmic motor patterns. However, in executing one of the most important functions of the nervous system, mainly voluntary movement, it is usually necessary to modulate or modify some of the reflexes. For example, when agonist muscles contract, the antagonist muscles are stretched. If the stretch reflex of these muscles is not suppressed and they contract, the movement becomes stiff, the net force is reduced, and some energy is wasted. Alpha-gamma coactivation can, in principle, solve this problem by relaxing the contraction of the intrafusal fibers of the antagonist muscle, allowing the muscle to be stretched without triggering the stretch reflex during a voluntary movement. Similarly, if we intentionally wash our hands in excessively hot water, the withdrawal reflex should be suppressed. Conversely, if we cautiously approach a shower to test how hot the water is, the threshold of the withdrawal reflex may have to be lowered.

The modulation of reflexes is implemented primarily by descending tracts, as will be discussed more fully in Section 12.2.5. These tracts, illustrated on the right-hand side of Figure 11.3, are somatotopically organized. Thus, medial pathways control axial muscles, that is, muscles close to the body axis, and are involved in the maintenance of posture and balance as well as coarse control of axial and proximal muscles. These pathways are predominantly under brainstem control and include the **vestibulospinal tract**, the **ventral reticulospinal tract**, the **tectospinal tract**, and the **ventral corticospinal** tract. On the other hand, lateral pathways control both proximal and distal muscles and are involved in most voluntary movements of arms and legs. These pathways are predominantly under cortical control and include the **lateral corticospinal tract**, the **lateral reticulospinal tract** and the **rubrospinal tract**. Similarly, axons from the premotor cortex terminate mostly in motor pools of proximal limb muscles, whereas axons from the primary cortex and supplementary motor area terminate mostly in the motor pools of the hand and digital muscles.

Presynaptic inhibition controlled by higher centers through their effects on GABAergic interneurons acting on the synaptic terminals of primary afferent fibers (Section 6.4), can direct sensory inputs to specific motoneuronal targets and modulate their effectiveness. Higher centers can also influence sensory input through the action of 5-HT and NE (Section 11.2.3) as well as extra-synaptic influences through release of histamine or peptides. The 5-HT and NE neuromodulatory systems inhibit the flexion reflex through their projections to the dorsal horn.

11.4 Reflexes Elicited by Stimulation

We discuss in this section reflexes that are elicited by electrical or mechanical stimulation.

SPOTLIGHT ON TECHNIQUE 11A: ELECTROMYOGRAPHY

Electromyography is the study of the electrical activity of normal and diseased skeletal muscle, using electrodes that record electrical activity of a muscle resulting from muscle action potentials. The recorded electrical activity is known as the **electromyogram** (EMG). The EMG could be elicited by voluntary contraction of the muscle or by electrical stimulation applied to a peripheral nerve innervating the muscle. For gross recording of electrical activity, the recording electrodes are normally metal surface electrodes placed on the skin over the belly of the muscle, whereas for more localized recording, needle electrodes are usually inserted inside the muscle. A needle electrode typically consists of a wire, insulated except for its tip, which is inserted inside a hypodermic needle of a diameter less than 1 mm. The electrical activity recorded is the voltage drop between the tip of the wire and the surrounding needle due to extracellular circulating currents of the muscle action potentials. A needle electrode would normally record electrical activity from a few motor units, but with proper placement, it could record the activity of a single motor unit. The electrodes are normally connected to a differential amplifier and to a loudspeaker, since the frequencies present in an EMG are within the audio range. "Listening" to the EMG is helpful in finding an optimal location for the recording electrodes and in diagnostic applications, as will be described later.

The upper trace of Figure 11.11 shows an EMG of a relatively forceful voluntary contraction resulting from the summation of action potentials from many motor units and lasting for about 100 ms. The amplitude of the recorded EMG depends on the placement of the recording electrodes. The upper trace of Figure 11.11 has a peak-to-peak amplitude of about 2 mV. The EMG is quantified by first taking the absolute value of the recorded voltage and then integrating, with respect to time, to obtain the area under the voltage vs. time trace. This area is

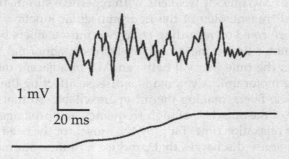

1 mV

20 ms

FIGURE 11.11
EMG of voluntary contraction.

taken as a measure of the strength of the activity and is shown as the lower trace in Figure 11.11.

Clinical EMG provides important information on the physiological status of skeletal muscle and its nerve supply. In cases of muscle paralysis, for example, it allows identification of the site of the lesion, or disorder, which could be in the higher brain centers, motoneurons, nerve fibers, neuromuscular junction, or the muscle itself. A typical application of EMG clinical measurement is that of the conduction speed of APs along motor nerve fibers in a peripheral nerve. An abnormally low speed would indicate some abnormality, or neuropathy, of the nerve fibers. To measure the conduction speed along the large motor nerve fibers in the leg, for example, an EMG is recorded from a foot muscle while applying an electrical stimulus successively to two positions on the leg and determining from the EMG the delay, or latency, between the stimulus and the response in each case. Dividing the distance between the two stimulation sites by the difference in the latencies from the two EMG records gives the required speed of conduction. The electrical stimulus is typically a few tens of volts in amplitude and 0.5–1 ms in duration, applied between two adjacent metal electrodes using an electrically isolated stimulator. To measure the conduction speed along the largest sensory fibers, the same procedure is applied, but now the nerve is stimulated at a distal point along a limb and successive recordings are made from the peripheral nerve at more proximal locations. Dividing the distance between the recording locations by the difference in the latencies of the corresponding responses gives the required conduction speed.

The EMG can be used to diagnose some diseases of the muscular system. Myasthenia gravis is an autoimmune disease that destroys acetylcholine muscle receptors, resulting in muscular weakness and eventually paralysis. In its early stages, it is manifested by drooping eyelids, double vision, a toneless voice, and difficulties in chewing and swallowing. The EMG quickly weakens with repetitive stimulation because of impaired transmission of the neuromuscular junction. In a sensitive test, performed by recording from two muscle fibers belonging to the same motor unit using a needle electrode, abnormal variation is observed in the time interval between APs of adjacent muscle fibers in the same motor unit. Myotonia is a disease affecting the membrane of the muscle fibers, making them hyperexcitable, so that the electrical activity of the muscle is of high frequency and prolonged duration, with a long relaxation time. Tapping the muscle or the needle electrode results in intense discharges that produce a thunderstorm-like sound in the loudspeaker. Muscular dystrophy is a genetic disease carried by a mother, and may also be carried by her daughters, but afflicts their male offspring only. It is a degenerative disease of the muscle, in which there is atrophy of some fibers, swelling of others, an increase in connective tissue separating muscle fibers, and deposition of fat between fibers that have become hypertrophied. With voluntary contraction, the action potentials are low in amplitude, short in duration, and high in frequency, producing a high-pitched whirring sound in the loudspeaker. With sustained contraction, the electrical activity alternately increases and decreases as the muscle begins to fatigue, at which time there is a marked reduction in electrical activity.

When a motor nerve is severed in an accident, for example, the severed part degenerates, so that the previously innervated muscle fibers are paralyzed and cannot be contracted voluntarily or reflexively. No electrical activity can be recorded from the affected muscle fibers following denervation. But after about a week or two, the denervated muscle fibers undergo small, random, and asynchronous contractions, which are not visible under the skin but which are associated with small fibrillation potentials that can be recorded with needle electrodes. If reinnervation does not occur, the muscle fibers atrophy, that is, become smaller and smaller in size and eventually die. If reinnervation occurs, as happens when the point of severance of the nerve is sufficiently far from the motoneuron cell body, the muscle fibers cease to atrophy, and the fibrillation potentials disappear gradually and are slowly replaced by normal muscle action potentials.

When a muscle is partially denervated, as can happen in poliomyelitis when some – but not all – of the motor neurons innervating a muscle are damaged, nondegenerated axons sprout new branches and innervate muscle fibers whose innervation has been lost. This results in "giant" motor units having a much larger number of muscle fibers than normal and producing correspondingly larger voltages in the EMG.

11.4.1 H-Reflex

The **H-reflex**, or **Hoffmann reflex**, is elicited by electrical stimulation of a muscle nerve, as illustrated in Figure 11.12. As the stimulus strength is increased from a low level, the largest Ia sensory fibers are stimulated first. This is because the largest fibers have the smallest internal resistance; so, for a given voltage stimulus applied externally to the mixed nerve, the voltage across the axonal membrane of the largest-diameter fibers will reach threshold first (Section 4.3.1). Two APs will be generated in the vicinity of the cathode-stimulating electrode and will travel in opposite directions (Figure 11.12a, top of diagram). The AP that reaches the sensory endings of the primary afferents will simply depolarize these endings temporarily and is extinguished (left-pointing dashed arrow). The AP that travels along an Ia fiber to the spinal

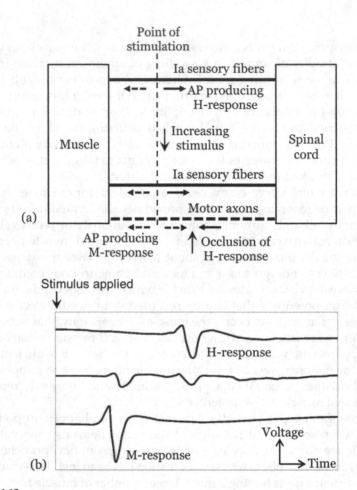

FIGURE 11.12
The H-reflex. (a) Stimulation of sensory and motor nerves with increasing stimulus strength;
(b) variation of H-response and M-response with stimulus strength.

cord reaches the synapse of this fiber on an α-motoneuron of the homonymous muscle. The APs generated in the α-motoneurons by the excited Ia fibers
will travel along the axons of the α-motoneurons to the muscle, causing it to
contract. The EMG recorded is the H-response, illustrated in the top trace of
Figure 11.12b. The response shown has a latency of about 20 ms. Of this, about
0.5 ms is the delay at the synapse between the Ia fibers and the α-motoneurons,
the rest being propagation time of the APs along the Ia fibers and the axons of
the α-motoneurons. When the stimulus strength is large enough to excite all
the Ia fibers in the nerve, the H-response reaches its maximum.

As the stimulus strength is increased, the motor axons begin to be stimulated directly by the externally applied stimulus, producing a response of
about 7 ms latency. This direct response is referred to as the M-response, or
Muscle response. As in the case of the Ia fibers, two APs are generated by the
stimulus, traveling in opposite directions along the motor axon (lower part

of the diagram in Figure 11.12a). The AP traveling in the normal direction toward the muscle is the **orthodromic** AP, whereas the AP traveling in the opposite direction is the **antidromic** AP. This action potential temporarily depolarizes the α-motoneuron.

As the M-response appears, the H-response gets smaller, as can be seen from the middle trace of Figure 11.12b. The reason is that the antidromic AP along the motor axon collides with the AP generated in the α-motoneuron by the Ia input, and which would otherwise contribute to the H-response (lower part of the diagram in Figure 11.12a). When two APs traveling in opposite directions along an axon collide, they are both annihilated because neither AP can propagate through the refractory zone of the other AP. This means that the motor axons that participate in the M-response cannot contribute to the H-response. As the M-response increases, the H-response will therefore get smaller (middle trace of Figure 1112b). Eventually, when the stimulus strength is large enough to activate all the motor fibers, the H-response disappears (bottom trace of Figure 11.12a).

11.4.2 Tonic Vibration Reflex

The **tonic vibration reflex (TVR)** is elicited by applying an electric vibrator to the muscle-tendon junction, the vibration being typically about 100 Hz in frequency and 1 mm or less in amplitude. The result is a slow contraction of the muscle that starts a few seconds after the application of vibration, builds up gradually to a level that is sustained as long as the vibration is applied, then gradually subsides over a few seconds after the vibration ceases. Recordings from α-motoneurons during the TVR reveal a slow depolarization with some epsps that are synchronized with the vibration and which are due to the monosynaptic terminations of the Ia afferents of the primary muscle spindles that are activated by the vibration. The slow depolarization is due to polysynaptic pathways mainly involving primary and secondary muscle spindles but also, and to a lesser extent, GTOs and cutaneous receptors.

The TVR can be at least partially suppressed voluntarily. Monosynaptic reflexes are suppressed during a TVR by presynaptic inhibition of the Ia afferents. The TVR is not a simple spinal reflex but is rather a complex reflex that involves supraspinal and intersegmental spinal mechanisms.

Summary of Main Concepts

- In humans, the vertebral column normally consists of 33 vertebrae, divided into five subdivisions: cervical, thoracic, lumbar, sacral, and coccyx.

- In humans, there are 12 pairs of cranial nerves and 31 pairs of spinal nerves.

- The gray matter of the spinal cord is divided transversely into ten Rexed laminae, labeled from I to X, dorsally to ventrally, except that lamina X surrounds the central canal of the cord. Longitudinally, ascending and descending fiber tracts surround the gray matter.
- Afferent fibers enter the spinal cord via the dorsal roots, and efferent fibers leave the spinal cord via the ventral roots.
- PIC plays a crucial role in shaping the firing patterns of motoneurons in a manner that depends on both the state of the motoneuron and the task being executed. PIC greatly amplifies the effects of synaptic input and can result in self-sustained firing of motoneurons.
- According to the size principle, small-size α-motoneurons innervating slow-twitch, fatigue-resistant muscle fibers are recruited first, by increasing synaptic input, whereas large-size α-motoneurons innervating fast-twitch, fatigable muscle fibers are recruited last. Derecruitment follows the reverse order.
- Interneurons, both excitatory and inhibitory, play a crucial role in controlling the activity of motoneurons. Renshaw cells mediate recurrent inhibition, whereas Ia inhibitory interneurons mediate reciprocal inhibition.
- The 5-HT and NE neuromodulatory systems strongly increase the intrinsic excitability of motoneurons, while they have inhibitory effects in the dorsal horn.
- A reflex is a largely involuntary and stereotyped action that occurs relatively fast in response to a certain stimulus.
- Examples of spinal reflexes are: the flexion reflex that moves a limb away from a harmful stimulus, the stretch reflex in response to muscle stretch, and the tendon-organ reflex in response to muscle force. Examples of reflexes elicited by stimulation are the H-reflex and the tonic vibration reflex.
- Spinal reflexes are modulated by supraspinal inputs, as may be required for particular types of movement.

12

Brain Motor Centers and Pathways

Objective and Overview

Whereas the preceding chapter was concerned with the lower hierarchical level of the somatomotor system, the present chapter discusses the higher levels of this system. The neuroanatomy and neurophysiology of some of the components of this system have been investigated in considerable detail, and responses of individual neurons or groups of neurons have been correlated with applied stimuli and with behavior. Functions have been ascribed to the various components of the higher levels of the somatomotor system based on a wealth of clinical observations and experimental results, including the effects of genetic modifications in animals. Yet, the manner in which the neuronal structures and properties of a given component bring about the ascribed functions is, in most cases, only poorly understood or has been postulated in rather general terms. This is even more so when it comes to the interactions of the various components of the somatomotor system to produce a given motor action. Hence, the aim of this chapter is to present various aspects of the higher levels of the somatomotor system that are meaningful and relevant to a general understanding of motor performance.

The chapter begins with explaining some general features and advantages of the hierarchical organization of the somatomotor system. The highest functional level of this hierarchy is then presented, namely, the motor association areas of the cerebral cortex that are concerned with the planning of movement or the formulation of the motor program. The main areas involved are considered and their primary roles explained.

The bulk of the chapter covers the middle hierarchical motor level whose function is to translate the planned movement into an operational form that can be executed by the lower level. The middle level comprises four components: the primary motor cortex, the basal ganglia, the cerebellum, and the brainstem nuclei together with the descending, extrapyramidal tracts that they give rise to. Each of these components is discussed, whenever feasible, in terms of its basic anatomy, afferent and efferent connections, neuronal organization and relevant neurophysiological properties, primary functions, and the main clinical disorders associated with the component in question. The motor brainstem nuclei considered are: the red nucleus, the reticular nuclei, and the vestibular nuclei; the extrapyramidal tracts involved in somatomotor activity being: the rubrospinal, reticulospinal, vestibulospinal, and tectospinal tracts. Some of the reflexes involving the head, neck, eyes, and the rest of the body are also considered.

Learning Objectives

To understand:

- Some general features of the hierarchical organization of the somatomotor system and its advantages
- The motor association areas and their involvement in the planning of movement
- Some basic anatomy and physiology of the primary motor cortex and the organization of the descending pathways
- The basic organization and physiological features of the basal ganglia and their correlation with clinical disorders
- The motor and nonmotor functions of the basal ganglia and the main clinical disorders of these ganglia
- The basic anatomy of the cerebellum, its afferent and efferent connections, distinctive neuronal organization and relevant physiology, primary functions, and the main clinical disorders associated with cerebellar lesions
- The contribution of the rubrospinal and tectospinal tracts to the control of movement
- The general anatomy and divisions of the reticular nuclei and their main functions as well as the main functions of the reticulospinal tract
- The basic anatomy of the vestibular nuclei, their afferent and efferent connections, and the function, pathways, and adaptation of the vestibulo-ocular reflex
- The main functions of the vestibulospinal tract

12.1 Hierarchical Organization

The somatomotor system, like the rest of the nervous system, is hierarchically organized, as this would offer many advantages. The complexity of the system precludes a centralized organization whereby a dominant center controls every detail of system behavior, as this would involve an enormous number of possible actions. The system becomes dauntingly complicated, unwieldy, inefficient, and sluggish. The hierarchical organization relieves the hypothetical center of a nonhierarchical organization of the mostly routine chores, and of having to specify every detail of system behavior. Maintenance of posture and stereotyped movements like reflexes or rhythmic motor patterns, for example, are handled at the lower levels in the hierarchy, effectively and without undue delay. The underlying circuitry at the lowest level can then be utilized by higher levels to execute more complex movements with more global and relatively simpler commands. Tasks are thus divided amongst different organizational levels, thereby allowing more effective and efficient planning, coordination, and control. Nevertheless, activity of lower levels is monitored by higher levels and directed as may be required.

Functionally, but not strictly anatomically, three levels of hierarchical organization of the somatic motor system can be distinguished: (i) the higher-level centers that are primarily concerned with the planning of movement, (ii) the middle-, or interpretation-level, centers that translate the planned movement into an operational form, and (iii) the lower, or execution level, centers that implement the required muscular activity. The highest levels are in the cerebral cortex and the lowest levels are the lower motoneurons in the brainstem and the spinal cord and their associated neuronal circuits. However, as will become evident from the discussion in this and the following chapter, this does not imply an exclusive specialization or serial progression between these aspects of movement. Rather, a given level is "predominantly" involved in one aspect of movement, but the activities of different levels are often intertwined with some shared inputs, and reciprocal, convergent, and divergent connections.

An important difference between the lower levels in the brainstem and spinal cord and the higher levels in the cerebral cortex, is that the neural circuits of the brainstem and spinal cord are largely "hard-wired" to generate complex motor patterns when triggered by higher centers or by external factors. On the other hand, cortical neural circuits and, to some extent, those of the middle level are generally "plastic", in the sense that their responses and functionality can change through learning or as a result of disuse or injury. Thus, if part of the cortical neuronal population that controls hand muscles is destroyed through loss of blood supply, for example, and the hand is not used, then the hand representation of the remaining undamaged part of the neuronal population is lost as neighboring areas controlling elbow and shoulder regions expand into the disused area. But if the hand is used, the undamaged hand areas expand into neighboring areas so as to restore hand use. Such reorganization evidently involves changes in the numbers and properties of synapses. A dramatic example of plasticity of the cerebral cortex is in full hemispherectomy, which involves the surgical removal of the frontal, parietal, temporal, and occipital lobes on one side as a treatment of severe, intractable epilepsy or, in some cases, a malignant brain tumor or as a result of severe brain injury due to an accident. After a recovery period of less than a few months, the patient quite remarkably is able to function almost normally. As to be expected, the younger the patient is, the fuller the recovery is.

12.1.1 Higher Levels

Little is known for certain about where exactly in the brain, and how, are somatic motor commands issued. A rather philosophical-type question is how a somewhat abstract notion such as an 'intention' to reach for an object becomes translated to electrical signals in the brain.

From a functional point of view, the cerebral cortex can be divided into **primary areas** and **association areas**. The primary areas are those where sensory signals are first received by the cortex or from which output signals of the cortex directly emanate. Examples of primary sensory areas are the **primary visual cortex** in the occipital lobe (Figure 1.9), the primary auditory cortex in the temporal lobe, and the **primary somatosensory cortex** in the postcentral gyrus of the anterior portion of the parietal lobe (Figure 12.1).

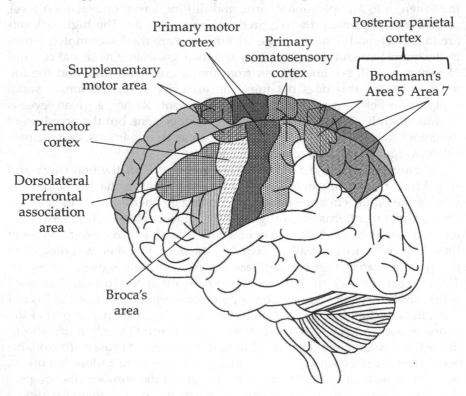

FIGURE 12.1
Motor areas of the cerebral cortex.

An example of a primary output area is the **primary motor cortex** in the precentral gyrus of the posterior portion of the frontal lobe, the primary motor cortex being separated from the **primary somatosensory cortex** by the **central sulcus** (Figure 12.1). The primary motor cortex is also referred to as the **somatomotor cortex, M1,** or **Brodmann's area 4.** In contrast, association areas, which constitute a considerably larger area of the cortex, are those areas where signals from different sensory modalities are integrated, or areas associated with "higher mental functions".

Movements are planned or programmed in the motor association areas of the cerebral cortex with the involvement of other brain regions, depending on the type of movement concerned. Planning of a movement generally involves: (i) knowledge of the surroundings and of the position of the body with respect to these surroundings, based on sensory inputs, which may include visual, auditory, and proprioceptive inputs; and (ii) a decision on what is judged to be the appropriate action, based on factors such as past experience (memory), motivation, and emotional state. Hence, planning of movement generally involves other, non-motor, cortical, and brain regions, which necessitates connections between these regions and motor association area, particularly parietal regions (for sensory inputs) and frontal regions (for higher mental functions). Anatomically, the motor areas are located between these two regions.

The somatomotor association areas comprise the following main structures (Figure 12.1), each of which may be divisible into functional units, based on their neuronal structure and connectivity with other regions of the cortex:

1. The **premotor cortex**, also known as the **premotor area (PMA)**, just anterior to the lateral part of the primary motor cortex. It uses information from other cortical regions to select movements, based on external events, such as visual cues, so that the movements are appropriate to the context of the intended action, particularly as regards the position of various body parts. For example, the premotor cortex is involved in the planning of the movement to cross the street while waiting for a pedestrian "go" signal. When this signal appears, the primary motor cortex becomes involved in initiating the movement. In line with its functions, the premotor area has reciprocal connections with several regions of the parietal cortex.

   Just anterior to the lower part of the premotor cortex in the dominant hemisphere, which is the left hemisphere in most people, is **Broca's area (Brodmann's areas 44 and 45)**, which extends ventrally into the inferior frontal gyrus of the frontal lobe. Broca's area is involved in speech production, which is a motor activity involving muscles of the vocal cords.

2. The **supplementary motor cortex**, also known as the **supplementary motor area (SMA)**, just anterior to the medial part of the

primary motor cortex. It is involved in the planning and coordination of complex movements, such as those requiring both hands, and in the planning of movements based on past experience as well as those initiated internally, that is, not in response to an external stimulus. The supplementary motor cortex and the premotor cortex are sometimes referred to as **Brodmann's area 6**, and they project to the primary motor cortex as well as to lower-level motor regions. The caudal part of the SMA, together with the primary motor cortex, receive inputs from the somatosensory cortex and the posterior parietal cortex. The rostral part of the SMA has reciprocal connections with the prefrontal cortex, that is, the anterior part of the frontal lobe implicated in complex cognitive functions and decision-making.

The SMA is also the site of the largest magnitude of the **readiness potential**, which is a slow, field voltage that is produced some 0.5–1.0 s before the onset of a voluntary movement, as recorded by an electromyogram (Spotlight on Free Will and Consciousness, Chapter 13).

3. The **posterior parietal cortex**, located directly posterior to the somatosensory cortex, also known as **Brodmann's areas 5 and 7**. It receives somatosensory, proprioceptive, and visual inputs and uses them to determine such aspects as the positions of the body and the target in space. It thereby produces internal models of the movement to be made, prior to the involvement of the premotor and motor cortices. These internal models are made available to the other motor association areas through the projection of the posterior parietal cortex to the premotor cortex and the rostral part of the SMA. Moving up the stairs, for example, depends on sensory inputs as to the height and width of the next step in the stairway and requires postural adjustments that shift the body center of gravity forwards.

The motor association areas are thus able to plan a movement, initiated internally or in response to some external cue, from its starting point to its destination, based on the position of the body in space, on the locations of targets and surrounding objects, and on past experience, motivation, and emotional state. Clearly, the prefrontal cortex, located rostral to motor association areas, plays a critical role in evaluating a given situation and reaching a decision as to the most appropriate choice of movement to make.

Focal lesions of premotor areas are generally manifested in impairment of the ability to choose the appropriate course of action. Lesions of the ventral premotor cortex impair the ability to use visual information about an object to control the hand so as to grasp an object in a manner that is appropriate for the object's size, shape, and orientation. Lesions of the dorsal premotor cortex, or supplementary motor area, impact the ability to learn and recall arbitrary sensorimotor transformations, such as temporal sequences of movements or conditional stimulus–response associations. Lesions of prefrontal areas associated with the supplementary motor area produce deficiencies in the initiation and termination of movements.

12.2 Middle Hierarchical Level

12.2.1 General

The middle level of the somatomotor system hierarchy comprises the primary motor cortex, the basal ganglia, the cerebellum, and motor nuclei of the brainstem. These structures are shown in gray in the block diagram of Figure 12.2, which depicts the main interconnections between the middle-level structures and between these and the rest of the nervous system.

Broadly speaking, the function of the middle level is to receive inputs from the higher centers and generate the required **motor program**, that is, the spatiotemporal pattern of action potentials that would activate the neurons of the lower levels so as to produce the desired movements. In generating the motor program, the middle-level structures utilize sensory inputs they receive from muscle, joints, eyes, and the vestibular apparatus.

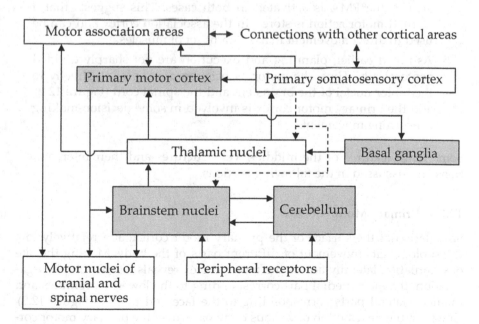

FIGURE 12.2
Block diagram of the main connections of the middle-level structures of the somatomotor system.

The distinction between planning and execution of movement is strikingly illustrated by the results of brain-imaging studies that seek to identify the areas of the brain that are involved in the performance of certain motor tasks. The following observations are noteworthy:

1. When subjects were requested to perform some finger movements from memory, activity was detected in cortical areas that included the somatosensory cortex, the posterior parietal cortex, parts of the prefrontal cortex, the premotor cortex, the supplementary motor cortex, and the primary motor cortex. But when the subjects were asked to mentally rehearse the movement without actually moving their fingers, no activity was detected in the primary cortex while other areas remained active.

2. Voluntary movements can be planned, but their execution can be delayed or cancelled altogether, unlike reflex actions.

3. When an activity is repeated often enough and is learned, so that it becomes almost automatic, the motor association areas become less involved, but the primary motor cortex becomes more involved as the activity becomes more precise.

4. After one learns to sign one's name with the dominant hand and then attempts to sign with one's toes, for example, the same hand area in the PMA is activated in both cases. This suggests that a learned motor action is stored in the association cortex and can be used to direct movement in a different set of muscles.

5. As noted earlier, planning and execution are not sharply divided anatomically. Thus, the motor association areas project directly to the motor nuclei of the brainstem and the spinal cord (Figure 12.2), and the primary motor cortex is involved in some decision-making concerning movement.

Some main features of the middle-level structures and their interconnections are discussed in the following sections.

12.2.2 Primary Motor Cortex

Stimulation of the surface of the primary motor cortex, at a relatively low threshold, elicits movement of different parts of the body. Moving the site of stimulation laterally across the motor cortex reveals a somatotopic organization, the most medial part corresponding to the lowest extremities and the most lateral parts corresponding to the face and tongue (Figure 12.3). However, the distribution of various body parts over the primary motor cortex is not uniform, resulting in a caricature of a human figure referred to as the **motor homunculus**, or little person. A relatively large cortical area

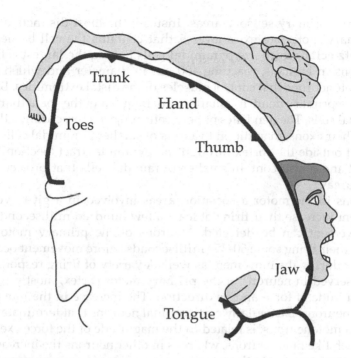

FIGURE 12.3
Motor homunculus.

representation is associated with finer movements, which involve small motor units and hence a larger number of motor neurons that need to be controlled. This is the case, for example, with:

1. The hand, including the four fingers and the opposable thumb, which underlies the amazing manual dexterity of humans.
2. Facial muscles, conveying a variety of facial expressions, which is important for social interaction.
3. The mouth and tongue, involved in vocalization.

It should be noted that the somatotopic organization depicted in Figure 12.3 is a diagrammatic representation of the responses in a frontal plane through the primary motor cortex. It does not imply a one-to-one mapping between cortical neurons and muscles. In fact, at the neuronal level, individual muscles and joints are represented at multiple sites in the primary motor cortex in a complex pattern, and cortical stimulation generally activates several muscles rather than individual muscles.

The primary motor cortex consists of six layers, like the rest of the neocortex. However, the cell-packed granular layer 4 is much less prominent than in the primary sensory areas. Instead, the most distinctive layer of the primary motor cortex is layer 5 that contains the cell bodies of the giant **Betz cells**, which are pyramidal cells having the largest cell bodies of neurons in humans, reaching 100 µm in diameter and which activate lower motoneurons that control muscles of the distal extremities. Betz cells have more primary dendritic shafts branching out of the soma than typical pyramidal cells. The dendrites of Betz cells project to practically all cortical layers. Their axons, in addition to axons of smaller pyramidal cells of layer 5, project outside the cortex through the pyramidal tract (Section 12.2.2.1). Layers 2 and 3 also contain smaller pyramidal cells that project to other cortical areas.

Neurons in the motor association areas involved in a given voluntary movement increase their firing at least a few hundred milliseconds before any movement can be detected. Neurons of the primary motor cortex increase their firing some 50–150 milliseconds before movement occurs and generally during the movement as well. A variety of firing responses have been observed in neurons of the primary motor cortex, mostly related to the magnitude of force and its direction. The increase in the rate of firing of some neurons, particularly corticospinal neurons that terminate directly on spinal motoneurons, is related to the magnitude of the force exerted at a joint, much like motoneurons, whereas in other neurons the increase in the rate of firing is related to the rate of change of force. Some neurons increase their firing rate in relation to the direction of force as well as its magnitude, whereas the activity of other neurons corresponds to kinematic variables, such as position and velocity of movement. Although individual neurons may have a maximum firing rate in a particular direction, the variation of firing rate of these neurons with direction is not highly selective. However, the average increase in the firing rate of a population of these neurons strongly correlates with the actual direction of force. Thus, the direction of the force is encoded not by the rate of firing of single neurons, but rather in the average increase of firing in a population of neurons. In tasks that require precise control of force, some neurons in the primary motor cortex increase their firing when the force decreases rather than increases. These neurons are believed to provide a more closely controlled derecruiting of motor units in the muscles involved. In many cases, neurons of the primary cortex are found to receive strong sensory input from the limb whose muscles they project to.

The primary motor cortex receives inputs from the basal ganglia and the cerebellum, via various thalamic nuclei, as will be elaborated later for outputs from these regions. It has interconnections with the somatosensory cortex, the motor association areas, and the frontal cortex.

In humans, lesions of the primary motor cortex disturb the dexterous execution of movements and cause deficits ranging from muscle weakness and discoordination to paralysis when upper motoneurons are completely destroyed.

SPOTLIGHT ON TECHNIQUES 12A:
RECORDING FROM THE BRAIN

Brain electrical activity is most conveniently recorded by surface electrodes placed on the scalp – a recording known as the **electroencephalogram** (**EEG**), referred to in Section 8.1.5. The EEG arises mainly from the synaptic inputs to the large pyramidal, cortical cells. An excitatory synaptic input to the apical dendrites produces an inward current, resulting in a positive charge intracellularly and a negative charge extracellularly (Figure 12.4a); current then flows through the main dendrite and back through the membrane and extracellularly to the apical dendrites, thereby completing the current loop, as required by conservation of charge. This extracellular current flow from lower regions, nearer to the cell soma, toward the apical dendrites, results in a negative voltage being recorded by a surface electrode near the apical dendrites, with respect to a distant reference electrode (Figure 12.4a). On the other hand, excitatory synaptic input close to the soma produces an extracellular current that flows toward the apical dendrites, resulting in a positive voltage being recorded by a surface electrode near the apical dendrites, with respect to a distant reference electrode (Figure 12.4b).

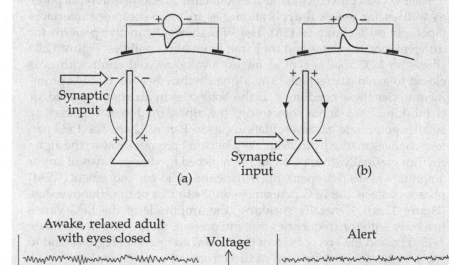

FIGURE 12.4

(a) Voltage recorded with excitatory synaptic input applied to apical dendrites of pyramidal cells; (b) voltage recorded with excitatory synaptic input applied close to the soma of pyramidal cells; EEG recorded in awake, relaxed state (c) and in the alert state (d).

Because the pyramidal cells are oriented parallel to one another and perpendicularly to the surface of the cortex, the voltages resulting from the extracellular currents of individual pyramidal cells are additive and are reinforced by the underlying rhythmic cortical activity (Section 8.1.5). The resulting voltages are large enough to be recorded by electrodes placed on the scalp. Nevertheless, the relatively high resistance of the bony skull attenuates the magnitude of the recorded voltages to the range 10–100 μV. To record these small voltages with minimal noise, active electrodes are used that combine a low-noise, Ag/AgCl electrode with an integrated circuit (IC) amplifier. To allow comparative interpretation of the EEG for clinical purposes between different subjects, the placement of EEG recording electrodes is standardized according to the 10–20 electrode system, so called because the spacing is based on intervals of 10 and 20 percent of the distances on the scalp from side to side and from front to back. The original system allowed for 19 electrodes, but has since been extended to 70 electrodes, if necessary. The reference electrode is commonly attached to one earlobe, and the ground electrode is attached to the mastoid, which is the back part of the temporal bone, on the same side of the head. The electrodes are usually embedded in a cap that is worn over the head.

The EEG has proved valuable for both clinical and research purposes as well as for practical applications, as in brain-computer interfaces (Spotlight on Techniques 13A). The EEG shows distinctive patterns for different, normal states and for some brain abnormalities. Figure 12.4c shows an EEG that is typical for an awake, relaxed adult with eyes closed to avoid distractions. The alpha rhythm (Section 8.1.5) is prominent under these conditions. If the subject is in an attentive state, or is thinking hard about something, the alpha rhythm is replaced by smaller amplitude, faster oscillations, as in Figure 12.4d. The EEG pattern changes markedly in sleep. As a person becomes drowsy, the alpha rhythm gradually disappears and is replaced by slower waves of larger amplitude as sleep deepens. But during the rapid-eye-movement (REM) phases of sleep, the EEG pattern is similar to that of the attentive state (Figure 12.4d). Generally speaking, the amplitude of the EEG varies inversely with the frequencies that are present, these being in the range 1–150 Hz, and EEG records from the frontal and parietal regions tend to show higher frequencies than records from the occipital regions.

The EEG is extensively used in the diagnosis and differentiation of different types of epilepsy, a condition in which groups of neurons in the brain become hyperexcitable and fire in near-synchrony producing distinctive wave patterns that can spread over a large area of the cortex and produce motor, sensory, or behavioral manifestations. The EEG is also useful for diagnosing superficial brain tumors, brain inflammation (encephalitis), stroke, some brain disorders (encephalopathies), and sleep disorders. The absence of an EEG is considered evidence of **biological death**.

Where a more accurate localization of the source of an abnormal brain activity is required, an invasive procedure is sometimes used to record brain electrical activity directly from the surface of the cerebral cortex, avoiding the skull. Such a recording is an **electrocorticogram (ECoG, Spotlight on Techniques 13A)**. An interesting minimally invasive procedure for brain electrical recording and stimulation uses a **stentrode**, which consists of a small metallic mesh – the stent – of a few millimeters in diameter, having small metal electrodes on its surface. The stentrode is inserted through blood vessels of the brain and maneuvered into the desired position under X-ray guidance, for example, as in the insertion of a stent through cardiac catheterization.

The currents that give rise to the EEG also produce magnetic fields, which are recorded in **magnetoencephalography (MEG)**. The magnetic fields are exceedingly small, of the order of tens of femtoteslas (fT, where femto denotes 10^{-15}). By comparison, the strength of the earth's magnetic field is of the order of tens of microteslas. Special devices, referred to as SQUIDs (superconducting quantum interference devices) are used for measuring these tiny magnetic fields and are set in a large, fixed "helmet" that covers a large area of the head. Recently, there have been attempts to reduce the "helmet" to a normal size that would allow unrestricted head movements.

There are important differences in the nature of EEG and MEG recordings. First, EEG recordings arise from both the radial components of current, which are normal to the cortical surface, as well as the tangential components of current, which are parallel to the cortical surface. Because of the difference in the spatial orientation of the magnetic fields associated with these current components, MEG detects only the tangential components. MEG is therefore most sensitive to activity in the walls of the cortical sulci, and is not sensitive to activity in the top of the gyri. Second, MEG primarily detects intracellular currents associated with synaptic activity because the magnetic fields associated with extracellular volume currents tend to cancel out. Third, magnetic fields decay with distance more rapidly than electric fields, so that MEG is more sensitive to superficial cortical activity. A distinct advantage of MEG is a more precise localization of the source of activity, such as an epileptic focus, because of a better spatial resolution, of 2–3 mm, compared to a spatial resolution of 7–10 mm for EEG. This is because magnetic fields are less distorted than electric fields by the skull, scalp, and CSF. Both EEG and MEG have millisecond temporal resolution and are therefore able to detect events that last for 10 ms or less. The fast time resolution and precise spatial localization of MEG have made it useful for functional brain imaging (Spotlight on Techniques 12B).

12.2.2.1 Descending Pathway

The major descending pathway from the cortex is the **pyramidal tract**, the longest and one of the largest tracts of the central nervous system, comprising over 1 million axons in humans. The pyramidal tract derives its name not from the fact that many of its cells of origin are cortical pyramidal cells but because, as it courses downward from the cortex, it forms a pyramidal-like bulge of triangular cross-section that runs down the ventral surface of the medulla. About half the axons of the pyramidal tract originate from the primary motor cortex and about 30% from the premotor area and the supplementary motor area, the remaining 20% of axons emanating from the somatosensory areas of the parietal lobe and the **cingulate gyrus** just above the corpus callosum. The anterior part of the cingulate gyrus (Brodmann's area 24) is involved in controlling facial muscles that partake in the expression of emotions. Most of the axons of the pyramidal tract are of fairly large diameter, of more than about 10 µm, and approximately 3% are of about 20 µm diameter and arise from Betz cells.

The pyramidal tract can be divided into parts:

1. The **corticobulbar tract** that ends in the brainstem (hence, "bulbar") and constitutes about 70% of the pyramidal tract.
2. The **corticospinal tract** comprising the remaining 30%. About 90% of the axons of the corticospinal tract cross over to the contralateral side of the body in the lower part of the medulla, at what is known as the **pyramidal decussation** (Figure 12.5), then travel down the spinal cord as the **lateral corticospinal tract**, and terminate in the motor nuclei at the various levels of the spinal cord. The axons that do not cross over travel down the spinal cord as the **ventral corticospinal tract** and most of them cross over to the contralateral side shortly before reaching their destination motor nuclei of the spinal cord.

The axons of the corticobulbar tract have one of three destinations:

1. Some of these axons cross over to the opposite side of the body and terminate in the motor nuclei of cranial nerves that innervate muscles of the face, jaw, tongue, and pharynx, as well as some of the neck muscles, and muscles that move the eye.
2. Some of the axons end, or have collaterals that end, in the motor nuclei of the midbrain, pons, and medulla, whose pathways will be discussed later.
3. Axons from the somatosensory areas end in the ventral posterior nucleus of the thalamus and in the dorsal column nuclei located at the junction of the spinal cord and the medulla, as indicated by

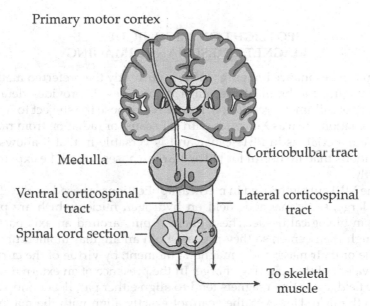

Primary motor cortex

Medulla

Corticobulbar tract

Ventral corticospinal tract

Lateral corticospinal tract

Spinal cord section

To skeletal muscle

FIGURE 12.5
The pyramidal tract and its subdivisions.

the dashed lines in Figure 12.2. This descending input modulates the flow of somatosensory information from the periphery to the brain.

The following should be noted concerning projections from the motor cortex:

1. Only a very small fraction of the motor axons of the corticospinal tract terminate ipsilaterally, the vast majority terminate contralaterally. The ipsilateral terminations mostly control muscles at or close to the midline of the body that are involved in the rotation of the trunk.

2. The terminations of the motor axons of the corticobulbar tract are more bilateral, except for the terminations affecting the tongue and the lower facial muscles, which are mainly contralateral.

3. Most of the terminations in the motor nuclei of cranial and spinal nerves are on interneurons. Only axons from the primary motor cortex terminate monosynaptically on α-motoneurons, particularly those that innervate muscles of the distal extremities, that is, the hand and fingers. This serves the dexterous use of the hand and fingers in humans, as the muscles involved can be directly controlled by the upper motoneurons of the primary cortex. Connections to γ-motoneurons and interneurons serve both to coordinate complex movements and to modify reflex responses, as may be required by a given movement.

SPOTLIGHT ON TECHNIQUES 12B:
MAGNETIC RESONANCE IMAGING

Magnetic resonance imaging (MRI) is currently the preferred method for imaging the brain in the majority of cases. It provides detailed images of millimeter resolution, it does not expose the subject to harmful radiation, such as X-rays (as with CT scans) or radiation from radioactive material (as in PET scans), and is versatile in that it allows for variations that are useful for particular purposes, as will be explained shortly.

The MRI currently used in medical applications is based on the effect of a large, static magnetic field on hydrogen nuclei, which are plentiful in biological tissue. These nuclei "spin" around an axis passing through their center, so they possess both an angular momentum, by virtue of their mass, and a magnetic moment, by virtue of the current equivalent of the spinning charge. In the presence of an external magnetic field, these tiny magnets tend to align either parallel or antiparallel to the field. However, they cannot exactly align with the magnetic field because their movement is restricted by being part of a liquid or a solid-like medium, as in biological tissue. Because the force due to the magnetic field is along an axis that is different from that of the angular momentum, the hydrogen nuclei will precess around the applied magnetic field, like a spinning top or gyro, at a frequency known as the **Larmor frequency** that is directly proportional to the strength of the applied field. If another magnetic field of frequency equal to the Larmor frequency is superposed on the static field, the hydrogen nuclei will absorb energy from the time-varying field, causing a "flip" from the parallel configuration to the higher-energy, antiparallel configuration. When the time-varying field, referred to as a radio-frequency or rf-field, is removed, the nuclei will revert back to their original state. In so doing, they induce a voltage in an appropriately positioned receiving coil. This decaying induced voltage, known as the **free induction decay (FID)**, is the basic nuclear magnetic resonance signal. The term resonance refers to the transition between the two states, corresponding to alternate absorption and release of energy, and the frequency at which this occurs is the **resonant frequency**.

In order to produce an image of a slice of biological tissue that is oriented perpendicularly to the applied strong magnetic field, three "gradient" magnetic fields, whose strength varies linearly with distance, are applied in pulse form along three orthogonal axes, one of which is parallel to the strong magnetic field. Using a rather elaborate sequence of pulses of the rf- and gradient fields, a two-dimensional image of a slice is produced using sophisticated signal processing based on Fourier transformations. The parameters of the applied pulses can be chosen for enhanced contrast based on solid vs. liquid form, fat vs. water content, and velocity of motion in the case of blood flow. In this way, some features of brain morphology can be enhanced such as myelination, which is useful in the diagnosis of multiple sclerosis. By measuring diffusion of water, which is facilitated along axonal membranes rather than across them, axon bundles are highlighted in a variant of MRI referred to as **diffusion tensor imaging (DTI)**. Software allows a multitude of operations on the reconstructed images, including coloring, zooming, 3-D reconstruction from images of contiguous slices, rotation, and smooth scrolling through sectional views.

An important feature of MRI is its potential for identifying substances at the molecular level and measuring their concentrations, as is routinely done in laboratory applications of nuclear magnetic resonance spectroscopy, where a molecule may be identified based on its resonant frequency as well as other factors. This is already being done for some substances such as metabolic intermediaries of neurotransmitters. MRI is being continuously improved. Stronger, uniform magnetic fields, exceeding 10 T, are being used for improved signal-noise ratio and generated using electromagnets whose windings are cooled to superconducting temperatures of a few degrees Kelvin. The structure of MRI machines is becoming more "open" and therefore more friendly to the subjects.

An important variation of MRI is **functional MRI (fMRI)** in which the MRI setup and procedure are modified to allow the performance of certain tasks during the imaging process, thereby identifying the brain structures involved in the given task. fMRI has proved valuable for functional brain-mapping, for showing how different parts of the brain interact during the performance of certain tasks, and for early detection of brain disorders such as strokes, tumors, epilepsy, Alzheimer's disease, and Parkinsonism, as well as for research into the relation between behavior and brain activity (Spotlight on Techniques 13B). Although different techniques can be used with fMRI, the most common is **blood oxygen level dependent (BOLD) contrast**. The underlying principle is that increased neuronal activity in a given region of the brain during the performance of a given task increases metabolic demand, which increases both blood flow and oxygen consumption. Oxygen, and hence oxygenated hemoglobin, is paramagnetic, that is, it is attracted to magnetic fields, whereas deoxygenated hemoglobin, is diamagnetic, that is, it is repelled from magnetic fields. Blood oxygenation is increased by an increase in blood flow and is reduced by an increase in metabolic activity, which consumes oxygen. Regions of increased neuronal activity will be more oxygenated because of the predominant increase in blood flow, compared to oxygen consumption, and the change in the level of oxygenation affects the MRI signal from these regions. Using statistical image-processing, these active brain regions are mapped onto the overall brain image. The spatial resolution is 2–3 mm, and the response time between the increase in neuronal activity and the detectable MRI signal is in seconds, not as fast as with MEG (Spotlight on Techniques 12A).

12.2.3 Basal Ganglia

The basal ganglia comprise four principal nuclei illustrated in Figure 12.6: (i) the **caudate nucleus** and the **putamen**, which together form the **dorsal striatum.** The caudate nucleus and the putamen are separated by the **internal capsule**, which is a fiber tract that carries signals to and from the cerebral cortex; the term striatum refers to the striated appearance of the white matter of these fibers and the slender gray matter bridges joining the caudate nucleus and the putamen, (ii) the **globus pallidus** – divided into the **globus pallidus external segment (GPe)** and the **globus pallidus internal segment (GPi)**, (iii) the **substantia nigra,** so named because it appears darker than neighboring regions due to the presence of the dark pigment neuromelanin in dopaminergic neurons; it is divided into the **substantia nigra pars compacta (SNc)**, so named because of its densely packed cells, and the **substantia nigra pars reticulata (SNr)**, so named because the axons passing through it give it a reticulated appearance, and (iv) the **subthalamic nucleus (STN)**, which is located between the thalamus and the substantia nigra. In addition,

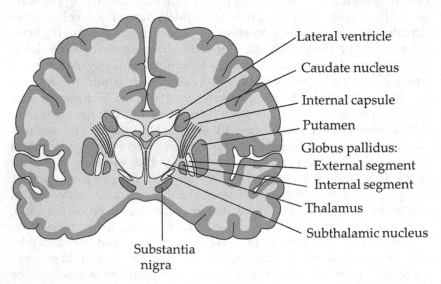

Lateral ventricle

Caudate nucleus

Internal capsule

Putamen

Globus pallidus:
  External segment
  Internal segment

Thalamus

Subthalamic nucleus

Substantia nigra

FIGURE 12.6
Nuclei of the basal ganglia.

the **ventral striatum** consisting of **nucleus accumbens** and the **olfactory tubercle** are anatomically part of the basal ganglia, but they belong to the limbic system and are not considered part of the motor system.

The main interconnections between these nuclei are shown in Figure 12.7. The basal ganglia do not have direct input or output connections with the spinal cord. Their primary input is from the cerebral cortex, and their primary output is to the thalamus and brainstem nuclei. The output nuclei of the basal ganglia are mainly the GPi and the SNr, although a direct connection has been reported from the STN to the pontine nuclei, and thence to the cerebellum. There is also evidence of a reciprocal connection between the cerebellar dentate nucleus and both the GPi and SNr (not shown in Figure 12.6). The excitatory connections represented by thin arrows in Figure 12.6 are glutamatergic, whereas the inhibitory connections represented by thin arrows are GABAergic. The thick arrows represent dopaminergic projections involving the neurotransmitter dopamine. These projections are excitatory or inhibitory, depending on the type of receptor in the target neurons in the dorsal striatum. The excitatory dopamine receptors are termed D1, whereas the inhibitory receptors are termed D2. Both receptors are G protein second-messenger systems (Section 6.3), involving an increase of cAMP in D1 receptors and a decrease of cAMP in D2 receptors. The basal ganglia contain about 80% of the brain's dopamine.

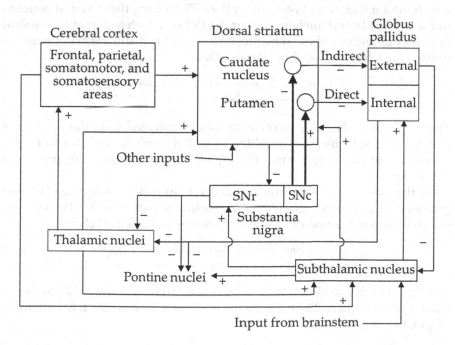

FIGURE 12.7
Block diagram of the main connections of the basal ganglia.

The output cells of the dorsal striatum are GABAergic medium spiny neurons, so named because their dendrites have an abundance of spines that receive inputs from glutamatergic cortical neurons and dopaminergic neurons of the SNc. The dopaminergic synapses are located at the bases of spines, where they selectively control the cortical inputs at the heads of spines. Glutamatergic inputs from thalamic neurons are received on dendritic shafts as well as spines. In addition, the medium spiny neurons receive inputs from axon collaterals of other medium spiny neurons, from large cholinergic interneurons and smaller GABAergic interneurons of the dorsal striatum. The medium spiny neurons have little background firing, unlike the output neurons of the globus pallidus and the SNr, which have a tonic firing that can be modulated up or down.

The cytoarchitecture of the other basal ganglia nuclei is quite different from that of the dorsal striatum. The output neurons of both the GPe and GPi are large and GABAergic. The GABAergic neurons of the SNr interdigitate with the more dorsal dopaminergic neurons of the SNc. The STN is densely packed with neurons, and its output neurons are glutamatergic.

There is a large convergence of cerebral cortical neurons onto neurons of the dorsal striatum and a similarly large convergence of these neurons on neurons of the globus pallidus. In humans, approximately 75 million neurons of the dorsal striatum converge on about 700,000 neurons of the globus pallidus.

Feedback loops exist between the cortex and the basal ganglia. One feedback loop is through the **direct pathway** involving striatal output neurons, which inhibit their target neurons in the GPi. In turn, these inhibit neurons in the **ventral lateral nucleus pars oralis (VLo)**, which is part of the **ventral lateral (VL)** nucleus of the thalamus. The result is disinhibition of neurons of the VLo, which excites the cortical cells:

$$\text{Cortex} \xrightarrow{+} \text{Striatum} \xrightarrow{-} \text{GPi} \xrightarrow{-} \text{Thalamus}$$
$$\underset{+}{\longleftarrow}$$

The feedback is therefore positive, as can be inferred from the product of the signs of the pathways around the loop. This positive feedback facilitates movement and is conducive to initiating movement or to selecting particular movements.

On the other hand, a negative feedback loop exists through an **indirect pathway** that involves striatal output neurons that inhibit GPe neurons, which inhibit in turn neurons of the subthalamic nucleus (STN):

$$\text{Cortex} \xrightarrow{+} \text{Striatum} \xrightarrow{-} \text{GPe} \xrightarrow{-} \text{STN} \xrightarrow{+} \text{GPi} \xrightarrow{-} \text{Thalamus}$$
$$\underset{+}{\longleftarrow}$$

Another negative feedback loop is the **hyperdirect pathway** that bypasses the dorsal striatum and GPe, which makes it faster and avoids the dopamine-responsive neurons of the dorsal striatum:

$$\text{Cortex} \xrightarrow{+} \text{STN} \xrightarrow{+} \text{GPi} \xrightarrow{-} \text{Thalamus}$$
$$\underset{+}{\longleftarrow}$$

The negative feedback via the indirect and the hyperdirect pathways hinders movement and could be used to terminate movement or prevent particular movements. It is believed that the indirect pathway normally predominates, but an increase in dopamine levels in the dorsal striatum, which occurs just before the onset of movement, shifts the balance in favor of the direct pathway. In addition, there are many feedback connections between the nuclei of the basal ganglia themselves, only some of which are shown in Figure 12.6.

The relative effects of the direct and indirect pathways correlate with the symptoms produced by well-known diseases of the basal ganglia. Parkinson's disease is characterized by reduced production of dopamine by neurons of the substantia nigra, which decreases dopamine levels in the dorsal striatum and results in two additive effects: (i) reduced excitation of the striatal target cells in the direct pathway, leading to less inhibition of GPi neurons, that is, more excitation of these neurons and, hence, more inhibition of the thalamic nuclei and less excitation in the cortex, and (ii) reduced inhibition of the striatal target cells in the indirect pathway, leading to more inhibition in the GPe, less inhibition of STN, more excitation of GPi neurons, and, again, more inhibition of the thalamic nuclei and less excitation of the cortex. Reduced activity in the motor cortex results in paucity of movement and impairment of movement initiation (**akinesia**), or slowness and reduced amplitude of movement (**bradykinesia**). Other symptoms associated with parkinsonism include: (i) rigidity due to simultaneous contraction of flexors and extensors, (ii) lack of facial expression, (iii) abnormal postures, such as flexion of head and neck, (iv) tremor at rest at about 4–5 Hz probably due to increased synchronization of oscillatory discharges in the GPe-STN loop, and (v) disturbed gait in the form of small, shuffling steps and a wide stance, without swinging of the arms. The disturbed gait probably safeguards postural stability in view of the slowness of postural reflexes. Other non-motor symptoms of parkinsonism include sleep disturbances, cognitive impairment, anxiety, and depression.

Whereas hypokinesia is associated with increased inhibition of the thalamus by the basal ganglia, a decrease in this inhibition results in **hyperkinesia**, or excessive, involuntary movement. These include rapid uncoordinated movement of body parts (**chorea**), twisting movements and sustained abnormal postures in the neck, trunk, and extremities (**dystonia**); and **hemiballism**, characterized by involuntary movements of the limbs on one side of the body. Hemiballism is caused by damage to the STN, which reduces excitation of the GPi and disinhibits the thalamus. The increased excitation and altered firing pattern of thalamocortical neurons cause these neurons to respond in an exaggerated manner or to discharge spontaneously, resulting in rapid, involuntary, and repetitive movements associated with violent flailing and swinging of the limbs and usually accompanied by a decrease in muscle tone.

In the hereditary disorder, **Huntington's disease**, striatal neurons of the indirect pathway degenerate, which reduces the inhibition of the GPe and, in turn, increases the inhibition of the STN. The resulting functional inactivation of the STN produces symptoms similar to those of hemiballism. In its later stages, Huntington's disease causes behavioral and personality disorders as well as cognitive impairment (**dementia**). Disruptions of the normal functioning of the basal ganglia can cause psychiatric disorders, such as depression and obsessive-compulsive disorders.

The basal ganglia have been implicated in a wide range of functions, as evidenced by the distribution of the inputs they receive. Practically all areas of the cerebral cortex project essentially topographically to the dorsal striatum, thence to other nuclei, and back through the feedback loops via the thalamus to the same cortical areas of origin of the given input to the basal ganglia. The dorsal striatum also receives: (i) feedback input from thalamic nuclei, (ii) dopaminergic input from the ventral tegmental area of the midbrain, which is believed to be part of the "reward" system in the brain, and (iii) serotonergic input from the raphe nuclei, which are a group of nuclei in the brainstem that are a major source of serotonin to the rest of the brain. Serotonin is a neurotransmitter that influences many brain functions, including mood, behavior, sleep, memory, and learning.

The following should be noted:

1. The basal ganglia are involved in a wide range of self-initiated or remembered movements, action selection related to reward or punishment, preparation for movement as well as its execution and sequencing, and control of some movement parameters, such as amplitude and velocity. However, it appears that the planning and execution functions are mediated by separate neuronal populations in the basal ganglia.

2. The basal ganglia are also involved in oculomotor activity as well, specifically in the control of saccadic eye movements, which are very rapid eye movements of velocity up to about 1000/s, that underlie visual fixation and rapid eye movements. The pathway involved is from the caudate nucleus to the SNr to the superior colliculus (not shown in Figure 12.6), where motoneurons controlling eye saccades are located.

3. The same regions of the putamen receive projections from both the cortical somatomotor areas concerned with a given movement as well as the cortical somatosensory areas involved in the movement. The putamen can thus integrate both the motor and sensory aspects of a given movement.

The basal ganglia are involved in non-motor functions as well. Inputs from the prefrontal cortex to the caudate nucleus are associated with: (i) cognitive tasks, such as organizing behavior responses and using verbal skills in problem solving, (ii) emotional stability, empathy, and social compatibility, and (iii) obsessive-compulsive disorders. The dorsal striatum also receives inputs from various structures of the limbic system implicated in motivation, working memory, disturbances of mood, and procedural learning, that is, acquisition of a skill through repetition.

A wide range of functions of the basal ganglia is postulated to fall under a broadened umbrella of goal-oriented behavior and reinforcement learning, which lends itself to computational modeling of the basal ganglia based on this formalism. The basic hypothesis is that, in response to a number of inputs competing for limited motor or cognitive resources, the basal ganglia select the most rewarding action. In this context, it is proposed that the dorsal striatum, receiving extensive sensory inputs from the cortex and thalamus, as well as dopaminergic inputs from the SNc, develops an internal model for estimation of the expected future reward. The GPe-STN loop is postulated to support reward-seeking exploratory activity. The dynamics of the thalamo-cortico-striatal loop are believed to underlie the ability to time intervals in the seconds-to-minutes range, which is important in addressing the temporal relationship between the stimulus and the rewarding action.

SPOTLIGHT ON TECHNIQUES 12C: DEEP BRAIN STIMULATION

Deep Brain Stimulation (DBS) is a surgical procedure in which electrodes are placed in specific regions of the brain and used to stimulate these areas so as to alleviate symptoms associated with certain diseases. It is most commonly used to control the tremor in Parkinson's disease but is also being used with other neurological and psychiatric disorders, including depression, obsessive-compulsive disorder, schizophrenia, and Tourette syndrome, characterized by involuntary, repetitive, and stereotyped movements and at least one vocalization.

The procedure is quite elaborate, involving 3-D MRI, stereotaxic surgery, simulations of the proposed trajectory of the electrode so as to avoid unwanted damage to neural and vascular tissue, and electrical recording and stimulation using a trial electrode. Eventually, two electrodes are implanted, usually bilaterally, in the optimal locations. In the case of Parkinsonism, the basal ganglia are targeted, particularly the globus pallidus and the subthalamic nucleus. The electrodes consist of four, small platinum–iridium contacts connected to a thin insulated lead wire that is connected in turn to a thin insulated wire extension that runs under the skin from the head down the neck and into the upper chest, where it is connected to electronic stimulators housed in the upper chest below the collarbone. The stimulators have their own power source, usually rechargeable by magnetic induction through the skin, and apply pulses of 1–3.5 V amplitude, current in the range of a fraction of a microampere to 100 μA, pulse width from few tens to few hundred microseconds, and frequency between 100 and 300 Hz. A hand-held controller allows the patient to wirelessly turn the stimulators on and off and adjust the stimulation parameters for maximum effectiveness, within the limits set by the physician.

The mechanism of action of DBS is not clearly understood. Electrical stimulation potentially induces myriad effects, including generation of action potentials in cell soma and axons, activation of voltage-gated channels, and release of neurotransmitters, neuromodulators, and other chemical agents. Although DBS is highly effective in alleviating the tremor associated with Parkinsonism, it does not seem to be effective in treating its other manifestations, such as disturbances in gait, speech, cognition, and mood.

Optical activation based on optogenetics (Spotlight on Techniques 7C) can be more selective and may provide an alternative to electrical stimulation of deep brain structures.

12.2.4 Cerebellum

12.2.4.1 Gross Anatomy

The cerebellum (or, little brain) constitutes about 10% of the brain by volume but contains more than half the number of neurons in the brain. It possesses several unique features that will be highlighted in this subsection. The cerebellum consists of two structures: (i) the **cerebellar cortex**, a tightly folded sheet of neural tissue, whose area is about 500 cm² in humans, and (ii) four deep nuclei on each side, embedded in the white matter that is partly enclosed by the cerebellar cortex (Figure 1.8). Unlike the cerebral cortex, the cerebellar cortex is continuous across the midline. Some deep fissures between folia allow grouping the folia into 10 lobules, designated by Latin names and Roman numerals I–X (Figure 12.8). Deeper fissures divide the cerebellar cortex into three lobes, which are from top to bottom: the anterior, posterior, and flocculonodular lobes (Figure 12.8). The fissure between the anterior and posterior lobes is the **primary fissure**, whereas the fissure between the posterior and flocculonodular lobes is the **posterolateral fissure**.

The medial part of the cerebellar cortex is the **vermis**, whereas the lateral parts are the cerebellar hemispheres. These are very prominent in apes and humans and can be divided into two parts, based on their connections to the cerebellar nuclei: an intermediate zone, or **pars intermedia**, next to the vermis, and a lateral zone, or **pars lateralis**.

From lateral to medial, the four cerebellar nuclei are: the **dentate, emboliform, globose,** and **fastigial** nuclei. The emboliform and globose nuclei are often grouped into a single **interposed nucleus**. The axons of Purkinje

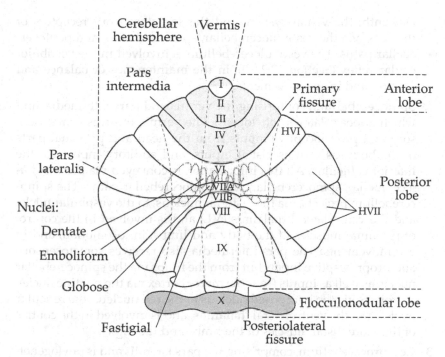

FIGURE 12.8
Overall structure of the cerebellum.

cells (Pc) are the sole output fibers of the cerebellar cortex and terminate topographically on the cerebellar nuclei: Purkinje cell axons from the pars lateralis terminate on the dentate nucleus, those from the pars intermedia terminate on the interposed nucleus, most of those from the vermis terminate on the fastigial nucleus, but those from the flocculonodular lobe, and some from the vermis, terminate in the vestibular nuclei, outside the cerebellum.

The cerebellum is connected to the dorsal aspect of the brainstem by three large fiber bundles on either side, referred to as the cerebellar peduncles, and identified as: the **inferior cerebellar peduncle**, or **restiform body**, the **middle cerebellar peduncle**, or **brachium pontis**, and the **superior cerebellar peduncle**, or **brachium conjunctivum**.

12.2.4.2 Afferent and Efferent Connections

Based mainly on their sources of input and their projections, the cerebellar hemispheres can be divided into three major subdivisions:

1. **Vestibulocerebellum**, comprising the flocculonodular lobe. Phylogenetically, this is the oldest part of the cerebellum, and which first appeared in fishes. It derives its input from the inner ear, or

labyrinth, the visual system, and the somatosensory receptors of the neck, via the vestibulocerebellar, spinocerebellar, and pontocerebellar paths. The vestibulocerebellum is involved in the vestibulo-ocular reflex (Section 12.2.5.3), in the maintenance of balance and posture and in eye movements.

2. **Spinocerebellum,** comprising the vermis and pars intermedia, and which appears later in phylogeny. The vermis receives somatosensory and proprioceptive inputs from the head and proximal parts of the body, as well as visual inputs, and auditory inputs via the inferior colliculus. All this information is conveyed via the vestibulocerebellar, spinocerebellar, and pontocerebellar paths. The spinocerebellum projects, via the fastigial nucleus, to the vestibular nuclei and the ventrolateral thalamus (VL) and is involved in the control of proximal muscles of the body and limbs, in locomotion, and in eye movements. The pars intermedia also receives somatosensory and proprioceptive inputs but from the limbs via the spinocerebellar nuclei, as well as inputs from the motor cortex via the pontine nuclei. It projects via the interposed nucleus to the red nucleus, the reticular nuclei, and the ventrolateral thalamus, and is involved in the control of the more distal muscles of the limbs and the digits.

3. **Cerebrocerebellum,** comprising the pars lateralis and is phylogenetically the most recent. It is much enlarged in apes and even more so in humans, paralleling the evolutionary enlargement of the prefrontal cerebral cortex. It receives its inputs from the association areas of the motor cortex, the prefrontal lobe, and the parietal lobe via the pontine nuclei. It projects via the dentate nucleus to the red nucleus, the reticular nuclei, and the ventrolateral thalamus. It is involved in many functions including the timing of voluntary movements, the execution of complex spatial and temporal sequences of movement, including speech, as well as in learning of motor skills and in certain cognitive functions, as will be elaborated later (Section 12.2.4.5).

The afferent and efferent connections to the cerebellum are summarized in Figures 12.9 and 12.10, respectively. The following should be noted about these connections:

1. In Figure 12.9 is shown an additional input from the **inferior olive,** or **inferior olivary nucleus,** which is a nucleus in the medulla oblongata that receives a wide variety of inputs: from the cerebral cortex, via the red nucleus, from the reticular formation, and from the spinal cord via the **spino-olivary tract.** This tract carries proprioceptive and somatosensory information from the periphery and originates from neurons in the posterior grey matter of the spinal cord. The inferior olive sends afferents to all parts of the cerebellum, as **climbing fibers** (CF, Section 12.2.4.3). The other main afferents to

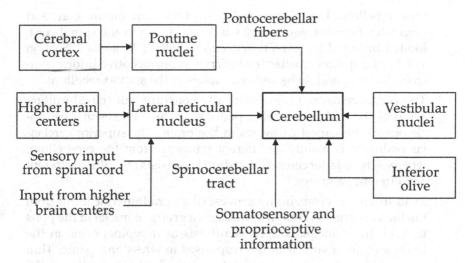

FIGURE 12.9
Main afferent connections of the cerebellum.

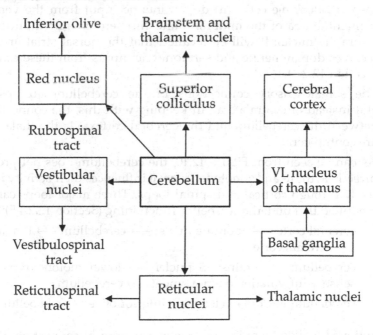

FIGURE 12.10
Main efferent connections of the cerebellum.

the cerebellum are **mossy fibers** (**MF**, Section 12.2.4.3). Both of these afferents are excitatory, with glutamate as the neurotransmitter.

2. The main input from the spinal cord to the cerebellum is directly via the spinocerebellar tract or indirectly via various reticular nuclei, particularly the lateral reticular nucleus, whose entire output is to

the cerebellum. The spinocerebellar tract has four subdivisions and originates from the neurons of Clarke's column in the spinal cord, located in Rexed lamina VII between vertebrae T1 and L3 (Section 11.1.3). The spinocerebellar tract carries proprioceptive information from the body and limbs and terminates in the spinocerebellum.

3. The pontocerebellar fibers practically constitute all the 20 million axons of the middle cerebellar peduncle, which makes this peduncle one of the largest pathways in the brain. The superior cerebellar peduncle is mainly an efferent pathway from the cerebellum, whereas the inferior cerebellar peduncle consists of multiple afferent and efferent pathways.

4. In addition, the cerebellum receives: (i) a noradrenergic input from the **locus coeruleus** complex, which is a group of nuclei in the pons that are the principal site for synthesis of norepinephrine in the brain and are involved in the responses to stress and panic. Thin afferent fibers from this complex terminate in the cerebellar nuclei and in the entire cerebellar cortex, where they generate prolonged ipsps in Purkinje cells; (ii) dopaminergic input from the ventral tegmental area of the midbrain; and (iii) serotonergic input from the raphe nuclei. It will be recalled that the dorsal striatum also receives dopaminergic and serotonergic inputs from these centers (Section 12.2.3).

5. The somatosensory connections of the cerebellum are ipsilateral instead of contralateral. In keeping with this, the connections between the cerebellum and the cerebral cortex, via the thalamus, are contralateral.

6. As can be seen from Figure 12.10, the cerebellum does not project directly to the α- and γ-motoneurons. It influences their activity indirectly through cortical and spinal loops. Three major loops can be identified that underlie cerebellar functioning (Section 12.2.4.5):

   i)   cerebral cortex $\longrightarrow$ pontine nuclei $\longrightarrow$ cerebellum $\longrightarrow$ thalamus $\longrightarrow$ cerebral cortex

   ii) cerebellum $\longrightarrow$ brainstem nuclei $\longrightarrow$ lower motoneurons $\longrightarrow$ sensory information on movement $\longrightarrow$ cerebellum

   iii) cerebellum $\longrightarrow$ red nucleus $\longrightarrow$ inferior olive $\longrightarrow$ cerebellum.

As mentioned earlier, there is evidence of a reciprocal connection between the cerebellar dentate nucleus and both the GPi and SNr (not shown in Figures 12.8 and 12.9). More recent physiological and fMRI studies have indicated primary and secondary somatotopic maps of the body, in the anterior and posterior lobes, respectively, namely in the pars intermedia region. There is no such mapping in the hemispheres but rather a topographic linking with frontal and parietal regions of the cerebral cortex.

12.2.4.3 Cellular Organization and Features

As noted earlier, the cerebellum has two main types of afferent fibers: MFs and CFs. In coursing through the sub-cerebellar-cortical region toward the cerebellar cortex, both of these types of fiber give off collaterals that excite cells in the cerebellar nuclei. It should be noted that, unless specified otherwise, the figures related to the number of cells and the dimensions of layers and cells that are quoted in this section refer to the cat cerebellum, which has been described in great detail.

The cerebellar cortex consists of three layers (Figure 12.11): (i) The deepest layer, the **granular layer**, varies in thickness with the curvature of the cortex, from as much as 500 μm at the top of the folia to as little as 100 μm at the bottom of the sulci; (ii) The most superficial layer, the **molecular layer**, varies much less in thickness with the curvature of the cortex, from about 400 μm at the top of the folia to about 250 μm at the bottom of the sulci, the thickness being about 350 μm over the rest of the molecular layer. (iii) The intermediate layer, the **Purkinje cell layer**, is 80–100 μm thick. The cerebellar cortex contains: (i) one type of excitatory, glutamatergic neurons, the granule cells (Gr), other than the less widespread unipolar brush cells, (ii) three main types of inhibitory interneurons: basket cells (Ba), stellate (St), and Golgi cells (Go), in addition to the less widespread Lugaro cells and candelabrum cells, and (iii) one type of inhibitory output neuron, the Purkinje cell (Figure 12.10). All the inhibitory cells are GABAergic. Most Golgi cells are both GABAergic and

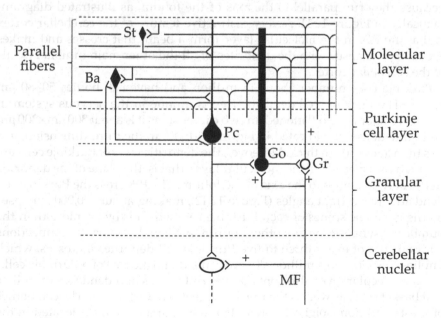

FIGURE 12.11
Cerebellar cytoarchitecture involving the mossy fiber (MF) input.

glycinergic. Remarkably, the cellular organization of the cerebellar cortex is highly regular, almost crystal-like, and is the same all over the cerebellar cortex, despite the projection of the cerebellum to different parts of the brain. This suggests that the cerebellar circuitry performs some essential computations that are used by different parts of the brain in implementing a variety of functions that are by no means exclusively motor. The basic cerebellar circuitry is also invariant throughout phylogeny, from the lowest vertebrates to humans, as explained later.

The cell bodies of granule cells, 5–8 µm across, are located in the granular layer. They number more than 50 billion in humans, or more than all the other neurons in the brain combined. Granule cells have 5 or 6 relatively short dendrites, less than 30 µm long, with claw-like terminations that synapse with **mossy fiber rosettes**, the bulbous terminals of MFs, and with the axon terminals of Golgi cells. All these neuronal elements are enclosed in a tight, glia-encapsulated structure referred to as a **cerebellar glomerulus**. As it approaches the cerebellar cortex, an MF branches into a number of folia, the branches being in sagittal, or transverse planes. In each folium, an MF divides into 2 or 3 preterminal branches, with each preterminal branch bearing about 10 mossy rosettes. Each MF rosette makes excitatory synapses with the dendrites of about 20 granule cells.

The thin axon of a granule cell, 0.1–0.2 µm in diameter, courses toward the surface of the folium and bifurcates in the molecular layer into two opposite T-branches, each about 3 mm long, referred to as **parallel fibers** (PFs) because they run parallel to the axis of the folium, as illustrated diagrammatically in Figure 12.11. A very distinctive feature of the cerebellar cortex is that the PFs in the molecular layer form a beam that crosses and makes excitatory synapses with the dendrites of all the other main inhibitory cells of the cerebellar cortex.

Purkinje cells number about 15 million and have cell bodies 50–80 µm across. The Purkinje cell dendritic tree is unique in the nervous system in that it is elaborate and flattened: in two dimensions it is about 300 µm×300 µm and quite dense, as indicated in Figure 7.1d; but in the third dimension, it is less than about 30 µm thick. Moreover, the dendritic trees of Purkinje cells are oriented transversely in the molecular layer, that is, the plane of the dendritic tree is at right angles to the axis of the folium. The PFs cross the Purkinje cell dendritic tree at right angles (Figure 12.12), making about 200,000 synapses on the dendritic spines of each Purkinje cell, which is by far more than the number of synapses on any other neuron. A PF makes synaptic connections with one out of every three to five Purkinje cell dendrites it crosses, which implies that close to a million PFs cross the dendritic tree of a Purkinje cell.

The molecular layer also contains the cell bodies and dendrites of stellate and basket cells, as well as the upper dendrites of Golgi cells. The cell bodies of the 4–5 million Golgi cells are 6–16 µm in diameter and are located in the granular layer. The dendrites of a Golgi cell extend radially for about 100 µm in all directions, mainly in the molecular layer and, to some extent, in the

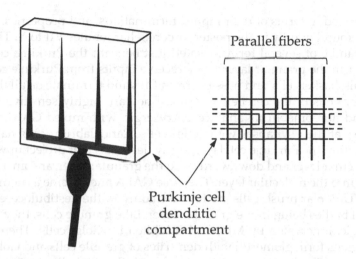

FIGURE 12.12
Innervation of Purkinje cells by parallel fibers.

granular layer. The upper dendrites of Golgi cells are interconnected through gap junctions. Golgi cell axons branch extensively in the granular layer and make inhibitory synapses with granule cells. Golgi cells receive excitatory inputs on their cell bodies and lower dendrites in the granular layer from some MFs and from ascending axons of granule cells (Figure 12.10).

Stellate cells, so called because of their star-like shape, are approximately 16 times as numerous as Purkinje cells. Those that are located more superficially in the molecular layer have cell bodies 5–9 μm in diameter, few dendrites, and short axons. More deeply located stellate cells are larger, have more elaborate dendritic trees and axons that extend transversely for up to a few hundred μm. The axons of stellate cells terminate on smooth Purkinje cell dendrites. The cell bodies of basket cells, which are approximately six times as numerous as Purkinje cells, are located in the lower part of the molecular layer. Their axons extend in the transverse direction and give off collaterals that reach a block of up to 10 Purkinje cells transversely on either side of the basket cell and up to 6 Purkinje cells on either side of the transversely oriented axon. Each Purkinje cell receives input from 20–50 basket cells, partly on the smooth Purkinje cell dendrites close to the cell body, but mostly in the form of a basket-like network, hence the name basket cell, around the lower cell body and initial segment of the Purkinje cell. Collaterals of Purkinje cell axons terminate on Golgi and basket cells (Figure 12.11) and to a lesser extent on stellate cells as well as on other Purkinje cells. The somas of basket cells are contacted by axons of other basket cells, and the somas of stellate cells are contacted by axons of other stellate cells.

In addition to the aforementioned main cell types, three other types of cells have been identified more recently in the cerebellum: Lugaro cells, candelabrum cells, and unipolar brush cells. These cell types have not yet been well characterized in terms of their inputs, terminations, and properties. Lugaro cells are found mainly in the posterior cerebellar lobules VII to X. Their cell bodies can be of several forms, located just beneath the Purkinje cell layer or deeper in the granular layer. They receive inputs from Purkinje cell axon collaterals, basket cells, and possibly from MFs and terminate on all the other inhibitory cells of the cerebellar cortex. They are highly sensitive to serotonin and norepinephrine and are GABAergic, with mixed GABA/glycine co-release at their synapses with Golgi cells. Candelabrum cells have their somas within the Purkinje cell layer, spiny dendrites that project upwards to the molecular layer and downwards into the granular layer, and an axon that projects into the molecular layer. They use GABA and glycine as neurotransmitters. Unipolar brush cells are found mainly in the vestibulocerebellum, their cell bodies being in the granular layer. Like granule cells, they are glutamatergic, innervated by MFs, and inhibited by Golgi cells. Their axonal terminations form glomeruli with dendrites of granule cells and Golgi cells.

Like MFs, CFs also branch as they approach the cerebellar cortex. CFs are so called because they wrap vine-like around the dendrites of Purkinje cells. A CF contacts up to 10 Purkinje cells making 500–1000 synapses on each cell, but a Purkinje cell is contacted by only one CF. CFs also contact Golgi, basket, and stellate cells. The connections of CFs are illustrated diagrammatically in Figure 12.13, where two types of cerebellar nuclear cells are shown: the larger excitatory cells, also contacted by MFs (Figure 12.8), and the smaller GABAergic

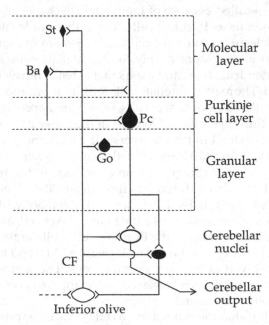

FIGURE 12.13
Cerebellar cytoarchitecture involving the climbing fiber (CF) input.

inhibitory cells. CF collaterals contact both types of cells. The inhibitory nuclear cells terminate in the **inferior olive**, which is the nucleus of origin of the CFs. Both types of nuclear cells are inhibited in turn by axons of Purkinje cells. The larger excitatory nuclear cells are also known as nuclear projection cells because their axons project to the targets of the cerebellar nuclei.

Several types of inhibitory connections are indicated in Figures 12.11 and 12.13. Thus, the overall connection of cerebellar afferents, nuclear cells, and cerebellar cortex is that of feedforward inhibition (Figure 7.3a), as illustrated in Figure 12.14a, which applies to MF and CF afferents and to both types of nuclear cells (shown in gray). Thus, cerebellar afferents excite both types of nuclear cells, with these being inhibited by Purkinje cells that are excited by the cerebellar afferents. Two other main feedforward inhibitory connections are those of: (i) granule cells, Purkinje cells, and basket and stellate cells (Figure 12.13b), and (ii) MFs, granule cells, and Golgi cells (Figure 12.13c). The feedforward inhibition involving stellate cells is through the Purkinje cell dendrites and would tend to counteract the excitatory parallel-fiber input at the more local dendritic level. On the other hand, the feedforward inhibition involving basket cells is through the lower cell body and initial segment of the Purkinje cell and would be much more effective in controlling the overall responses of Purkinje cells.

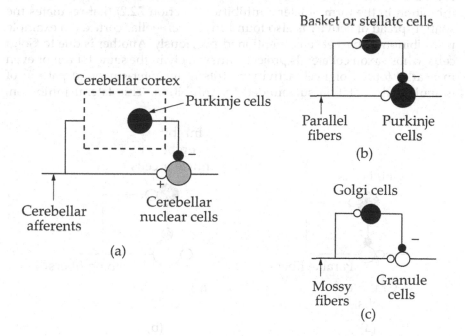

FIGURE 12.14
Feedforward inhibitory connections involving: (a) the whole of the cerebellar cortex, where the cerebellar nuclear cells represent both excitatory and inhibitory cells; (b) Purkinje cells and basket or stellate cells, and (c) granule cells and Golgi cells.

Two main feedback inhibitory connections can be identified: (i) one in the cerebellar cortex, involving granule cells and Golgi cells (Figure 12.15a), and (ii) the other outside the cerebellar cortex, involving the inferior olive and the cerebellar nuclei, through the CF afferents (Figure 12.15b). Two additional feedback paths have been identified involving cerebellar nuclei: (i) inhibitory cerebellar nuclear cells project back to the cerebellar cortex where they terminate on Golgi cells and disinhibit granule cells, and (ii) axon collaterals of some cerebellar nuclear output cells project back as MFs to the cerebellar cortex, terminating on granule cells that excite Golgi cells that, in turn, inhibit Purkinje cells that project to the same nuclear cells, thereby providing a positive feedback loop.

The cerebellum also has recurrent feedback mediated by axon collaterals, whereby the axon collaterals of a given type of neuron innervate another type of neuron, which in turn affects the neurons of the first type. Thus, Purkinje cell collaterals inhibit basket or stellate cells, which in turn disinhibits Purkinje cells. Purkinje axon collaterals also inhibit Golgi cells, which disinhibit granule cells and excite Purkinje cells.

It will be noted that negative feedback in a given neuronal population tends to regulate the level of activity of the given population and prevent excessive activity. Feedforward inhibition, with its disynaptic delay, narrows the window for excitatory inputs to exercise their effect. Feedforward inhibition in the form of lateral inhibition (Section 7.2.2) that regulates the spatial spread of activity is also found in the cerebellar cortex. An example is inhibition by basket cells, mentioned previously. Another is due to Golgi cells, whose axon collaterals project transversely in the same folium or even in nearby folia. Golgi cell activity results in a center-surround pattern of granule cell excitation surrounded by inhibition. This lateral inhibition,

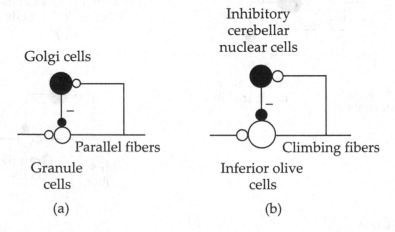

FIGURE 12.15
Feedback inhibitory connections involving: (a) parallel fibers and Golgi cells; (b) climbing fibers and inhibitory cerebellar nuclear cells.

together with the feedback inhibition depicted in Figure 12.15a can give rise to self-sustained oscillations.

It is worth noting that the primitive cerebellar cortices of lower vertebrates, such as the lamprey, which is an ancient eel-like fish, have the basic neuronal circuit elements, namely, MFs, CFs, granule cells, PFs, Purkinje cells, and short-axon stellate cells, all connected in the same manner as in higher vertebrates. In higher vertebrates, lateral inhibition is introduced by long-axon stellate cells, then primitive Golgi cells appear that lack the upper dendrites. In the fully developed cerebellar cortices of birds and mammals are found basket cells, typical Golgi cells, and Purkinje axon collaterals. Moreover, the number of cells increases with phylogenetic development, the cells acquire more elaborate dendritic arborizations, particularly Purkinje cells, the number of PFs per Purkinje cell increases markedly, and axonal connections of inhibitory interneurons become more extensive.

Na^+ channels are expressed in the axon hillock of Purkinje cells, in the soma, and most probably in the dendritic trunk immediately adjacent to the soma. The rest of the dendrites are devoid of Na^+ channels. K_v3 potassium channels are also expressed in Purkinje cell somas, and $K_v3.4$ potassium channels are expressed in Purkinje cell dendrites (Section 7.3.3). Voltage-gated Ca^+ channels are abundant in the dendrites, including dendritic spines.

Purkinje cells generate two types of spikes, referred to as simple spikes and complex spikes, illustrated in Figure 12.16, as recorded intracellularly in the soma. The simple spikes are generated in the soma/axon hillock due to PF stimulation. These spikes are about 0.5 ms wide and 80–100 mV in amplitude. Applying a depolarizing current pulse to the soma of magnitude just above threshold results in tonic firing till the end of the pulse. Further increase in the magnitude of the current pulse results in repetitive bursting of the simple spikes. This repetitive bursting is probably due to the interplay of Ca^{2+} and K^+ channels as described in Section 7.3.

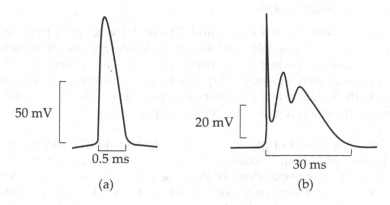

FIGURE 12.16
(a) Simple spikes of Purkinje cells; (b) complex spikes of Purkinje cells.

As noted in Section 7.3.1, Purkinje cell somas have Na⁺ that are blocked during the upstroke of the action potential but are unblocked during repolarization and are thus available for generating a new action potential. This enables Purkinje cells to fire at high frequencies of up to several hundred spikes per second. The high-frequency firing is further aided by K_v3 potassium channels in Purkinje cell somas, which help repolarize the cell.

The complex spikes arise from the intense depolarization caused by the simultaneous activation of the 500–1000 synapses that a CF makes on a Purkinje cell. The complex spikes generate in the Purkinje cell soma and axon hillock a burst of 3–5 action potentials at a frequency of about 500/s. Activation of Ca^{2+} channels in fine dendrites and in dendritic spines, as occurs during PF activity, results in low-threshold, prolonged, plateau-like, all-or-none responses that can last up to several hundred milliseconds. The larger dendrites support Ca^{2+} dendritic spikes of slow onset of 10–20 ms duration and 30–60 mV amplitude. CF synapses activate Ca^{2+} channels in the main dendrites.

CFs have a low rate of spontaneous activity of a few APs per second, but fire single APs in temporal relation to specific inputs to the inferior olive. Neurons of the inferior olive have electrotonic coupling through gap-junctions, so that clusters of these neurons fire synchronously. The CFs from such clusters terminate in thin sagittal, strips that extend across several folia of cerebellar cortex. Purkinje cells within a sagittal strip inhibit a common group of cells in the cerebellar nuclei, which project in turn to a specific part of the inferior olive. Interestingly, the inhibitory cells of the cerebellar nuclei, which terminate in the inferior olive (Figure 12.12), are able to regulate the electrotonic coupling between olivary neurons by shunting depolarizing currents in dendritic processes of these neurons.

The sagittal organization of MF inputs, CF inputs, and Purkinje cell projections is noteworthy.

12.2.4.4 Cerebellar Plasticity

The dual excitatory effects of PFs and CFs on Purkinje cells have aroused much interest as a possible mechanism underlying some motor-learning tasks. It is now established that synaptic plasticity in the form of LTP, LTD, and changes in intrinsic excitability is widespread throughout the cerebellum, in both excitatory and inhibitory synapses. The following are the main features of the emerging view on cerebellar plasticity:

1. The early results of paired activation of MFs and CFs and the peculiarities of LTD in Purkinje cells were mentioned in Section 6.5.2.2. Two different populations of Purkinje cells have since been identified, based on their expression of **zebrin II**, also known as **aldolase C**, an enzyme in the glycolysis pathway. One population, having a

high expression of zebrin II, and referred to as zebrin II+, have thin myelinated axons and an average at-rest firing frequency of about 60 spikes/s. These Purkinje cells are predominant in the flocculonodular lobe (lobule X), in the most caudal part of the vermis (lobule IX), and in the central zone (lobules VI and VII). In the rest of the cerebellum are Purkinje cells that have little or no expression of zebrin II. These zebrin II- cells have thick myelinated axons, an average at-rest firing frequency of about 90 spikes/s, and occur in sagittal strips that alternate with sagittal strips of zebrin II+. It is believed that LTD is more prominent at synapses between PFs and zebrin II- Purkinje cells. In zebrin II+ Purkinje cells, either the synapses with PFs undergo LTP, or the intrinsic excitability of these cells is enhanced.

2. CFs also terminate on interneurons of the cerebellar cortex. If the purpose of CF activity is to induce synaptic plasticity, then such plasticity will also be induced in interneurons on which the CFs terminate. LTP has been reported at the synapses of PFs and molecular layer inhibitory interneurons as a result of paired activation of PFs and CFs, and LTD has been reported in Golgi cells as a result of high-frequency activation of PFs.

3. LTP/LTD occurs at the MF-granule cell synapses; the balance between these two effects depends mainly on the dynamics of $[Ca^{2+}]_i$, as discussed in Section 6.5.2.2, but is also believed to be influenced by cholinergic neuromodulation shifting the threshold of the transition between LTP and LTD, by nitric oxide (NO) affecting the probability of release of presynaptic vesicles, and by the effects of Golgi cell inhibition. Long-term modification of the intrinsic excitability of granule cells also occurs through changes in $[Ca^{2+}]_i$ and in persistent Na^+ and Ca^{2+} currents.

4. The general sequence of events at the output of the cerebellar nuclei following a CF or an MF input is: (i) excitation of cerebellar nuclear output cells due to CF and MF collaterals, (ii) a powerful inhibition due to the activation of Purkinje cells terminating on the nuclear cells, and (iii) disinhibition, often prolonged, as the Purkinje cells are in turn inhibited by the inhibitory interneurons of the cerebellar cortex. Following the removal of the hyperpolarization, cerebellar nuclear output cells show a prominent rebound depolarization, which is further aided by the hyperpolarization-activated current I_h (Section 7.3.5). A high-frequency burst of MF activity that precedes the rebound depolarization induces LTP. Cerebellar nuclear cells are also contacted by CF collaterals and either LTP or LTD is induced depending on the dynamics of $[Ca^{2+}]_i$. Moreover, the intrinsic excitability of cerebellar nuclear cells may be affected by Purkinje cell activity or by a high-frequency burst of MF activity.

12.2.4.5 Cerebellar Disorders

Although the cerebellum is involved in other than motor functions, as mentioned later, cerebellar disorders are clearly manifested as motor abnormalities related to force, direction, and rate of movement as well as coordination and precision of movement. The following are movement disorders that have a clinical designation:

1. **Ataxia**: lack of proper muscular coordination of movement, as in walking with a staggering or reeling gait that is wide-based so as to help maintain balance. This gait is similar to that of drunkenness, which is not surprising, as alcohol is believed to interfere with the proper functioning of cerebellar granule cells.

2. **Dyssynergia**: decomposition of synergistic, multijoint movement, that is, lack of proper cooperation between closely related muscle groups. It is manifested, for example, by great difficulty, when lying on one's back, in touching one knee with the heel of the opposite leg, then sliding the heel down the leg to the ankle.

3. **Hypotonia**: reduced muscle tone, where muscle tone refers to residual or background muscle tension, as well as resistance to passive stretch. It is manifested, for example, by excessive swinging of the leg, pendulum-like, following a patellar tendon tap.

4. **Asthenia**: muscle weakness, that is, reduced force of contraction.

5. **Dysdiadochokinesia**: inability to move rapidly. It is manifested, for example, by inability to rapidly move the hand alternately between pronation (palm downwards) and supination (palm upwards).

6. **Dysmetria**: or abnormality in the range of movement. It is the inability to stop a movement at the appropriate time, resulting in overshooting or undershooting the intended target.

7. **Decomposition of movement**: or the breaking up of a normally smooth movement into a series of unitary movements. Instead of smoothly stepping over an object, for example, the leg is first raised, then moved forward, and finally lowered, in a series of distinct movements.

8. **Intention tremor**, which becomes manifest during movement, particularly when the direction of movement is sharply changed. This is in contrast to tremor at rest due to a disorder in the basal ganglia.

9. **Dysarthria**, in which speech is slurred or is irregular in volume or rhythm, due to inability to control properly the flow of air past the vocal cords.

10. **Nystagmus**, manifested in difficulty to fixate the gaze on a target, so that the eyes will drift from the target, which is then corrected by a rapid eye movement, or saccade.

In addition, lesions in the pars lateralis have been shown to delay the initiation of a voluntary movement.

There is mounting evidence for the involvement of the cerebellum in cognitive functions. Cerebellar lesions result in what is collectively described as the **cerebellar cognitive affective syndrome (CCAS)**, manifested as (i) deficits in linguistic and social skills, (ii) executive function disorder, or impairment of the ability to analyze, plan, organize, schedule, and complete tasks; (iii) affective disorder, involving anxiety, depression, and extreme shifts in mood; (iv) disorders of attention and emotional control; and (v) other behavioral and psychotic disorders, including autism and schizophrenia.

12.2.4.6 Cerebellar Function

It is evident from the block diagram of Figure 12.16 and from the aforementioned manifestations of cerebellar disorders, that the cerebellum is not directly involved in the initiation of movement but rather in the control and coordination of movement, so that the movement is executed smoothly, precisely, and with proper coordination of the various participating muscles. The cerebellum is involved in three crucial aspects of movement: (i) planning of movement, in conjunction with the higher motor centers; once the signal for an intent to move is received from these centers, the cerebellum, through its connections to the primary motor area, is involved in the programming of the timing, direction of movement, and its force; (ii) shaping and modification of the motor signals directed to the α- and γ-motoneurons, through the influence exerted on the motor nuclei of the brainstem; and (iii) monitoring, through direct input from the periphery, of the execution of an intended movement. This enables the cerebellum to effectively exercise both feedback and feedforward control of movement. Feedback control is exerted through comparing the planned, or intended, movement with the movement that is actually being executed and generating the appropriate signals for correcting any "error", or deviation from the intended movement. As mentioned earlier, this mode of control is appropriate for slow movements and postural adjustments. The cerebellum can also exert feedforward control by modifying the motor signals, before they reach their destination, in accordance with the sensory context, such as the position and state of body parts and in the light of past experience.

The cerebellum has been shown to be involved in various motor-learning tasks, such as eyeblink conditioning, and adaptation of the vestibulo-ocular reflex (Section 12.2.5.3). The cerebellum seems to play a crucial role in the learning of motor skills involving a sequence of movements, as in playing the piano. It appears that the cerebellum is most active during the early phase of the learning of a motor sequence, during which cerebellar activity adjusts movement kinematics, based on sensory feedback, so as to accurately produce the desired motor output and reduce the error during motor learning. The prefrontal cortex and the basal ganglia are particularly important for motor learning. Once the motor sequence has been well-learned, the dorsal striatum and motor cortical areas are critical for long-term storage and execution of the learned sequence. The basis for motor learning in the cerebellum is synaptic plasticity of the PF-Pc synapses in the cerebellar cortex and synapses of MFs in the cerebellar nuclei. The synapses between PFs and interneurons of the cerebellar cortex that are also innervated by CFs may also undergo synaptic plasticity that contributes to motor learning.

Through the fastigial nucleus and the thalamus, the vermis is linked to the basal ganglia and the limbic system (Section 1.3), thereby influencing the responses of this system. The cerebellum also affects autonomic activities such as blood pressure, heart rate, respiration, and gastrointestinal motility.

It is believed that the characteristic regular cellular organization of the cerebellar cortex underlies a most salient feature of all the functions of the cerebellum, namely, its ability to control timing to a high degree of resolution, of the order of a few milliseconds. Moreover, the patterns of cerebellar activity at these high time resolutions could be used to compensate for delays that occur in signal transmission and processing in the nervous system by adding a predictive component to the movement (Section 13.1.3).

12.2.5 Brainstem Nuclei and Descending Tracts

This section focuses on the motor nuclei of the brainstem and their descending motor fiber tracts, collectively known as **extrapyramidal tracts**, to distinguish them from the pyramidal tract descending from the cerebral cortex (Section 12.2.2.1). The extrapyramidal tracts comprise the rubrospinal, the reticulospinal, the vestibulospinal, and the tectospinal tracts.

12.2.5.1 Red Nucleus

The red nucleus is located in the **tegmentum**, or rostral midbrain, next to the substantia nigra (Figure 12.17). It is divided into two parts: the rostral **parvocellular** division, which in humans constitutes most of the red nucleus, and the caudal **magnocellular** division. The red nucleus receives input from the cerebral cortex, via the corticobulbar tract, and from the cerebellar dentate and interposed nuclei. Most of its axons project from the parvocellular division to the inferior olive, the thalamus, and reticular nuclei. The magnocellular division projects to the spinal cord as the **rubrospinal tract**, adjacent to the lateral corticospinal tract and the lateral reticulospinal tract (Figure 11.3).

In humans, the rubrospinal tract is of less importance than in lower animals because of the development of a direct corticospinal projection on motoneurons. The rubrospinal tract is involved in controlling muscles of the shoulder and the upper arm, facilitating flexion in the upper extremities, as in arm swinging during walking. It is not involved in the leg muscles, as the tract terminates in the superior thoracic region of the spinal cord.

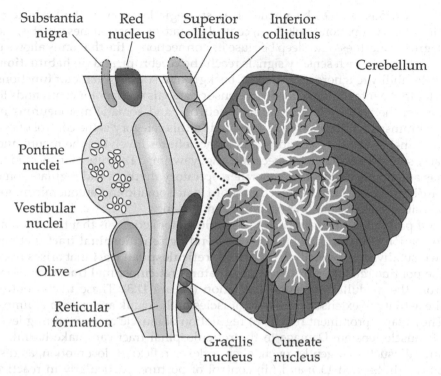

FIGURE 12.17
Brainstem nuclei. The midsagittal plane crosses the reticular formation and the cerebellar cortex. The other neuronal structures are located symmetrically on the two sides of the midsagittal plane.

12.2.5.2 Reticular Nuclei

The **reticular formation** is an extensive network of neurons that extends from the rostral midbrain to the caudal medulla (Figure 12.17). It is a heterogeneous assemblage of neuronal clusters that is pervaded by bundles of axons crossing the brainstem. Rostro-caudally, it is divided into three parts: the **pontine reticular formation** in the region of the lower pons, the **mesencephalic reticular formation** above this central region, and the **medullary reticular formation** below the central region. Sagittally, the reticular formation is divided into three columns: the raphe nuclei that form a ridge in the middle of the reticular formation, adjacent to which is the **medial reticular formation**, then the **lateral reticular formation**. The medial reticular formation contains gigantocellular nuclei, having cells of large size, whereas the lateral reticular formation contains parvocellular nuclei, having cells of small size.

The diverse functions of the reticular formation are divided into two broad categories: modulatory functions and premotor functions. Both of these functions are predicated on the integration by the reticular formation of inputs from various sources by virtue of its strategic location in the brainstem. Modulatory functions include: (i) control of state of consciousness, that is, the degree of alertness and sleep because its connection to the thalamus allows it to influence which sensory signals reach the cerebrum, and (ii) **habituation**, or the ability to ignore meaningless, background stimuli. Premotor functions refer to the integration of feedback sensory signals with motor commands to control the activity of visceral motoneurons and somatic motoneurons in the brainstem and spinal cord. Included in this category is the control of eye movements, facial expressions, and some reflexes involving the mouth and face, such as swallowing, sneezing, and yawning. The reticular formation has autonomic centers that generate respiratory rhythms and regulate some cardiac functions and centers that coordinate somatic and autonomic motor activities such as vomiting, laughing, and crying.

Of particular interest related to somatic motor activity is that the reticular formation is the origin of a major descending **reticulospinal tract** that has two subdivisions: a **medial**, or **ventral**, **reticulospinal tract** that arises from the pontine reticular formation and a **lateral reticulospinal tract** that arises from the medullary reticular formation (Figure 11.3). These tracts control the activity of extensor and flexor muscles of the trunk and proximal limbs. They play a prominent role in: (i) regulation of muscle tone, or resting level of muscle tension. Damage to the reticulospinal tract can make harmless stimuli, such as a gentle touch, elicit a flexor reflex; (ii) locomotion, as discussed in Section 13.4; and (iii) control of posture, particularly in reactive postural adjustments, as in moving from a sitting to a standing position, or in anticipatory postural adjustments in relation to voluntary activity. For example, when we voluntarily push or pull on something, posture-controlling muscles of the trunk and legs are activated before the muscles involved in the voluntary action, so as to maintain the proper body posture, based on past experience and judgment. Subsequent postural adjustment involves reactive, or feedback, control. It is also believed that the reticulospinal tract mediates the "startle reaction" in response to an unexpected, loud sound.

12.2.5.3 Vestibular Nuclei

The vestibular system serves several important functions. Together with the visual and proprioceptive systems, it maintains balance, provides critical information for spatial orientation and navigation, participates in postural adjustments – particularly of the head – contributes to the perception of self-motion, and is involved in a number of supportive and protective reflexes.

The vestibular receptors are found in the **vestibular labyrinth** of the inner ear and comprise two **otolith organs** – the **utricle** and **saccule** – and three **semicircular canals**. The otolith organs provide information on the static position of the head relative to the gravitational axis and respond to linear accelerations, whereas the semicircular canals respond to the angular acceleration of the head arising from self-induced rotations or from external forces. Both the linear and angular accelerations can be in any direction. These vestibular receptor organs project, through approximately 20,000 nerve fibers on each side, via cranial nerve VIII, to the vestibular nuclei and also directly to the cerebellum.

There are four vestibular nuclei on each side, located in the rostral medulla and caudal pons (Figure 12.17): superior, lateral, medial, and inferior. The lateral vestibular nucleus is also known as **Deiters' nucleus**. Cerebellar afferents to the vestibular nuclei project ipsilaterally from the vermis of the anterior lobe to the lateral vestibular nucleus, from the flocculonodular lobe to the other three vestibular nuclei, and bilaterally from the fastigial nucleus to the lateral and inferior vestibular nuclei. The vestibular nuclei integrate a broad range of visual and somatosensory inputs, including inputs from the spinal cord, particularly neck proprioceptive information, inputs from subcortical visual centers, and inputs from the cerebral cortex, including premotor head movement commands.

The main projections of the vestibular nuclei are to: (i) the cerebellum, mainly to the flocculonodular node from the medial and inferior vestibular nuclei, (ii) the thalamus, mainly the ventral posterior complex of the thalamus, and thence to the cerebral cortex, including areas in the parietal and temporal regions, motor and premotor regions, and frontal eye fields; (iii) the nuclei controlling extraocular muscles that mediate eye movements, namely, the **oculomotor nucleus**, the **abducens nucleus**, and the **trochlear nucleus**; these projections are mainly from the superior and medial vestibular nuclei; (iv) the spinal cord via the vestibulospinal tract, and (v) other vestibular nuclei on the same side or the opposite side.

The vestibular system is involved in the **vestibulo-ocular reflex (VOR)**, which is a composite reflex that keeps the gaze fixed on a target despite head movement, as in the case of head shaking or vertical head oscillations that occur with locomotion. The main pathways involved are shown in Figure 12.18. Essentially, movement of the head is detected by the vestibular labyrinths on both sides, which then send signals to the vestibular nuclei. These signals are processed in the nuclei controlling extraocular muscles in such a manner so as to cause movements of the two eyes that are equal and opposite to the head movements, thereby stabilizing the retinal images. Cerebellar inputs are superposed on this simple three-neuron arc for several purposes. Thus, there are many instances when the VOR should be suppressed, so that the gaze is locked to the movement of the head. The cerebellum allows this visual suppression of the VOR. Moreover, the VOR requires some adjustment, or calibration, to keep the gaze stable at different viewing distances because the eyes have to undergo some translational movement as well as rotational movement when the viewing distance is varied, since they cannot both be perfectly aligned with the axis of rotation at different viewing distances. Such errors would cause a slip in the retinal image and are detected by retinal ganglion cells and corrected by the CFs of the cerebellum through a pathway via the **accessory optic system**. A similar

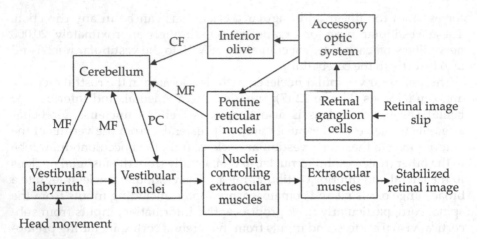

FIGURE 12.18
The main pathways involved in the vestibulo-ocular reflex.

effect occurs when one wears corrective lenses, for example, since these change the size of the visual image, which necessitates a change in the magnitude of the eye rotation that compensates for a given head rotation. The cerebellum is involved in this **VOR adaptation,** which is achieved through synaptic plasticity in the PF synapses, as well as in some neurons of the vestibular nuclei. The adaptation of the VOR can extend over a period of a few days and is considered a form of motor learning. It is also important during development as the size of the head changes. Moreover, the cerebellum is involved in distinguishing between head tilts and purely translational movements of the head because the otolith organs respond in the same manner to both of these types of movement, whereas the semicircular canals respond to the head tilts but not to the translational movement. The VOR is very well coordinated with other eye and head movements through extensive interconnections with various neuronal circuits that mediate these movements as well as with those involved in the perception of self-motion. But illusions can be experienced, as in the case of sitting in a stationary vehicle and watching another slowly moving vehicle, which can give the illusion of motion of one's vehicle. Impairment of the VOR is associated with some clinical symptoms, as in **oscillopsia,** which is manifested as difficulty in fixating on visual targets while the head is moving.

The vestibulospinal tract has two components: (i) the lateral vestibulospinal tract, which originates in the lateral vestibular nucleus and descends, ipsilaterally, the length of the spinal cord; and (ii) the medial vestibulospinal tract, which originates in the medial vestibular nucleus and extends bilaterally through mid-thoracic levels of the spinal cord. The APs along the axons of the lateral vestibulospinal tract monosynaptically excite the motoneurons that activate the antigravity muscles, which are extensor muscles of the trunk and legs and flexor muscles of the arms, and they disynaptically inhibit motoneurons that activate the antagonists to the antigravity muscles. This pathway mediates balance, contributes to the maintenance of posture, and is involved in the **vestibulospinal reflex** (VSR) that stabilizes the body. When the head and trunk are tilted together to one side, for example, this reflexively activates the trunk and leg extensors on the side to which the head is tilted, so as to stabilize the body. The VSR, together with other reflexes, is also involved in the "righting" of the body to prevent a fall when slipping.

The medial vestibulospinal tract is involved in the stabilization of head position, in the coordination of head and eye movements, and in the **vestibulocollic reflex** (VCR) that regulates head position by reflexive action on the neck muscles in response to stimulation of the vestibular organs. For example, if one trips, the forward pitch of the body and head stimulates the semicircular canals so that the VCR causes the neck muscles to reflexively pull the head up. The dorsiflexion of the head causes extension of the arms through other reflexes, such as the **cervicospinal reflex** that generates limb movements in response to activation of neck proprioceptors, in order to protect the head when falling. As in the case of the VOR, the vestibulospinal reflexes are suppressed when making voluntary head movements.

12.2.5.4 Tectospinal Tract

The **tectospinal tract** (Figure 11.3), also known as the **colliculospinal tract**, is an extrapyramidal motor tract that coordinates head, neck, and eye movements. The tract originates in the superior colliculus, which is situated rostrally, just below the thalamus (Figure 12.17). The superior colliculus, together with the inferior colliculus, comprise the **tectum**, or roof of the midbrain, in humans. The part of the midbrain between the tectum and tegmentum constitutes the **midbrain tegmentum**. The two colliculi on each side form four prominences referred to as the **corpora quadrigemina**.

The superior colliculus is a layered, multi-sensory structure that receives inputs from the retina, the visual, parietal, and frontal cortices, as well as inputs from the auditory and somatosensory systems. The superior colliculus has a major ascending projection to the frontal cerebral cortex through the medial dorsal nucleus of the thalamus. It projects to the nuclei controlling extraocular muscles that mediate eye movements, and to the cervical spinal cord through the tectospinal tract.

The inferior colliculus is the main midbrain nucleus of the auditory pathway and receives input from the auditory cortex, from the superior colliculus, and from several peripheral brainstem nuclei in the auditory pathway. It is involved in the binaural localization of sound, in frequency recognition, and pitch discrimination, as well as in the processing of sounds having complex temporal patterns.

The tectospinal tract is involved in reflexively directing the head and eyes to arresting visual and auditory stimuli that require immediate attention.

12.2.5.5 *C3-C4 Propriospinal System*

The neurons of the C3-C4 propriospinal system have their cell bodies at the C3-C4 segmental level and their axons descending down the spinal cord. They receive monosynaptic excitation from the corticospinal tract and from the other descending tracts and, to a much lesser extent, from peripheral afferents. They also receive (i) feedforward inhibition from interneurons of the spinal cord that are themselves excited by the descending inputs, and (ii) feedback inhibition from interneurons of the spinal cord that are themselves excited by peripheral afferents. Individual neurons of the C3-C4 propriospinal system project monosynaptically to motoneurons of muscles acting at different joints. They also have ascending projections to the lateral reticular nucleus, which in turn projects to the cerebellum. In addition, the C3-C4 propriospinal system has its own inhibitory interneurons that project to motoneurons and to the lateral reticular nucleus. Although the C3-C4 propriospinal system was initially described for the cat, there is evidence that a similar system exists in humans.

It is believed that the C3-C4 propriospinal system subserves complex motor synergies that involve coordination between muscles acting at different joints. It has also been suggested that this system may be involved in modifying the descending motor signals to cater for any sudden changes in the internal or external conditions that may have occurred since the formulation of the initial motor command.

Summary of Main Concepts

- The somatomotor system is hierarchically organized, which offers the important advantages of less complexity, as well as more efficient and faster operation.

- From a functional viewpoint, three hierarchical levels of the somatomotor system can be identified: (i) the higher-level centers that are primarily concerned with the planning of movement, (ii) the middle-level centers that translate the planned movement into an operational form, and (iii) the lower centers that implement the required muscular activity.

- The higher-level centers are the motor association areas, comprising: the premotor cortex, the supplementary motor cortex, and the posterior parietal cortex. These areas are concerned with the planning of movement, based on the position of the body in space, on the locations of targets and surrounding objects, and on past experience, motivation, and emotional state.

- The middle level of the somatomotor system hierarchy comprises the primary motor cortex, the basal ganglia, the cerebellum, and somatomotor nuclei of the brainstem.

- As exemplified by the motor homunculus, the area of primary cortex allocated to a given body region depends on degree of finesse of muscular control that is required in that region.

- The primary cortex differs in its anatomy from the sensory areas and is characterized by the presence of cell bodies of giant Betz pyramidal cells that control muscles of the distal extremities.

- Neurons of the primary motor cortex increase their firing some 50–150 milliseconds before movement occurs and generally during the movement as well. The increase in the rate of firing of these cortical neurons may be related to: the magnitude of the force of the movement, the rate of change of the force, or to the direction of the force. In the latter case, the average increase of firing in a population of neurons, rather than in single neurons, strongly correlates with the actual direction of force.

- About half the axons of the pyramidal tract originate from the primary motor cortex and about 30% from the premotor supplementary motor areas. The terminations of the motor axons of the corticospinal division of the pyramidal tract are practically all contralateral, whereas the terminations of the motor axons of the corticobulbar division are more bilateral.

- The basal ganglia consist of four principal nuclei: the dorsal striatum, comprising the caudate nucleus and the putamen; the globus pallidus; the substantia nigra, and the subthalamic nucleus. The primary input to the basal ganglia is from the cerebral cortex, and their primary output is to the thalamus and brainstem nuclei.

- There are three feedback loops between the cerebral cortex and the basal ganglia, through the thalamus. the direct pathway is a positive feedback loop that facilitates movement, whereas the indirect and hyperdirect pathways are negative feedback loops that hinder movement. Some of the well-known clinical disorders of the basal ganglia can be correlated with neuronal activity in these loops.

- Based mainly on their afferent and efferent projections, the cerebellar hemispheres can be divided into three major subdivisions: (i) vestibulocerebellum, comprising the flocculonodular lobe and is involved in the vestibulo-ocular reflex, in the maintenance of balance, posture, and eye movements; (ii) spinocerebellum, comprising the vermis and pars intermedia and is involved in the control of the more distal muscles of the limbs and the digits; and (iii) cerebrocerebellum, comprising the pars lateralis and is involved in many functions including the timing of voluntary movements, the execution of complex spatial and temporal sequences of movement, including speech, as well as in the learning of motor skills and in certain cognitive functions.

- There are two main types of afferents to the cerebellum: the MFs that generate simple spikes in Purkinje cells and the CFs that project from the inferior olive and which generate complex spikes in Purkinje cells.

- The cerebellar cortex has a distinctive, highly regular structure in which the axons of granule cells, the PFs, cross at right angles to the flat dendritic trees of Purkinje cells. Cerebellar circuitry has feedforward, feedback, and recurrent inhibition.

- The PF-Pc cell synapses undergo synaptic plasticity in conjunction with PF stimulation. This synaptic plasticity underlies motor learning that the cerebellum is involved in, together with the cerebral cortex and basal ganglia.

- In humans, the rubrospinal tract is less important than in lower animals and is involved in controlling muscles of the shoulder and the upper arm and facilitating flexion in the upper extremities, as in arm-swinging during walking.

- The functions of the reticular formation include control of state of consciousness, habituation, control of eye movements, facial expressions, and some reflexes involving the mouth and face – such as swallowing, sneezing, and yawning – autonomic respiratory rhythms, and some cardiac functions and centers that coordinate somatic and autonomic motor activities such as vomiting, laughing, and crying.

- The reticulospinal tract, through its two subdivisions, controls the activity of extensor and flexor muscles of the trunk and proximal limbs and plays a prominent role in the regulation of muscle tone, locomotion, and control of posture, particularly in reactive postural adjustments.

- The vestibulo-ocular reflex is a composite reflex that keeps the gaze fixed on a target despite head movement. The cerebellum is involved in the suppression of this reflex, in its calibration, and in its adaptation.

- The vestibulospinal tract mediates balance, contributes to the maintenance of posture, and is involved in the vestibulospinal reflex that stabilizes the body and in the vestibulocollic reflex that regulates head position by reflexive action on the neck muscles.

- The tectospinal tract is involved in reflexively directing the head and eyes to arresting visual and auditory stimuli that require immediate attention.

- The C3-C4 propriospinal system may be involved in modifying the descending motor signals to cater for any sudden changes in the internal or external conditions that may have occurred since the formulation of the initial motor command.

13

Control of Movement and Posture

Objective and Overview

Movement is an amazing feat of the neuromuscular system. Although it is normally executed seamlessly and effortlessly, it is nevertheless quite a complicated process, as will become evident from this chapter. Movement is sometimes classified as either voluntary or involuntary in the form of reflexes and rhythmic motor patterns, as in breathing and walking. However, voluntary and involuntary movements are not independent, since the same set of muscles can participate in voluntary, reflex, and rhythmic movements, as in the case of breathing muscles, for example. Moreover, movement generally entails some postural adjustments so as to maintain balance.

Although movement is caused by muscle activity, whose electrical and mechanical manifestations are amenable, at least in principle, to direct and precise measurement, the fundamental nature of motor control remains largely conjectural. Thus, it is not known for sure what exactly does the motor system command – is it muscle force, spatial coordinates of the end-point of movement, or what?

The objective of this chapter is to highlight some of the main features and problems of motor control. The chapter starts with examining some aspects of movement, mainly, lever action of muscles, co-contraction of antagonist muscles, feedforward and feedback control, motor coordination, and motor equivalence. Motor learning and memory are examined next and some of their features highlighted. Posture is then considered in terms of balance, maintaining an upright posture, and postural adjustments during movement. This is followed by examining locomotion, including analysis of gait, central pattern generators, and their regulation and control by descending and sensory inputs.

The chapter ends with a brief review of some hypotheses on the control of movement before elaborating on the equilibrium point hypothesis in some detail, as this hypothesis has some attractive features that merit special consideration.

Learning Objectives

To understand:

- Some aspects of movement, namely, lever action of muscles, co-contraction of antagonist muscles, feedforward and feedback control, motor coordination, and motor equivalence
- Some important and distinctive features of motor learning and memory
- What is involved in postural adjustments
- The phases of gait and the nature of central pattern generators for locomotion as well as their regulation and control by supraspinal inputs
- The main features of the equilibrium point hypothesis on the control of movement

13.1 Aspects of Movement

To appreciate some of the subtleties and complexities of the control of movement, consider the case of catching an incoming ball and throwing it back on a target. As the ball approaches, it is visually tracked with eye movement locked to head movement, which entails suppression of the vestibulo-ocular reflex (Section 12.2.5.3) and control of head movement through visual feedback. As the ball gets closer, the body and limbs are configured to catch the ball, and the hand is positioned so that the ball is judged to hit the palm of the opened hand. This involves our sense of proprioception, or **kinesthesia**, that is our ability to perceive the position of our body parts in space, without relying on vision. The joints of the limbs, shoulder, and trunk are stiffened by co-contraction of antagonists (Section 13.1.2) so that catching the ball will not cause the arms or the body to swing unduly. This is mainly based on judgment and experience. When the ball hits the palm, the fingers close around the ball in a near-reflexive movement reminiscent of the **palmar grasp reflex** that is observed in newborns, but which subsides later. The force applied by the fingers is enough to prevent the ball from slipping but is not too strong so as to waste energy and possibly cause discomfort. When the ball is to be thrown at the target, the body is appropriately configured, and the arm is swung with the ball firmly in grasp. At what is judged to be the right moment, the muscles of the shoulder and upper arm are activated, and the ball is released in the direction of the target. This requires not only proper coordination of all the muscles involved but also the maintenance of balance as the center of mass of the body shifts during the movements involved. The throwing of the ball and aiming at the target is an example of feedforward control (Section 13.1.3), as there is no feedback to correct for any error in the movement. Again, this is a matter of judgment and experience. The performance improves with practice, based on motor learning and memory (Section 13.2).

Various aspects of movement are considered in this section that are relevant to the muscle activation needed for the execution of a given movement.

13.1.1 Lever Action

Skeletal muscles usually exert the force they develop through some type of lever action, where a lever system consists ideally of a rigid member that can rotate about a pivot point, or fulcrum, with two forces acting on the rigid member at two different points and tending to rotate the rigid member in opposite directions. The forces can be denoted as: (i) the load, or force acting on the system, and (ii) the effort, or force that is used to oppose the load force.

Depending on the relative positions of the two forces and the fulcrum, lever systems are divided into three classes (Figure 13.1): (i) **Class I levers**, in which the load and effort are applied on opposite sides of the fulcrum, as in a seesaw or an oar in a rowboat; (ii) **Class II levers**, having the fulcrum at one end, and the load closer to the fulcrum than the effort, as in a wheelbarrow or a bottle opener used to remove a metal cap; and (iii) **Class III levers**,

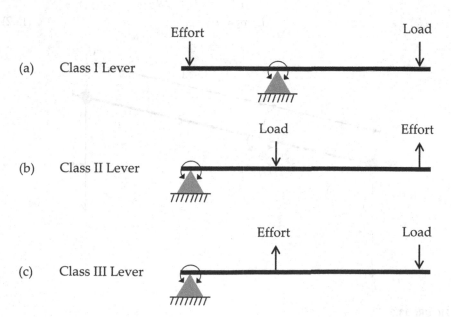

FIGURE 13.1
Classes of levers; (a) class I; (b) class II; and (c) class III.

having the fulcrum at one end, but with the effort closer to the fulcrum than the load, as in the case of a fishing rod when pulling a fish out of the water.

The principal feature of a lever system is that the ratio of the values of the load and effort, and the ratio of the relative velocities at which these forces move, depend on the distances between the fulcrum and the points of application of these forces. This can be illustrated with respect to Figure 13.2, where it is assumed that the rigid member can only rotate about the fulcrum and is prevented from any lateral movement. Under equilibrium conditions, the moments of L and F about the fulcrum P are equal and opposite, neglecting the weight of the rigid member. Thus:

$$L \times d_L \cos\theta = F \times d_F \cos\theta \qquad (13.1)$$

where the right-hand side and left-hand side represent the moments of F and L, respectively, about the fulcrum. $\cos\theta$ cancels out from both sides of Equation 13.1. The moment of a force about a point of rotation can be interpreted as: (i) the product of the magnitude of the force and the perpendicular distance from the point of rotation to the line of action of the force, or (ii) the product of the distance from the point of application of the force to the point of rotation, along the length of the rigid member, and the magnitude of the component of the force perpendicular to the rigid member.

If in Figure 13.2 L and F are assumed to move at this instant with velocities v_L and v_F, respectively, in opposite directions, then conservation of power, or the rate at which energy is delivered or expended, requires that:

$$L \times v_L = F \times v_F \qquad (13.2)$$

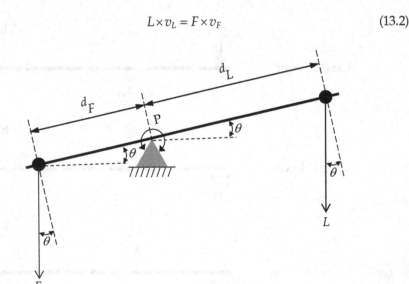

FIGURE 13.2
Diagram for deriving the expression for lever action in terms of the ratio of the values of the load and effort and the ratio of the relative velocities at which these forces move.

It follows from Equations 13.1 and 13.2 that:

$$\frac{L}{F} = \frac{v_F}{v_L} = \frac{d_F}{d_L} \qquad (13.3)$$

If $d_F < d_L$, then $v_F < v_L$ and $F > L$. Since a larger velocity means that a larger distance is traveled in the same interval of time, then under these conditions, larger effort F acting over a short distance can move a smaller load L over a larger distance. The lever system is said to possess a **mechanical disadvantage** equal to F/L. Conversely, if $L > F$, the lever system is said to have a **mechanical advantage** equal to L/F.

Although the three classes of levers are found in the human body, Class III levers are the most common, in which case muscles are positioned to exert a larger force than the load, but the distance and the velocity of muscle movement are magnified at the load. A Class I lever is involved in the forward and backward movement of the head (Figure 13.3a). The fulcrum is the **atlanto-occipital joint** between the skull and the top vertebra of the spinal cord. The effort is exerted by the muscles in the back of the neck, and the load is the force that tilts the head forwards. A Class II lever is involved in the lifting of the body by plantar flexion (Figure 13.3b), where the fulcrum is the ball of the foot, the load is the weight of the body acting through the leg bones, and the effort is exerted mainly by the gastrocnemius muscle. This is the movement of a high-jumper at takeoff from the ground and involves a mechanical advantage. A Class III lever is involved in the familiar effort exerted by the biceps when flexing the elbow while holding a weight (Figure 13.3c).

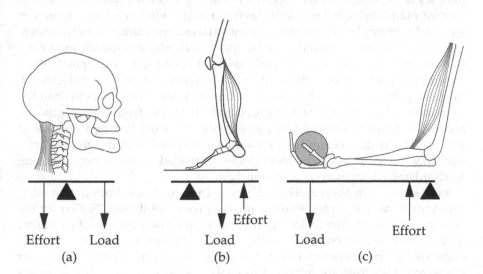

FIGURE 13.3
Examples of classes of levers in the body; (a) class I; (b) class II; and (c) class III.

13.1.2 Co-contraction of Antagonist Muscles

Because muscles can only pull and cannot push, motion in opposite directions at a given joint requires the action of two opposing sets of muscles, the agonists and the antagonists for the motion involved (Section 9.3.4). Thus, the agonists and the antagonists are alternately activated in the execution of rhythmic movements. In the simplest form of walking, for example, the hips are alternately flexed and extended by different sets of muscles, which are the agonists and antagonists for flexion or extension. Inhibition of antagonist muscles when the agonist muscles are activated is referred to as reciprocal inhibition (Section 11.2.2.2).

It may seem natural to assume that when movement in a given direction is required, only the agonists should be activated, with the antagonists inactivated. For a given net or resultant force, inhibiting the antagonists minimizes the force that needs to be developed by the agonists and hence reduces the energy expended. As mentioned previously, training to increase the force developed at a given joint, such as the knee, involves learning to deactivate the antagonists. However, reciprocal inhibition applies in practice only to relatively few types of movement. For example, in executing a slow voluntary movement of a limb, without an opposing load, the antagonists are inhibited while the agonists are active.

In most cases, contraction of antagonists is required for several reasons: (i) to terminate the movement, particularly fast movements, where activation of antagonists decelerates the movement and brings it to an end; (ii) to stabilize a joint, as when stabilizing the elbow joint by co-contraction of the biceps and triceps muscles; stabilization of the spine by co-contraction of antagonist muscles is important for the control of posture, and joint stability is important for maintaining balance and staying upright when walking on uneven terrain, for example; (iii) to increase joint stiffness and damping of the movement; this is useful in stabilizing the movement and in limiting joint rotation under a given load, particularly when the load is unanticipated; thus, in jumping from a moderate height to land with both feet on the ground, the knee joint is stiffened by co-contraction of antagonists, so as to limit the movement of the knee joint and prevent the buttocks from hitting ground; and (iv) to learn or execute fine movements, particularly when reversal of movement direction is required. This is in accordance with the general principle that any process can be more finely controlled by using two opposing mechanisms rather than only one mechanism.

As discussed in Section 11.2.2.2, the spinal cord has built-in circuitry, involving inhibitory interneurons, for reciprocal inhibition. For co-activation of antagonists, this reciprocal inhibition must be suppressed by higher centers. This is an example of higher centers having to suppress "wired" responses in lower centers under certain conditions, bearing in mind that the wired responses are intended to enable faster execution of movement and to simplify control by higher centers in most cases.

13.1.3 Feedforward and Feedback Control

Basically, feedforward control is open-loop, whereas feedback control is closed-loop, which means that in feedback control the action is influenced by its consequences, as fed back in a manner that can modify the on-going action. In open-loop control the action is not modified by its consequences, but it can still be modified by willfully changing the command during the action independently of the consequences. **Ballistic movements** are extremely rapid movements that once initiated cannot be modified in any way.

Feedback control, or more accurately, negative feedback control, is diagrammatically illustrated in Figure 13.4. The desired movement could be the tracking of a spot, that is arbitrarily moving on a screen, by an identical spot that can be moved on the screen by an operator using a joystick. The actual movement is sensed and converted by the feedback path to a form that can be compared with that of the desired movement. The difference, which is an error, is applied through the forward path to effect the actual movement. If, for example, the actual movement is less than it should be, the sensed movement through the feedback path will also be less. The difference between the desired movement and the sensed movement, which is the error, increases, which will in turn increase the actual movement through the forward path and bring it back closer to the desired movement.

The loop formed by the forward path and the feedback path in the direction of signal propagation, which is clockwise in Figure 13.4, is the **feedback loop**. Clearly, to keep the error small, then in going around the feedback loop, the ratio of sensed movement to error should be high. This ratio is the **loop gain** and is the product of the gain of the forward path, which is normally relatively large, and the gain of the feedback path, which could be less than unity. When the loop gain is large, only a very small error is sufficient to make the sensed movement nearly equal to the desired movement and the tracking of the target will be very good.

The negative feedback system works very well as long as the loop gain is large and delays around the loop are negligibly small. Under certain

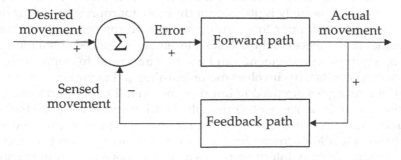

FIGURE 13.4
Block diagram of a feedback control system.

conditions, however, the negative feedback system could show instability. Suppose, for example, that the loop gain is high, but there is appreciable delay in the feedback path. If the desired movement suddenly increases to a new steady value, for example, the change in the sensed movement will be delayed. In the meantime, the error will be relatively large and positive, causing an overcorrection in the actual movement. This overcorrection will, after a certain delay, cause the error to go negative, so the actual movement will swing in the opposite direction, which, after a delay, will give rise to a large positive error, and so on. Theoretically, there will be oscillations of increasing amplitude. In practice, the amplitude of oscillations will be limited by system nonlinearities, or, if there is sufficient damping in the system, the actual movement will oscillate with decreasing amplitude about the new steady value before settling to this value. This is reminiscent of the intention tremor observed in some cerebellar disorders (Section 12.2.4.5). The oscillatory behavior could be eliminated by reducing the loop gain, at the expense of less accurate tracking, or by executing very slow movements, which severely limits the type of movement that can be executed.

In the somatomotor system, there is an inevitable delay in the generation, transmission, and processing of the sensory signal in the feedback path, the delay being about 200 ms for the response to a visual stimulus. This places a limitation on the maximum loop gain and hence the tracking accuracy that can be achieved and necessitates taking some measures to improve the overall performance, as will be discussed shortly. In feedforward control, sensory feedback is absent, which means that movement proceeds according to the initial command for the desired movement. The somatomotor system uses both feedforward and feedback control. In the initial part of a movement, sensory feedback is not yet available anyway, so the movement inevitably begins under feedforward control. If the movement is relatively slow, sensory feedback is used to make "on the fly" corrections to the ongoing movement. If the movement is ballistic, the movement will be completed under feedforward control according to the formulated command. However, even in such cases, sensory input is important for two reasons: (i) if the movement is being executed for the first time, the sensory information about the position and state of the body is utilized for the most appropriate planning of the ballistic movement, prior to its initiation, and (ii) the sensory information about the consequences of the ballistic movement, or indeed about the execution of any type of movement, can be stored and used to improve similar, future movements. This involves motor learning and memory.

Various strategies are used to improve the overall motor performance: (i) slowing down the movement – this is the familiar speed-accuracy trade-off; when accuracy of movement is called for, the movement is executed more slowly so as to allow sensory feedback to exercise its full effect; a variant of this slowing down is intermittency, whereby a movement is momentarily suspended for a "rest" interval, which allows sensory feedback to catch up; and (ii) adding a predictive component to the motor command, derived from the motor command and the current state of the body. Being predictive, this component is time-advanced in nature, which tends to compensate for the feedback delays. When picking up a box of relatively large size but unknown content, we automatically anticipate, or predict, that it is heavy and apply a relatively large force to lift it, not waiting for sensory feedback. If the force turns out to be too large or too small, based on sensory feedback, the force is adjusted accordingly. The cerebellum is believed to be involved in adding the predictive component.

Problem 13.1

Show that in a linear system based on Figure 13.4, $\dfrac{Y}{X} = \dfrac{G}{1+HG}$, where X is the input, Y is the output, G is the gain in the forward path, and H is the gain in the feedback path, so that the loop gain is HG. Note that if $HG \gg 1$, the overall gain Y/X is nearly $1/H$ independently of G.

Problem 13.2

Derive the overall gain in Figure 13.4 if the forward path is a G-H feedback system and the feedback path is a G_f-H_f feedback system. Show that if both are very high, the overall gain reduces to $H_f/(1+HH_f)$.

ANS.: $\dfrac{Y}{X} = \dfrac{G}{1+HG+GG_f/(1+H_jG_f)}$

13.1.4 Motor Coordination

The coordinated contraction of muscles in a given movement is one of the most dazzling aspects of motor control. Almost any type of movement requires some muscular coordination. This is because when a muscle contracts, force is generated at both ends of a muscle, so that, in order for only one end to move, the other end must be stabilized by another group of muscles. Thus, even if the elbow is flexed by contraction of the biceps, the shoulder is stabilized by contraction of other muscle groups. Another reason for muscular coordination is the role of the antagonists in the execution of movement, as explained in Section 13.1.2.

A striking feature of motor control is **motor variability**, exemplified by the fact that a motor task is not executed by following exactly the same trajectory, even in hundreds of repetitions. Some variability is to be expected because of random variations, or "noise", in the characteristics and activities of constituent neural elements, such as the responses of sensory receptors, the excitability of motoneurons, and variability in input activity from higher centers affecting the recruitment and derecruitment of motor units, etc. However, a more fundamental cause of motor variability, at the system level, is **motor redundancy**, which arises because a given movement can in general be executed by a theoretically infinite number of combinations of motions of individual joints. Consider, for example, moving the arm to touch the forehead, which involves mainly three joints: shoulder, elbow, and wrist. The shoulder joint has three axes of rotation: flexion-extension, abduction-adduction, and internal-external rotation (Section 9.3.4). The elbow joint has one axis of rotation (flexion-extension), and the wrist has three axes of rotation: flexion-extension, abduction-adduction, and pronation-supination of forearm. The arm is said to have seven **degrees of freedom**, which means that seven independent parameters are required to specify the configuration of the arm in space. Clearly, the final state of touching the forehead can be achieved by a theoretically infinite number of combinations of these seven degrees of freedom. That is, there is no unique choice of joint angles to effect the required movement. Evidently, the strategy used by the nervous system leads to somewhat different choices of these parameters in successive repetitions of a given movement, resulting in motor variability.

The calculation of the coordinates of an endpoint, such as fingers on the forehead, from the angles of the joints involved is referred to as a **forward kinematic transformation**, whereas the reverse problem of calculating the joint angles that will result in the required coordinates of the endpoint is referred to as an **inverse kinematic transformation**. The calculation of joint angles and angular velocities resulting from a set of torques applied to the joints involved is referred to as a **forward dynamic transformation**, whereas the reverse problem of calculating the joint torques that will result in the required movement trajectory is referred to as an **inverse dynamic transformation**. Because of motor redundancy, the number of parameters that account for the degrees of freedom is larger than the number of equations that can be derived from applying these inverse transformations to the given movement in three dimensions. The problem is said to be **ill-posed**, and the solution is mathematically indeterminate. To obtain a unique solution, some additional constraints have to be applied, based on minimizing some cost function such as the time for a movement, or the time integral of muscle force. However, it is unlikely, for several reasons, that the motor system controls movement by solving such sets of mathematical equations derived from Newtonian mechanics: (i) the computational load is rather massive, considering the nonlinear characteristics of the constituent neuromuscular elements, and the resulting movement will be slow, given the time frames of neuronal responses; (ii) the neuronal organization of the nervous system does not seem to lend itself to performing such computations; and (iii) having the higher centers compute the time profiles of torques to be generated at all the joints involved is not compatible with the hierarchical organizational principles of the motor system. A more plausible strategy is that of **motor synergies**.

The concept of motor synergies was advanced by the famous Russian neurophysiologist, Nicolai Bernstein. In one of his classical investigations of motor control in the 1920s, Bernstein studied the arm movements of professional blacksmiths as they hammered chisels. Such a movement may be repeated by a blacksmith hundreds of times daily over a number of years. Bernstein examined in detail the arm trajectories of individual blacksmiths in successive repetitions. He observed the variability in arm movement between repetitions but noted that the variability in the end movement of the hammer hitting the chisel is smaller than the variability in the motions of the individual joints involved; that is, the joints compensated for errors in each other's movement so as to result in a smaller variability in the end movement.

A motor synergy is used to denote either a group of muscles that are activated by a single command to execute a particular movement, or the pattern of muscle co-contraction that is involved in this particular movement. In either case, the central idea is that the higher hierarchical centers do not specify the detailed command to each muscle. Only a single command is issued and the synergy executes the command in the desired manner. Synergies are task-specific and are learned, not hardwired like reflexes, and a given muscle can be part of more than one synergy. In the case of Bernstein's blacksmiths, the synergy controls the trajectory of the arm in each repetition, allowing variability in the movements of the individual joints, but in such a manner that the errors in the movements of individual joints compensate for one another, resulting in less variability in the end movement. The compensation can be partly guided by sensory feedback, both visual and proprioceptive, but is also governed by the synergy that is established through the learning process, utilizing communication between motoneurons controlling the individual muscles. This communication is effected through spinal and supraspinal mechanisms and is reflex-like, as is proprioceptive feedback.

Synergy circumvents the degrees-of-freedom problem and relieves the higher motor centers from having to compute and command the time courses of individual muscle forces involved. The variabilities in the motions of individual joints associated with the synergy are subtractive, resulting in a smaller variability in the desired end movement.

13.1.5 Motor Equivalence

Motor equivalence refers to the similarity in the result of a given motor activity, irrespective of how this activity is carried out. One's signature, for example, retains the same general structure when using the preferred hand on paper, or on a blackboard, or when using the non-preferred hand, or with the pen gripped between the teeth, or using the foot in the sand. The differences between writing on paper on a horizontal desk and writing on a vertically oriented blackboard, for example, are quite significant: (i) different sets of muscles are involved, the finger muscles when using paper, and the arm muscles when using the blackboard; (ii) writing on the blackboard requires larger muscle forces and joint movements because of the larger size of the letters and the generally larger frictional forces opposing movement; and (iii) gravity perturbs writing on the blackboard by aiding downstrokes and opposing upstrokes.

Motor equivalence has led to the concept that purposeful movement is represented in the brain in an abstract manner rather than in terms of specific commands to a particular set of muscles. The abstract representation is translated to specific commands as required for a particular manner of execution of the motor activity. These abstract representations are referred to as **generalized motor programs** or as **motor schema**. Such an abstract representation is yet another manifestation of hierarchical control and is highly advantageous. It provides a high degree of flexibility by allowing a relatively easy adaptation of the required motor activity to implementation by different effectors – or by the same set of effectors but at different speeds, spatial extents, and conditions as, for example, when writing with extremely cold hands. It also more readily allows modification of movement as may be required in new circumstances and relieves the higher centers from having to cope with perturbations during execution. Such flexibility would not be practical with a preset motor program for every type or form of movement.

It is highly unlikely that the generalized motor program is represented in terms of the usual kinematic and kinetic variables, such as muscle forces or joint torques, angles, or velocities. In the case of handwriting, for example, the variables could be the relative spacing and orientation of the strokes for the various letters.

An interesting instance of motor control and coordination is **handedness**, that is, the preference to use one hand or arm for faster or more precise motor activity. It is estimated that about 90% of humans are right-handed, about 10% are left-handed, and about 1% change hand preference between tasks. That is not to say that the dominant hand achieved some "superiority" over the non-dominant hand. There is evidence of some degree of specialization between the two hands. The dominant hand, controlled by the contralateral hemisphere, is more adept in tasks that require quick, coordinated multi-joint movements, whereas the non-dominant hand performs better in tasks that involve stabilization of position and load compensation. This is a common experience when hammering a nail, for example, or cutting a loaf of bread. The dominant hand is part of the trajectory of the desired movement, while the non-dominant hand stabilizes the object in position, against the load imposed by the movement. It is believed that having both hands perform equally well in all types of motor tasks is redundant and wasteful of neural tissue, whereas a division of labor preserves function with a smaller totality of nervous tissue. The same argument has been applied to the specializations of the two cerebral hemispheres. Redundancy has of course the advantage of added reliability in case of impairment of function of one of the executing entities, but the nervous system allows, to a large extent, for this contingency through learning of new tasks by the unimpaired entity. Even after losing the dominant hand, a person can learn to write with the other hand.

SPOTLIGHT ON FREE WILL AND CONSCIOUSNESS

It was mentioned in Section 12.1.1 that when a voluntary movement is made, a readiness potential is produced in the supplementary motor area some 0.5 s before the intended muscular activity. The question may be asked as to the relative timing of the willful decision to make the movement. Intuitively, it may be assumed that the willful decision comes first, followed by the readiness potential, then the muscular activity.

Neuroscientist Benjamin Libet and his coworkers sought to answer this question using an experimental setup in which subjects were asked to flex their wrists at a time of their own choosing (Libet et al., 1983). The readiness potential was recorded by an EEG (Spotlight on Techniques 12A), and muscular activity by an EMG (Spotlight on Techniques 9A). The timing of the conscious act to flex the wrist was determined by having a spot of light move around a circle with markings on the screen of a cathode-ray oscilloscope. The subjects later reported the position of the spot on the circle when they decided to make the movement. The results of this experiment, published in 1983, were somewhat surprising. The readiness potential occurred first, followed about 350 ms later by the conscious will to act, and after another 200 ms or so by the actual movement. In other words, the neuronal processes that planned the movement began about 350 ms *before* the subject made the conscious decision to flex the wrist.

The same conclusion, but with a wider time margin, was reached by psychologist John-Dylan Haynes and his coworkers using fMRI (Spotlight on Techniques 12B) (Soon et al., 2008). As letters were displayed on a screen in front of them at intervals of 0.5 s, subjects were free to press, at a time of their own choosing, either a button with their right hand or a button with their left hand. They would then identify the letter they saw when they decided which button to press. According to the results, published in 2008, the researchers could predict from the fMRI image in the frontal and parietal cortex, and with about 70% accuracy, which button will be pressed. This prediction could be made up to 10 s *before* the reported time at which the decision to press a button was made.

Needless to say, such results generated much controversy and debate among scientists and philosophers about the existence of free will. Both Libet and Haynes determined in later experiments that subjects could still freely abort or override their decisions, within a certain time interval after making them, which indicates that some freedom of choice can still occur.

Free will is intimately related to consciousness, for it is based on the conscious perception of being able to choose between alternatives.

Consciousness encompasses self-awareness as individuals, volition and free will, and the ability to form a subjective view of the world, which is unique to every individual. Subjective phenomena include sensations, emotions, ideas, thoughts, opinions, attitudes, personal values, and past experiences. Consciousness is at the very essence of our existence as human individuals, societies, and species and is the cornerstone of culture and civilization. As such, the nature of consciousness has engaged philosophers and thinkers since the times of the ancient Greeks. It continues to be a subject for debate and research among philosophers, psychologists, and neuroscientists. Many theories have been advanced, either dealing explicitly with consciousness, or implicitly in terms of subjective vs. objective, or the mind-brain problem; but no real progress has been made so far, despite all efforts over the years. It is as if understanding consciousness is outside the frame of current thinking and needs a major breakthrough.

13.2 Motor Learning and Memory

Humans engage in a great deal of motor learning over a lifetime. Toddlers have to learn to stand upright, to walk, and to talk. Children at school learn how to write and possibly play sports and do artwork. Later in life, many people learn how to ride a bicycle, or a motorcycle, or drive a car. Many vocations and hobbies involve motor learning, from manual labor to surgery, arts, crafts, playing musical instruments, dancing, and sports. All this learning is subserved by a highly adaptive motor system, superbly dexterous hands, and exquisite senses. Moreover, as we grow, our body parts become heavier, and our limbs get longer, so that the kinematics and dynamics of movement change, and the motor system must adapt to these changes.

Motor learning is generally classified as **procedural**, **implicit**, or **nondeclarative**, meaning that it is not readily expressible in words; it can occur without consciously thinking about it. What was learned can be performed "habitually" without conscious awareness. The following should be noted about motor learning:

1. Motor learning can occur with a single trial, as for example, when learning to hold a delicate object with the right amount of force; or it can take years of practice, if it involves intricate or unnatural movements, as in professional gymnastics, or learning to play the violin, which involves precise control of individual fingers.

2. Sensory feedback is generally required for motor learning, but different sensory modalities are more important in some motor tasks than in others. Thus, proprioception is more important than vision in dynamic tasks that depend on forces acting at joints, whereas vision is more important in kinematic tasks that depend on arm trajectories.

3. It is generally true that gross motor skills involving groups of large muscles, as in swimming and bicycling, are essentially retained after a period of non-use. On the other hand, there is a retention loss of fine motor skills involving groups of small muscles, as in playing the piano or calligraphy, after a period of non-use, although performance levels can be largely recovered over a shorter period than that of the initial learning.

4. Motor learning seems to occur at the level of the generalized motor program. Thus, when a new mode of writing with the dominant hand is learned, the learning is manifested when writing with the non-dominant hand. This would indicate that the motor association areas are involved in the learning process.

It is well established that the prefrontal cortex, primary motor cortex, the basal ganglia, and the cerebellum engage in motor learning, although the exact role of each of these regions is not clear. As stated previously, the cerebellum seems to play a crucial role in the learning of motor skills involving a sequence of movements. The cerebellum is most active during the early phase of the learning of a motor sequence, during which cerebellar activity adjusts movement kinematics, based on sensory feedback, so as to accurately produce the desired motor output and reduce error during motor learning. The prefrontal cortex and the basal ganglia are particularly important for motor learning. Once the motor sequence has been well learned, the dorsal striatum and motor cortical areas are critical for long-term storage and recall of the learned sequence. Synaptic plasticity is believed to underlie the learning process.

Learning a motor skill is not by itself very useful if the learned skill cannot be stored and retrieved later. Where and how memories are stored and how they are retrieved remains conjectural. It is generally assumed that memories are stored in synapses. Macromolecules have also been implicated in memory storage and retrieval. In his classical experiments conducted early in the twentieth century, the American neuropsychologist Karl Lashley showed that rats retained the memory of a learned motor task even after 98% of the cortical areas related to the task were removed, irrespective of which areas were spared. This suggested a distributed nature of memory, as in a hologram, which is a two-dimensional recording on a photographic medium of an interference pattern produced by a laser beam, part of which is applied directly to the medium and the other part is reflected from a three-dimensional object. When a laser beam of the same frequency is applied to the medium, a hologram appears as a three-dimensional image of the object.

Holograms of different objects can be stored on the same medium using laser beams of different frequencies. An important aspect of holography is that any small part of the hologram can be used to reproduce the image of the whole object, which means that the information on the object is stored in a distributed manner. What neuronal mechanisms can give rise to distributed memory also remains conjectural.

SPOTLIGHT ON TECHNIQUES 13A: BRAIN-MACHINE INTERFACE

A **brain-machine interface (BMI)**, also commonly referred to as a **brain-computer interface (BCI)** is a system for recording brain neuronal activity representing intentions of movement and converting this activity to signals for real-time control of external devices, such as a cursor on a computer screen, a wheelchair, a robotic arm, a prosthesis, or an exoskeleton. The purpose of a BMI is to improve the quality of life of individuals with sensorimotor disabilities.

The components of a BMI system perform four basic types of operations: (i) recording of neuronal activity, (ii) processing and decoding of the recorded signals, generally using artificial intelligence (AI) procedures, to extract salient information about the intended movement, (iii) generation of control signals that are appropriate for the effector being utilized, and (iv) providing sensory feedback for improving overall performance.

Neuronal activity is most commonly recorded electrically using electrodes on the scalp (EEG, Spotlight on Techniques 12A), or on the surface of the cerebral cortex (epidural ECoG), or under the dura mater (subdural ECoG), or intracortically, that is, penetrating the cortex. A sophisticated example of the latter is the silicon-based, slanted **Utah electrode array (UEA)** that is made using integrated-circuit (IC) manufacturing techniques. The square array can have up to 256 tine electrodes, insulated except for their tips, spaced 0.4 mm apart, and varying in length, diagonally across the array, from 1 to 1.5 mm so as to record from various depths. Further developments have halved the spacing of the tines to 0.2 mm, replaced wiring connections by wireless ones, and added optical waveguides for optogenetic applications (Spotlight on Techniques 7C). The challenge in using electrodes is to enable good quality recording, of acceptable magnitude and signal-to-noise ratio, over a prolonged period of many years in the presence of degrading factors, such as corrosion of contacts, deterioration of insulation, and inflammatory response of neural tissue. The recordings are mainly from the primary motor cortex but could also be combined with signals from other cortical areas, such as the premotor cortex and the posterior parietal cortex. A new technology is that of using ultra-thin, conductive nanowires, each about one fifteenth the diameter of a human hair, for recording as well as stimulation. 100,000 wires can be deployed at various depths in multiple regions of the brain.

It should be noted that the decoding operations do not decipher the motor commands, since the nature of these commands is largely unknown. The operations essentially extract, using statistical methods, information that is relevant to the intended movement. The BCI user must learn, through training, to generate cortical activity patterns that will effectively control the external devices. The training may include computer-assisted learning and immersive virtual reality, operant conditioning, and sensory feedback. Significantly, some of the subjects who undergo intensive training for BMI experience improved neurological function on top of the ability to use the BMI more effectively. Visual and haptic feedback through specially provided sensors is generally used to improve performance during task execution. The haptic feedback can be quite sophisticated in that it can sense force, acceleration, grip pressure, and slip.

An interesting extension of BMI is to use it in the reverse direction, that is, to feed external signals into the brain to influence thoughts and perception.

13.3 Posture

13.3.1 Balance

The control and maintenance of posture is another truly amazing feat of the motor system. Not only can we stand upright, but we can also control our posture, that is, the relative position of our body parts, during movement. Witness a waiter weaving swiftly between tables in a crowded restaurant while balancing with one hand a tray full of food and drinks. Or consider how one may regain one's balance when stumbling.

From a mechanical viewpoint, standing upright is akin to balancing a pendulum having a multi-jointed stem, in an inverted vertical position, with its loaded end up. This is an inherently unstable structure that must be stabilized by the sensorimotor system. Before we consider how this is achieved, it is necessary to define some terms that are commonly used in discussing stability and equilibrium:

Center of mass (COM): It is the point around which the mass of a body is equally distributed in all directions. Hence, when a body moves under the influence of external forces, the COM is the point that moves as if the entire mass of the body were concentrated at that point and all the external forces were applied at that point. The **center of gravity (COG)** is the point where the weight of the body, due to the force of gravity, may be considered to act. The COG and COM are the same point as long as the gravitational force is uniform over the body, which is the case for all practical purposes. For a person in a normal standing position, the COM is located anterior to a level in the upper third of the sacrum (Figure 11.1).

Center of Pressure (COP): It is the point of application of the resultant of the forces acting on the body at the support surfaces. Normally, when a person is standing barefoot on a flat surface, the support forces act on the body at the surfaces of contact of the ball and heel of each foot. Assuming equal distribution of the body's weight between the two feet, the COP will be in the midsagittal plane (Figure 13.5a).

Base of Support (BOS): It comprises the contact area between the body and the surface of support plus any intervening space. For a person standing on a flat surface, the BOS is the contact area of the two feet as well as the intervening area, as illustrated in the lower part of Figure 13.5a.

In a normal standing position on a flat surface, with the weight equally distributed between the two feet, the COP is in the middle of the BOS, mediolaterally, and is vertically below the COM (Figure 13.5a). Under these conditions, there is no net translational force on the body, nor is there a net torque acting to rotate the body. The body is in stable equilibrium. If the person leans to one side so that one foot just lightly touches the flat surface, the weight is now supported by the other foot, the COP will act on this foot, and the COM will be shifted sideways, vertically above the COP. In Figure 13.5b the COP is just at the edge of the BOS. The body is still balanced, that is, in postural equilibrium, but it is not a stable equilibrium because leaning further in the same direction will cause the COM to move outward, beyond the BOS (Figure 13.5c). There will be a resultant torque that rotates the body in the direction of the bend, causing it to topple sideways, as balance is lost. Hence, *when only gravity acts on the body, stability requires that the vertical projection of the COM falls within the BOS, on the COP. The further this projection is from the edge of the BOS in a given direction, the greater is the stability in that direction, that is, the more difficult it would be to lose balance.*

When forces other than gravity act on the body, the body can be in a stable position, with the vertical projection of the COM falling outside the BOS. In Figure 13.5d, where the person is holding to a rope attached to a wall, the vertical projection of the COM falls outside the BOS, but the tension in the rope provides a counter torque that balances the body. A gymnast doing a handstand with the body in a horizontal position is balancing his body by

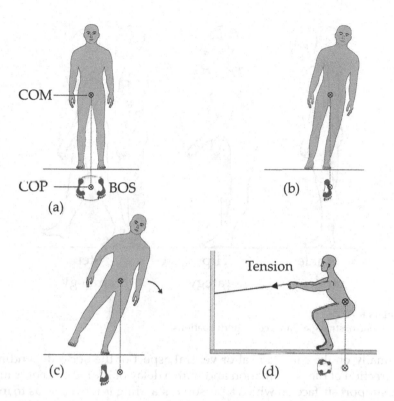

FIGURE 13.5
Relations between the center of mass (COM), the center of pressure (COP), and the base of support (BOS) under various conditions: (a) standing upright and still; (b) maximal leaning without toppling; (c) toppling if leaning further than in (b); (d) holding a rope attached to a wall.

the forces developed by his core and leg muscles, while the vertical projection of the COM falls outside the BOS provided by the hands.

Maintaining balance is not only a matter of keeping the vertical projection of the COM within the BOS. When one walks on a narrow support, such as a plank spanning a trench or a log over a stream, one automatically extends the arms laterally for better balance. This does not change the vertical projection of the COM but increases the moment of inertia of the body about the axis of rotation, thereby making it more difficult for any rotational forces to topple the body sideways. This is also the reason why tightrope walkers carry a long balancing pole.

Balance control is also influenced by the emotional state. Fear of falling, for example, affects postural tone stiffness, balance, and responses to postural disturbances.

There are three basic balance control strategies that are induced by perturbations (Figure 13.6). The first is the **ankle strategy**, which involves sequential activation of the ankle, thigh, and hip muscles, progressing distally to

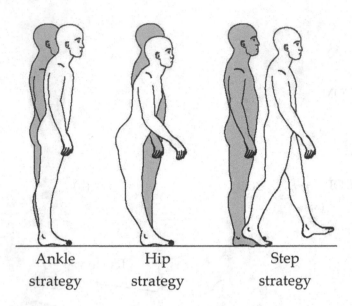

Ankle   Hip   Step
strategy   strategy   strategy

FIGURE 13.6
Balance control strategies induced by perturbations.

proximally on the same dorsal or ventral aspect of the body, depending on the direction of the perturbation and with a delay of 70–130 ms. For example, if the support surface on which a person is standing is moved so as to induce a forward sway (Figure 13.6a), this would result in the activation of the ankle plantar-flexors (gastrocnemius, soleus, and plantaris), the knee flexor (biceps femoris) and the hip extensors (mainly the gluteus maximus and the hamstring muscles). Balance is restored through rotation of the body about the ankle joints, which can rotate in all directions. The ankle strategy is limited by the torque that can be exerted by the foot in contact with the supporting surface and would be used for slow and small perturbations and when standing on uneven terrain.

For fast or larger amplitude perturbations, or when the support surface is narrow, so that little ankle torque can be applied, the **hip strategy** is used (Figure 13.6b), where the hip joint can also move in all directions. The posture shown in the figure involves sequential activation of the trunk muscles (abdominals) and thigh muscles (quadriceps), progressing proximally to distally, with about the same delay of 70–130 ms. This strategy rotates the trunk forward and downward, and the legs backward, which stabilizes the body with respect to the BOS and decreases the moment of inertia about the ankle, thereby allowing a given ankle torque to produce a larger angular acceleration of the body. The hip strategy is limited by the horizontal force that is produced by friction between the feet and the supporting surface. When the ankle and hip strategies are insufficient to maintain balance, a step is taken (Figure 13.6c) which maintains balance by enlarging the BOS.

13.3.2 Upright Posture

Upright posture involves propping up the bony skeleton so that it does not collapse to the ground under gravity. But the upright posture is inherently unstable, like the inverted pendulum previously mentioned, so it has to be actively maintained. This is chiefly effected by tonic activity of the so-called antigravity muscles, which are the extensor muscles of: (i) the ankle (the triceps surae plantar flexors, or calf muscles), (ii) the knee (the quadriceps muscles), (iii) the hip (mainly the gluteus maximus), and (iv) the neck, to prevent the head from falling forward, in addition to the paraspinal muscles of the back, which hold up the spine. When joints are fully extended vertically by the extensor muscles, a major support against gravity is provided through mechanical contact between bones, as in the knee joint, or between bone and ligament, as in the hip joint.

The tonic activity of the antigravity muscles is maintained through sensory, negative feedback involving the vestibular, visual, and somatosensory systems. The contribution of the somatosensory system is mainly through proprioceptors participating in the tonic stretch reflex (Section 11.3.3). Thus, if joints tend to flex under the influence of gravity, the extensor muscles will stretch and develop a counteracting force through the tonic stretch reflex. Another important contribution of the somatosensory system is through cutaneous receptors in the sole of the foot, that is, in the **glabrous skin** (the skin devoid of hair) on the plantar side of the foot. These receptors are of four types (**Merkel discs, Meissner corpuscles, Pacinian corpuscles, and Ruffini endings**) and provide information on the distribution of pressure on each foot from the supporting surface, as well on the magnitude, direction, and rate of change of this pressure, and on the geometry and compliance of the supporting surface. This information is of considerable importance for balance and movement, as evidenced by the relatively large area of projection of the foot receptors in the somatosensory cortex.

The vestibular system provides information about head position with respect to gravity as well as the direction and speed of movement (Section 12.2.5.3). The visual system provides information on the position of the body and its movement with respect to the environment. Information provided by a single sensory modality, however, is often insufficient or ambiguous, so more than one sensory modality is used. The vestibular system, for example, does not provide information about the trunk, independently of the head. The visual system may provide ambiguous information, as when sitting in a stationary vehicle and watching another vehicle move, which gives the sensation of movement of one's vehicle. Moreover, one sensory modality may be more important in a given situation. The visual system is of prime importance when walking on uneven terrain and avoiding obstacles.

When a person stands still in an upright position, a posture referred to as a **quiet stance**, the COM and COP are not stationary but are continually moving in a random-like manner. They show alignment only when their motion is averaged over time and not at any particular instant. The horizontal movement of the COM about its average value is referred to as **postural sway**. Its amplitude is normally about a few millimeters but is increased by closing the eyes, cooling the soles or standing on foam, and by head placement in different positions, reflecting, as expected, the effects of the visual, somatosensory, and vestibular systems, respectively. Several factors can contribute to postural sway such as breathing, heartbeat, and somatosensory variability. Inevitable variations in the forces of extensor muscles, and their relative contributions on each side of the body, would affect the COP, which would produce a torque that moves the COM. Moreover, since the upright posture is maintained by negative feedback, the dynamics of negative feedback would introduce damped oscillations in the COM in response to ongoing perturbations.

13.3.3 Postural Adjustments

Retaining posture during movement is an all-important function of the sensorimotor system that is even more challenging than stabilizing an inherently unstable upright posture. It is accomplished through postural readjustment, an all-too-common example of which is **gait initiation**, that is, the transition from standing to walking. As one leg, the **swing** leg, is lifted the BOS is reduced to the area of contact between the supporting surface and the foot of the other leg, the **stance leg**, so that the projection of the COM is now outside the BOS (Figure 13.7). There results a torque that tumbles the

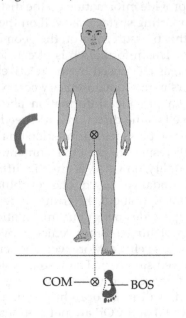

$$COM—\otimes \quad BOS$$

FIGURE 13.7
Gait initiation.

body toward the swing leg. But even before the heel of the swing leg is lifted off the ground, an **anticipatory postural adjustment** occurs in the form of a motor synergy that adducts the hip on the swing side, with some flexion of the hip and knee on the stance side. This shifts the COP toward the stance side, which tends to produce a counter movement to that resulting from lifting the foot on the swing side. Nevertheless, the body falls laterally toward the swing side, and the fall is brought to a stop when the foot of the swing leg is placed on the ground. This enlarges the base of support and produces a ground reaction force whose peak may be as large as twice the body weight. The effect is to brake the body fall and have the projection of the COM within the base of support, eventually restoring balance.

Movement perturbs posture in several ways. When a joint is flexed, the extensor, antigravity muscles connected to this joint must be inhibited, that is, the stretch reflex suppressed. Movement alters body geometry and generally affects the COM, COP, and BOS, which in turn influences balance. Moreover, particularly with fast movements, **interaction torques** are produced at joints, which affect the forces developed by postural muscles. These interaction torques are quite complex and involve: (i) **inertial torques**, contributed to each joint due to the movement of every other joint, (ii) **centripetal torques**, due to rotary motion about a joint, and (iii) **coriolis torques**, due to joints moving at different angular velocities. All the foregoing effects entail some postural adjustment during movement.

Postural adjustments may be initiated by the subject prior to the movement, as in the anticipatory postural adjustments mentioned earlier in connection with gait initiation. These anticipatory adjustments essentially depend on experience with expected perturbations to posture. Unexpected perturbations are handled in several ways. First, the elasticity of muscles, tendons, and other tissues naturally generates forces that oppose movement. Second, the phasic stretch reflex (Section 11.3.3) is activated; this reflex is largely monosynaptic and has a time delay of a few tens of milliseconds. Third, corrective reactions are triggered by sensory inputs arising from the perturbation; these corrections are essentially pre-programmed. For example, when a strong wind blows from the front, one would automatically: (i) move one leg forward, the other back, with slanting ankles, which enlarges the BOS, (ii) lean forward in the direction of the wind, which moves the COM forward, close to the edge of the BOS nearest to the direction of the wind, and furthest away from the opposite edge of the BOS, which enhances stability, and (iii) bend the knees and stiffen them by co-contraction of antagonist muscles, which adds strength and sturdiness to our posture and lowers the COM somewhat, thereby reducing the moment of the wind force that tends to knock one over.

In making postural adjustments, the motor system uses motor synergies quite effectively because of the advantages of this mode of motor control (Section 13.1.4). These postural synergies, whether anticipatory or reactive, are inherently part of the synergies of any planned voluntary movement.

As previously mentioned, these synergies involve motor learning and are subject to adaptation because of changes in the environment or in the task to be executed. Higher motor centers must therefore be involved in these synergies, including the cerebellum, basal ganglia, reticular formation, and cortical motor areas. The primary motor cortex and the motor association areas are expected to be more involved in anticipatory postural adjustments than automatic postural reactions. Where postural adjustments are influenced by the emotional state, the limbic system would be involved.

Because any disturbance to normal posture will generate muscle forces that will restore this posture, the question arises as to how voluntary movement can occur without opposition from these postural stabilizing forces? This is known as the **posture-movement paradox**. The equilibrium-point hypothesis addresses this problem by combining posture and movement in a single scheme (Section 13.5.4).

13.4 Locomotion

Locomotion is an essential part of daily living that we take for granted. Seemingly effortless, it is in fact a highly complex motor activity that involves carefully tuned coordination between different muscle groups. We will start with an analysis of **gait**, which is the manner in which we walk, followed by a description of the basic neuronal processes underlying its execution and control.

13.4.1 Phases of Gait

The initiation of gait was discussed in the preceding section in connection with Figure 13.7, where the emphasis was on stability in the mediolateral direction. In addition to the previously mentioned anticipatory postural adjustments that shift the COP mediolaterally, similar adjustments shift the COP backwards, through inhibition of the ankle plantar-flexors, followed by activation of ankle dorsi-flexors. This produces a torque that helps propel the body forward and launch the initial step with the intended step length and initial velocity.

The energy expended in walking along a horizontal path (level walking) varies quadratically with speed, up to about 100 m/min (6.6 km/hr) (Ralston, 1976) (see Problem 13.3). This energy seems to be essentially gender-independent when expressed per kg of body weight, particularly if based on lean body weight. Interestingly, humans seem to adjust step length and stepping rate for minimal energy expenditure per unit distance (Problem 13.4).

In both running and walking, movement of the arms helps propel the body forward and enhances body stability. The trunk is normally stabilized during locomotion, but there may be a small oscillation of the head, in which case, the gaze is stabilized by the vestibulo-ocular reflex (Section 12.2.5.3).

In steady walking, human locomotion is divided into two phases, the **swing phase** and the **stance phase**. The swing phase begins with the toe of the swing leg leaving the ground and ends when the heel of this foot strikes the ground. This phase is therefore characterized by having the swing foot off the ground during the phase. The stance phase begins when the heel of one foot contacts ground at the end of the swing phase and ends when the toe of this foot leaves the ground at the beginning of the next swing phase. The foot is therefore in contact with the ground throughout this phase.

The sequence of a stance phase and a swing phase is a **gait cycle** (GC), and the distance moved in a gait cycle is a **stride**. A **step** is the distance moved between the heel of one leg striking ground and the next strike by the heel of the opposite leg. In walking, at least one foot is always on the ground, but in running, there is a brief period, the **float phase**, when both feet are off the ground and at no time is the body weight supported by both legs.

In normal walking, the swing phase lasts for about 40% of the gait cycle, and the stance phase, the remaining 60%. The gait cycle duration decreases with increasing speed in walking and running, primarily due to a shortening of the stance phase, with much less change in the duration of the swing phase. The speed of locomotion in humans can vary from a very slow speed, to about 4–to 4.8 km/hr in normal walking, to about 6 km/hr in brisk walking, and up to about 45 km/hr toward the end of a 100 m dash in a sprint race. The gait changes with the speed of locomotion, from that of slow walking to that of fast running.

We will consider next the main movements of flexion and extension of the hip, knee, and ankle joints in the swing and stance phases. The angles of the hip, knee, and ankle joints that change during locomotion are indicated in Figure 13.8. The precise orientation of the thigh, leg, and foot during the various stages of the swing and stance phases will, of course, differ between individuals. What will be described is a "typical" gait cycle with reference to the movement of the right limb.

The swing phase can be divided into three stages. In the **initial swing stage** (Figure 13.9a), the hip of the swing foot is extended by about 10° at the beginning of this stage and is flexed to about 20°. The knee is flexed to 40°–60°, and the ankle goes from about 20° of plantar flexion to a neutral position. In the **midswing stage**, the hip is flexed to about 30° and the leg is extended so that the flexion at the knee is reduced to about 30°. The ankle is dorsiflexed. In the **terminal swing stage**, the hip remains flexed at approximately the same angle, the leg is almost fully extended, and the ankle is in a neutral position.

In addition to these movements, there is, in the swing phase, rotation of the pelvis toward the stance leg and a counteracting rotation of the spine in the opposite direction and a slight lateral tilt of the pelvis toward the swing leg. In running, the foot gets much more elevated than in walking. This raises

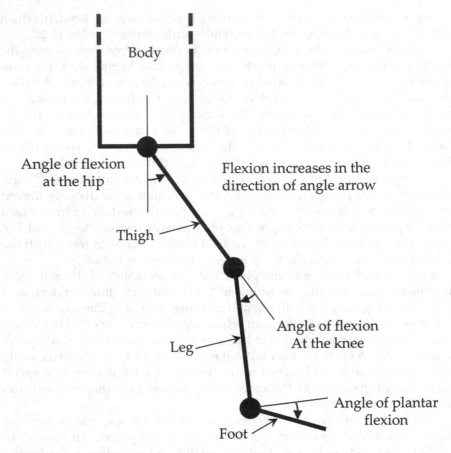

FIGURE 13.8
Changes in joint angles during locomotion.

the COM of the leg and reduces the torque needed to swing the leg forward. Whereas in walking, the heel first strikes the ground in the stance phase, the ball of the foot first strikes the ground in running.

The stance phase can be divided into five stages (Figure 13.9b): The **initial contact stage** is a short interval that marks the beginning of double support of the body by both legs. The hip is flexed by about 30°, the leg is almost fully extended, and the ankle is moved from a neutral position into plantar flexion. In the **loading response stage**, the body absorbs the impact of the foot with the ground by rolling forward in pronation. The COM is at its lowest level. The hip is extended to about 10° of flexion, the leg is flexed to 15°–20°, and ankle plantar flexion is increased to 10°–15°, with the foot being flat on the ground. In the **midstance stage**, the COM is at its highest level, the body is supported by one single leg because the other leg is in the swing phase, and the body begins to move from force absorption at impact to forward

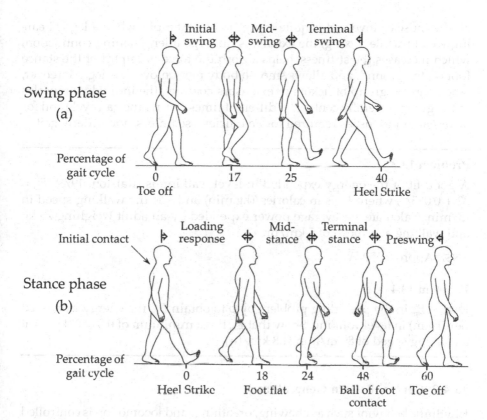

FIGURE 13.9
Phases of locomotion; (a) swing phase; (b) stance phase.

propulsion. The hip moves from 10° of flexion to extension. The knee reaches maximal extension before the beginning of flexion. The ankle is dorsiflexed and slightly supinated, that is, the foot is rolled outward. During the **terminal stance stage**, the heel leaves the ground (heel-off), so that the weight on that foot is supported by the ball of the foot, underneath the heads of the metatarsal bones. In this phase, the body weight is divided over the metatarsal heads. The hip is extended to about 10°–15°, the knee is slightly flexed by a few degrees, and the ankle supinates and plantar flexes. In the toe-off, **or pre-swing, stage**, the toes leave the ground, the big toe being the last part of the foot to leave the ground, marking the beginning of the swing phase. The hip becomes less extended, the knee is flexed 35°–40°, and plantar flexion of the ankle increases to about 20°.

In addition to these movements, there is in the stance phase, rotation of the pelvis away from the stance leg, and a counteracting rotation of the spine in the opposite direction.

Whilst it is generally true that flexors contract during flexion and extensors during extension, the pattern and sequencing of the contractions of all the muscles involved is quite complex. For example, when a leg is bearing weight while flexing, the extensors undergo a lengthening contraction, which increases the stiffness, helps absorb the force of impact of the stance foot on the ground, and allows smooth body motion over the leg. Moreover, when a given group of flexors or extensors contract, the individual muscles in the group generally contract at different times during the gait cycle and for different durations, as required for an efficient, seamless, and orderly gait.

Problem 13.3

A good fit of the energy expended in level walking is (Ralston, 1976): $E_w = 32 + 0.005v^2$, where E_w is in calories/(kg.min) and v is the walking speed in m/min. Calculate the average power expended by an adult weighing 75 kg and walking at a speed of 5 km/hr.

ANS.: Approx. 350 W.

Problem 13.4

Divide E_w in the preceding problem by v to obtain E_m, the energy expended per (kg.m) in level walking. Show that E_m has a minimum of 0.8 cal/(kg.m) at a walking speed of 80 m/min (4.8 km/hr).

13.4.2 Central Pattern Generators

Rhythmic behavior such as chewing, breathing, and locomotion is controlled by specialized neuronal circuits, referred to as **central pattern generators** (**CPGs**), which are mainly responsible for generating the rhythm and pattern of the rhythmic activity. More complex CPGs are involved in organized, stereotyped behavior such as swallowing, coughing, and sneezing and which require coordination of many different muscles. In some continuous rhythmic behavior, the periodicity is generated by neurons having pacemaker properties, as in the case of the heartbeat.

In general, rhythmic activity can result from: (i) intrinsic pacemaking involving a sodium persistent current, $I_{Na,p}$, K⁺-dominated leak currents (K_{leak}), and Ca^{2+}-activated nonselective cationic current (CAN) I_h (Section 7.3.3); (ii) network dynamics involving excitatory and inhibitory neuronal populations that are coupled in a certain manner; and (iii) a combination of these two mechanisms. Thus, both mechanisms are implicated in the control of breathing, for example, resulting in medullary neurons firing in bursts, where the inter-burst interval determines the rate of breathing, and the frequency of firing within a burst determines the force of contraction of the breathing muscles and hence the depth of breathing. In many cases, CPGs utilize circuitry that underlies various reflexes.

As a result of animal experiments and observations of humans with spinal injuries, it was concluded that the CPGs for leg movement during locomotion are located in the lumbosacral spinal cord. They are, nevertheless, controlled by descending inputs from the motor cortex and the brainstem, and their activity is modulated by sensory inputs from the periphery. The CPG for leg movement is distributed longitudinally over several segments of the spinal cord and is concentrated transversely in laminae VII, VIII, and X (Figure 11.2).

There is evidence that the CPG network for each limb comprises multiple interconnected modules that control the movement of each joint, these modules being oriented rostro-caudally in the order of: hip, knee, and ankle modules. A CPG module, enclosed by the dashed rectangle in Figure 13.10 consists of two layers: a rhythm-generating layer and a pattern-generating layer. The latter is responsible for coordinating: (i) flexor-extensor activity at each joint, (ii) activity of the CPGs of the joints on one side, and (iii) activity of the CPGs of the limbs on the two sides. The more rostral, rhythm-generating module of the hip appears to be dominant, that of the knee being less dominant, and that of the ankle being less dominant still. Intersegmental connections ensure coordination between the joint modules on one side. Coordination between the CPGs of the limbs on either side is achieved through commissural interneurons whose axons cross the midline from one side of the spinal cord to the other via the ventral white commissure (Figure 11.3). Axons of these interneurons can be confined to the same segment, can ascend or descend over several segments, or can bifurcate into ascending and descending fibers.

The interneurons of the pattern-generating layer excite motoneurons monosynaptically, whereas interneurons of the rhythm-generating layer are several synapses away from the motoneurons and project directly to the interneurons of the pattern-generating layer. It is not known for certain how

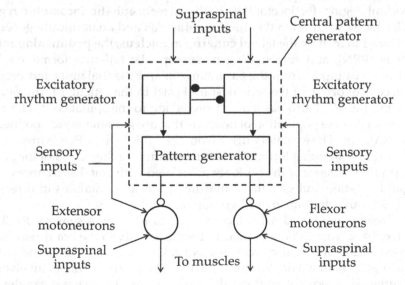

FIGURE 13.10
Diagrammatic Illustration of central pattern generation.

the rhythm is generated in the mammalian CPGs. It is believed that each CPG has a separate excitatory rhythm generator for each side. The excitatory designation means that no inhibition is involved, since it has been shown that rhythmic activity persists when all fast inhibitory transmission is blocked. The two rhythm generators mutually inhibit one another, so that their activation alternates, that is, when one rhythm generator is active, the other is inhibited. At least four classes of interneurons, some excitatory and some inhibitory, have been implicated in the mammalian CPGs, based on their expression of distinct transcription factors, that is, proteins that are involved in the transcription of DNA to RNA. One group of interneurons that are involved in reciprocal inhibition, between the rhythm generators and reciprocal inhibition between agonists and antagonists within the pattern generators, are the inhibitory Ia interneurons (Section 11.2.2.2). They are referred to in this context as rIa-INs (reciprocal Ia inhibitory interneurons), which are themselves inhibited by Renshaw cells (Section 11.2.2.1).

Although the details of the mammalian rhythm generators are unknown, their behavior is governed by intrinsic neuronal characteristics and synaptic connectivities and properties. The following neuronal characteristics are particularly relevant: (i) endogenous bursting behavior (Section 7.3.3), (ii) postinhibitory rebound (Section 7.3.5), (iii) spike frequency adaptation (Section 7.3.5), and (iv) plateau potentials (Sections 7.4.4 and 7.4.5).

13.4.3 Extra-Spinal Influences

The CPGs are under the control of the motor cortex as well as through the reticulospinal, rubrospinal, and vestibulospinal tracts (Section 12.2.5). An important region for locomotion is the **mesencephalic locomotor region** (**MLR**), located ventral to the inferior colliculus and anatomically generally considered to include the lateral **cuneiform nucleus**, the **pedunculopontine nucleus** (**PPN**), and surrounding mesencephalic reticular formation. The MLR receives inputs from the basal ganglia and the thalamus and projects to the cells of origin of the reticulospinal tract in the reticular formation. In primates, the MLR also has a direct projection to the spinal cord. The MLR contains a diverse population of neurons that are glutamatergic, cholinergic, or GABA-ergic. The excitatory drive from the MLR to the CPGs is involved in initiating and stopping locomotion and determining the period of the generated rhythm. Lesions of the MLR are associated with gait disturbances, such as gait hesitation and gait ataxia manifested as abnormalities in direction, amplitude, and rhythmicity of locomotion.

The cerebellum coordinates locomotion, as it receives sensory feedback and feedback from CPGs and modulates the activity in the brainstem nuclei (Section 12.2.5), which also receives feedback from CPGs. Cerebellar lesions disturb gait, as in ataxia (Section 12.2.4.5). The motor cortex is involved in the initiation of locomotion through its connections to the basal ganglia and influences locomotion through its descending pathways as well as through direct outputs to the spinal cord. Cortical input is important for visual direction of locomotion through information provided by the visual cortex via the posterior parietal cortex.

Sensory input to the CPGs is important for modifying locomotor activity, as when encountering obstacles, for changing the amplitude of the motor output, and for regulating phase changes between flexor and extensor activation.

13.5 The Equilibrium Point Hypothesis

13.5.1 General Views of Motor Control

A central question in the control of movement is: what is it that the CNS controls during movement? Is it muscle force, muscle velocity, movement trajectory, joint stiffness, damping, or other mechanical variables? Figure 13.11 illustrates a simplified functional, anatomically based description of muscular control. It is seen that, in principle, the following modes of control are possible:

1. **Force control**, through activation of α-motoneurons by higher centers. The negative feedback involving GTOs will tend to keep the force constant, as discussed in connection with Figure 13.2.3.

2. **Length control**, through activation of static γ-motoneurons. The contraction of intrafusal fibers increases the static response of the muscle spindle receptors (Section 9.4.2.2), which activates α-motoneurons, causing the muscle to contract. The length of the muscle decreases, and the firing of intrafusal fibers is reduced to the level that maintains the required length. The negative feedback tends to keep the muscle length constant, as discussed in connection with Figure 11.9 for the lower loop involving the static response of muscle spindle receptors.

3. **Velocity control**, through activation of dynamic γ-motoneurons. The sequence of events and the mode of control are analogous to that involving static γ-motoneurons but is based on the upper loop in Figure 11.9, incorporating the dynamic response of muscle spindle receptors.

4. Another possible mode of control by higher centers is through **α-γ linkage**, that is coactivation of α- and γ-motoneurons. Compared to direct activation of γ-motoneurons, as in aforementioned items 2 and 3, which in turn activates α -motoneurons through feedback, the process is faster, since muscle activation begins with the direct

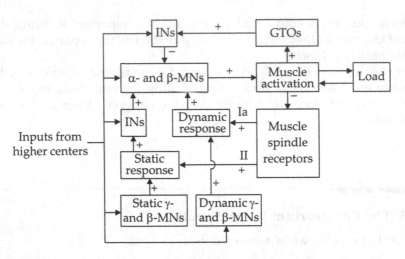

FIGURE 13.11
Block diagram of muscular control.

activation of α-motoneurons. The role of the feedback becomes essentially the correction of deviation of the movement from that intended. Moreover, since activation of α-motoneurons controls force, whereas activation of static and dynamic γ-motoneurons controls muscle length and velocity, respectively, then it is possible, at least in principle, to control muscle stiffness (ratio of force to length) and muscle viscous damping (ratio of force to velocity), as may be required by a given movement.

Several hypotheses have been advanced over the years concerning control of movement. According to Merton's **servo-hypothesis** introduced in the 1950s, movement is controlled by activation of γ-motoneurons, as in aforementioned items 2 and 3. Hence, the hypothesis is also known as the **γ-model**. There is no convincing experimental evidence for this hypothesis, nor is the loop gain sufficiently high for the negative feedback system to act effectively as a length or velocity controller or servomechanism. A later modification is that movement is initiated by α-γ linkage as **servo-assisted control** in which the stretch reflex mainly compensates for deviations in muscle length due to load conditions.

The **α-model** was introduced by Bizzi and his associates in the 1970s, according to which muscle force is the primary controlled variable. The model suffers from some basic difficulties in that the predicted muscle force-length characteristics are essentially a family of lines passing through a point on the length axis representing the resting length of the muscle, and the control becomes unstable for a load whose force increases with muscle length.

Beginning with the 1980s, a more elaborate hypothesis of motor control, subsequently referred to as the **EMG-force control hypothesis,** emerged as a result of contributions from a number of investigators and from work in robotics and adaptive control. The electromyogram (EMG) is an extracellular recording of the action potentials of muscle resulting from the activation of muscle fibers by α-motoneurons (Spotlight on Techniques 9A). The EMG can therefore be considered to encode the force developed by the muscle. The timing, amplitude, duration, and overall pattern of the EMG are, in general, the result of both descending motor commands and the activity of peripheral reflex loops. A central idea of the EMG-force control hypothesis is, again, that the nervous system is primarily concerned with the specification of forces required for a given motor activity. These forces are computed by the nervous system from **inverse dynamics**, that is, from the kinematic variables required, such as position, velocity, and acceleration. This is the inverse of the usual problem in dynamics, in which the forces are used to calculate the kinematic variables. The system parameters required for inverse dynamics computations are derived from **internal models** constructed by the nervous system, taking into account the mechanical properties of muscles and body parts, their interaction with the environment, and the motor learning gained from experience. The computation of forces includes a feedforward, or predictive, element to speed up the movement and make it more effective, and the computation is modified in accordance with feedback on the progression of the movement, compared to that planned, and on any unanticipated influences.

A different approach, known as the **equilibrium-point (EP) hypothesis**, or **λ-model**, was advanced principally by Feldman in 1966 and has been elaborated since. It is an example of **threshold-control theory (TCT)**. In this view, the motor system does not compute inverse dynamics or kinematics. Rather, the higher-centers command basically changes the threshold of the tonic stretch reflex λ, that is, the muscle length at which the muscle begins to generate an active force, as explained below. The EP hypothesis has several attractive features: (i) the central command signal depends on internal, physiological variables, (ii) the attainment of the final, desired endpoint generally relies on sensory feedback control, (iii) in executing a movement, reflex circuitry, motor synergies (Section 13.1.4), and biomechanical properties of muscle are exploited, and (iv) the posture-movement paradox is resolved by combining posture and movement in a single scheme. The EP hypothesis has been steadily gaining ground, supported by arguments that it provides the framework for explaining aspects of motor control that other hypotheses can explain only vaguely or not at all.

An illustrative car analogy can be invoked to illustrate the basic concept of the EP hypothesis. Consider a futuristic car whose driver, analogous to the highest level of motor hierarchy, has only to issue a command to instruct the car to proceed to a specified destination. The car is assumed to have the intelligent technology that enables it to compare the destination to its starting position, plan the best route, taking into consideration all relevant factors, such as road and traffic conditions, compute all required actions, such as steering, acceleration, and braking, then proceed accordingly, avoiding obstacles and other cars while abiding by traffic laws and accepted norms of good driving. In principle, such a scenario is possible, although overly complex and hardly practicable. This situation is analogous to the motor system having to compute the inverse dynamics. Consider next a more realistic situation in which a driver uses some tools such as the steering wheel, accelerator, and brake, analogous to the λs of the muscles, to guide the car to different destinations without the need to compute beforehand all the required actions for every required destination. In steering the car, the driver utilizes learned skills, judgment, and sensory feedback based on vision, hearing, and body sensations.

The EP hypothesis will be presented in some detail in what follows and used to shed some light on important aspects of motor control.

13.5.2 Basics of the EP Hypothesis

The λ-model originated from experiments on the unloading of elbow flexors and extensors, the biceps and triceps muscles, from different initial elbow angles. Figure 13.12 illustrates the nature of the observed results. An open circle represents initial values of elbow angle and torque exerted by the flexors when a certain load was applied to the arm. After this initial point was established, a certain amount of the load was removed, causing the elbow angle to decrease, which yielded a new set of steady-state values of elbow angle and torque, as represented by a filled circle. The initial values of elbow and load were then restored and another amount of load was removed, giving a new set of steady-state values represented by another filled circle, and so on. The filled circles for the same set of initial conditions were joined by a curve, such as that labeled IC_1 in Figure 13.12.

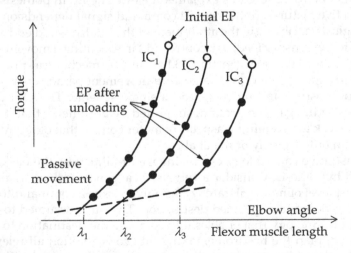

FIGURE 13.12
Invariant characteristics obtained for the unloading of elbow flexors and extensors.

The subject then intentionally moved the elbow to a new starting point, and the procedure was repeated, giving a new curve, such as that labeled IC_2, and so on. A family of curves was thus obtained, whereby each curve represented involuntary movement of the arm of the subject, as dictated by unloading from the same initial state, and without voluntary intervention by the subject. However, the change to a different curve was due to a voluntary movement. The individual curves IC_1, IC_2, etc. are nonintersecting but successively displaced along the horizontal axis, with some change in shape. Each curve represents the spring-like, force-length relation of active muscle *with* attendant afferent input from the peripheral receptors involved. Evidently, the shift from one curve to the next, due to voluntary intervention by the subject, is effected by higher centers through changing some variant parameter of the system. This variant was identified as the threshold muscle length λ at which an individual curve departs from the passive torque-elbow angle characteristic, obtained when the elbow was rotated with the elbow muscles fully relaxed. In other words, λ is the length at which the flexor muscle, in this case, begins to generate active force for the individual curve in question. Each individual curve is referred to as an **invariant characteristic** (IC) because λ is invariant for that characteristic.

λ is specified by central commands to α-motoneurons, γ-motoneurons, and associated interneurons. The shape of an IC, which is essentially an active force-length characteristic, is influenced by all the peripheral inputs that can act on the α-motoneurons either directly or through interneurons. These inputs include those from proprioceptors of all the muscles and joints involved in the movement as well as cutaneous receptors that may be activated by changes in muscle length or joint position. The pathways involved by these peripheral inputs need not be exclusively spinal; they could include supraspinal centers as well. However, the peripheral inputs that determine the shape of a given IC are involuntary and are commonly referred to as reflex effects. In other words, the shape of the IC is, in general, determined not by the stretch reflex alone, but by all reflexes that involve α-motoneurons, γ-motoneurons, and associated interneurons.

It follows that the EP depends on: (i) λ, as the starting point for development of active force, (ii) how the muscle force varies with muscle length along the IC, and (iii) the force-length, or torque-angle characteristic of the load. Figure 13.13 illustrates the movement of an initial EP on IC_2 to EPs on IC_1 for an isometric, isotonic, or elastic load.

The fact that λ, in this simple case of a single muscle under static conditions, is the only parameter that is needed to describe control by higher centers, dramatically simplifies this control in various situations, as will be elaborated later. It means that higher centers do not have to compute any kinematic variables, these being determined by λ, the IC, and the load on the muscle.

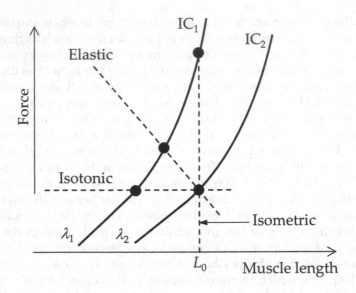

FIGURE 13.13
Movement of an initial equilibrium point (EP) from one invariant characteristic to another, for isometric. isotonic, and elastic loads.

13.5.3 Elaboration of the EP Hypothesis

The ICs of Figure 13.12 are obtained under static conditions and the discussion so far has been restricted to the steady state. The next step in the development of the EP hypothesis is to remove the static restriction and to identify the anatomical substrates involved.

Muscle activity is the result of recruitment of motor units due to activation of α-motoneurons. Muscle activation begins when the membrane voltage threshold is reached in the α-motoneuron having the lowest threshold among the α-motoneurons of the pool innervating the given muscle. This threshold voltage $v_{thr}(t)$ is, in general, a function of time due to the effects of membrane voltage variations on spike generation in the α-motoneuron. The voltage excursion v_m from a given instantaneous membrane voltage $v_m(t)$ to $v_{thr}(t)$ depends also on $v_m(t)$, which depends in turn on all the synaptic inputs to the α-motoneuron from both descending and peripheral influences. In this context, the threshold concept can be generalized in terms of a new muscle length threshold λ^* that includes time variation as well as additional factors. Thus,

$$\lambda^* = \lambda - \mu\upsilon + \rho + \varepsilon \qquad (13.4)$$

where all the quantities are, in general, functions of time and have the following interpretations:

λ^*: composite, time-dependent muscle length threshold at which muscle activation begins

λ: effect on λ^* due to inputs from higher centers. In the static case considered previously and ignoring other effects included in Equation 13.4, $\lambda = \lambda^*$.

μ: parameter that includes the effect of dynamic sensitivity of the muscle spindle receptors in the form of the dependence of the afferent input from these receptors on the velocity of muscle contraction, $v = dx/dt$, x being the muscle length. The minus sign associated with the term μv in Equation 13.4 accounts for the fact that μ is a positive quantity and v is positive when the muscle is lengthening, but the effect of μv for positive v to depolarize the membrane and bring it closer to threshold, thereby reducing the threshold λ^*. This will be explained later in connection with Figure 13.14.

ρ: effect on λ^* of peripheral inputs from other than the muscle under consideration, such as: (i) reciprocal inhibition from antagonist muscles mediated by Ia inhibitory interneurons, and (ii) cutaneous inputs from pressure-sensitive receptors in the fingertips during grasping.

ε: effect on λ^* of changes in $v_{thr}(t)$ arising from the intrinsic properties of the neuronal membrane. These could be due, for example, to inactivation properties of the ionic channels associated with spike generation in the α-motoneuron.

A muscle is active if:

$$x - \lambda^* > 0 \qquad (13.5)$$

This could occur if: (i) x is increased by stretching the muscle, or (ii) λ^* is reduced due, for example, to a reduction in λ, as when the elbow flexors are

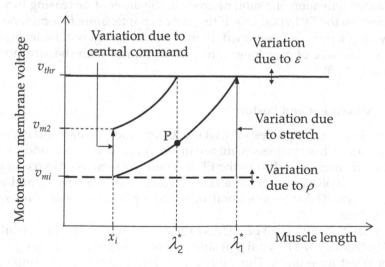

FIGURE 13.14
The thresholds for muscle activation at the level of the α-motoneuron.

voluntarily activated by elbow flexion. Moreover, the **degree of muscle activation**, A, is an increasing function of $(x - \lambda^*)$:

$$A = f(x - \lambda^*), \quad x > \lambda^* \tag{13.6}$$

As A increases, more motor units are recruited, the frequency of firing of α-motoneurons already recruited increases (Section 10.2.4), and the amplitude of the EMG increases. *Motor activity is the result of the nervous system minimizing $(x - \lambda^*)$ in the new state.* The minimum value of $(x - \lambda^*)$ is that required to overcome external forces. This is an expression of the *principle of minimal action*, which is referred to later in the discussion of motor redundancy (Section 13.5.6).

Figure 13.14 illustrates Equation 13.4 at the level of the α-motoneuron. Starting from an initial state (v_{mi}, x_i), a muscle stretch, including velocity effects, increases the depolarization and brings the membrane to threshold at the muscle threshold length λ_1^*. An increase in λ from the initial state (v_{mi}, x_i) brings the membrane voltage to v_{m2}, so that a muscle stretch now brings the membrane to threshold at a muscle length $\lambda_2^* < \lambda_1^*$. If the initial length and membrane voltage correspond to point P, then a depolarization due to a central command will bring the membrane voltage to v_{thr} at a muscle length λ_2^*. The central command has therefore changed the muscle length threshold from λ_1^* to λ_2^*.

It is seen that the variation of voltage with stretch is essential for the transformation from membrane voltage to muscle length, as in λ_1^* and λ_2^*, as well as the transformation of the central command to a change of threshold from λ_1^* to λ_2^*.

Note that activation of α-motoneurons in the order of increasing threshold is inherent in the EP hypothesis. If the same synaptic input to α-motoneurons innervating a muscle is assumed, then the recruitment will be in the order of increasing size of α-motoneurons, in accordance with the size principle (Section 11.2.1.3).

13.5.4 Movement and Posture

Inherent in the EP hypothesis is that movement and the maintenance of posture are not distinct processes but are in fact controlled by one and the same mechanism, namely, a shift in the EP to a new position. As discussed previously, this involves in general: a central command to a new IC and a shift along the new IC due to peripheral, or reflex, inputs as a result of interaction with the load.

There are two important problems to be considered: (i) postural stability, or the maintenance of a bipedal gait, and (ii) how posture-maintaining reflexes do not resist movement. Thus, during walking, running, or jumping, the center of body mass may fall outside the base of support, particularly when stumbling, in which case a leg may be automatically moved forward to prevent falling. How are these posture-stabilizing reactions evoked? According to the EP hypothesis, the posture-stabilizing mechanisms are now reset to an area on the ground that is outside the initial base of support. The present position becomes a deviation from the newly specified one, and the same posture-stabilizing mechanisms now generate forces that move the joints to the new stable position.

As for the second problem of resistance of posture-maintaining reflexes to movement, it is clear that the stretch reflex, which is strongly involved in maintaining muscle length, and hence posture, will resist any attempt to change muscle length during movement. The classical view is that such reflexes must somehow be suppressed or inhibited during movement, by such mechanisms as α-γ linkage or reciprocal inhibition. According to the EP hypothesis, the shifting of muscle activation thresholds generates forces that move the joint to the new position, as in posture stabilization just mentioned. Threshold shifting concomitantly involves both the central commands for the movement and the associated reflexes. There is no artificial distinction between the two influences; they work together and not against one another.

It should be noted from the preceding discussion of threshold control that this control is feedforward in nature, that is, the change in threshold precedes muscle activity, as in the upward shift of P in Figure 13.14 to threshold. As mentioned previously, feedforward control is anticipatory and avoids the destabilizing effects of delays in feedback loops.

13.5.5 Agonist-Antagonist Co-contraction

Consider the situation at a single joint according to the EP hypothesis. Assuming that the joint muscles are completely relaxed in the absence of a load, the torque-angle characteristic will be as indicated by the solid line in figure 13.15a. The threshold angle indicated by r_1 is the same for flexors and extensors, and is also the equilibrium angle under these conditions. If the elbow is passively extended, the elbow angle increases and the torque developed by the flexor increases in the positive direction along the upper part of the curve. If the elbow is passively flexed, the elbow angle decreases and torque is developed by the extensor. Since this torque is in a direction opposite to that of the flexor, it is negative, and the lower part of the curve is traced. Note that in drawing this IC, the threshold r for muscle activation has been assumed along the horizontal axis, for simplicity, and not at the point of departure from the passive characteristic, as in Figure 13.12.

Suppose that the elbow is voluntarily flexed to a new position, again in the absence of a load. The IC will move to the left to a new threshold value and elbow angle r_2. The threshold for the extensor moves in the same direction as that of the flexor, and to the same extent, so as to have the same threshold for both muscles. This must be the case, so that passive flexion activates the flexor, whereas passive extension from the same elbow angle activates the extensor.

If a tonic load L is applied when the common threshold is r_1, the elbow will be extended and new equilibrium point EP_1 is established at an angle θ_1 (Figure 13.15a). If central commands shift the threshold to r_2, the new equilibrium point EP_2 is established at an angle θ_2.

Because a flexor and an extensor are involved at the joint, the central commands can activate both muscles simultaneously, corresponding to agonist-antagonist co-contraction. Activating the flexor moves the threshold and the IC to the left, so that a larger positive torque is developed at the same joint angle. Activating the extensor moves the threshold and the IC to the right, so that a negative torque of larger magnitude is developed at the same joint angle. In Figure 13.15b the flexor threshold is shifted from r_1 to $(r_1 - c_1)$, whereas the extensor threshold is shifted to $(r_1 + c_2)$, where c_1 and c_2 are such that the same equilibrium angle θ_1 is maintained. The algebraic sum of the two shifted ICs gives a composite IC that intersects the load line at a new equilibrium point having an angle θ_1. The slope of the composite IC exceeds the slope at EP_1, which means that the stiffness of the joint is increased at the same elbow angle.

The co-contraction command, which moves the thresholds of the agonists and antagonists in opposite directions by facilitating or inhibiting them together, is known as the c command. On the other hand, the r, or reciprocal, command moves the thresholds in the same direction by facilitating

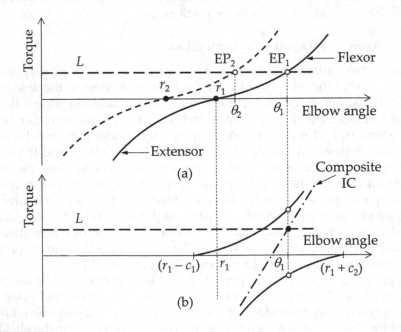

FIGURE 13.15
Agonist-antagonist co-contraction; (a) torque-angle characteristic at no load for two elbow positions; (b) torque-angle characteristic when the flexor muscles and the extensor muscles are co-activated.

the agonist and inhibiting its antagonist. By applying r and c commands together, the nervous system can adjust the equilibrium angle and stiffness for a given load.

Under dynamic conditions, the location of the thresholds along the horizontal axis is affected by the quantities ρ and ε in Equation 13.4. The speed of the movement is affected by the rate at which the r command is changed and by the contribution of the μv term in Equation 13.4, which depends on the degree of activation of the dynamic γ-motoneurons.

Let us generalize the movement of a single-joint in terms of what has been discussed so far concerning the EP hypothesis. Let the EP be P_1 on IC_1 having a threshold R_1^* (Figure 13.16), which can now be denoted as the **joint referent**, rather than λ_1^*, which can be reserved for the threshold of a single muscle. Moreover, R for the joint can include both r and c commands discussed in connection with Figure 13.15. Let the joint be moved to a new referent position R_2^*. Both R_2^* and the shape of IC_2 are in general determined by: (i) central commands and other factors included in Equation 13.4, and (ii) r and c commands that influence both the velocity and the stiffness, as required by the subject and in the light of past experience with the same or similar movements. Note that so far, this is feedforward control.

Once R_2^* is specified, the final equilibrium point is determined by the shape of IC_2 and the load characteristic. P_2' is the EP for an isometric load, whereas P_2''' is the EP for an isotonic load. These set two limits for the EP on IC_2. For a load with a spring-like torque-angle characteristic, the EP is P_2''. The R_2^* determination of the final EP does not involve computation of mechanical variables by the nervous system.

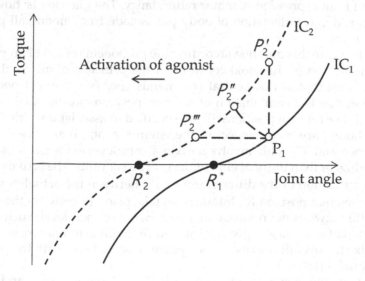

FIGURE 13.16

Generalization of the torque-angle characteristic for a single joint for three types of loads, as in Figure 13.13.

When the new referent R_2^* is specified, the quantity $(x - \lambda_1^*)$, the difference between the actual length x of the agonist and its threshold length λ_1^*, is a maximum, many α-motoneurons are recruited, and the agonist is strongly activated, with maximal acceleration of the joint. As the agonist shortens, its activation is reduced, taking into account contributions from other actors such as recurrent inhibition and inhibition from the antagonist. Some of the α-motoneurons are derecruited accordingly. The motion of the elbow activates the antagonist through stretch, in addition to the reduced reciprocal inhibition from the agonist, thereby further de-activating the agonist and decelerating the movement. The sequence of agonist and antagonist muscle activity in general affects R_2^* and the shape of the IC. At the final EP, the overall muscular activity at the joint is minimized, in accordance with the principle of minimal action.

If the destination EP deviates for some reason from the desired end position of the joint, a new referent position is specified by the nervous system, and the process is repeated. This adjustment, if required, is based on sensory feedback.

13.5.6 Motor Redundancy

If we reach for a book on a shelf, the final position of the hand can be reached through virtually an infinite number of possible positions of the arm, forearm, and hand, corresponding to different angles of the wrist, elbow, and shoulder joints. The existence of a relatively large number of degrees of freedom that can be chosen from in reaching the desired destination of the hand is an expression of motor redundancy. The question is: how is the actual position or orientation of body parts made from among all possible combinations?

The answer to this question according the proponents of the EP hypothesis is a combination of threshold control and the principle of minimal action. The basic concept is that central commands specify a referent body configuration that is a combination of referent positions for the various joints involved. The referent positions: (i) are specified in association with the position of body parts with respect to the environment, (ii) are based on past experience, and (iii) could involve more than one level in the nervous system that are higher than the final effector level of motor units. The activity at each level is a function of the difference $(Q - R^*)$ between the actual position Q and the referent position R^*. Interactions take place: (i) between the various levels, (ii) between neuromuscular components of each level, such as agonists and antagonists at a given joint, and (iii) with external forces, so as to minimize the overall activity in the system in accordance with the principle of minimal action.

In locomotion, the referent position is shifted continuously and repetitively with each step, the motion being produced by the system minimizing the difference between the actual body position and the referent position in each step. As a result, the same body configuration is produced but in a different part of the surrounding space. The referent position is appropriately modified when encountering an obstacle.

Summary of Main Concepts

- Lever systems are divided into three classes, depending on the relative positions of the load and effort forces with respect to the fulcrum. Skeletal muscles in the body most commonly exercise their action through Class III lever systems having the fulcrum at one end, with the muscle effort closer to the fulcrum than the load. The muscle exerts a larger force than the load, but the distance and the velocity of muscle movement are magnified at the load.

- Co-contraction of agonist and antagonist muscles often occurs in practice in order to terminate a movement, to stabilize a joint, or to increase joint stiffness.

- The somatomotor system uses both feedforward and feedback control. The performance of a feedback system is limited by the gain and the inevitable delays involved but is improved by introducing a predictive component.

- Motor variability is exemplified by the fact that different trajectories are followed in successive repetitions of the same motor task. A fundamental cause of motor variability is motor redundancy, which arises because a given movement can, in general, be executed by a theoretically infinite number of combinations of motions of individual joints.

- The problem of motor redundancy is circumvented by motor synergies, by which a single command is issued to execute a particular movement by the concerted action of a group of muscles. The motor synergies are task-specific, are learned, and are executed in such a manner that the errors in the movements of individual joints compensate for one another, resulting in less variability in the end movement.

- Motor equivalence refers to the similarity in the result of a given motor activity, irrespective of which set of muscles is used to carry out this activity. This suggests that a purposeful movement is represented in the brain in an abstract manner, referred to as a generalized motor program, or motor schema, rather than in terms of specific commands to a particular set of muscles.

- Motor learning is procedural learning that it is not readily express-ible in words. It seems to occur at the level of the generalized motor program and involves the prefrontal cortex, primary motor cortex, the basal ganglia, and the cerebellum. Synaptic plasticity is believed to underlie motor learning.

- The cerebellum plays a crucial role in the learning of motor skills involving a sequence of movements. The cerebellum is most active during the early phase of the learning of a motor sequence, during which cerebellar activity adjusts movement kinematics, based on sensory feedback. Once the motor sequence has been well learned, the dorsal striatum and motor cortical areas are critical for long-term storage and recall of the learned sequence.

- It is generally assumed that memories are stored in synapses. Macromolecules have also been implicated in memory storage and retrieval. Motor learning is holographic in nature, in that it is distrib-uted, and a small part can be used to reproduce the whole.

- When only gravity acts on the body, stability requires that the verti-cal projection of the center of mass falls on the center of pressure and that the center of pressure should lie within the base of support.

- Upright posture has to be actively maintained. This is chiefly effected by tonic activity of the antigravity, extensor muscles of the ankle, the knee, the hip, and the neck, in addition to the paraspinal muscles of the back, which hold up the spine. The tonic activity of the anti-gravity muscles is maintained through sensory, negative feedback involving the vestibular, visual, and somatosensory systems. The contribution of the somatosensory system is mainly through pro-prioceptors participating in the tonic stretch reflex and also through cutaneous receptors in the sole of the foot.

- Posture is retained during movement by postural adjustments, which could be anticipatory or preprogrammed. The three basic balance control strategies that are induced by perturbations are: the ankle strategy, the hip strategy, and taking a step. In making pos-tural adjustments, the motor system uses motor synergies, so that these postural synergies are inherently part of the synergies of any planned voluntary movement.

- In steady walking, human locomotion is divided into two phases: the swing phase, during which the swing foot is off the ground, and the stance phase, during which the foot is in contact with the ground. Each of these phases consists of a number of stages involving flexion and extension movements of the hip, knee, and ankle joints.

- Locomotion is controlled by central pattern generators located in the lumbosacral spinal cord. These pattern generators are controlled by descending inputs from the motor cortex and the brainstem, and their activity is modulated by sensory inputs from the periphery. It seems that the central pattern generator network for each limb comprises multiple interconnected modules that control the movement of each joint.

- According to the EP hypothesis, the higher-centers command basically changes the threshold of the tonic stretch reflex λ, that is, the muscle length at which the muscle begins to generate an active force, as explained below. The EP has several attractive features: (i) the central command signal depends on internal, physiological variables, (ii) the attainment of the desired endpoint generally relies on sensory feedback control, (iii) in executing a movement, reflex circuitry, motor synergies, and biomechanical properties of muscle are exploited, and (iv) the posture-movement paradox is resolved by combining posture and movement in a single scheme.

Bibliography and References

General (Relevant to more than one chapter)

Books

Aidley DJ (1998) *The Physiology of Excitable Cells*, 4th ed, Cambridge University Press, Cambridge, UK.

Bear MF, Connors BW, Paradiso MA (2016) *Neuroscience: Exploring the Brain*, 4th ed, Wolters Kluwer, Philadelphia, PA.

Bower JM, Beeman D (1998) *The Book of GENESIS*, 2nd ed, Springer-Verlag, New York.

Gandevia SC, Proske U, Stuart DG (2002) *Sensorimotor Control of Movement and Posture*, Springer Science+ Business Media, New York.

Hille B (2001) *Ion Channels of Excitable Membranes*, 3rd ed, Sinauer Associates, Inc., Sunderland, MA.

Jack JJB, Noble D, Tsien RW (1983) *Electric Current Flow in Excitable Cells*, 2nd ed, Clarendon Press, Oxford.

Johnston D, Wu SMS (1995) *Foundations of Cellular Neurophysiology*, The MIT Press, Cambridge, MA.

Kandel ER, Schwartz JH, Jessel TM, Siegelbaum Steven A, Hudspeth AJ (2013) *Principles of Neural Science*, 5th ed, McGaw-Hill, New York.

Latash ML (2008) *Neurophysiological Basis of Movement*, 2nd ed, Human Kinetics, Champaign, IL.

Levitan IB, Kaczmarek LK (2015) *The Neuron: Cell and Molecular Biology*, 4th ed, Oxford University Press, New York.

Nicholls JG, Martin AR, Fuchs PA, Brown DA, Diamond ME, Weisblat DA (2011) *From Neuron to Brain*, 5th ed, Sinauer Associates, Inc., Sunderland, MA.

Nieuwenhuys R, Voogd J, Van Huijzen C (2008) *The Human Central Nervous System*, 4th ed, Springer, Berlin.

Plonsey R, Barr RC (2007) *Bioelectricity: A Quantitative Approach*, 3rd ed, Springer, New York.

Poznanski RR (ed) (1999) *Modeling in the Neurosciences: From Ionic Channels to Neural Networks*, CRC Press, Boca Raton, FL.

Purves D, Augustine GJ, Fitzpatrick D, Hall WC, LaMantia AS, Mooney RD, Platt ML, White LE (2018) *Neuroscience*, 6th ed, Oxford University Press, New York.

Saterbak A, San KY, McIntire LV (2007) *Bioengineering Fundamentals*, Prentice Hall, Upper Saddle River, NJ.

Shepherd GM (2004) *The Synaptic Organization of the Brain*, 5th ed, Oxford University Press, Oxford.

Stein RB (1980) *Nerve and Muscle: Membranes, Cells, and Systems*, Plenum Press, New York.

Voet D, Voet JG, Pratt CW (2016) *Fundamentals of Biochemistry: Life at the Molecular Level*, 5th ed, John Wiley and Sons, Hoboken, NJ.

URLs

Brain facts and figures: http://faculty.washington.edu/chudler/facts.html

---

Chapter 1 Introduction: Background Material

Book

Capellos C, Birlski BHJ (1980) *Kinetic Systems: Mathematical Description of Chemical Kinetics in Solution*, R.E. Krieger Publishing Company, Huntington, NY.

Q&A and Review Articles

Allen NJ, Barres BA (2009) Neuroscience: glia - more than just brain glue. *Nature* 457(7230):675–677.

Brown A (2003) Axonal transport of membranous and nonmembranous cargoes: A unified perspective. *J Cell Biol* 160(6):817–821.

Pizzo P, Pozzan T (2007) Mitochondria–endoplasmic reticulum choreography: Structure and signaling dynamics. *Trends Cell Biol* 179(10):511–517.

Chapter 2 The Cell Membrane in the Steady State

Ion Hydration Special Issue (2006) *Biophys Chem* 124(3):169–302.

Review Articles

Bezanilla F (2000) The voltage sensor in voltage-dependent ion channels. *Physiol Rev* 80(2):555–592.

Gouaux E, MacKinnon R (2005) Principles of selective ion transport in channels and pumps. *Science* 310(5753):1461–1465.

Original Articles

Doyle DA, Cabral JM, Pfuetzner RA, Kuo A, Gulbis JM, Cohen SL, Chait BT, MacKinnon R (1998) The structure of the potassium channel: Molecular basis of K^+ conduction and selectivity. *Science* 280(5360):69–77.

Sabah NH (1999) Origin of the resting potential. *IEEE Eng Med Biol Mag* 18(5):100–105.

Sabah NH (2000) Rectification in biological membranes. *IEEE Eng Med Biol Mag* 19(1):106–113.

Sabah NH (2000) Reactance of biological membranes. *IEEE Eng Med Biol Mag* 19(4):89–95.

Chapter 3 Generation of the Action Potential

Review Article

Meves H (1984) Hodgkin-Huxley: Thirty years after, in: Kleinzeller A (ed) *Current Topics in Membranes and Transport*, vol 22, The Squid Axon, pp279–329.

Original Articles

Hamil OP, Marty A, Neher E, Sakmann B, Sigworth FJ (1981) Improved patch-clamp techniques for high-resolution current recording from cells and cell-free membrane patches. *Pflügers Arch* 391(2):85–100.

Hodgkin AL, Huxley J (1952) A quantitative description of membrane current and its application to conduction and excitation in nerve. *J Physiol* 117(4):500–544.

Kuyucak S, Chung SH (1994) Temperature dependence of conductivity in electrolyte solutions and ionic channels of biological membranes. *Biophys Chem* 52(1):15–24.

Llano I, Webb CK, Bezanilla F (1988) Potassium conductance of the squid giant axon. *J Gen Physiol* 92(2):179–196.

Chapter 4 Propagation of the Action Potential

Book Chapter

Taylor RE (1963) Cable theory, in: Natsuk WL (ed) *Physical Techniques in Biological Research, Volume VI, Electrophysiological Methods, Part B*, Academic Press, New York, pp219–262.

Review Articles

Waxman SG (1975) Integrative properties and design principles of axons. *Int Rev Neurobiol* 18:1–40.

Waxman, SG (1980) Determinants of conduction velocity in myelinated nerve fibers. *Muscle & Nerve* 3(2):141–150.

Original Articles

Coppin CML, Jack JJB (1972) Internodal length and conduction velocity of cat muscle afferent nerve fibers *J Physiol* 222(Suppl):91–93.

Dun FT (1970) The length and diameter of the node of Ranvier. *IEEE Trans Biomed Eng* BME 17(1):21–24.

Goldman L, Albus JS (1968) Computation of impulse conduction in myelinated fibers; theoretical basis of the velocity-diameter relation. *Biophys J* 8(5):596–607.

Paintal AS (1966) The influence of diameter of medullated nerve fibres of cats on the rising and falling phases of the spike and its recovery. *J Physiol* 184(4):791–811.

Rushton WAH (1951) A theory of the effects of fibre size in medullated nerve. *J Physiol* 115(1):101–122.

Sabah NH (2000) Aspects of nerve conduction. *IEEE Eng Med Biol Mag* 19(6):111–118.

Stein RB, Pearson KG (1971) Predicted amplitude and form of action potentials recorded from unmyelinated nerve fibres. *J theor Biol* 32(3):539–558.

Tasaki I (2004) On the conduction velocity of nonmyelinated nerve fibers. *J Integr Neurosci* 3(2) 115–24.

Chapter 5 The Neuromuscular Junction

Review Articles

Bowman WC (2006) Neuromuscular block. *Br J Pharmacol* 147(Suppl 1):S277–S286.

Engel AG, Ohno K, Sine SM (2003) Sleuthing molecular targets for neurological diseases at the neuromuscular junction. *Nat Rev Neurosci* 4(5):339–352.

Unwin N (2000) The Croonian Lecture 2000: Nicotinic acetylcholine receptor and the structural basis of fast synaptic transmission. *Phil Trans R Soc Lond B* 355(1404):1813–1829.

Original Articles

Land BR, Salpeter EE, Salpeter MM (1981) Kinetic parameters for acetylcholine interaction in intact neuromuscular junction. *Proc Natl Acad Sci USA* 78(11):7200–7204.

Severs NJ (2007) Freeze-fracture electron microscopy. *Nature Protocols* 2(3):547–576.

Takeda T, Sakata A, Takahide M (1999) Fractal dimensions in the occurrence of miniature end- plate potential in a vertebrate neuromuscular junction. *Prog Neuro-Psychopharmacol Biol Psychiat* 23(6):1157–1169.

Unwin N (2005) Refined structure of the nicotinic acetylcholine receptor at 4 Å resolution. *J Mol Biol* 346(4):967–989.

Van der Kloot W, Cohen IS (1984) End-plate potentials in a model muscle fiber. *Biophys J* 45(5):905–912.

Chapter 6 Synapses

Review Articles

Citri A, Malenka RC (2008) Synaptic plasticity: Multiple forms, functions, and mechanisms. *Neuropsychopharmacology Reviews* 33(1):18–41.

Connors BW, Long MA (2004) Electrical synapses in the mammalian brain. *Annu Rev Neurosci* 27:393–418.

Grant BD, Donaldson JG (2009) Pathways and mechanisms of endocytic recycling. *Nat Rev Mol Cell Biol* 10(9):597–608.

Hnasko TS, Edwards RH (2012) Neurotransmitter co-release: Mechanism and physiological role. *Annu Rev Physiol* 74:225–243.

Hormuzdi SG, Filippov MA, Mitropoulou G. Monyer H, Bruzzone R (2004) Electrical synapses: A dynamic signaling system that shapes the activity of neuronal networks. *Biochim Biophys Acta* 1662(1–2):113–137.

Kerchner GA, Nicoll RA (2008) Silent synapses and the emergence of a postsynaptic mechanism for LTP. *Nat Rev Neurosci* 9(11):813–825.

Malenka RC (2003) Synaptic plasticity and AMPA receptor trafficking. *Ann NY Acad Sci* 1003(1) 1–11.

Miaczynska M, Pelkmans L, Zerial M (2004) Not just a sink: Endosomes in control of signal transduction. *Curr Opin Cell Biol* 16:400–406.

Nimchinsky EA, Sabatini BL, Svoboda K (2002) Structure and function of dendritic spines. *Annu Rev Physiol* 64(1):313–353.

Park M (2018) AMPA receptor trafficking for postsynaptic potentiation. *Front Cell Neurosci* 12:361.

Rizzuto R, Pozzan T (2006) Microdomains of intracellular Ca^{2+}: Molecular determinants and functional consequences. *Physiol Rev* 86(1):369–408.

Segal M (2005) Dendritic spines and long-term plasticity. *Nat Rev Neurosci* 6(4):277–284.

Zucker RS, Regehr WG (2002) Short-term synaptic plasticity. *Annu Rev Physiol* 64(1):355–405.

Original Articles

Kauer JA, Malenka RC (2006) LTP: AMPA receptors trading places. *Nat Neurosci* 9(5):593–594.

Mazzolini M, Anselmi C, Torre V (2009) The analysis of desensitizing CNGA1 channels reveals molecular interactions essential for normal gating. *J Gen Physiol* 133(4):375 386.

Sun Y, Olson R, Horning M, Armstrong N, Mayer M, Gouaux E (2002) Mechanism of glutamate receptor desensitization. *Nature* 4179(6886):245–253.

Toni N, Buchs P-A, Nikonenk I, Povilaitite P, Parisi L, Muller D (2001) Remodeling of synaptic membranes after induction of long-term potentiation. *J Neurosci* 21(16):6245–6251.

Chapter 7 Neurons

Book Chapter

Chen-Izu Y, Izu LT, Nanasi PP, Banyasz T (2012) From action potential-clamp to "Onion-Peeling" technique – Recording of ionic currents under physiological conditions, in: Shad FK (ed) *Patch Clamp Technique*, IntechOpen, London, UK, pp143–162.

Review Articles

Agnati LF, Leo G, Zanardi A, Genedani S, Rivera A, Fuxe K, Guidolin D (2006) Volume transmission and wiring transmission from cellular to molecular networks: History and perspectives. *Acta Physiol* 187(1–2):329–344.

Antic SD, Zhou WL, Moore AR, Short SM, Ikonomu KD (2010) The decade of the dendritic NMDA spike. *J Neurosci Res* 88(14):2991–3001.

Bean BP (2007) The action potential in mammalian central neurons. *Nat Rev Neurosci* 8(6):451–465.

Bennett MVL, Zukin RS (2004) Electrical coupling and neuronal synchronization in the mammalian brain. *Neuron* 41(4):495–511.

Enyedi P, Czirják G (2010) Molecular background of leak K currents: Two-pore domain potassium channels. *Physiol Rev* 90(2):559–605.

Faber DS, Pereda AE (2018) Two forms of electrical transmission between neurons. *Front Mol Neurosci* 11:427.

Faber ESL, Pankaj S (2007) Functions of SK channels in central neurons. *Clin Exp Pharmacol Physiol* 34(10):1077–1083.

Farrant M, Nusser Z (2005) Variations on an inhibitory theme: Phasic and tonic activation of GABAA receptors. *Nat Rev Neurosci* 6(3):215–229.

Guru A, Post RJ, Ho YY, Warden MR (2015) Making sense of optogenetics. *Int J Neuropsychopharmacol* 18(11):1–8.

Häusser M, Spruston N, Stuart GJ (2000) Diversity and dynamics of dendritic signaling. *Science* 290(5492):739–744.

Hibino H, Inanobe A, Furutani K, Murakami S, Findlay I, Kurachi Y (2010) Inwardly rectifying potassium channels: Their structure, function, and physiological roles. *Physiol Rev* 90(1):291–366.

Llinás RR (2014) Intrinsic electrical properties of mammalian neurons and CNS function: A historical perspective. *Front Cell Neurosci* 8:320.

Markram H, Toledo-Rodriguez M, Wang Y, Gupta A, Silberberg G, Wu C (2004) Interneurons of the neocortical inhibitory system. *Nat Rev Neurosci* 5(10):793–807.

Matus A (2005) Growth of dendritic spines: A continuing story. *Curr Opin Neurobiol* 15(1) 67–72.

McBain CM, Fisahn A (2001) Interneurons unbound. *Nat Rev Neurosci* 2(1):11–23.

Robinson RB, Siegelbaum SA (2003) Hyperpolarization-activated cation currents: From molecules to physiological function. *Annu Rev Physiol* 65:453–480.

Russell JT (2011) Imaging calcium signals *in vivo*: A powerful tool in physiology and pharmacology. *Br J Pharmacol* 163:1605–1625.

Szobota S, Isacoff EY (2010) Optical control of neuronal activity. *Annu Rev Biophys* 39:329–4.

Vacher H, Mohapatra DP, Trimmer JS (2008) Localization and targeting of voltage-dependent ion channels in mammalian central neurons. *Physiol Rev* 88(4):1407–1447.

Waters J, Schaefer A, Sakmann B (2005) Backpropagating action potentials in neurones: Measurement, mechanisms and potential functions. *Prog Biophys Mol Biol* 87(1):145–170.

Yang W, Yuste R (2017) *In vivo* imaging of neural activity. *Nat Methods* 14(4):349–359.

Original Articles

Dolphin AC (2006) A short history of voltage-gated calcium channels. *Brit J Pharmacoll* 147(Suppl 1):S56–S62.

Francavilla R, Villette V, Martel O, Topolnik L (2019) Calcium dynamics in dendrites of hippocampal CA1 interneurons in awake mice. *Front Cell Neurosci* 13:98.

Gutman AM (1991) Bistability of dendrites. *Int J Neur Syst* 1(4):291–304.

Hay E, Hill S, Schürmann F, Markram H, Segev I (2011) Models of neocortical layer 5b pyramidal cells capturing a wide range of dendritic and perisomatic active properties. *PLoS Comput Biol* 7(7):e1002107.

Liu S, Michael T, Shipley MT (2008) Olfactory Intrinsic conductances actively shape excitatory and inhibitory postsynaptic responses in bulb external tufted cells. *J Neurosci* 28(41):10311–10322.

Magee JC (1999) Dendritic I_h normalizes temporal summation in hippocampal CA1 neurons. *Nat Neurosci* 2(6):508–514.

Metz AE, Jarsky T, Martina M, Spruston N (2005) R-type calcium channels contribute to afterdepolarization and bursting in hippocampal CA1 pyramidal neurons. *J. Neurosci* 25(24):5763–5773.

Oláh S, Füle M, Komlósi G, Varga C, Báldi R, Barzó P, Tamás G (2009) Regulation of cortical microcircuits by unitary GABA-mediated volume transmission. *Nature* 461(7268):1278–1282.

Rinke I, Artmann J, Stein V (2010) ClC-2 voltage-gated channels constitute part of the background conductance and assist chloride extrusion. *J Neurosci* 30(13):4776–4786.

Shah M, Haylett DG (2000) Ca^{2+} channels involved in the generation of the slow afterhyperpolarization in cultured rat hippocampal pyramidal neurons. *J Neurophysiol* 83(5):2554–2561.

Uhlhaas PJ, Pipa G, Lima B, Melloni L, Neuenschwander S, Nikolic D, Wolf Singer W (2009) Neural synchrony in cortical networks: History, concept and current status. *Front Integr Neurosci* 3:17.

Williams SR, Stuart GJ (2000) Site independence of EPSP time course is mediated by dendritic I_h in neocortical pyramidal neurons. *J Neurophysiol* 83(5):3177–3182.

Yu Y, Maureira C, Liu X, McCormick D (2010) P/Q and N channels control baseline and spike-triggered calcium levels in neocortical axons and synaptic boutons. *J Neurosci* 30(35):11858–11869.

Chapter 8 Neuronal Firing Patterns and Models

Books

Bower JM, Beeman D (1998) *The Book of GENESIS*, 2nd ed, Springer-Verlag, New York.

Gerstner W, Kistler WM, Naud R, Paninski L (2014) *Neuronal Dynamics: From Single Neurons to Networks and Models of Cognition*, Cambridge University Press, Cambridge, UK.

Izhikevich EM (2007) *Dynamical Systems in Neuroscience*, The MIT Press, Cambridge, MA.

Reeke GN, Poznanski RR, Lindsay KA, Rosenberg JR, Sporns O (eds) (2005) *Modeling in the Neurosciences: From Biological Systems to Neuromimetic Robotics*, 2nd ed, Taylor & Francis Group, Boca Raton, FL.

Book Chapters

Rall W, Agmon-Snir H (1998) Cable theory for dendritic neurons, in: *Methods in Neuronal Modeling: From Ions to Networks*, 2nd ed, Koch C, Segev I (eds) The MIT Press, Cambridge, MA, pp27–92.

Segev I, Burke R (1998) Compartmental models of complex neurons, in: *Methods in Neuronal Modeling: From Ions to Networks*, 2nd ed, Koch C, Segev I (eds) The MIT Press, Cambridge, MA, pp93–136.

Review Articles

Burkitt AN (2006) A review of the integrate-and-fire neuron model: I. homogeneous synaptic input. *Biol Cybern* 95(1):1–19.

Burkitt AN (2006) A review of the integrate-and-fire neuron model: II. inhomogeneous synaptic input and network properties. *Biol Cybern* 95(2):97–112.

Original Articles

Agmon-Snir H, Segev I (1993) Signal delay and input synchronization in passive dendritic structures. *J Neurophysiol* 70(5):2066–2085.

Fardet T, Ballandras M, Bottani S, Métens S, Monceau P (2018) Understanding the generation of network bursts by adaptive oscillatory neurons. *Front Neurosci* 12:41.

Goldfinger MD (2005) Rallian "equivalent" cylinders reconsidered: Comparisons with literal compartments. *J Integrative Neurosci* 4(2):227–263.

Hay E, Hill S, Schürmann F, Markram H, Segev I (2011) Models of neocortical layer 5b pyramidal cells capturing a wide range of dendritic and perisomatic active properties. *PLoS Comput Biol* 7(7):e1002107.

Hayar A, Karnup S, Shipley MT, Ennis M (2004) Olfactory bulb glomeruli: External tufted cells intrinsically burst at theta frequency and are entrained by patterned olfactory input. *J Neurosci* 24(5):1190–1199.

Howell FW, Dyhrfjeld-Johnsen J, Maex R, Goddard N, De Schutter E (2000) A large-scale model of the cerebellar cortex using PGENESIS. *Neurocomputing* 32–33:1041–1046.

Hu H, Vervaeke K, Storm JF (2002) Two forms of electrical resonance at theta frequencies, generated by M-current, h-current and persistent Na^+ current in rat hippocampal pyramidal cells. *J Physiol* 545(3):783–805.

Izhikevich EM (2001) Resonate-and-fire neurons. *Neural Netw* 14(6–7):883–894.

Kim H, Jones, KE (2011) Asymmetric electrotonic coupling between the soma and dendrites alters the bistable firing behaviour of reduced models. *J Comput Neurosci* 30:659–674.

Lindsay KA, Rosenberg JR, Tucker G (2003) Analytical and numerical construction of equivalent cables. *Math Biosci* 184(2):137–164.

Radulescu AR (2010) Mechanisms explaining transitions between tonic and phasic firing in neuronal populations as predicted by a low dimensional firing rate model. *PLoS one* 5(9):1–14.

Rall W (1962) Theory of physiological properties of dendrites. *Ann NY Acad Sci* 96(4):1071–1092.

Traub RD, Jefferys JGR, Miles R, Whittington MA, Toth K (1994) A branching dendritic model of a rodent CA3 pyramidal neurone. *J Physiol* 481(1):79–95.

Traub RD, Kopell N, Bibbig A, Buhl EH, LeBeau FEN, Whittington MA (2001) Gap junctions between interneuron dendrites can enhance synchrony of gamma oscillations in distributed networks. *J Neurosci* 21(23):9478–9486.

Chapter 9 Skeletal Muscle

Book Chapter

Banks R (2018) Muscle spindles and tendon organs, in: *Reference Module in Biomedical Sciences*, Elsevier, Amsterdam, pp1–10.

Review Article

Berchtold MW, Brinkmeier H, Müntener M (2000) Calcium ion in skeletal muscle: Its crucial role for muscle function, plasticity, and disease. *Physiol Rev* 80(3):1215–1265.

Original Articles

Krivickas LS, Dorer DJ, Ochala J, Frontera WR (2011) Relationship between force and size in human single muscle fibres. *Exp Physiol* 96(5):539–547.

Mileusnic M, Brown IE, Lan N, Loeb GE (2006) Mathematical models of proprioceptors. I. Control and transduction in the muscle spindle. *J Neurophysiol* 96(4):1772–1788.

Mileusnic M, Loeb GE (2006) Mathematical models of proprioceptors. II. Structure and function of the Golgi tendon organ. *J Neurophysiol* 96(4):1789–1802.

Chapter 10 Functional Properties of Muscle

Book Chapter

Brenner B, Mahlmann E, Mattei T, Kraft T (2005) Driving filament sliding: Weak binding cross-bridge states, strong binding cross-bridge states, and the power stroke, in: Sugi H (ed) *Sliding Filament Mechanism in Muscle Contraction: Fifty Years of Research*, Springer, New York, pp75–91.

Review Articles

Alcazar J, Robert Csapo R, Ara I, Luis M. Alegre LM (2019) On the shape of the force-velocity relationship in skeletal muscles: The linear, the hyperbolic, and the double-hyperbolic. *Front Physiol* 10:769.

Aromataris EC, Rychkov GY (2006) ClC-1 chloride channel: Matching its properties to a role in skeletal muscle. *Clinical and Experimental Pharmacology and Physiology* 33(11):1118–1123.

Bullard B, Pastore A (2011) Regulating the contraction of insect flight muscle. *J Muscle Res Cell Motil* 32(4–5):303–313.

Severs NJ (2000) The cardiac muscle cell. *BioEssays* 22(2):188–199.

Original Articles

Azizi E, Brainerd EL, Roberts TJ (2008) Variable gearing in pennate muscles. *PNAS* 105(5):1745–1750.

Chin L, Yue P, Feng JJ, Seow CY (2006) Mathematical simulation of muscle cross-bridge cycle and force-velocity relationship. *Biophys J* 91(10):3653–3663.

Chapter 11 Spinal Cord and Reflexes

Book

Kernell D (2006) *The Motoneurone and its Muscle Fibres*, Oxford University Press, Oxford.

Review Articles

Alvarez FJ, Fyffe REW (2007) The continuing case for the Renshaw cell. *J Physiol* 584(1):31–45.

ElBasiouny SM, Schuster JE, Heckman CJ (2010) Persistent inward currents in spinal motoneurons: Important for normal function but potentially harmful after spinal cord injury and in amyotrophic lateral sclerosis. *Clin Neurophysiol* 121(10):1669–1679.

Heckman CJ, Johnson M, Mottram C, Schuster J (2008) Persistent inward currents in spinal motoneurons and their influence on human motoneuron firing patterns. *Neuroscientist* 14(3):264–275.

Heckman CJ, Mottram C, Quinlan K, Theiss R, Schuster J (2009) Motoneuron excitability: The importance of neuromodulatory inputs. *Clin Neurophysiol* 120(12):2040–2054.

Hodson-Tole EF, Wakeling JM (2009) Motor unit recruitment for dynamic tasks: Current understanding and future directions. *J Comp Physiol B* 179(1):57–66.

Jankowska E (2001) Spinal interneuronal systems:Iidentification, multifunctional character and reconfigurations in mammals. *J Physiol* 533(1):31–40.

Kanning KC, Kaplan A, Henderson CE (2010) Motor neuron diversity in development and disease. *Annu Rev Neurosci* 33(1):409–40.

Rudomin P (2009) In search of lost presynaptic inhibition. *Exp Brain Res* 196(1):139–151.

Chapter 12 Motor Brain Centers and Pathways

Review Articles

Chakravarthy VS, Joseph D, Bapi RS (2010) What do the basal ganglia do? A modeling perspective. *Biol Cybern* 103(3):237–253.

De Zeeuw CI, Ten Brinke MM (2015) Motor learning and the cerebellum. *Cold Spring Harb Perspect Biol* 7:a021683.

Fujita H, Sugihara I (2013) Branching patterns of olivocerebellar axons in relation to the compartmental organization of the cerebellum. *Frontiers in Neural Circuits* 7:3.

Klein AP, Ulmer JL, Quinet SA, Mathews V, Mark LP (2016) Nonmotor functions of the cerebellum: An introduction. *Am J Neuroradiol* 37(6):1005–1009.

Lang EJ, Blenkinsop TA (2011) Control of cerebellar nuclear cells: A direct role for complex spikes? *Cerebellum* 10(4):694–701.

Mapelli L, Pagani M, Garrido JA, Egidio D'Angelo E (2015) Integrated plasticity at inhibitory and excitatory synapses in the cerebellar circuit. *Front Cell Neurosci* 9:169.

Milardi D, Quartarone A, Bramanti A, Anastasi G, Bertino S, Basile GA, Buonasera P, Pilone G, Celeste G, Rizzo G, Bruschetta D, Cacciola A (2019) The cortico-basal ganglia-cerebellar network: Past, present and future perspectives. *Front Syst Neurosci* 13:61.

Pierrot-Deseilligny E (2002) Propriospinal transmission of part of the corticospinal excitation in humans. *Muscle & Nerve* 26(2):155–172.

Prestori F, Mapelli L, D'Angelo E (2019) Diverse neuron properties and complex network dynamics in the cerebellar cortical inhibitory circuit. *Front Mol Neurosci* 12:267.

Ramaswamy S, Markram H (2015) Anatomy and physiology of the thick-tufted layer 5 pyramidal neuron. *Front Cell Neurosci* 9:233.

Sanes JN, Donoghue JP (2000) Plasticity and primary motor cortex. *Annu Rev Neurosci* 23(1):393–415.

Tsubo Y, Isomura Y, Tomoki Fukai T (2013) Neural dynamics and information representation in microcircuits of motor cortex. *Front Neural Circ* 7:85.

Young NA, Collins CE, Kaas JH, (2013) Cell and neuron densities in the primary motor cortex of primates. *Front Neural Circ* 7:30.

Zang Y, De Schutter E (2019) Climbing fibers provide graded error signals in cerebellar learning. *Front Syst Neurosci* 13:46.

Original Articles

Chun SS, Brass M, Heinze H-J, Haynes J-D (2008) Unconscious determinants of free decisions in the human brain. *Nat Neurosci* 11(5):543–545.

Deans MR, Gibson JR, Sellitto C, Connors BW, Paul DL (2001) Synchronous activity of inhibitory networks in neocorctex requires electrical synapses containing connexin36. *Neuron* 31(3):477–485.

Hoge GJ, Davidson KG, Yasumura T, Castillo PE, Rash JE, Pereda AE (2011) The extent and strength of electrical coupling between inferior olivary neurons is heterogeneous. *J Neurophysiol* 105(3):1089–1101.

Libet B, Gleason CA, Wright EW, Pearl DK (1983) Time of conscious intention to act in relation to onset of cerebral activity (readiness-potential): The unconscious initiation of a freely voluntary act. *Brain* 106(3):623–642.

Chapter 13 Control of Movement and Posture

Books

Abernethy B, Hanrahan SJ, Kippers V, Mackinnon LT, Pandy, MG (2013) *The Biophysical Foundations of Human Movement*, 3rd ed, Human Kinetics, Champaign, IL.

Hamilton N, Weimar W, Luttgens K (2012) *Kinesiology: Scientific Basis of Human Motion*, 12th ed, The McGraw-Hill Companies, New York.
Rose J, Gamble JG (eds) (2005) *Human Walking*, 3rd ed, Lippincott Wiliams and Wilkins, Philadelphia.
Sternad D (ed) (2009) *Progress in Motor Control A Multidisciplinary Perspective*, Springer Science+Business Media, New York.

Book Chapters

Feldman AG, Goussev V, Sangole A, Levin F (2007) Threshold position control and the principle of minimal interaction in motor actions, in: Cisek P, Drew T, Kalaska JF (eds) *Progress in Brain Research*, vol 165, Elsevier B.V., Amsterdam, pp267–180.
Feldman AG, Levin MF (2009) The equilibrium-point hypothesis – Past, present and future, in: Sternad D (ed) *Progress in Motor Control A Multidisciplinary Perspective*, Springer Science+Business Media, New York, pp699–726.
Ralston HJ (1976) Energetics of human walking, in: Herman RM, Grillner S, Stein PSG, Stuart DG (eds) *Neural Control of Locomotion*, Plenum Press, New York, pp77–98.

Review Articles

Dhawale AK, Smith MA, Ölveczky BP (2017) The role of variability in motor learning. *Annu Rev Neurosci* 40:479–498.
Grillner S (2006) Biological pattern generation: The cellular and computational logic of networks in motion. *Neuron* 52 (5):751–766.
Kiehn O (2016) Decoding the organization of spinal circuits that control locomotion. *Nat Rev Neurosci* 17(4):224–230.
Latash ML (2008) Evolution of motor control: From reflexes and motor programs to the equilibrium-point hypothesis. *J Hum Kinet* 19(19):3–24.
Rossignol S, Dubuc R, Gossard JP (2006) Dynamic sensorimotor interactions in locomotion. *Physiol Rev* 86(1):89–154.
Sainburg RL (2005) Handedness: Differential specializations for control of trajectory and position. *Exerc Sport Sci Rev* 33(4):206–213.
Whelan PJ (2010) Shining light into the black box of spinal locomotor networks 2010. *Philos Trans R Soc Lond B Biol Sci* 365 (1551):2383–2395.
Yiou E, Caderby T, Delafontaine A, Fourcade P, Honeine JL (2017) Balance control during gait initiation: State-of-the-art and research perspectives. *World J Orthop* 8(11):815–828.

Original Articles

Latash ML (2010) Motor synergies and the equilibrium-point hypothesis. *Motor Control* 14(3):294–322.
Wing AM (2000) Mechanisms of motor equivalence in handwriting. *Curr Biol* 10(6):R245–R247.

Index

Page references to figures are italicized; page references to tables are in boldface.

Printed in the United States
by Baker & Taylor Publisher Services